Plastics

Plastics

Microstructure and Applications

Third Edition

N J Mills

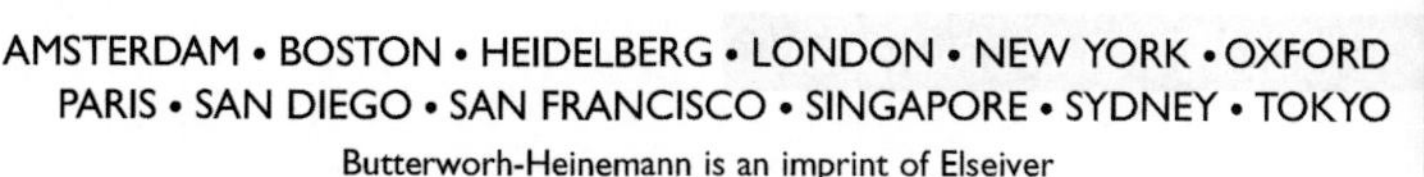

Butterworth-Heinemann is an imprint of Elsevier
Linacre House, Jordan Hill, Oxford OX2 8DP, UK
30 Corporate Drive, Suite 400, Burlington, MA 01803, USA

First edition published by Arnold 1986
Second edition by Arnold 1993
Third edition published by Butterworth Heinemann 2005
Reprinted 2006

British Library Cataloguing in Publication Data
A catalogue record for this book is available from the British Library

Library of Congress Cataloging-in-Publication Data
A catalog record for this book is available from the Library of Congress

ISBN–13: 978-07506-5148-6
ISBN–10: 0-7506-5148-2

For information on all Butterworth-Heinemann publications
visit our website at books.elsevier.com

Transferred to Digital Printing 2009

Contents

Preface

This book is intended for students of engineering and materials science degree courses, and for scientists and engineers as an introduction to the properties and applications of plastics. The mechanical design of plastics products is emphasised and physical properties in terms of microstructure are explained in detail. The sales of plastics are growing, partly at the expense of traditional materials, and partly via the development of new markets. When plastics are substituted for other materials, products should be redesigned to suit polymer processing. Therefore, the merits and limitations of these processes must be understood. Processing has permanent consequences on the microstructure of the product; these must be anticipated, and used to advantage if possible. This interlocking nature of different aspects of plastics technology provides a challenge to engineers.

By exploring the relation between the properties of plastics and their microstructure we begin to see the possibilities and the limitations of this class of materials. The responses of polymers to the environment differ from other materials, so the particular pitfalls must be recognised. The book emphasises concepts, and links between polymer engineering and other areas of science and technology. The derivation of key equations is included, since the assumptions made should be recognised. To keep the length manageable, details of polymer properties or processing routes should be sought from the sources given in the Further Reading section. Questions are given for each chapter in Appendix D, to give the student confidence in polymer engineering approach.

Major changes were made in revising the second edition. In this edition, a new chapter, Chapter 1, introduces properties of plastics through practical exercises, to help students see the relevance of more academic chapters. Computer modelling has revealed the mechanics of many types of composites, so the emphasis of Chapter 4 has shifted to modelling. Applications, product design and process technology have moved on; consequently, the case studies in Chapter 14 were updated. A new Chapter 15 introduces sport and biomaterials with case studies, since increasing numbers of students are enrolled in courses related to these areas. The material has been thoroughly updated, and the principles of polymer structure–property relationships set out more clearly.

Materials science and engineering degree courses traditionally had a considerable practical content, to apply the principles of the subject, and develop practical skills. Experimental work on polymer microstructure, measurements of mechanical and physical properties, and use of polymer process equipment, are of great benefit in developing understanding. The basics of microstructure and processing are usually covered in the first year of a course. Detailed consideration of mechanical properties is best left until after simple elastic materials have been studied. Polymer selection, covered in Chapter 13, can be integrated with the selection of other materials. There is sufficient range of topics in Chapters 10 and 11 to suit options on electrical, chemical or optical properties. The case studies in Chapters 14 and 15 illustrate the compromises needed in the design of complex products. The references to primary research journals could be used as the basis of literature search exercises; now relatively easy via online journals.

It is assumed that the reader has an elementary knowledge of the mechanics of materials. However, appendices given at the end summarise the necessary principles and provide the heat and fluid flow theories relevant to plastics.

Birmingham, March 2005

Chapter 1

Introduction to plastics

Chapter contents

1.1 Introduction

This chapter encourages the reader to familiarize themselves with plastics. It aims to open the reader's eyes to design features in familiar products, and to relate these features to polymer processes. This prepares them for polymer selection exercises in Chapters 13–15. The dismantling exercises can be adapted to suit different courses; for students on a biomaterials course, blood sugar monitors, asthma inhalers, or blood apheresis units can be dismantled. For those on a sports/materials course, the components of a running shoe could be considered (see also further reading). Product examination can be tackled at different levels. The level described here is suitable at the start of a degree course. Later, when most of the topics in the book have been studied, more complex tasks can be tackled – improving the design of an existing product, with reselection of materials and processing route.

There are some polymer identification exercises, using simple equipment. This would make the reader familiar with the appearance of the main plastics. Professional methods of polymer identification, such as differential scanning calorimetry, Fourier transform infrared (FTIR) spectroscopy and optical microscopy, may be dealt with later in degree courses.

This book explores the characteristic properties of polymers and attempts to explain them in terms of microstructure.

1.2 Dismantling consumer products

Using familiar products, the aim is to note component shapes, to see how they are assembled, and measure the variation in thickness. Recycling can also be considered; the ease of dismantling depends on whether the product was intended to be repaired, or to be scrapped if faulty. Screws may be hidden under adhesive labels, and the location of snap-fit parts may be difficult to find.

1.2.1 Plastic kettle

A new plastic kettle can be bought for less than £30, or a discarded one used. Preferably use a cordless kettle, which can be lifted from the powered base. The following four activities can be extended if necessary, by consideration of aesthetics, weight, and ease of filling and pouring.

Briefly touch the kettle's outer surface when the water is boiling

Although the initial temperature of the kettle's outer surface may be 90 °C, the low thermal conductivity of the plastic body compared with that of your finger, means that the skin surface temperature takes more than a minute to reach an equilibrium value, and this value is c. 50 °C. *With a dry finger* touch the kettle's outer surface for less than 5 s. If you have access to a digital thermometer with a fine thermocouple probe, tape the thermocouple to the outer surface of the kettle and check the temperature. What can you deduce about the thermal conductivity of the plastic? Chapter 5 explores the balance between thermal conduction through the plastic and convection from its outer surface.

Measure the thickness of the body at a range of locations

Dismantle the kettle and make a vertical section through the body with a hacksaw. Use callipers to measure the body thickness at a range of locations, and mark the values on the plastic.

Over what range does the thickness vary? Reasons for the nearly constant section thickness are given in Chapter 13. Figure 1.1 shows a typical section. Check how the colouring is achieved. If there is no paint layer on the outside, the colour must be integral (for pigments, www.specialchem4polymer.com).

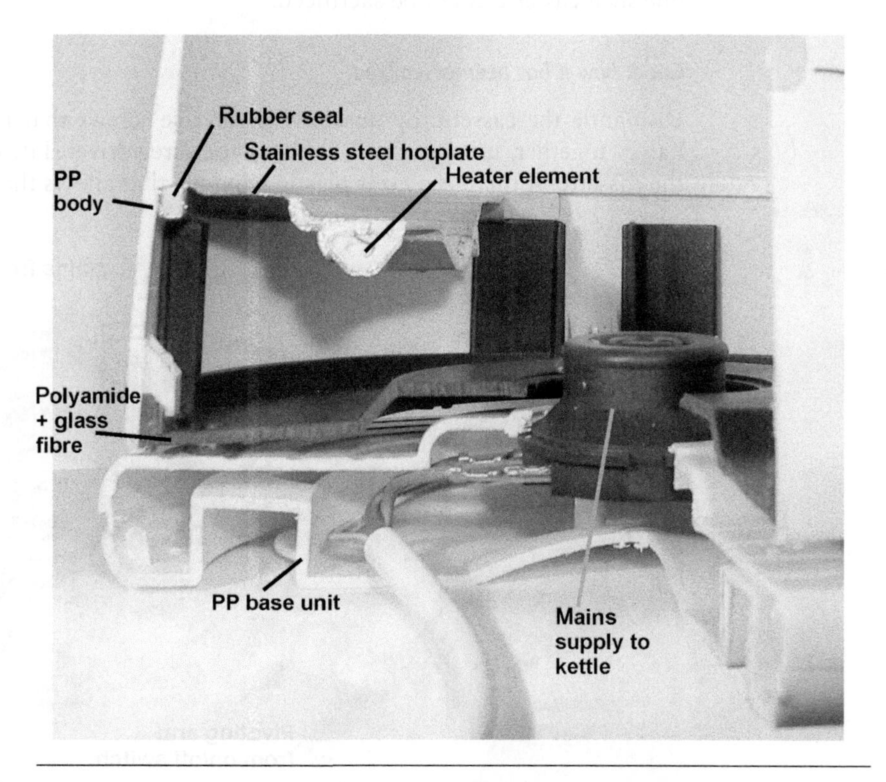

Figure 1.1 Section of a plastic kettle and powered base unit (most of electrical heater was removed from the kettle).

Examine the electrical insulation in the base unit

Note how the electrical conductors are insulated from the parts that are handled. A metal-bodied kettle must have separate insulation wherever mains-connected parts are attached; however plastic is an electrical insulator. Figure 1.1 shows the mains power connections in the base. The coloured insulation (live, neutral and earth) of the braided copper wires is plasticised PVC.

Examine the linking mechanism for the heater switch

Identify the mechanism that connects the on/off switch to the internal contact switch that applies mains power to the heater unit. Identify the thermostat that detects the boiling of the kettle, and note how it switches off the power. Figure 1.2 shows a typical arrangement.

1.2.2 VHS video cassette

Video cassettes are becoming obsolete with the increasing use of DVDs, so one such cassette could be sacrificed.

Check how it has been assembled

Dismantle the cassette by unscrewing the five screws that fasten the two halves together, using a small Phillips-type screwdriver. Lift off the top of the cassette. If there is a clear plastic window, that allows the tape levels to

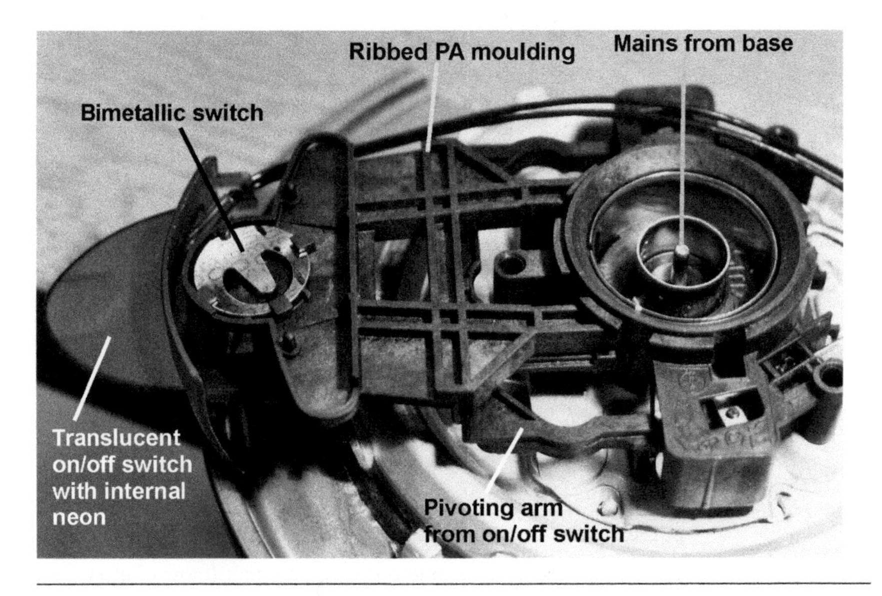

Figure 1.2 Underside of the heater unit inside the kettle, showing the power switch and switch mechanism.

be seen, check how it is attached to the main body. Count the number of parts. After dismantling, see how easy it is to reassemble!

Identify the plastic springs that lock the spools

When a cassette is removed from a recorder, the tape spools are locked to prevent the unwinding of the tape. When it is inside the recorder, a pin presses through a flap at the base of the cassette, causing a lever to operate on two plastic mouldings (Fig. 1.3). They engage with slots in the rim of the tape spools. Check, by pressing with a finger, that the springs can be easily bent. They are made of the engineering thermoplastic polyoxymethylene.

Measure the tensile strength of the tape

Unwind some of the 13 mm wide coated PET tape and measure the thickness (0.02 mm) with callipers. A loop of tape can withstand a tensile force of about 60 N before it yields and about 80 N before it fails in tension. Check this by using a spring balance on a loop of tape, and calculate its tensile strength (approximately 150 MPa). It must bend around cylinders of diameter 5 and 6 mm (Fig. 1.4), so it must have a very low bending stiffness. It must resist wear as it is dragged over the stationary metal cylinders. It must be dimensionally stable, so that the coating is not damaged.

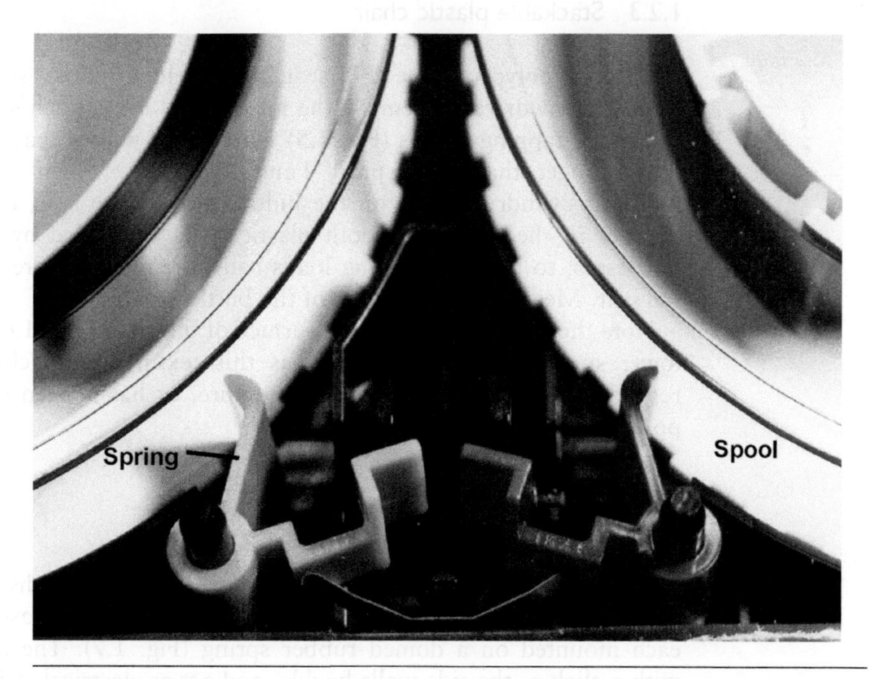

Figure 1.3 Mechanism that locks the spools when the tape is not being played, as seen inside a video cassette.

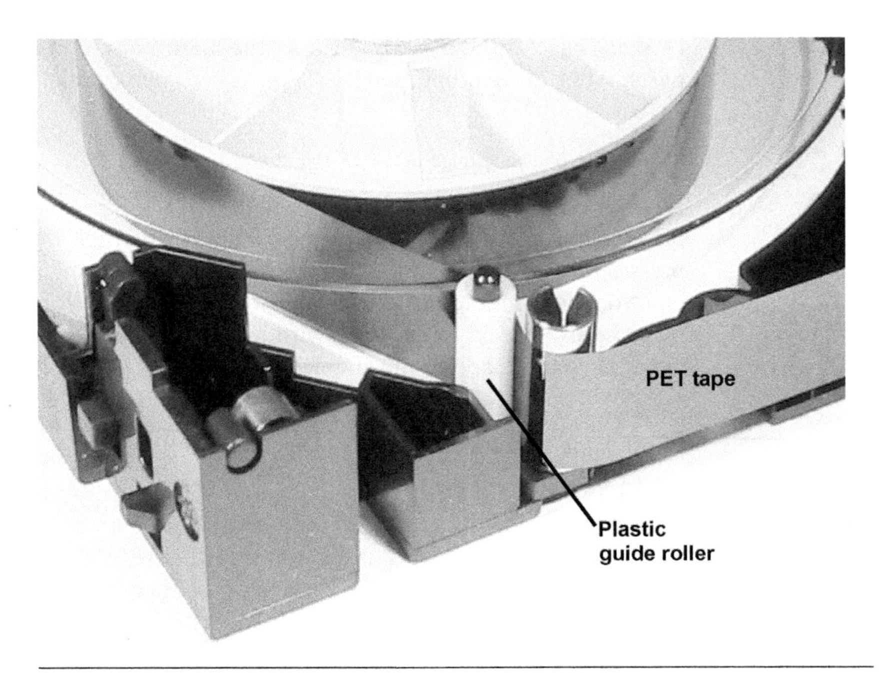

Figure 1.4 PET tape in a VHS cassette passes round a plastic guide roller and a fixed metal cylinder.

1.2.3 Stackable plastic chair

This has a polypropylene (PP) seat, with welded tubular-steel legs. Use a Phillips screwdriver to remove the four screws that attach seat to the legs. These self-tapping screws (Fig. 1.5) with sharp, widely spaced threads, are much longer than the typical 4 mm thick seat. When screwed into a moulded cylindrical boss on the hidden side of the seat, the threads cut grooves in the initially smooth plastic. It is supported by four or more buttresses, to prevent bending loads causing failure, where the boss joins the seat. Measure the thickness of the buttresses.

Note the texture on the upper surface of the chair (Fig. 1.6), whereas the lower surface is smooth. How has this texture been achieved? Is it a reproduction of the mould surface texture, or has it been produced by a post-moulding operation?

1.2.4 Telephone handset

An old handset from an office may be available for dismantling. The numbers for dialling are printed on separate thermoplastic mouldings, each mounted on a domed rubber spring (Fig. 1.7). The domes depress with a click as the side walls buckle, and act as electrical switches. A layer

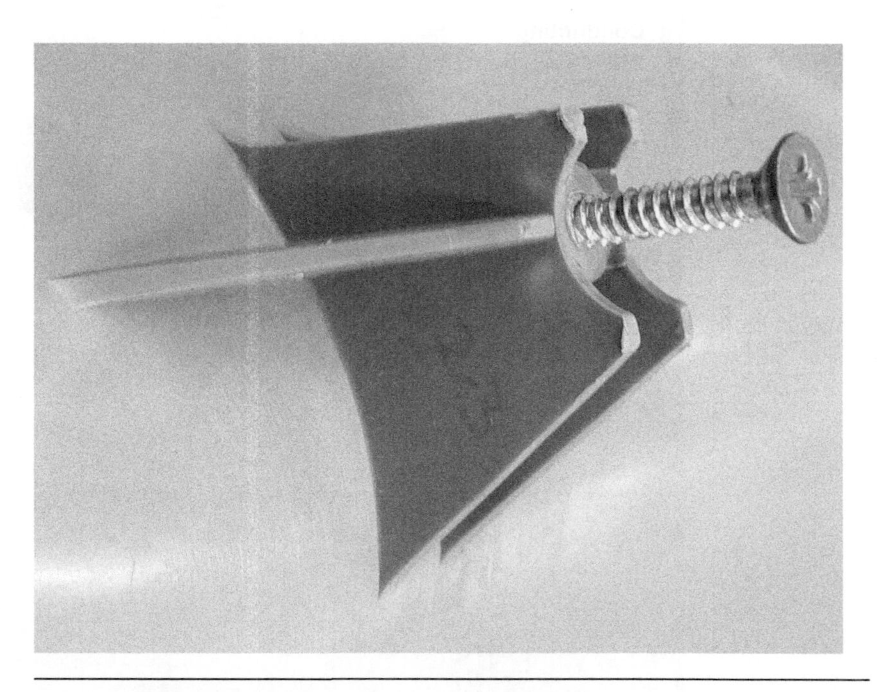

Figure 1.5 Self-tapping screw for attached tubular metal legs, and the boss with buttresses under the seat of a PP stacking chair.

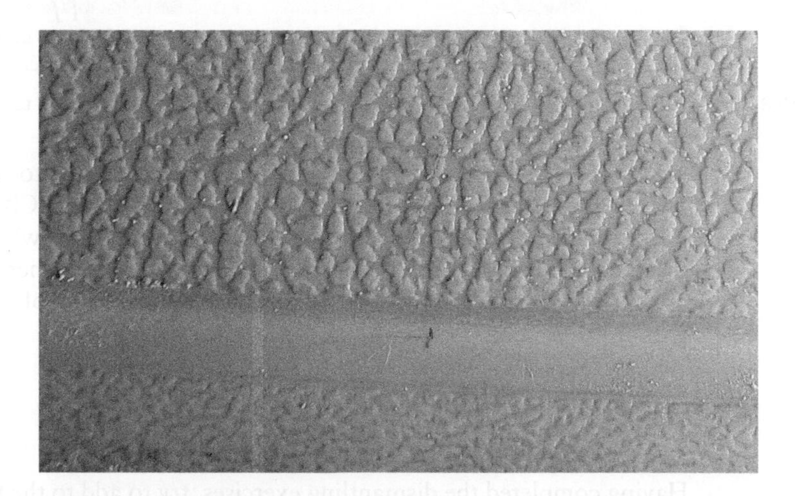

Figure 1.6 Texture on the upper surface of a PP chair.

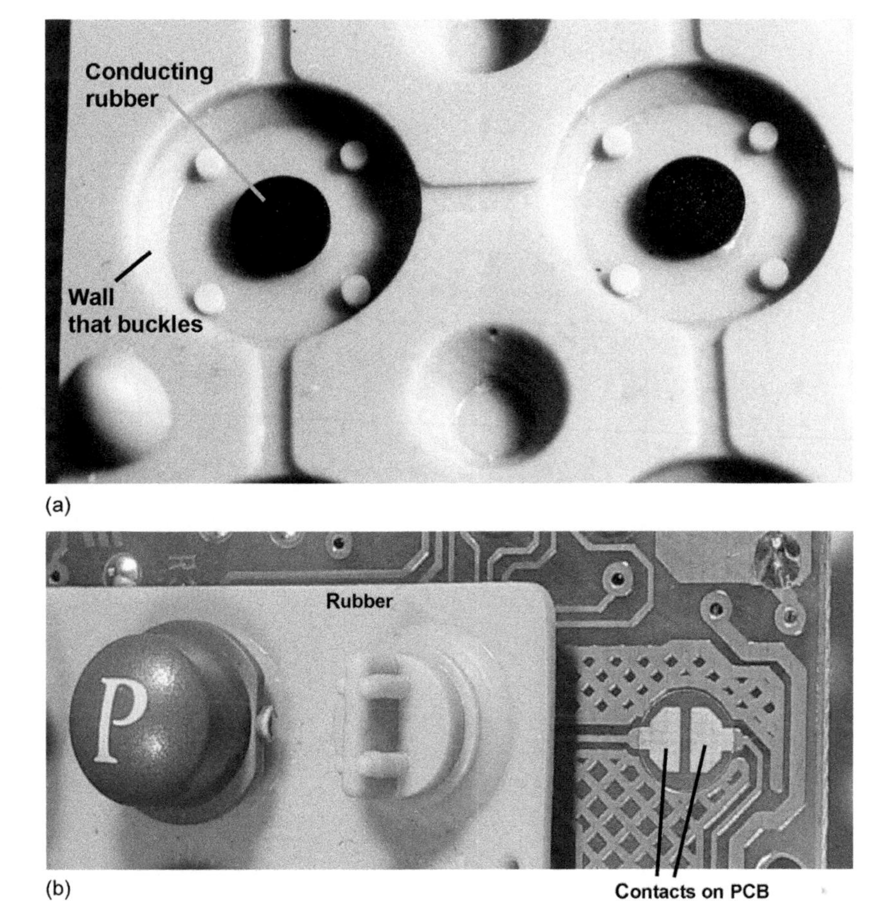

Figure 1.7 Views from both sides of an injection-moulded rubber switch from a telephone.

of carbon-black filled rubber on the base comes into contact with copper tracks on the printed circuit board (PCB). The PCB consists of a polyester resin plus woven fibreglass (GRP) composite, which is also an insulator. The copper tracks on the PCB lead to holes where components are mounted; the PCB must tolerate the temperature of molten solder without distortion.

1.2.5 Summary

Having completed the dismantling exercises, try to add to the following list. *Plastics have advantages over metals of being*

1. self-coloured, by adding about 0.1% of dispersed pigment. There are no painting costs, and the product maintains its colour if scratched.

2. electrical insulators. There is no need for insulating layers between live parts and the body of product, and assembly is simplified.
3. thermal insulators. This conserves energy, and touching a kettle body will not cause scalds.
4. of low density, so lightweight products can be made.
5. impact resistant, with a high yield strain, so thin panels do not dent if locally loaded.

Plastics have advantages over ceramics or glass of being

1. tough, so that the impacts are unlikely to cause brittle fractures.
2. low melting point, so the energy costs for processing are low.
3. capable of being moulded into complex shapes with the required final dimensions (they are 'net-shape', with no final machining stage).

1.3 Mechanical and optical properties of everyday products

Several disposable plastic products are considered, to illustrate mechanical and optical properties.

1.3.1 Crazing and fracture of a biro

Find a Bic biro (or a similar ballpoint pen) with a transparent polystyrene body. Hold it up towards a light source and bend it, using the thumbs as the inner and the forefingers as the outer loading points. Make sure that the curved portion is away from you and not aimed at anyone else. Deform the biro by about 10 mm and hold this for about 30 s, then release the load. The biro should return to its original shape, showing that large elastic strains can occur. Tilt the biro against the light and look for parallel reflective planes (Fig. 1.8a). These are called *crazes*.

Continue the loading until the body fractures. Although the ink tube will trap the broken pieces of the body, it is likely that a small piece(s) of PS might detach (Fig. 1.8b). Do not do the experiment without the ink tube, as pieces can fly off at speed. The strain energy released by the fracture is enough to create more than one fracture surface.

1.3.2 Ductile yielding of low-density polyethylene strapping

Low-density polyethylene (LDPE) strapping, cut from 0.42 mm thick film, is used to hold four packs of drink cans together. If pulled slowly with the hands, parts of the strapping undergo tensile necking followed by cold drawing of the thin region (Fig. 1.9). Mark parallel lines at 5 mm intervals

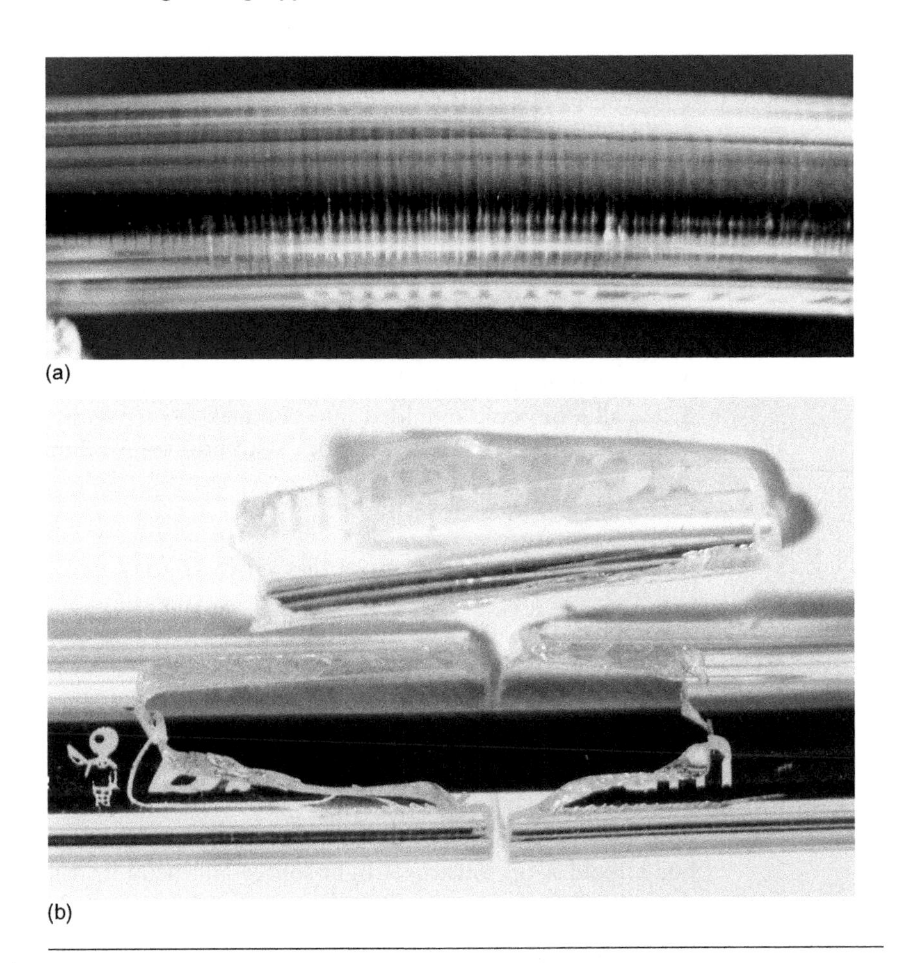

(a)

(b)

Figure 1.8 (a) Crazes in, (b) broken pieces of, a PS Biro after a bending experiment.

across the LDPE before the experiment. Note the extension ratio in the neck, and how the shoulder of the neck moves into the un-necked region.

1.3.3 Optical properties of a CD and polyethylene film

This requires a laser pointer and a CD. Observe safety precautions: do not aim the laser beam at anyone's eyes. Aim it, at approximately normal incidence, at the side of the CD that appears silvered. When the beam hits the tracks near the centre of the disc, a diffraction pattern is created (Fig. 1.10). This pattern is a two-dimensional analogue of X-ray diffraction from a three-dimensional crystal.

If the laser beam hits the main part of the disc, there are just two diffraction peaks, in addition to the directly reflected beam. These are caused by the

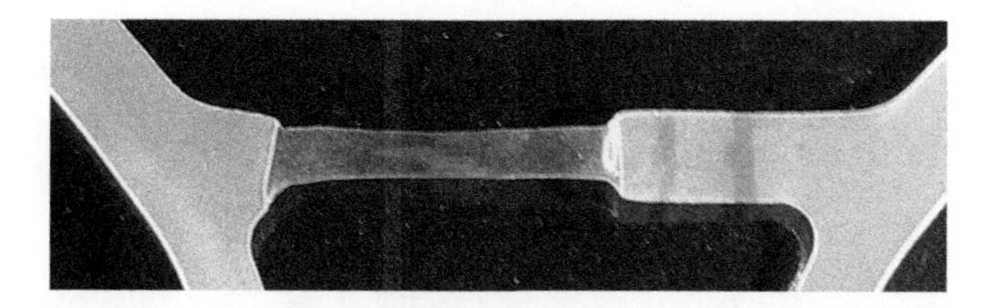

Figure 1.9 Necking and cold drawing of LDPE strapping from a four-pack of drink tins.

Figure 1.10 Diffraction pattern from a laser pointer, when shown on the track near the end of a music CD; the direct reflection has the cross pattern.

regular track spacing in the radial direction. As the circumferential pits are irregularly spaced along each track, this part of the disc acts as a one-dimensional diffraction grating. The diffraction pattern is used to keep the reading head on the track (for more details see Chapter 14). If you scratch off part of the label and the underlying metallized layer, the CD will be transparent in this region. Hence the material, polycarbonate, is transparent.

Macro-bubbles, used inside cardboard boxes for the shock-resistant packaging of goods (Fig. 1.11), are manufactured from 200 mm wide tubular polyethylene film approximately 0.05 mm. The tube is inflated with air then welded at approximately 100 mm intervals. Place a macro-bubble on top of a printed page with a range of font sizes, and note the smallest font size that you can read. High-density polyethylene (HDPE) bubbles scatter

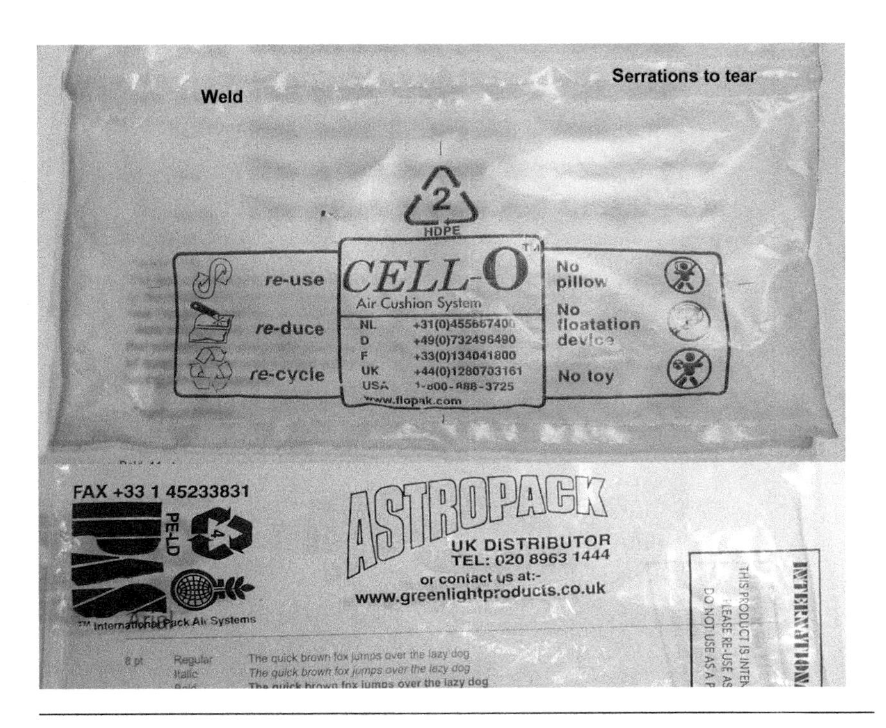

Figure 1.11 LDPE and HDPE macro-bubbles (deflated) on top of a test page with a range of font sizes.

light more than LDPE, so it is more difficult to read the text. If a HDPE bubble is lifted by about 20 mm, it is impossible to read the text.

1.3.4 Degradation of polymers in sunlight

Visit a beach and collect plastic articles that have been there for a couple of years. Apart from foam and hollow air-filled products, there will be polyethylene (PE) or polypropylene (PP) products, which are less dense than water. Note how colours have faded, the surface has become opaque, and the product has started to crack.

1.3.5 Viscoelasticity of a foam bed

Acquire some 'slow recovery' foam such as *Confor* (samples are often given away by bedding showrooms). Compress the surface with one hand for a minute, and then observe how long it takes for the indentations in the foam to disappear. Repeat the exercise after the foam has been placed in a refrigerator (5 °C) when it will be much stiffer, or after it has been placed in an oven at 60 °C (when it will be much less stiff and will recover quickly). This shows that the strong viscoelastic response only occurs in a temperature range where the polymer is leathery; close to its glass transition temperature.

1.4 Identifying plastics

Make a collection of food packaging: a milk bottle, a carbonated drink bottle, a supermarket carrier bag, a near-transparent lidded container for food, a margarine container.

Note how plastic bottles have replaced glass for soft drinks, milk, ketchup, etc. One-trip plastic bottles are essential for the sales of bottled water, while they have replaced metal cans for many products. Even containers apparently made from paper (such as Tetrapak) rely on an inner polyethylene layer to protect the paper from the liquid contents.

Use the methods below to identify which plastics are used in one or more products.

1.4.1 Recycling marks

Recycling marks on products (Fig. 1.11) allow the common plastics to be identified (Table 1.1). Sometimes numbers are used in place of the abbreviation for the polymer name.

1.4.2 Product appearance, if unpigmented

Translucent products are semi-crystalline, e.g. PE. Some thin (<1 mm) or highly oriented products appear transparent, in spite of being semi-crystalline (e.g. PET bottles), since the crystals are too small to scatter light. Some thicker PP products appear translucent, but thin mouldings, especially if the PP is nucleated, will appear nearly transparent.

Table 1.1 Recycling marks for polymers

No.	*Legend*	*Polymer*
1	PET	Polyethylene terephthalate
2	HDPE	High-density polyethylene
3	PVC	Polyvinyl chloride
4	LDPE	Low-density polyethylene
5	PP	Polypropylene
6	PS	Polystyrene

Transparent mouldings thicker than 1 mm will be one of the glassy polymers (PVC, PS, PC, etc.). If a thin film of molten, unpigmented plastic is opaque to light it is likely to be filled.

1.4.3 Density

An electronic densitometer can measure the density of small (<10 g) pieces, using Archimedes principle. The pieces are first weighed in air, then again, while suspended in water. Table 1.2 gives the densities and melting points of the main polymers. They are arranged in classes, in order of increasing density. The density of semi-crystalline plastics increases with crystallinity, so a range is given. If a significant amount of a reinforcing or toughening material is added, the density changes, making it more difficult to identify the polymer.

1.4.4 Melting temperatures

Table 1.2 shows the temperature T_m at which the crystalline phase melts, or, for non-crystalline polymers, the glass transition temperature T_g at which the glass changes into a melt. Samples can be dragged across the surface of metal hotplates, set to a range of temperatures. However, when the polymer is just above T_m, some polymers leave a streak of melt, while others of higher viscosity just deform. Therefore, transition temperatures can be overestimated.

1.4.5 Young's modulus

Estimate the order of magnitude of the Young's modulus of a flat part of the product by flexing it. This works best if a standard sized (say 100 mm long, 20 mm wide, 2 mm thick) beam is cut from the product and loaded in three-point bending, since the bending stiffness varies with the cube of the thickness. LDPE is of a much lower Young's modulus (c. 100 MPa) than most other plastics (1–3 GPa), and the surface can be marked with a finger nail.

1.5 Product features related to processing

The aim is to recognise design features associated with processes. The diagrams in Chapter 5 show the major processes. Both the product shape and surface marks provide clues for process identification.

Table 1.2 Polymer densities and transition temperatures

Abbreviations	*Polymer*	*Density ($kg\,m^{-3}$)*	*T_g (°C)*	*T_m (°C)*	*Event if bent through 90°*
Semi-crystalline plastics					
P4MP	Poly (4-methyl-pentene-1)	830	25	238	Semi-brittle
PP	Polypropylene	900–910	−10	170	Whitens
LDPE	Low-density polyethylene	920–925	−120	120	Ductile
MDPE	Medium density polyethylene	935–945	−120	130	Ductile
HDPE	High-density polyethylene	955–965	−120	140	Ductile
PA 6	Polyamide 6	1120–1150	50	228	Ductile
PA 66	Polyamide 6,6	1130–1160	57	265	Ductile
PET	Polyethylene terephthalate	1336–1340	80	260	Ductile
POM	Polyoxymethylene (Acetal)	1410	−85	170	Semi-brittle
PVDC	Polyvinylidene chloride	1750	−18	205	
PTFE	Polytetrafluoro ethylene	2200	−73	332	Ductile
Glassy plastics					
PS	Polystyrene	1050	100		Brittle
SAN	Styrene acrylonitrile copolymer	1080	100		
ABS	Acrylonitrile butadiene styrene copolymer	990–1100	100		Whitens
PC	Polycarbonate	1200	145		Ductile
PVCu	Polyvinyl chloride unplasticised	1410	80		Ductile
PMMA	Polymethyl methacrylate	1190	105		Brittle

T_m, crystal melting temperature; T_g, glass transition temperature.

1.5.1 Blow mouldings

These are hollow containers, usually with an opening of smaller diameter than the body. Both ends of the moulding may be cut off to produce a tubular product, or one end cut off for a bucket-shaped container. The wall thickness varies with position, and there is a weld line across the closed end of the container (Fig. 1.12). Sometimes near-parallel lines are visible on the inner surface. These indicate the extrusion direction when the parison (tubular preform) emerged from a die. Look for the weld location on the base of an HDPE milk bottle; this aligns with the external surface line from the mould split. Section the milk bottle vertically, in a plane perpendicular to the weld line, and measure the thickness at the weld line compared with elsewhere. Note the threads in the neck are corrugated.

Stiff HDPE tool boxes can be created by allowing the two sides of the blow moulding to come into contact at some locations and form welds (Fig. 1.13). The 0.4 mm thick hinge region is created by pressing the HDPE with metal bars, and the 'click shut' catches are also part of the blow moulding.

1.5.2 Extruded products

These have a constant cross section. Examples are domestic gutters or down-pipe, or (replacement) window frames, made from PVC. Look through a length of an extrusion, towards a window, for markings parallel to the extrusion direction, which have come from the die. The outer surface is in contact with a sizing die, whereas the inner surface cools in air and can

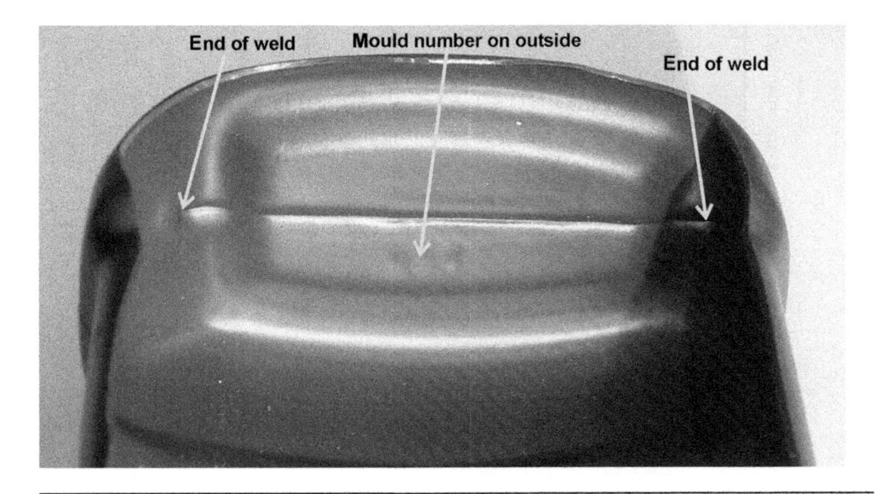

Figure 1.12 Sectioned blow-moulded bottle, showing the weld line at the base.

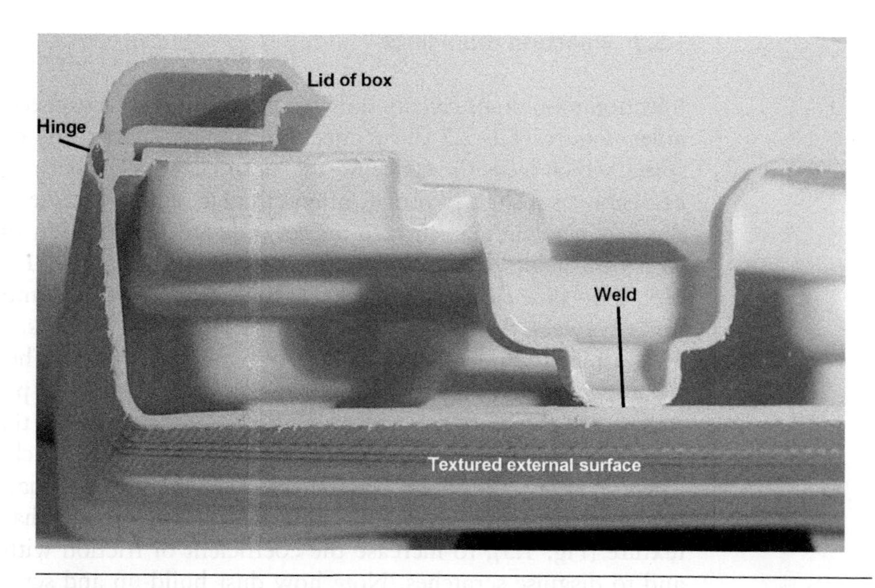

Figure 1.13 Section through a blow-moulded HDPE tool box (45 mm thick). Both lid and base of the box are hollow, with reinforcing welds at intervals interior.

Figure 1.14 Extruded HDPE pipe, with corrugations at 18 mm intervals in the outer layer, for buried electric cables.

change shape slightly. The pipe wall provides the bending stiffness and resistance against weathering. Pipes for cable TV in the UK are green with a corrugated exterior, but a smooth inner wall. Figure 1.14 shows a pipe for electrical cable, with an outer red corrugated layer bonded to a smooth inner black layer (details of the process are shown in Fig. 13.2). Such pipes offer maximum resistance to crushing by soil loads for a given weight of polymer.

1.5.3 Injection mouldings

Injection mouldings can contain T-junctions (where *ribs* meet a surface) and holes. Figures 1.1–1.7 show injection moulded parts. The point where the *sprue*, which feeds the melt into the mould cavity, has been removed should be visible as a slightly rough, often circular region. On the concave side of the product, circular surface marks indicate the location of *ejector pins*, which push the cold moulding from the mould. Figure 1.15 of a moulded box, shows the ejector pin marks on the inside of the box, and a moulded-in hinge between the two halves.

Consider the polypropylene seat of a stackable chair. The seat sides are 'bent over', providing a place to grip the seat. These sides provide bending stiffness to the seat. You can prove this if you can cut off the side parts; if you lean back in the chair, it flexes excessively at the back/seat junction. There is more about the bending creep of plastics in Chapter 7 and the bending stiffness of beams in Chapter 13. The seat surface has a moulded-in texture (Fig. 1.5), to increase the coefficient of friction with your clothes, and to disguise scratches. Note how dust build-up and scratches spoil the appearance of the hidden side, which is smooth.

1.5.4 Thermoformed products

These tend to be curved panels, or shallow containers. They have a variable thickness, since only the convex side contacts a metal die. They can be as thin as 0.1 mm, since a sheet of melt is stretched before contact with the cold mould, or as thick as 10 mm. There will be no signs of any injection point or ejector pins. Typical examples are disposable coffee cups (Fig. 1.16a), margarine containers, baths and shower trays.

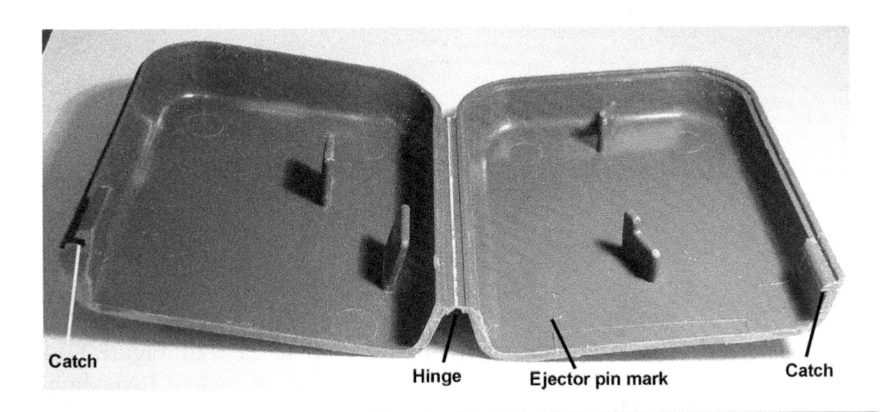

Figure 1.15 Section through an injection-moulded PP box for a micrometer. The moulded-in hinge has whitened in use.

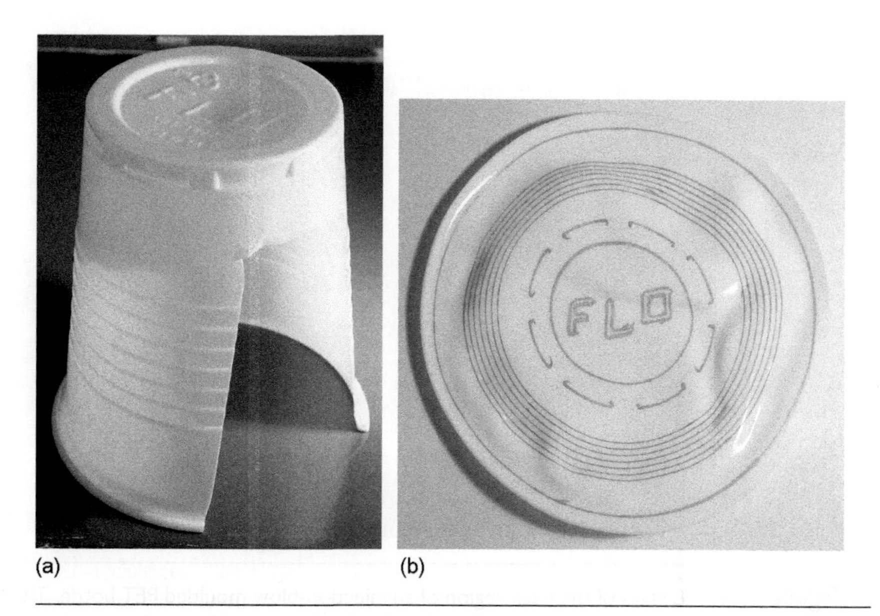

(a) (b)

Figure 1.16 Section through a thermoformed PS disposable coffee cup, with shallow corrugations in the 0.2 mm thick sidewall. The corrugations were outlined in felt-tip then the cup heat reverted to 0.8 mm thick sheet.

Use a sharp pair of scissors to cut a section through the cup, and note the corrugations which increase the bending stiffness of the wall. The corrugations also provide grip and reduce heat transfer to the fingers. Note the variation in wall thickness.

Either use a hot air blower (for paint stripping), or put the cup on a layer of aluminium foil in an oven at 120 °C and note the gradual shape reversion to a nearly-flat sheet (Fig. 1.16b). The thermoforming process involved the elastic stretching of a sheet of polymer melt, and this orientation was frozen into the cup when it cooled. On reheating, the plastic attempts to return to its original shape.

1.5.5 Blown film

The blown film process creates a continuous tube of film, usually less than 0.25 mm thick, which is flattened and rolled up. It can be cut into lengths and welded to produce products such as supermarket carrier bags or protective bubbles (Fig. 1.11). For a carrier bag, determine where the film has been welded and where it has been cut, or folded. Support the handles on a spring balance, then use sand (or tins) as the loading medium and determine the tensile strength of the polyethylene in the handle region.

Figure 1.17 Details of the neck region of an injection-blow moulded PET bottle. The bottle on the left has been heat treated at 120 °C, while the neck of a preform is shown on the right.

1.5.6 Injection-blow moulded bottles

Compare a PET carbonated drink bottle with an HDPE milk bottle. The moulded neck threads of the PET bottle (Fig. 1.17) have T-sections. The internal pressure of 4 bar in the carbonated drink bottle can only be resisted by a lightweight bottle if the polymer is oriented to increase its strength. However, the milk bottle is under no internal pressure, so a lower cost material (HDPE) and process can be used. Gas diffusion is covered in Chapter 11, while the stress analysis of a pressure vessel is covered in Appendix C. Note the location of the injection sprue in the centre of the base of the PET bottle. Try placing an empty PET bottle in an oven at 120 °C for 1 h. Note how it shrinks both in length and diameter, showing that the PET had been biaxially oriented. The base and the neck become milky in appearance, due to the crystallisation of these initially amorphous regions. The shrinkage in these regions is relatively low. The main part of the bottle remains clear, since it was already semi-crystalline.

1.6 Summary

Hopefully you are now familiar with the appearance and some typical properties of the commodity thermoplastics, and can recognise how some products have been made. You are now ready to study the microstructure and processing of polymers in more detail, and to find out how the properties can be related to the microstructure.

Chapter 2

Molecular structures and polymer manufacture

Chapter contents

2.1 Introduction

2.1.1 Size scale

Polymers can be considered on a range of size scales, from that of the molecules to that of the final products (Fig. 2.1). This chapter considers the molecular level: the regularity of the arrangement of monomer units in polymer chains, the characterisation of that arrangement, and the effects on the polymer crystallinity.

Chapter 3 discusses polymer chain shapes and packing in both crystalline and amorphous forms, while Chapter 6 explains the effect of polymer processing on the microstructure on a millimetre length scale. You should be able to synthesise these views, and use models on appropriate scales to explain the mechanical and physical properties of products.

2.1.2 Categories of polymers

Polymers can be subdivided into three main categories. *Thermoplastics*, consisting of individual long-chain molecules, can be reprocessed; products can be granulated and fed back into the appropriate machine. *Thermosets* contain an infinite three-dimensional network, which is only created when the product is in its final form, and cannot be broken down by reheating. *Rubbers* contain looser three-dimensional networks, where the chains are free to change their shapes. Neither thermosets nor rubbers can be reprocessed. Some polymers, such as polyurethanes, can be produced in both thermoplastic and thermoset variants.

2.1.3 Commodity and engineering thermoplastics

The relative importance of thermoplastics can be judged from their annual consumption (Table 2.1). The first six in the table are regarded as

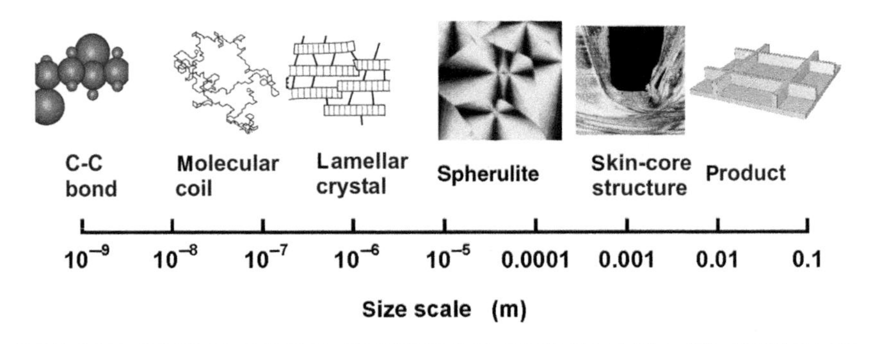

Figure 2.1 Range of size scales in polymer science.

Table 2.1 Thermoplastic consumption in Europe in 2003 (US prices in February 2005)

Thermoplastic	*Abbreviation*	*Consumption (%)*	*Price ($/lb)*
Polyethylene, low density	LDPE	21	0.8–1.0
Polyethylene, high density	HDPE	14	0.8–0.9
Polypropylene	PP	21	0.7–0.9
Polyvinyl chloride	PVC	15	0.4–0.8
Polyethylene terephthalate	PET	10	0.9
Polystyrene	PS	8	0.7–0.9
Acrylonitrile butadiene styrene	ABS	2.1	0.8–1.1
Polycarbonate	PC	1.2	1.4–1.8
Polyamide	PA	0.8	1.4–1.7
Polymethyl methacrylate	PMMA	0.8	1.2–2.2
Acetal	POM	0.5	1.3–1.5

Sources: Consumption, www.apme.org, % of total 38 million tonnes; prices, *www.plasticstechnology.com*.

commodity thermoplastics. Many manufacturers compete to supply these. Prices change quite rapidly, in response to the price of crude oil, so the table indicates relative prices. The low density of thermoplastics, ranging from $900\,\mathrm{kg\,m^{-3}}$ for polypropylene (PP) to $1400\,\mathrm{kg\,m^{-3}}$ for polyvinyl chloride (PVC), means that the material costs are low in volume terms. The remaining thermoplastics in Table 2.1 are called *engineering thermoplastics* because of their superior mechanical properties, but the distinction is a fine one. They are produced on a smaller scale and have prices about twice that of commodity thermoplastics. Finally, there are speciality plastics which only sell a few thousand tonnes per annum. An example is polytetrafluoro ethylene (PTFE) which has unique low friction properties.

Thermoplastics can be divided into amorphous and semi-crystalline solids. The amorphous polymers are glassy at temperatures lower than T_g (the glass transition temperature) and rubbery liquids at higher temperatures. Semi-crystalline thermoplastics have an amorphous phase, and a crystalline phase with a melting temperature T_m. The transition temperatures of the main thermoplastics are listed in Table 1.2.

2.1.4 Thermosets and rubbers

The crosslinking reaction, which occurs in the production of thermosets, also provides good adhesion to other materials. Therefore, epoxy and polyester resin matrices are used for fibre-reinforced composites, amino resins are used for bonding chipboard, while phenolics are used for bonding fibres in brake pads, and sand for metal casting. These specialised products do not fit in well with the discussion of thermoplastic properties in this book. The consumption of thermosets is almost static, reflecting a loss of some markets to thermoplastics with a high temperature resistance.

Rubber consumption is dominated by tyre production. In these, conveyor belts, and pressure hoses, thin layers of either steel wire or polymeric fibre reinforcement take the main mechanical loads. These layers, with rubber interlayers, allow flexibility in bending, whereas the reinforcement limits the in-plane stretching of the product. The applications are dominated by natural rubber and styrene butadiene copolymer rubber (SBR). Other rubbers have specialised properties: butyl rubbers have low air permeability, nitrile rubbers have good oil resistance, while silicone rubbers have high and low temperature resistance. Rubbers play a relatively small role in this book, but the rubbery behaviour of the amorphous phase in semi-crystalline thermoplastics is important.

2.2 Bonding and intermolecular forces in polymers

2.2.1 Covalent bonds

The single covalent bond consists of an electron shared between two atoms. There are three main ways of covalent bonding, of molecules such as methane (Fig. 2.2):

a) Ball and stick models emphasise bond directions and the distances between atom centres, but have unrealistically small atoms.
b) Space-filling models emphasise molecular packing, but do not allow the bond directions to be easily seen.
c) Electronic shell models emphasise the number of electrons in the outer shell.

Carbon has four electrons in its outer shell (quantum number $n = 2$); the shell would be full if it contained eight electrons. The outer shell (quantum number $n = 1$) of hydrogen contains one electron; it would be full if it contained two electrons. In methane, CH_4, the carbon atom forms covalent

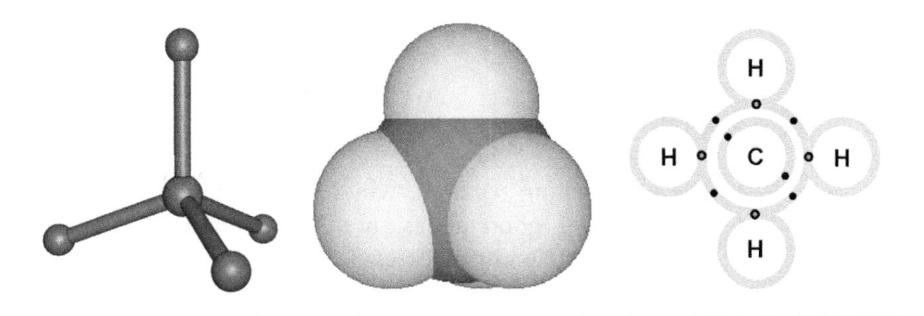

Figure 2.2 Models for methane: Ball and stick, space-filling and molecular orbitals.

bonds to four hydrogen atoms; each atom now has a full outer electron shell. Covalent bonds, indicated as C—H, are directional. In the methane molecule, the lines joining the centres of the H atoms to the C atom are at 109° 28′ to each other; the H atom centres are at the corners of a tetrahedron, with the C atom at the centre. The majority of bonds in polymer chains are covalent single bonds. They do not allow long-range electron movement, so polymers are electrical insulators, and consequently can transmit light.

Many monomers, and some polymers, contain covalent double bonds, written as C=C to indicate that two electrons are shared between the atoms. The first electron forms a σ (single) bond but the second forms the less stable π bond. The double bond prevents the rotation of the C atoms (or other atom pairs such as —C=O) relative to each other.

If a covalent bond is broken, each atom has an unpaired electron known as a *free radical*, shown as a dot as in C•. Free radicals are extremely reactive and consequently their lifetimes are measured in milliseconds.

2.2.2 Van der Waals forces

Van der Waals forces are a weak form of attraction; electron oscillations in one atom induce electron movement in neighbouring atoms, thereby attracting them. These forces, responsible for holding neighbouring polymer chains together, are not shown in diagrams of polymer structures. In any of the condensed polymer states (melt, glassy or crystalline) van der Waals forces cause neighbouring polymer molecules to pack closely together. Polymer molecules can only separate in solution; there is no gaseous state.

Van der Waals forces are easier to quantify in molecular solids, such as solid methane, where they are the sole intermolecular force. The potential energy E of two methane molecules, with their centres a distance R apart, is given by

$$\frac{E}{E_0} = \left(\frac{R_0}{R}\right)^{12} - 2\left(\frac{R_0}{R}\right)^{6} \tag{2.1}$$

where the constants $R_0 = 0.43\,\text{nm}$, $E_0 = 0.0127\,\text{eV}$. The R^{-12} term is a short-range repulsion force, whereas the R^{-6} term is an attractive force. In the potential energy versus distance graph (Fig. 2.3), a potential energy minimum of depth E_0 occurs at R_0. We expect the interaction between sections of neighbouring polymer molecules, such as $—CH_2—$ groups, to have an energy versus separation curve that is similar in shape to the one shown in Fig. 2.3. At room temperature, the *thermal energy* per carbon atom is of order kT, which is of the same order as the depth of the potential well. Consequently, the occupied energy level will be close to the top of the potential well. This diagram can be used to explain the high thermal

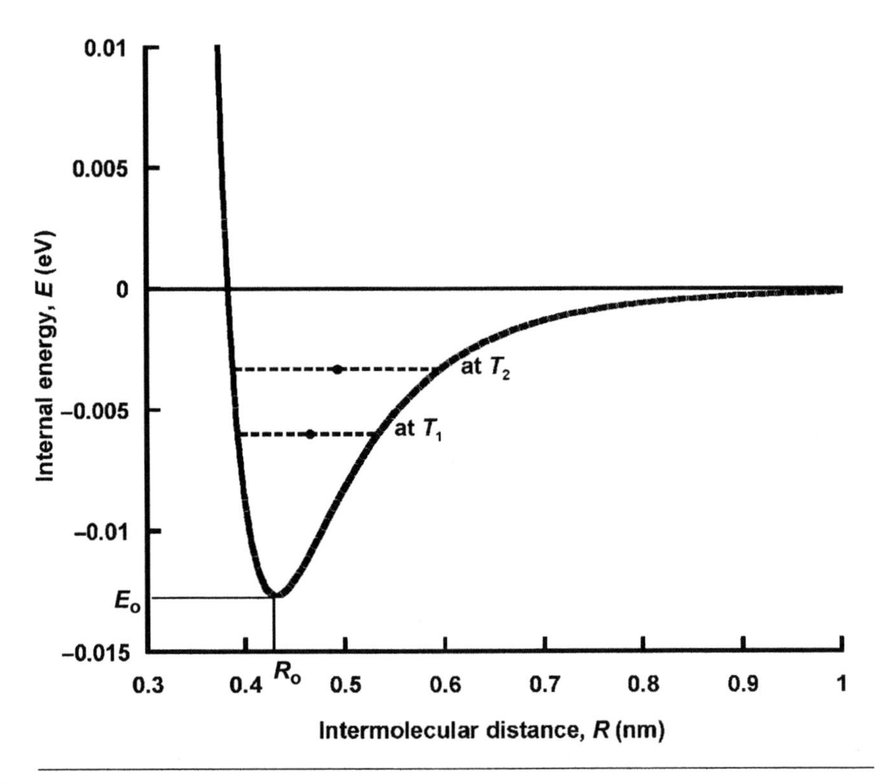

Figure 2.3 Variation of the internal energy E of a pair of methane molecules with the intermolecular distance R.

expansion coefficients of polymers; the mean separation of two CH_2 groups at a temperature T_1 occurs at the midpoint of the horizontal line shown. When the temperature is raised to T_2, the mean separation increases. The skew nature of the potential well explains the thermal expansion.

The energy E_0 needed to separate two methane molecules, is only about 1% of that needed to break a covalent bond. Table 2.2 compares the densities and melting points of crystalline forms of carbon and hydrocarbons. Continuous covalent bonds run in three dimensions in diamond (Fig. 3.3), two dimensions in graphite (a sheet of linked hexagons) and one dimension (along the chain) in polyethylene. The stronger the covalent bonding, the higher the density and melting point.

2.2.3 Hydrogen bonds

Hydrogen bonds exist in certain polymers, in particular polyamides. These bonds, intermediate in strength between covalent bonds and van der Waals forces, are responsible for the anomalous properties of water compared

Table 2.2 Forms of carbon and their properties

Material	*Covalent bonds in n dimensions*	*Density at 20 °C (kg m^{-3})*	*Melting point (°C)*
Diamond	3	3510	3820
Graphite	2	2250	3800
Polyethylene	1	1000	137
Methane	0	543 (at −200 °C)	−180

with H_2S. In polyamides, the hydrogen atom covalently bonded to nitrogen transfers part of this bond to the carbonyl group (—C=O) on the neighbouring polymer chain. Hydrogen bonds will be shown in crystalline structural models by a series of dots as in

$$—N—H\cdots\cdots O=C—$$

2.2.4 Ionic bonds

Ionic bonds, in which electrons are donated to, or received from other atoms, occur in a few polymers. Du Pont *Surlyn* ionomers are copolymers of ethylene and methacrylic acid. Part of the methacrylic acid is neutralised with zinc or sodium ions. Ionic bonds are relatively strong and ionomers contain clusters of ions, which act rather like crosslinks. Some ionomers are blended with other polymers to improve toughness.

2.3 Polymerisation

2.3.1 Naming addition polymers

The main types of polymerisation reaction are *addition* and *step-growth* polymerisations. Commodity plastics are all made by addition polymerisation, in which a vinyl monomer (one containing a C=C double bond) is converted into the polymer by the opening of the double bond. For example, the polymerisation of ethylene can be written

$$n\ CH_2{=}CH_2 \rightarrow \left[CH_2-CH_2 \right]_n$$

where the integer n is the degree of polymerisation. Addition polymer names consist of the prefix, poly- plus the monomer name. Table 2.3 gives some examples and the corresponding names.

Table 2.3 Structures of addition polymers

Generic structure	*Side group (X)*	*Polymer name*
$-[CX_2-CX_2]-$	H	Polyethylene
	F	Polytetrafluoro ethylene
	CH_3	Polypropylene
$-[CH_2-CX_2]-$	C_6H_5	Polystyrene
	Cl	Polyvinyl chloride
	CN	Polyacrylonitrile
$-[CH_2-CHX]-$	Cl	Polyvinylidene chloride
	F	Polyvinylidene fluoride
$-[(CH_2)_n-O]-$	N = 1	Polyoxymethylene
	$n = 2$	Polyoxyethylene

Addition polymerisation has three stages: initiation, propagation and termination. The reaction is usually *initiated* by the thermal decomposition of an unstable initiator molecule, such as a peroxide, to produce two free radicals. The free radical on the initiator fragment, shown as I•, attacks the covalent π bond in a monomer, leaving a free radical on the monomer.

$$I\bullet + CH_2{=}CH_2 \rightarrow I{-}CH_2{-}CH_2\bullet$$

Initiator decomposition is slow compared to the succeeding *propagation* steps, in which a monomer adds to the growing chain, with the free radical transferring to the chain end

$$\sim\sim\sim CH_2{-}CH_2\bullet + CH_2{=}CH_2 \rightarrow \sim\sim\sim CH_2{-}CH_2{-}CH_2{-}CH_2\bullet$$

Chain growth ceases when the free radical is either destroyed, as in the *termination* reaction when two chains link

$$\sim\sim CH_2{-}CH_2\bullet + \sim\sim CH_2{-}CH_2\bullet \rightarrow \sim\sim CH_2{-}CH_2{-}CH_2{-}CH_2 \sim\sim$$

or *chain-transfer* occurs to continue polymerisation of another chain

$$\sim\sim\sim CH_2{-}CH_2\bullet + H_2 \rightarrow \sim\sim\sim CH_2{-}CH_3 + H\bullet$$

The degree of polymerisation is controlled by the termination step. This may occur naturally as a result of a reaction with impurities in the monomer, or with a specific *chain transfer agent* such as a thiol compound, containing the weak S–H bond. The polymerisation reaction is irreversible.

2.3.2 Naming step-growth polymers

More complex polymer structures can be made by *step-growth* polymerisation; usually two monomer structures alternate in the chain. The alternative name is *condensation* polymerisation, since a by-product of water or other small molecule is often produced. Each monomer molecule has reactive groups at both ends. For example a diol can react with a dibasic acid.

$$HO-R-OH + HO-\overset{\overset{O}{\|}}{C}-R'-\overset{\overset{O}{\|}}{C}-OH \rightleftharpoons HO-R-O-\overset{\overset{O}{\|}}{C}-R'-\overset{\overset{O}{\|}}{C}-OH + H_2O$$

R and R′ represent unspecified chemical groups, while the two-way arrows indicate that the polymerisation process is reversible. To move the reaction equilibrium to the right, water must be removed from the reactor. The product still has reactive groups at both ends; in the general reaction step, an n mer and an m mer equilibrate with a $n+m$ mer. As the degree of polymerisation becomes large, this particular reaction produces a polyester.

The name of a step-growth polymer consists of poly- plus the name of linking group formed in the polymerisation. These are generic (family) names, so the groups R and R′ must be specified to identify the polymer. However, as relatively few polymers are commercialised, it is often unnecessary to spell out the details of the groups R and R′. Thus, the polycarbonate of 2,2 bis(4-hydroxyphenol) propane (the technical name is *Bisphenol A*) with *phosgene* ($COCl_2$) is referred to simply as polycarbonate (PC), because no other polycarbonate is sold on any scale. Table 2.4 lists the most common linking groups.

The polyurethane reaction, between a di-isocyanate and a diol

$$HO-R_2-OH + O{=}C{=}N-R_1-N{=}C{=}O \rightleftharpoons \left[O-\overset{\overset{O}{\|}}{C}-\underset{\underset{H}{|}}{N}-R_1-\underset{\underset{H}{|}}{N}-\overset{\overset{O}{\|}}{C}-O-R_2 \right]_n$$

produces no by-products, which is an advantage if the polymer is made in a mould.

While the reaction proceeds, there is equilibrium between polymer molecules of different degrees of polymerisation. Those with $n = 1, 2, 3, \ldots 10$ are referred to as oligomers. When a fraction p of the end groups have reacted, the mean degree of polymerisation is

Table 2.4 Linking groups in step-growth polymers and example structures

Linking group	Structure	Example polymer	Structure
Amide	$-NH-C(=O)-$	Polyamide 6	$[-NH-C(=O)-(CH_2)_5-]$
Carbonate	$-O-C(=O)-O$	Polyamide 6,6	$[-NH-C(=O)-(CH_2)_4-C(=O)-NH-(CH_2)_6-]$
Ester	$-C(=O)-O$		
Ether	$-O-$	Polycarbonate	$[-C_6H_4-C(CH_3)_2-C_6H_4-O-C(=O)-O-]$
Sulphone	$-S(=O)_2-$	Polyether sulphone	$[-C_6H_4-S(=O)_2-C_6H_4-O-]$
Urethane	$-NH-C(=O)-O-$	Polyethyleneterphthalate	$[-(CH_2)_2-O-C(=O)-C_6H_4-C(=O)-O-]$

$$\bar{n} = \frac{1}{1-p} \tag{2.2}$$

Consequently, the reaction must be taken very close to completion, with $p > 0.999$, to obtain a useful high polymer. This means that the starting reagents must be pure and present in stoichiometric (exact ratio according to the molecular formula) amounts. It may be necessary to prepare and purify an intermediate monomer to allow the reaction to proceed to a high polymer. The polymerisation takes place as a batch process over a period of hours. In contrast with addition polymerisations, large amounts of heat are not evolved, so the polymer does not have to be suspended in a heat transfer medium such as water. The polymer must be in the melt state to allow the rapid diffusion of reactive groups towards each other; thus, polymerisation may need to be completed at a high temperature.

2.3.3 Molecular weight distribution

Molecular weights are measured in atomic mass units, with hydrogen = 1 unit, carbon = 12 units, etc. The polymer molecular weight M is related to the *degree of polymerisation* n and the repeat unit molecular weight M_r by

$$M = n\,M_r \tag{2.3}$$

It is impossible to manufacture a truly monodisperse polymer in which every molecule has the same value of M. Either the *molecular weight distribution* (MWD), or statistical averages of the MWD, are measured to characterise polymers. The *mean* and *standard deviation* are familiar statistical measures. An equivalent of the mean is used to characterise polymers, but the standard deviation is not used because the distribution shapes are skew rather than 'normal'.

If f_i is the frequency of occurrence of molecules with degree of polymerisation i, the mean degree of polymerisation $\bar{n}$ is given by

$$\bar{n} = \frac{\sum_{i=1}^{\infty} f_i i}{\sum_{i=1}^{\infty} f_i} \tag{2.4}$$

The *number average molecular weight* M_N is defined as the product of $\bar{n}$ with the repeat unit molecular weight M_r

$$M_N = \bar{n} M_r \tag{2.5}$$

If a polyethylene has $n = 400$, then as $M_r = 28$, $M_N = 11\,200$. The moderate value of n may give a false impression that really large molecules are not present.

The weight average molecular weight M_W is defined by

$$M_W = M_r \frac{\sum_{i=1}^{\infty} f_i i^2}{\sum_{i=1}^{\infty} f_i i} \tag{2.6}$$

The ratio M_W/M_N is used to characterise the width of MWDs. For a range of polymers produced by a particular route, the MWD often has the same relative width, so the ratios M_W/M_N and M_Z/M_W remain constant. M_Z is an average of i^3 calculated by an equation similar to Eq. (2.6). For example, most commercial PVCs have $M_W/M_N = 2$. In such a case the measurement of a single molecular weight average is enough to specify the whole MWD.

Molecular weight averages are measured using properties that directly relate to the molecular size. The ratio, of osmotic pressure of a dilute polymer solution to the solution concentration, is proportional to the number of molecules present per unit volume, hence inversely proportional to M_N. M_W can be measured from the intensity of light scattered from dilute polymer solutions. Gel permeation chromatography (GPC) is an *indirect* method of determining the MWD. In this technique, sometimes called size exclusion chromatography, a constant flow rate of solvent passes through columns filled with swollen crosslinked polymer gel. Columns containing different pore sizes are placed in series, to separate polymers over a wide molecular weight range. The sample of dilute polymer solution is injected into the flow stream at a known time, then the polymer concentration at the far end of the columns, detected by refractive index measurement, is plotted against the elution volume. The largest molecules are the quickest to pass through the columns, because they are unable to diffuse into the smaller passages in the gel. GPC instruments are calibrated using samples of so-called 'monodisperse' polystyrene, having $M_W/M_N = 1.05$. Figure 2.4 shows that the range of molecular weights in a polyethylene is extremely wide, extending from oligomers with $n < 10$ to very large molecules with $M > 10^6$. The distribution appears to be close to the Gaussian or Normal distribution, but the molecular weight scale is logarithmic; hence it is referred to as a log-normal distribution.

The theoretical form of the MWD can be calculated for step-growth polymerisations, since the reaction is at equilibrium. The number fraction f_i of molecules with i repeat units is equal to the probability that a molecule chosen at random has i repeat units. Since each step in forming a chain is mutually exclusive, we can multiply the probabilities that the first unit is

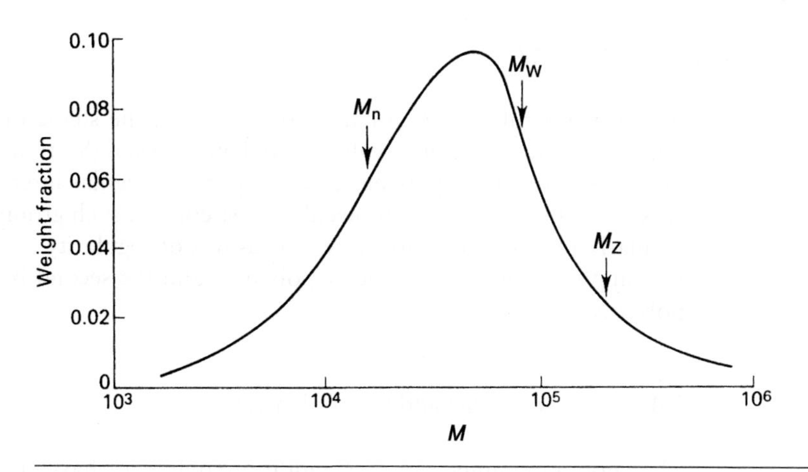

Figure 2.4 The molecular weight distribution of a polyethylene, determined using GPC. The various molecular weight averages are shown.

polymerised, that the second is polymerised and so on. These probabilities are equal to p, the average extent of reaction, while the probability that the ith unit is not polymerised is $(1 - p)$. Consequently

$$f_i = p^{i-1}(1 - p) \tag{2.7}$$

When this theoretical distribution is substituted in Eqs (2.4) and (2.6), we find that

$$M_N = \frac{M_r}{1-p} \quad \text{and} \quad M_W = M_r\left(\frac{1+p}{1-p}\right) \tag{2.8}$$

As commercially available polymers have $p \cong 1$, the ratio of $M_W/M_N \cong 2$. There is no such simple theory for addition polymerisation. If there are a number of different types of polymerisation sites on a catalyst used for free radical addition polymerisation, the MWD can be very broad, while 'single-site' catalysts can produce narrow MWDs.

We will see later that the desirable mechanical properties, such as resistance to cracking, improve as M_N increases. On the other hand, the ease of fabrication of polymers by melt processing decreases rapidly as M_W increases. Commercial polymers therefore have MWDs that are the best compromise for a particular process and application area. There is a trend to manufacture polymers of narrower MWD, to improve mechanical properties without sacrificing processability.

For *quality control* purposes, rather than measuring molecular weight averages, properties that correlate with molecular weight can be measured. Examples are the dilute solution intrinsic viscosity, and the melt viscosity under specific conditions (see the melt flow indexer in Section 7.1.1).

2.4 Chain regularity

As a general rule, for a polymer to crystallise, the shape of its molecules must repeat at regular intervals. In a polymer crystal (Section 3.4.2) a group of atoms (part of the polymer chain) repeats at regular intervals across the crystal. Consequently, the molecules must contain such groups of atoms, in regularly repeating positions. Two measures of regularity are explored; the first applies to most commodity polymers and the second is important for polyethylene.

2.4.1 Stereoregular addition polymers

The monomer units for most addition polymers have asymmetric side groups; one exception is polyethylene. We consider the vinyl monomer $H_2C{=}CHX$, where the side group X represents Cl, CH_3, etc. During polymerisation the monomer units add head to tail, so X side groups occur on alternate C atoms in the polymer. Figure 2.5 shows part of the polymer molecule, with its backbone in the fully extended form, a planar zigzag of C atoms. As the C—C—C bond angle is 112°, if any C—C bond rotates, the backbone C atoms will no longer all lie in one plane. In the view shown, the side groups X appear to be on one side of the chain or the other.

Catalysts control the position of the monomer unit added, relative to the end of the growing chain. The monomer side group can be on the same side as the last one – a *meso* (m) placement – or on the opposite side – a *racemic* (r) placement. The stereoregularity of polymer chains is often idealised as being of one of three types.

Isotactic: the side groups X are all on the same side of the chain
Syndiotactic: the side groups alternate from side to side
Atactic: the side groups have random positions

These words are based on the Greek roots *iso-* = same, *a-* = not, *tactos* = form. Isotactic chains contain exclusively mm monomer placements; syndiotactic chains contain exclusively mr or rm placements, while atactic chains contain 25% rr, 25% mm and 50% mr placements. Special catalysts are necessary to produce isotactic polypropylene, which is highly crystalline, whereas atactic polystyrene is produced without the use of special catalysts. It is possible to produce isotactic polystyrene and atactic polypropylene, but these have very limited markets.

Stereoregularity is more accurately described by the statistical distribution of sequences of neighbouring side group placements. In the simple (Bernoullian) form of chain growth statistics, the probability α of an m

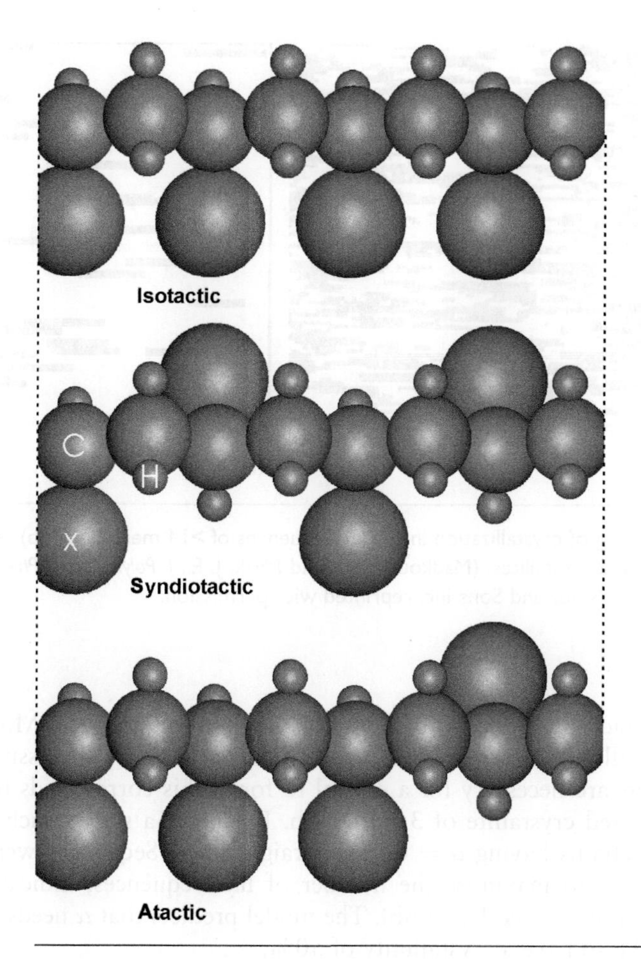

Figure 2.5 Views of part of fully extended vinyl polymer chains, showing the various stereoisomers.

placement is independent of the previous monomer placement. Consequently, the probability of two consecutive m placements (mm) is α^2, while that of mr placements is $\alpha(1 - \alpha)$. Nuclear magnetic resonance (NMR) can determine the proportions of *mmmm*, *mmmr*, etc. monomer unit pentads in a polymer. For a commercial PVC, the triad populations obey Bernoullian statistics with a fraction of $mr + rm$ triads $= 0.51 \pm 0.02$. Consequently, the probability of n successive racemic placements is 0.71^n. Hence, there is a 0.06 probability that eight racemic units occur in a row. Since the crystallinity of PVC is about 10% it can be inferred that such sequences are regular enough to fit into a crystal lattice.

Stereoregularity control, by choice of catalyst and polymerisation conditions, is most important for polypropylene, of the commodity polymers.

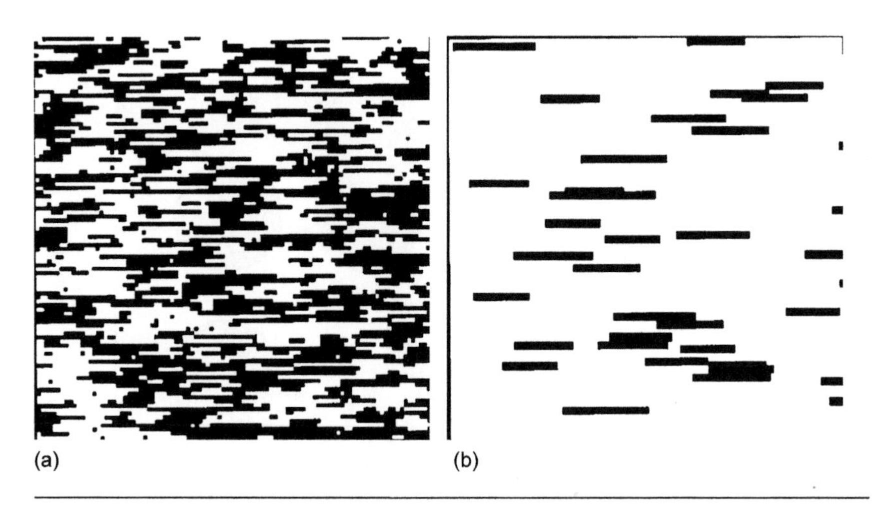

Figure 2.6 Modelling of crystallization in PP: (a) Sequences of >14 meso units; (b) sequences shifted to create crystallites. (Madkour, T. M. and Mark, J. E., *J. Polym. Sci. B Phy.* **35**, 2757, 1997) © John Wliley and Sons Inc. reprinted with permission.

A typical isotactic PP has 96–99% of *mmmm* pentads. Modelling of the crystallinity of isotactic PP assumed that at least 14 successive *meso* placements are necessary for a crystal to form; this corresponds to the smallest observed crystallite of 3 nm length. Figure 2.6a shows such sequences, in 100 chains having $\alpha = 0.9$, as straight lines. Sequences were slid by one another to maximise the number of like sequences, indicating the likely crystalline areas (Fig. 2.6b). The model predicts that α needs to be 0.95 for the PP to have a crystallinity of 50%.

2.4.2 Copolymerisation

Ethylene has a symmetrical monomer, so the concept of tacticity does not apply. Consequently, the crystallinity of polyethylene is controlled either by chain branching or by copolymerisation. Copolymers are classified into random and block copolymers (Fig. 2.7) depending on whether the monomer locations are random, or whether long blocks of each monomer exist. Polyethylene copolymers are random. The figure suggests that the local composition of a random copolymer is the same as that of the monomer mixture. However, in a batch copolymerisation, monomers tend to add to the end of a growing chain at different rates. The monomer ratio drifts as the polymerisation proceeds, so polymer formed at the end of the polymer-

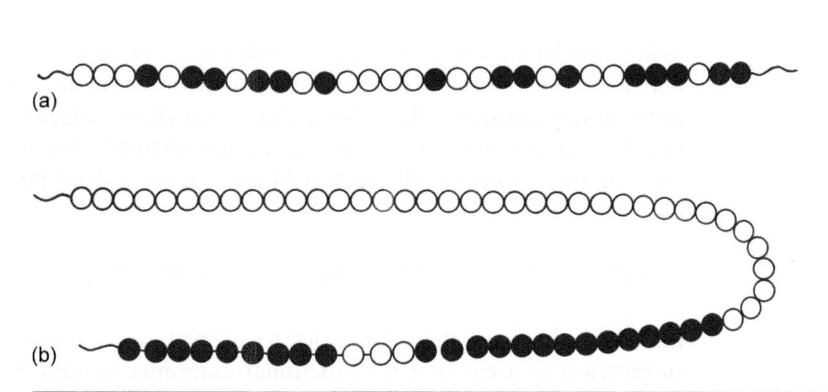

Figure 2.7 Copolymers of two monomers: (a) Part of a random copolymer containing 60% M_1; (b) part of a block copolymer in which there is a 90% probability that each monomer is joined to another of the same kind.

isation can differ markedly in composition from that formed at the start. However, for polyethylene copolymers made by continuous polymerisation, the composition is more stable.

2.4.3 Block copolymers

The nomenclature *poly* (M_1–*b*–M_2) is used where M_1 and M_2 are the monomer names; for example poly (styrene-b-butadiene). To make block copolymers, the polymer chains must have the ability to propagate (*living polymers*) when the first monomer is replaced by the second. In conventional addition polymerisation the chain termination and transfer processes make the lifetime of a growing polymer chain too short. Consequently, special ionic polymerisation catalysts were developed. A fixed number of di-anions such as $^{-}[C_6H_5CHCH_2CH_2CHC_6H_5]^{-}$ are introduced into an inert solvent. These propagate from both ends if a suitable monomer is introduced. As there are no termination or transfer reactions, once the first monomer has been consumed, a second monomer can be introduced to produce a triblock copolymer such as styrene–butadiene–styrene. Each block has a precisely defined molecular weight. These materials undergo phase separation (Chapter 4) and act as *thermoplastic rubbers*.

2.5 Branched and crosslinked polymers

2.5.1 Chain branching

When ethylene (ethene) is copolymerised with small proportions of higher alkenes (olefins), the resulting short-chain branches modify the polymer crystallinity (Section 3.4.1). Long-chain branched molecules (Fig. 2.8) can occur as a result of a side reaction; for example when a propagating polyethylene molecule abstracts a H atom from a dead polyethylene molecule

$$-CH_2-CH_2\bullet + -CH_2-CH_2- \rightarrow -CH_2-CH_3 + -CH\bullet -CH_2-$$

and the side chain then continues to propagate. Alternatively, if a small proportion of a tri- or tetra-functional monomer is used in a step-growth polymerisation, this produces single or double branches in the resulting polymer.

In *graft copolymers*, the polymer backbone consists of one monomer and the branches of another. For example, polybutadiene contains carbon–carbon double bonds that can be attacked by a free radical initiator

$$I\bullet + \sim CH_2-CH{=}CH-CH_2\sim \rightarrow \sim CH_2-\dot{C}H-CHI-CH_2\sim$$

If styrene is available, polystyrene branches can be grafted on to the polybutadiene backbone. The grafting efficiency is not high, as separate polystyrene molecules will also be formed. Once the polystyrene concentration reaches 2%, phase separation occurs, with spheres of polystyrene

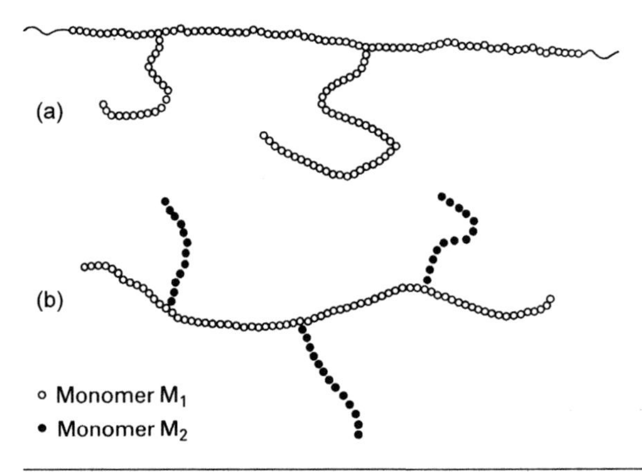

Figure 2.8 (a) Long-chain branching as found in LDPE polymerised at high pressures; (b) graft copolymerisation of a monomer M_2 onto a backbone of monomer M_1.

forming in the polybutadiene matrix. In the final composite material, the graft copolymer is concentrated at the polystyrene polybutadiene phase boundaries, where it aids the mechanical properties.

2.5.2 Thermosets

When the density of branch points is increased in a polymer, there is a progression, from a collection of branched molecules, through a single infinite tree molecule containing no closed rings (Fig. 2.9a), to a three-dimensional network molecule (Fig. 2.9b). When a single tree molecule forms, the *gel point* occurs; if a solvent is added the majority of the polymer forms a swollen gel, rather than dissolving. Both thermosets and rubbers are

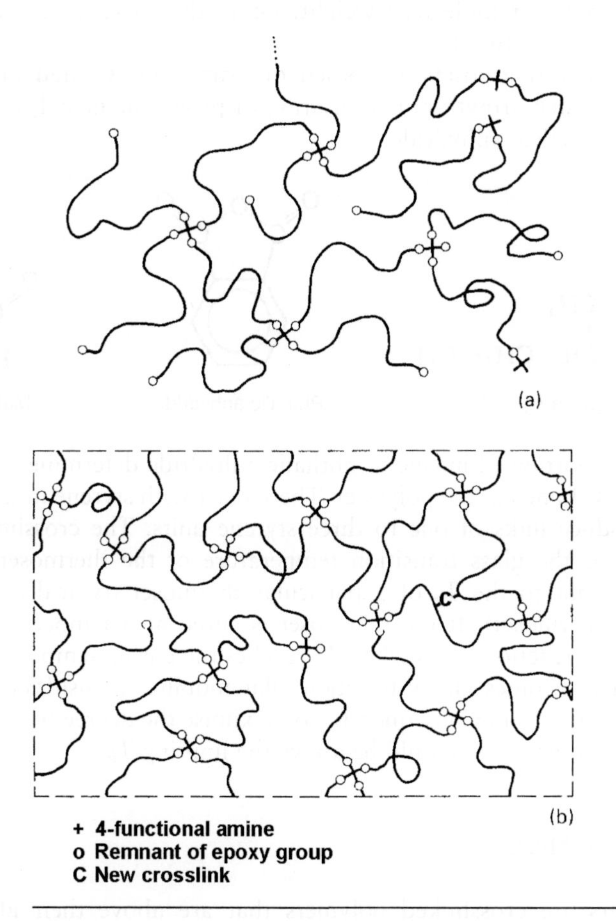

Figure 2.9 (a) Part of an infinite tree molecule that forms during the crosslinking of a thermoset; (b) part of an infinite three-dimensional network molecule, showing the effect of another crosslink (C).

examples of infinite, three-dimensional network molecules. The chemical structure can be illustrated by the epoxy thermoset system. There are two components, a prepolymer of molecular weight 1000–2000 with reactive epoxy groups at each end, and a multifunctional amine 'hardener'. In the crosslinking reaction

$$R_1-\underset{\mid}{\overset{O}{\overbrace{C-CH_2}}}\hspace{-2.5em}_{H} \quad + \quad H_2N-R_2-NH_2 \rightarrow R_1-\underset{H}{\overset{OH}{C}}-\underset{H}{\overset{H}{C}}-\underset{H}{N}-R_2-\underset{H}{\overset{H}{N}}$$

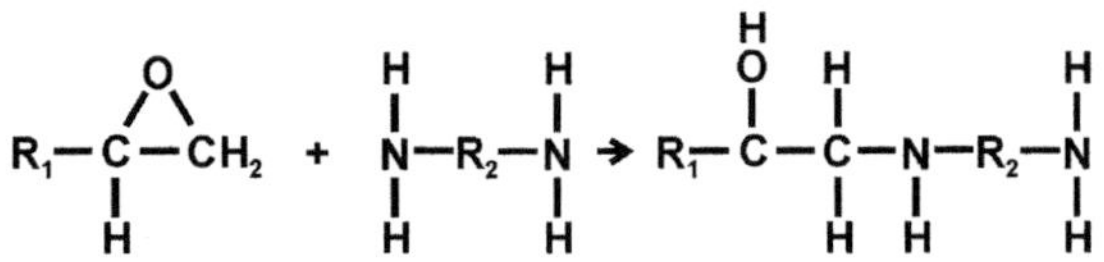

the epoxy ring is opened without any by-product being produced. Figure 2.9b shows that if stoichiometric quantities of amine hardener are used, each of these molecules is linked to four others by a network chain of the prepolymer molecular weight. Other thermoset systems produce less well-defined networks.

Polyester thermosets are based on partly unsaturated linear polyester from the step-growth polymerisation of propylene glycol, phthalic anhydride and maleic anhydride.

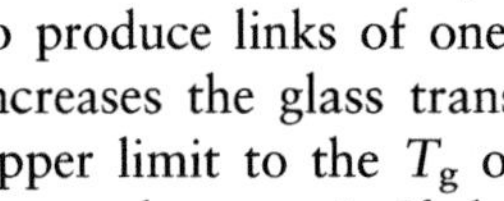

$HO-\overset{CH_3}{\overset{|}{C}H}-CH_2-OH$ — Propylene glycol

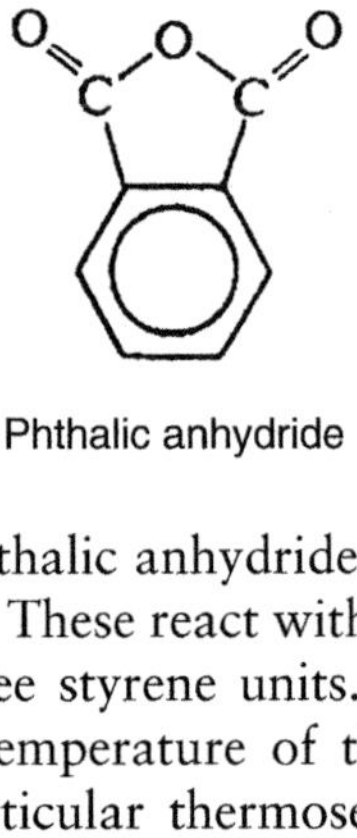

Phthalic anhydride

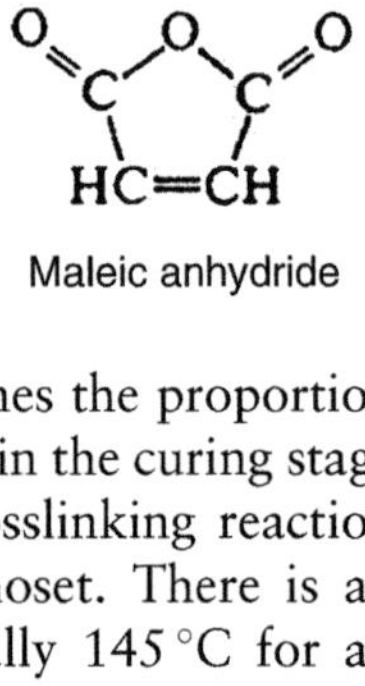

$\overset{O=C-O-C=O}{HC=CH}$ — Maleic anhydride

The proportion of maleic to phthalic anhydride determines the proportion of C=C bonds in the polyester. These react with styrene in the curing stage to produce links of one to three styrene units. The crosslinking reaction increases the glass transition temperature of the thermoset. There is an upper limit to the T_g of a particular thermoset (typically 145 °C for an epoxy thermoset). If the thermoset is cured at a temperature below this limit, the reaction stops when T_g reaches the curing temperature; when the polymer becomes glassy the molecular mobility is insufficient for further reaction to occur. Consequently, to maximise the degree of crosslinking, the curing temperature should be above the limiting T_g.

2.5.3 Rubbers

Rubbers are crosslinked polymers that are above their glass transition temperatures at room temperature. If a crystalline phase can form, its melting point (T_m) must also be below room temperature. In contrast with

Table 2.5 Rubber structures

Rubber	*Polymer*	*Structure*	T_g (°C)	T_m (°C)
Natural	Polyisoprene	$[-CH_2-C(CH_3)=CH-CH_2-]$	−73	25
	Polybutadiene	$[-CH_2-CH=CH-CH_2-]$		55
Butyl	Polyisobutylene	$[-CH_2-C(CH_3)_2-]$	−70	0
Silicone	Polydimethyl siloxane	$[-O-Si(CH_3)_2-]$	−123	−70

thermosets, most rubbers are prepared by crosslinking a high molecular weight polymer (polyurethane rubbers are an exception). The repeating units of rubbers (Table 2.5) neither contain rigid phenyl rings in the main chain, nor as side groups, as they would hinder chain rotation. Polar groups, which would increase the strength of intermolecular forces, are also absent. The polymers contain unsaturated carbon double bonds, to allow crosslinking. If, as for silicone rubber, the repeat unit appears saturated, it is copolymerised with 1–2% of a monomer that produces unsaturated groups.

The original crosslinking process for natural rubber, called *vulcanisation*, involved mixing in 2–3% of sulphur plus an accelerator. On heating to 140 °C the sulphur reacts with C=C bonds on neighbouring polyisoprene chains to form sulphur crosslinks $C-(S)_n-C$. Typically, 15% of the crosslinks are monosulphide ($n = 1$), 15% are disulphide and the rest are polysulphide with $n > 2$. The polysulphide crosslinks are partially labile, which means that they can break and reform with other broken crosslinks when the applied stresses are high. This leads to permanent creep in compressed rubber blocks. To avoid such *permanent set*, 'efficient' vulcanisation systems have been developed that produce only monosulphide crosslinks.

The rubber network can be characterised by the network chain molecular weight M_c. The network chain is the length of polymer chain between neighbouring crosslinks, and M_c is its number average molecular weight. M_c grams of rubber contains 1 mole of (Avogadro's number N_A of) network chains. 1 m^3 of rubber has a mass 1000ρ grams where ρ is the rubber density in kg m^{-3}; it contains N network chains, where N is the density of network chains. The number of network chains per gram of rubber is the same in both examples, so N is related to the network chain molecular weight by

$$\frac{M_c}{1000\rho} = \frac{N_A}{N} \tag{2.9}$$

If the crosslinks are four-functional, each connects four network chains. Figure 2.9 (point c) shows that when a crosslink is introduced between two network chains, the result is four network chains, i.e. each new crosslink increases the number of network chains by two. Consequently, the crosslink density is $N/2$. Hence, by Eq. (2.9), M_c is inversely proportional to the crosslink density.

2.6 Technology and economics of manufacture

2.6.1 Monomer manufacture

Monomer prices are determined by the raw material and energy costs, the number of stages in their manufacture and the scale of manufacture. The main raw material is the naphtha fraction from crude oil distillation, which boils in the range 20–200 °C; it contains hydrocarbon molecules with 4–12 °C atoms. *Cracking* of a mixture of naphtha and steam at a temperature of about 850 °C for 0.5 s, produces a complex mixture of products, some of which are used for plastics manufacture. These are separated by fractional distillation (Fig. 2.10). The gases are liquefied by cooling to −140 °C at pressures up to 40 bar, then a series of fractional distillation towers strip off the products; roughly 30% of ethylene (ethene), 15% of propylene (propene), with 20% fuel gas, 20% gasoline and 9% of four carbon atom hydrocarbons. Aromatic compounds (benzene, toluene, xylene, etc.) produced in a separate part of the petrochemical complex, are also used for monomer production.

In contrast with ethylene monomer production, many stages are needed to produce the monomers adipic acid and hexamethylene diamine for nylon 6,6 (Fig. 2.11). None of these stages is 100% efficient, although new catalysts have increased the efficiency of some stages. Since energy is consumed in each reaction stage, and the capital cost of the many reactors is high, the cost per tonne of nylon 6,6 is four or five times that of polyethylene. Moreover the scale of production is smaller by a factor of 50; the implications of this are explored later.

2.6.2 Polymerisation processes

The large-scale manufacture of addition polymers is usually via a continuous process, with monomer addition and polymer removal occurring at a

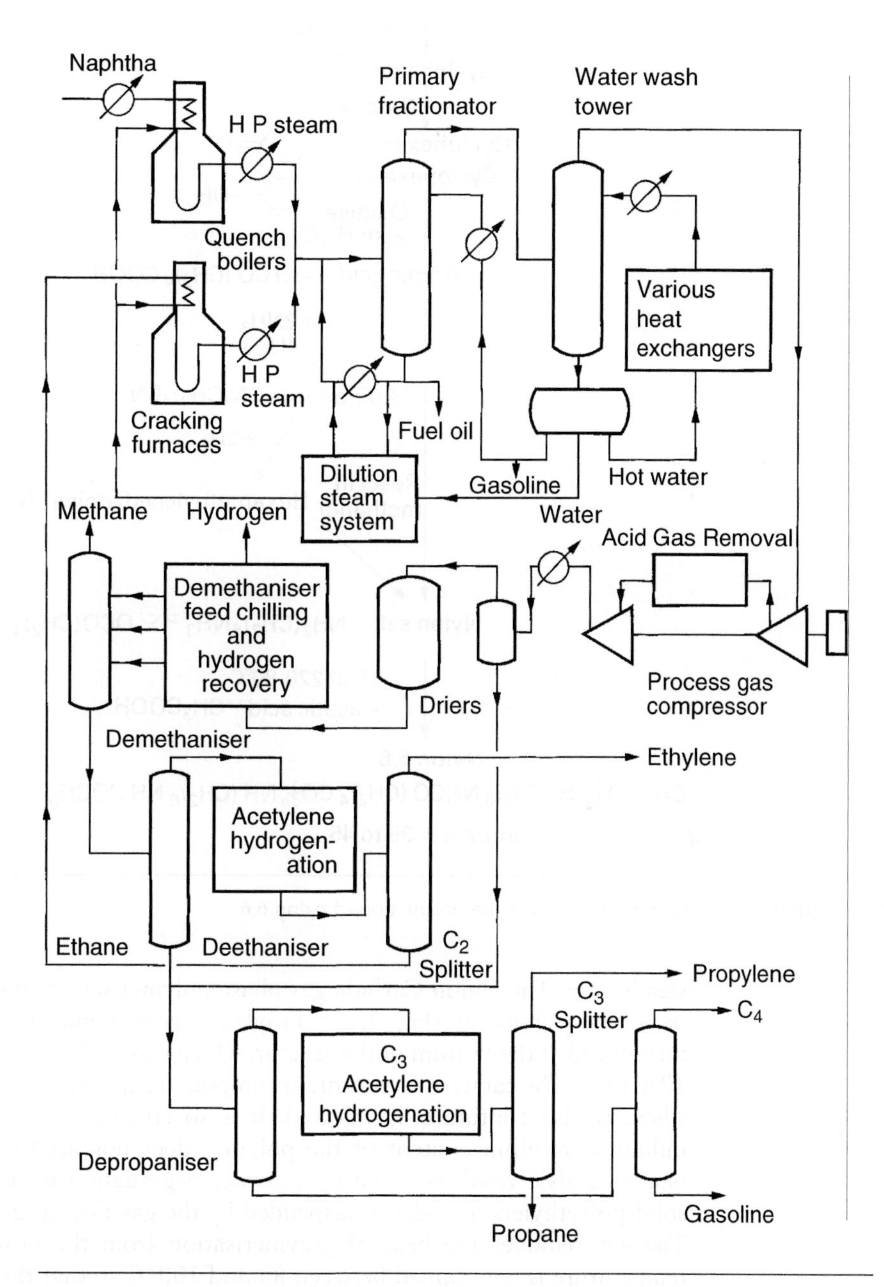

Figure 2.10 Flow chart for the separation of the products from cracking naphtha.

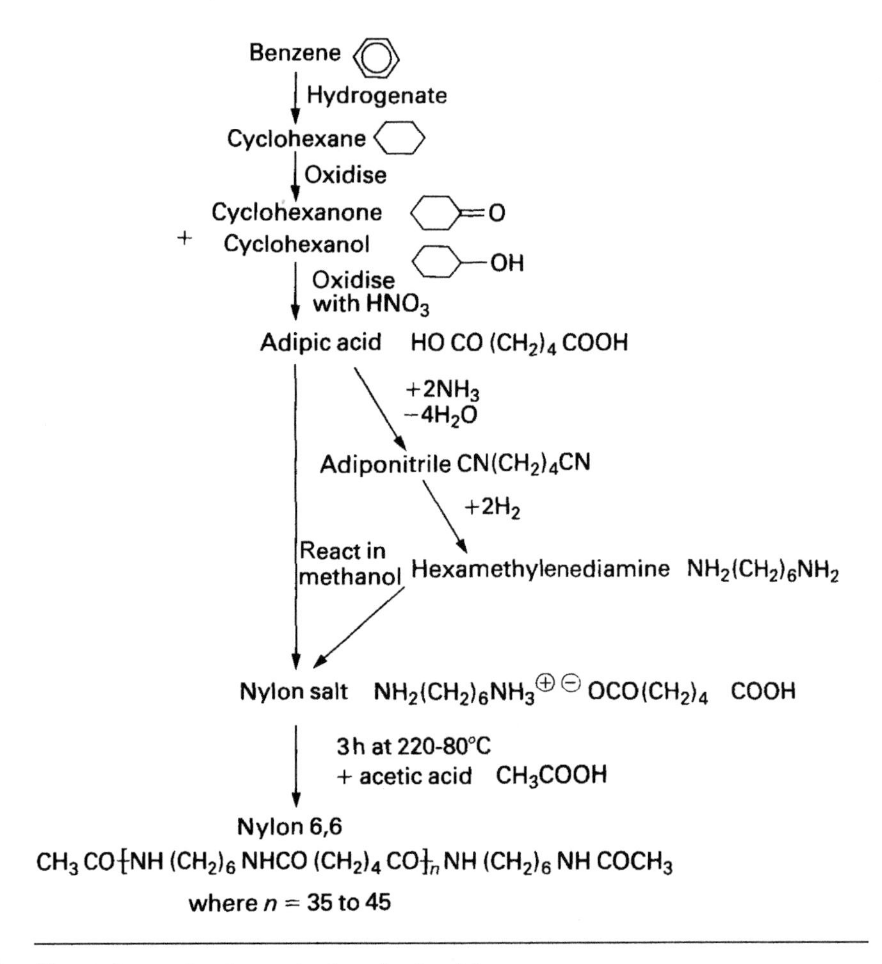

Figure 2.11 Chemical route for the production of nylon 6,6.

steady rate. The Union Carbide gas phase polymerisation of ethylene illustrates the technology (Fig. 2.12). Ethylene gas at about 20 bar pressure is introduced at the bottom of the reactor, which can be 2.5 m in diameter and 12 m high. The catalyst, a chromium compound supported on finely divided silica, is also continuously injected. It is so efficient that the 1 part per million chromium content of the polymer does not need to be removed (some catalyst residues speed up polymer degradation or are toxic). The solid polyethylene powder is suspended by the gas flow as a fluidised bed. The gas removes the heat of polymerisation from the powder, and the temperature is maintained between 85 and 100 °C so that the polyethylene particles are solid and do not stick together. There is a mixture of monomer, polymer and a small proportion of growing chains in the reactor at any time, because chain propagation is much faster than initiation. Only about 2% of the ethylene is polymerised each time it passes through the reactor;

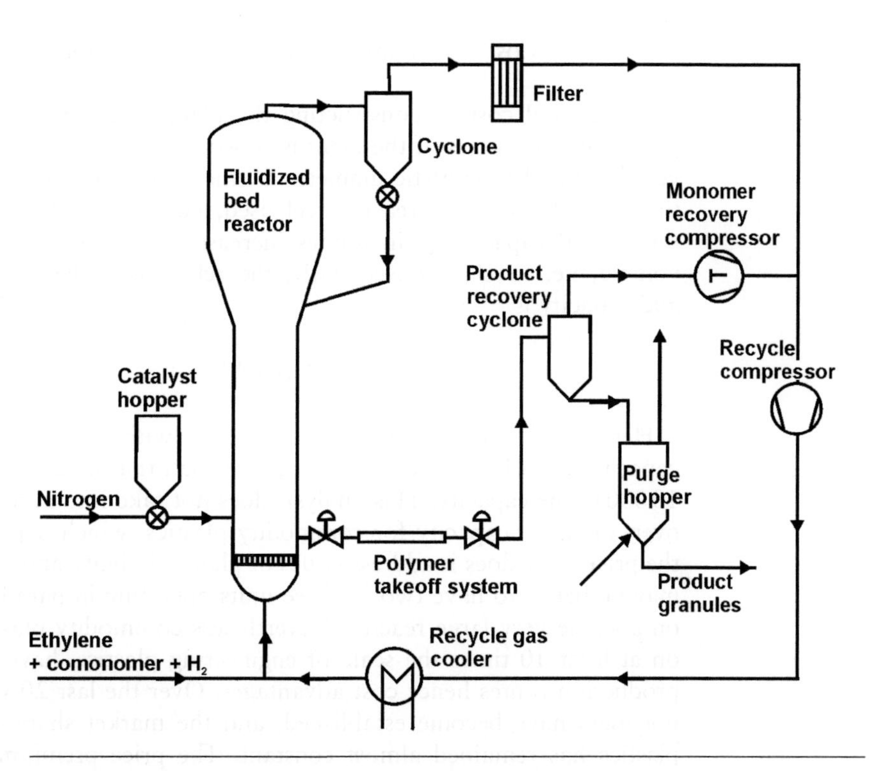

Figure 2.12 Union Carbide process for gas phase polymerisation of ethylene.

the remaining monomer is compressed and cooled before being returned to the reactor. The solid polyethylene particles sink down the reactor as they grow in size; after 3–5 h they are about 0.5 mm in diameter and can be removed from the reactor. The polymer needs to be purged with nitrogen before being conveyed to storage in an inert gas stream.

The reactor can produce polyethylene homopolymer, or copolymers with butane, octane, etc. so the overall crystallinity of the product can be controlled. The molecular weight is controlled by additions of hydrogen, and the width of the molecular weight distribution can be changed by modifying the catalyst.

2.6.3 The economics of scale

The polyethylene process just described was originally operated as a pilot plant with a reactor 0.1 m in diameter, producing 50 tonnes per year. The largest operating reactor has a 4.5 m diameter reactor and is capable of producing 100 000 tonnes per year. With increased scale, the production cost per tonne of polymer decrease considerably. The production costs contain main elements.

$$\text{Cost} = \text{monomer} + \text{energy} + \text{share of capital cost} \tag{2.10}$$

The capital cost of constructing the plant is depreciated over a fixed period of say 5 years. If the plant is constructed of steel pipes and pressure vessels of a characteristic diameter D, then the capital cost £C is proportional to the surface area of steel used, i.e. to D_2. On the other hand, the annual capacity Q in tonnes increases in proportion to the reactor volume, i.e. to D_3. Consequently, the relationship between capital cost and capacity is

$$C \propto Q^{0.66} \tag{2.11}$$

Hence, polyethylene produced in a plant with a 100 000 tonne annual capacity has a lower manufacturing cost than that made in a plant with a 10 000 tonne capacity. This analysis does not take into account temporary over- or under-capacity for commodity plastics, which depresses or raises the price. Nor does it address issues of plant reliability and maintenance; it may be better to have two or three units operating in parallel then to rely on a single very large reactor. Nevertheless commodity plastics, produced on at least 10 times the scale of engineering plastics, have much simpler production routes hence cost advantages. Over the last 20 years, few new polymers have become established, and the market share of commodity plastics has remained almost constant. The price premium for PET and ABS, relative to the commodity plastics, has sunk. Even if a new polymer offers specific property advantages, it faces stiff competition from commodity plastics upgraded by adding reinforcing fillers and toughened by rubbery additions.

2.7 Grades and applications of commodity plastic

Each commodity thermoplastics is available in many grades, either adapted to the requirements of a process, or to the demands of a product type. The major variants will be described and some of the jargon explained. A few of the important applications will be described, while others appear throughout the book.

2.7.1 Polyethylenes

The original ICI process for polymerising ethylene, developed in the 1940s, produces low density polyethylene (LDPE) with a density in the range 910–935 kg m^{-3}. Ethylene, compressed to pressures between 1400 and 2400 bar at a temperature between 200 and 250 °C, is above the critical point where liquids and gases can be distinguished. When it is polymerised using a free radical catalyst, side reactions cause the formation of both short- and long-chain branches on the molecules. Short ethyl ($—CH_2CH_3$) or *n*-butyl ($—CH_2CH_2CH_2CH_3$) branches replace a hydrogen atom on between 1.6 and 3% of the C atoms in the chain. The infrequent long-chain branches create comb-shaped molecules (Fig. 2.8b), with different flow properties to linear polyethylenes. If the polymerisation pressure is increased, the molecular weight increases while the number of branches decreases.

From 1955 onwards, Ziegler catalysts were used to produce high-density polyethylene (HDPE), with densities in the range 955–970 kg m^{-3}. The Union Carbide process, described in Section 2.6.2 is a related process, operating in the gas phase. Polymerisation occurs at low to medium pressures of 1–200 bar, with less side reactions, hence fewer short-chain branches (0.5–1% of the C atoms) and no long-chain branches. Subsequently, the range of polymer densities has been extended by copolymerisation with butene or 1-hexene, which produce ethyl and *n*-butyl short-chain branches, respectively. The copolymers are classified as medium density polyethylenes (MDPE) if the density is approximately 940 kg m^{-3} and linear low-density polyethylenes (LLDPE) if in the range of 918–935 kg m^{-3}.

Metallocene single-site catalysts allow use of higher copolymer content than do Ziegler catalysts. Typical comonomers are 1-hexene and 1-octene, which produce an *n*-hexyl branch. Very low-density polyethylenes can be made, with no long-chain branches and a narrow molecular weight distribution ($M_W/M_N = 2$), hence a lower level of melt elasticity. Sometimes the polymers with densities < 910 kg m^{-3} are called polyolefin elastomers. Metallocene polyethylenes have uniformly thin crystalline lamellae (Chapter 3), and slightly lower melting points than Ziegler PEs. Chain branches cannot easily be accommodated in polymer crystals, so the percentage crystallinity decreases as the amount of comonomer increases (Fig. 2.13). To estimate the volume fraction crystallinity, it is quicker to measure density than to use X-ray methods. It is assumed that the crystalline phase density is constant as the % crystallinity varies, i.e. the defect population inside the crystals that affects the density, is constant. As this assumption is not quite valid, crystallinity values vary by a few %, depending on the method of measurement.

2.7.1.1 *Density and melt flow index*

The mass of crystals in 1 m^3 of polymer is the product of the volume fraction crystallinity, V_c and the crystal density ρ_c. The mass of amorphous material is product of the amorphous volume fraction crystallinity $(1 - V_c)$ and the amorphous density ρ_a. Consequently, the polymer density ρ is linearly related to V_c by

$$\rho = V\rho_c + (1 - V_c)\rho_a \tag{2.12}$$

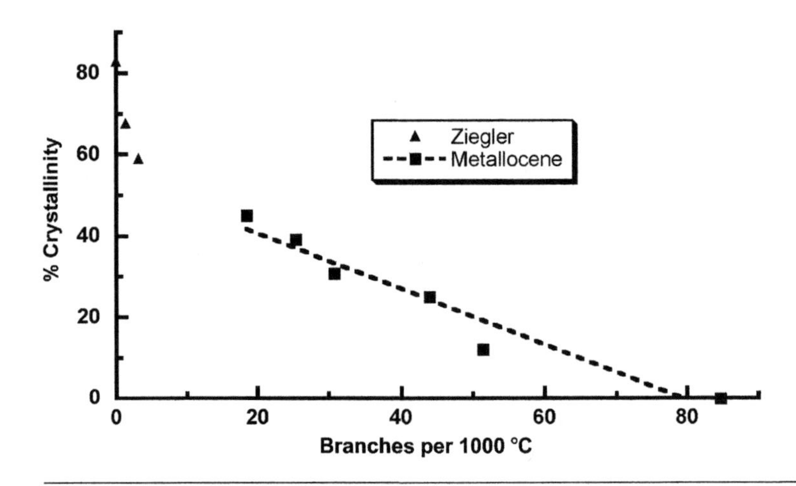

Figure 2.13 Crystallinity of polyethylene copolymers vs. the number of short branches per 1000 °C atoms. C4 comonomers for metallocene; C4 and C6 comonomers for Ziegler. (Data from Mirabella FM and Bafna A, *J. Polym. Sci. B* **40**, 1637, 2002).

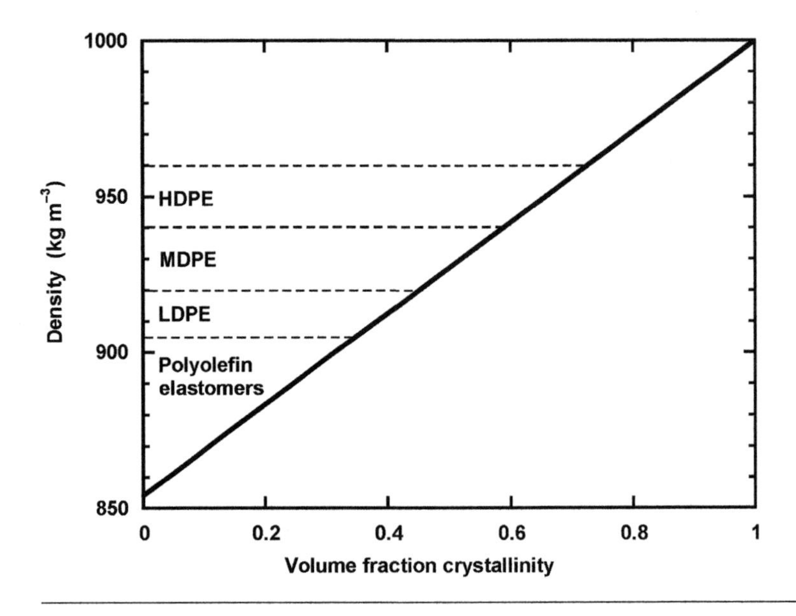

Figure 2.14 Variation of polyethylene density with crystallinity, with boundaries of PE classes.

where for polyethylene at 25 °C $\rho_c = 1000\,\mathrm{kg\,m^{-3}}$ and $\rho_a = 854\,\mathrm{kg\,m^{-3}}$. Figure 2.14 shows the variation of density versus crystallinity with the regions of HDPE, LDPE and polyolefin copolymer elastomers.

The *melt flow index* (MFI) is used as a production control assessment of the average molecular weight. The MFI is an inverse measure of melt viscosity

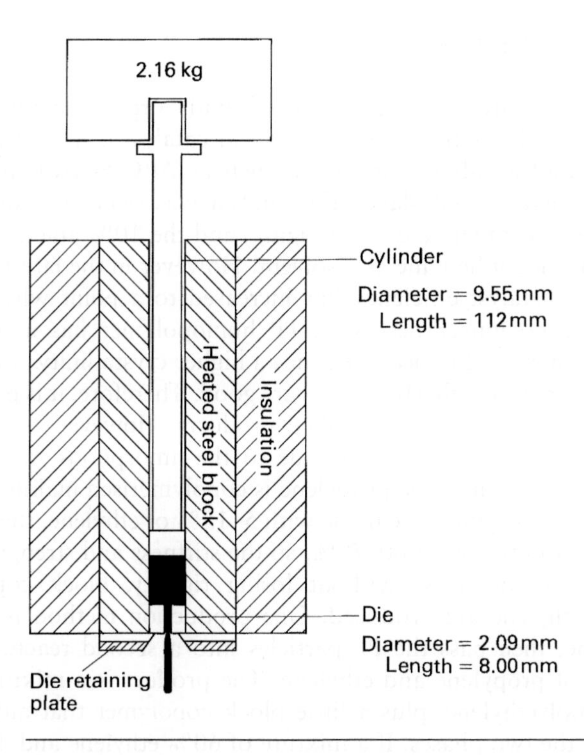

Figure 2.15 Melt Flow Indexer used for the quality control of polymer molecular weight.

under specific conditions, so as the MFI increases, the polymer melt should flow more easily. It is measured by heating the polyethylene to 190 °C and applying a fixed pressure of 3.0 bar to the melt via a piston of diameter 9.5 mm using a weight of 2.16 kg (Fig. 2.15). The MFI is defined as the output rate, in gram per 10 min from a standard die of 2.1 mm diameter and 8.0 mm length. Section 3.3.3 explains that the zero-shear rate melt viscosity is a function of the weight average molecular weight M_W. As the MFI is measured in the non-Newtonian region, the correlation between 1/MFI and M_W is not perfect; nevertheless it is better than the correlation with M_N or M_Z.

Polyethylenes with MFI > 5 are used for injection moulding whereas grades with MFI < 0.5 are used for extrusion and blow moulding. Ultra high molecular weight polyethylene (UHMWPE) with M_W of $2–4 \times 10^6$ is so viscous that its MFI cannot be measured. It is used for bearings in machinery, or in joint replacements in the human body (Chapter 15). About 80% of LDPE is used as film, mainly for packaging. Minor percentages are used for cable insulation, injection moulding and the extrusion coating of cardboard or other materials. The more rigid HDPE is used for the blow moulding of containers, and for the extrusion of pipe for gas and water supplies. A proportion is used for film, either on its own, or blended with LDPE. Blends of such similar polymers are compatible.

2.7.2 Polypropylene

Propylene is invariably polymerised with organometallic catalysts, to achieve high isotactic content. A typical catalyst is a mixture of titanium chloride and an aluminium alkyl, such as $Al(C_2H_5)_2Cl$. In the first and second generation PP plants, the catalyst was removed from the polymer by solvent treatment and centrifuging, and the 10% atactic polypropylene removed using *n*-heptane as a solvent. However in the latest process generation, the catalyst efficiency has increased to a point where these costly stages are no longer necessary. PP homopolymer has a higher Young's modulus than HDPE, because it has a higher crystallinity, and the mechanical properties of high MFI grades are good. Therefore, it is preferred for the injection moulding of thin-walled products. However, the homopolymer becomes brittle at about 0 °C due to the amorphous fraction becoming glassy. To overcome this, propylene is copolymerised mainly with ethylene. If random copolymers are made with 5–15% of ethylene, the crystallinity is reduced from the usual 60–70%, so the stiffness and strength are reduced. To improve toughness without losing rigidity, block copolymers with 5–15% ethylene are produced. One production method is to polymerise propylene, then pass the PP particles into a second reactor containing a mixture of propylene and ethylene. The product is a mixture of polypropylene, polyethylene, plus a little block copolymer that aids the bonding between the two phases. If a mixture of 60% ethylene and 40% propylene by weight is used in the second stage, a rubbery copolymer with a glass transition of −60 °C is formed. There have been significant improvements in the toughness of polypropylenes in recent years (Galli *et al.*, 2001). At the same time the optical clarity has increased, due to decreases in the spherulite size, and the Young's modulus has increased, due to increased crystallinity.

The higher melting point of polypropylene (170 °C) compared with HDPE (135 °C) makes it more suitable for fibre applications, and 30% of the production goes into carpets and ropes. About 20% is used in the form of oriented polypropylene film for wrapping potato crisps, cigarettes, etc. where good clarity is required.

2.7.3 Polyvinyl chloride

Most of the PVC intended for melt processing is made by suspension polymerisation in a batch reactor (Fig. 2.16), a large stirred pressure vessel with a water cooling jacket for temperature control.

A suspension of droplets of vinyl chloride monomer (VCM) in water, of size 30–150 μm, is formed by agitation with a stirrer. It is stabilised by a colloidal layer of partially hydrolysed polyvinyl acetate or other water-soluble polymer. When the suspension is polymerised at a temperature in the range 50–70 °C, PVC molecules form in the VCM droplets. As PVC is insoluble in VCM, it precipitates in the form of primary particles, initially

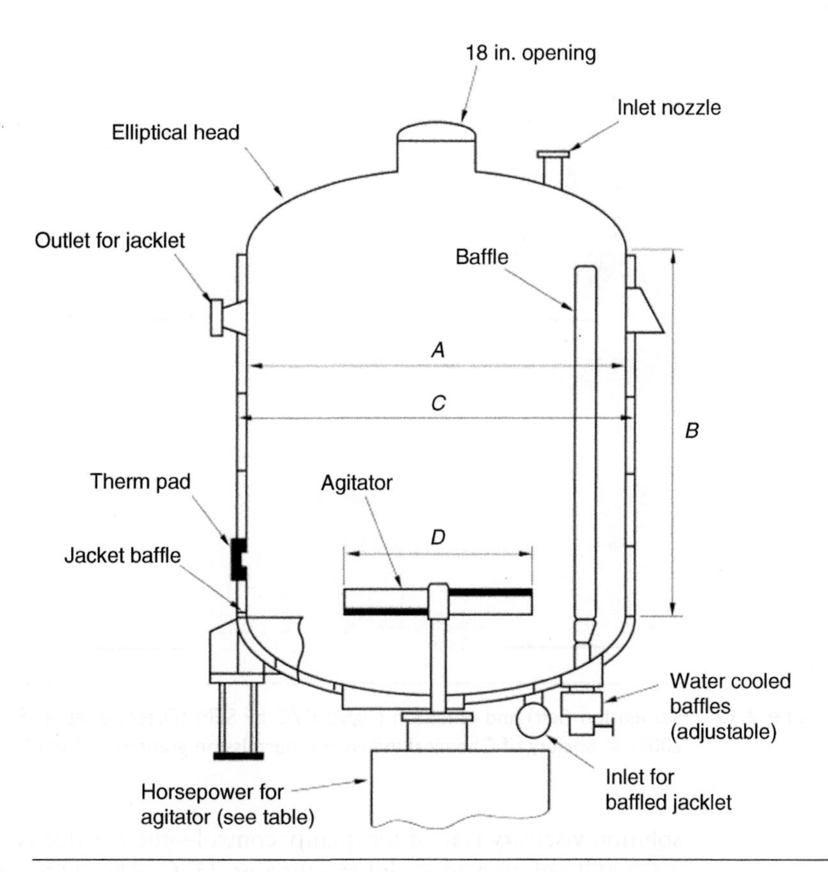

Figure 2.16 Batch reactor for PVC suspension polymerisation (Saeki Y *et al.*, *Prog. Polym. Sci.* **27**, 2055, 2002).

0.1–0.2 μm in diameter. As the percentage conversion increases over a 5–6 h period, the number of primary particles per droplet increases and the particles coalesce into aggregates of irregular shape (Fig. 2.17). The reaction is stopped at 85–95% conversion and the autoclave is discharged. Treatment with steam removes the remaining VCM from the PVC, which is then centrifuged and dried in a cyclone. The resulting powder particles are 'bumpy' spheres, of diameter about 150 μm; the internal porosity (Fig. 2.17) is used to absorb stabilisers, lubricants and plasticisers in the dry blending stage of processing. The colloid, that stabilised the suspension, also controls the porosity of the PVC grains.

It is also possible to polymerise 'emulsion' PVC. While as an emulsion in water, the spherical particles have diameters of 0.1–1 μm. On drying, these agglomerate to grains of mean size 30–60 μm (Fig. 2.17), smaller than the 100–160 μm grain size of suspension PVC.

The melt viscosity of PVC depends on the degree to which the particle structure is retained, as well as on the molecular weight. Consequently, a

Figure 2.17 Suspension (left) and emulsion (right) PVC by SEM (Diego, B. *et al.*, *Polym. Intl.* **53**, 515, 2004) © Society of Chemical Industries, permission granted by John Wiley & Sons.

solution viscosity is used for quality control—the *K*-value is the viscosity of a 0.5 g/dl solution in cyclohexanone at 25 °C. M_W increases with the *K*-value; PVCs with *K*-values of 55–62 are used for injection moulding and the extrusion of thin foil, whereas large diameter pipe, which requires a maximum toughness, uses PVC with *K*-values of 66–70.

Unplasticised PVC (UPVC) has a better resistance to sunlight and chemicals than other commodity plastics. The resistance can be improved by adding mineral fillers, while the toughness can be improved by adding a rubber phase. UPVC products are widely used on the exterior of houses (gutters, wall cladding, window frames) as well as waste-water systems.

PVC is unique among the commodity plastics in that about 50% of the production is sold with plasticiser incorporated. Plasticisers are high boiling point liquids which swell the PVC and reduce the glass transition temperature. Figure 2.18 shows how the shear modulus changes with temperature at different plasticiser contents. The glass transition temperature, at which the modulus falls most rapidly, is reduced to below room temperature at 40% plasticizer content. The 10% crystallinity of PVC prevents such a material being a sticky liquid at room temperature; instead it is a rubbery solid.

For a plasticiser to be compatible, so that large amounts dissolve in PVC, it should contain polar ester groups or polarisable benzene groups. It must also have a low vapour pressure at room temperature and not diffuse out of

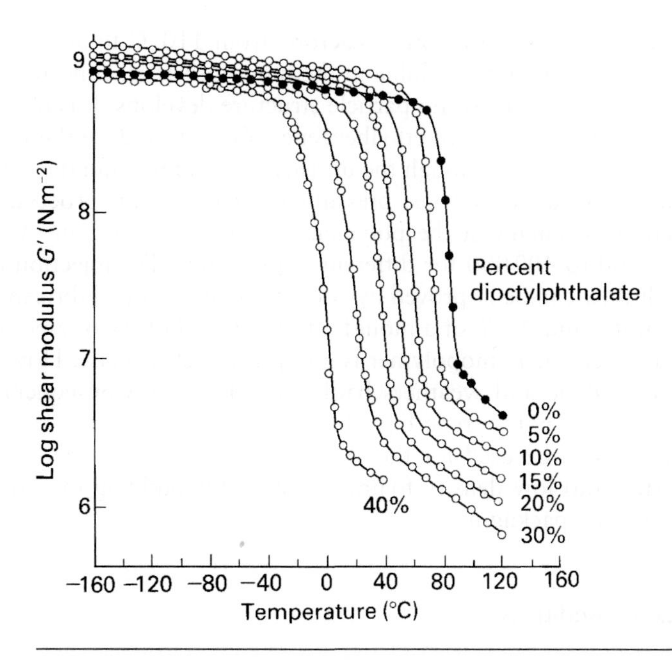

Figure 2.18 Shear modulus of PVC vs. temperature at different contents of dioctylphthalate plasticiser (taken with permission from Koleske JC and Wartman LH, *Polyvinylchloride*, Macdonald and Co., 1967).

the polymer rapidly. Most PVC plasticisers are esters, products of reactions between long-chain alcohols and dibasic acids. An example is dibutyl phthalate, formed by reacting butanol and phthalic anhydride. We return to plasticisers in the blood bag case study in Chapter 15.

Plasticised suspension PVC is first melt compounded and then extrusion coated onto copper wire to provide electrical insulation. It can be calendered between parallel rolls to produce flexible sheeting for flooring. Emulsion PVC can be converted into a *plastisol* by mixing it with a high proportion (50–70%) of a plasticiser. The plastisol, containing swollen 1 μm particles suspended in the plasticiser, is a liquid at room temperature with a medium viscosity of 20–50 $\mathrm{N\,s\,m^{-2}}$. It can be rotationally cast in moulds, coated onto wallpaper or cloth, or used to dip-coat metal products. A subsequent heating to between 150 and 175 °C causes the rest of the plasticiser to be absorbed in the PVC and the particles to fuse into a homogeneous solid.

2.7.4 Polystyrene and toughened derivatives

Styrene homopolymers are produced by a free radical polymerisation, that proceeds to completion as the styrene/polystyrene mixture is taken through a series of gradually hotter reactors, from 110 °C to about 200 °C. Unlike PVC, the polymer is soluble in the monomer at all concentrations, and the product is atactic, so no particle structure develops. It is usual to dilute the system with 3–12% of ethylbenzene solvent to reduce the melt viscosity at the later stages. Although the unreacted monomer and the ethylbenzene are flashed off under reduced pressure at the end of the process, some of these materials remain in the polymer. The molecular weight M_W varies from 100 000 to 400 000 for different applications. For injection moulding, the melt flow can be improved by adding about 1% of a lubricant such as butyl stearate, and 0.3% of a mould release agent (a wax or zinc stearate).

Polystyrene homopolymer is an optically clear glass. However, it is relatively brittle and weathers badly out of doors. Consequently, toughened versions of polystyrene have been developed (Chapter 4). As an oriented film, it can be used for the electrical insulation of capacitors. When converted into low density foam, it is used for building insulation and shock absorbing packaging.

2.7.5 Additives

A plastic consists of a polymer plus additives. Additives of many kinds, will be discussed where the appropriate properties are described. Fibres or rubber additives will modify the mechanical properties (Chapter 4), or plasticizers can change the state of PVC from a glass to a rubber. Stabilisers for melt processing are particularly necessary for polymers like PVC. Various additives delay polymer degradation (Chapter 10). Optical properties (Chapter 11) are modified by the addition of inorganic pigments or organic dyes. An additive can have multiple effects; carbon black increases the tensile modulus of rubbery polymers, it changes the colour, and, by absorbing UV radiation, it improves the outdoor weathering behaviour.

The physical form of the polymer affects the ease of additive dispersion. Powder blending is a low energy process, so polymers like PVC, that need large proportions of additives, are sold in powder form. Pigments or stabilisers, needed in proportions of 1% or less, are often dispersed in an extruder. Granules of a masterbatch, which contain 50% of the additive mixed into the relevant polymer, are mixed with granules of the polymer. Universal masterbatches are becoming more common; the host polymer is compatible in small amounts with the major plastics.

Chapter 3

Microstructure

Chapter contents

3.1 Introduction

This chapter describes the microstructures of the main types of polymer, concentrating on features used later to explain physical properties. The order of magnitude of elastic moduli for rubbers, glassy polymers and polymer crystals will be related to their molecular mobility and inter-molecular forces. These values will be used in Chapter 4 to predict the moduli of semi-crystalline polymers.

For any particular polymer, the microstructure passes through two or more states as the temperature is increased. Figure 3.1 shows what happens to five typical polymers, chosen to be in different states at 20 °C. The first three are linear polymers, and the other two are network polymers. The main transitions occur at the glass transition temperature T_g and the melting temperature of the crystalline phase T_m. Once the microstructures of typical states are understood, the reader can apply these to any polymer for which T_g and T_m are known.

3.2 Modelling the shape of a polymer molecule

3.2.1 Conformations of the C—C bond

The consequences of directional covalent bonding on the shape of polymer molecules will be explored, to generate the typical polymer shapes that exist in each of the microstructural states. The four covalent single bonds from a carbon atom, point towards the corners of a tetrahedron, with the carbon atom at the centre. The angle between any two of the bonds is 109.5°, and the C—C inter-atomic distance is 0.154 nm. In a polymer chain, every C—C bond can potentially rotate on its axis; whether it does or not depends on the temperature. Figure 3.2 defines the rotation angle ϕ of bond C_2—C_3 from the relative positions of bonds C_1—C_2 and C_3—C_4.

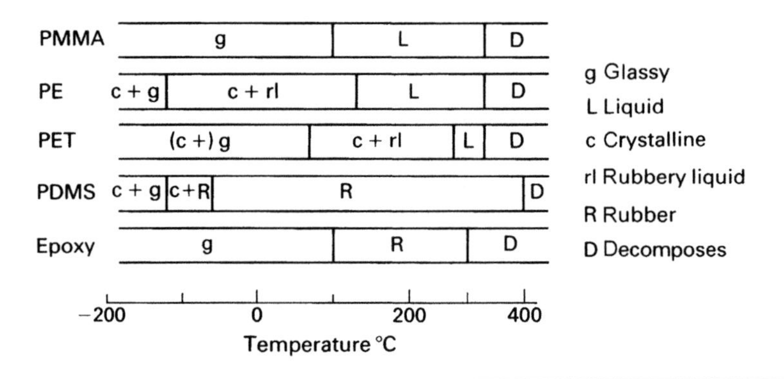

Figure 3.1 Changes in the state of five typical polymers with changes in temperature.

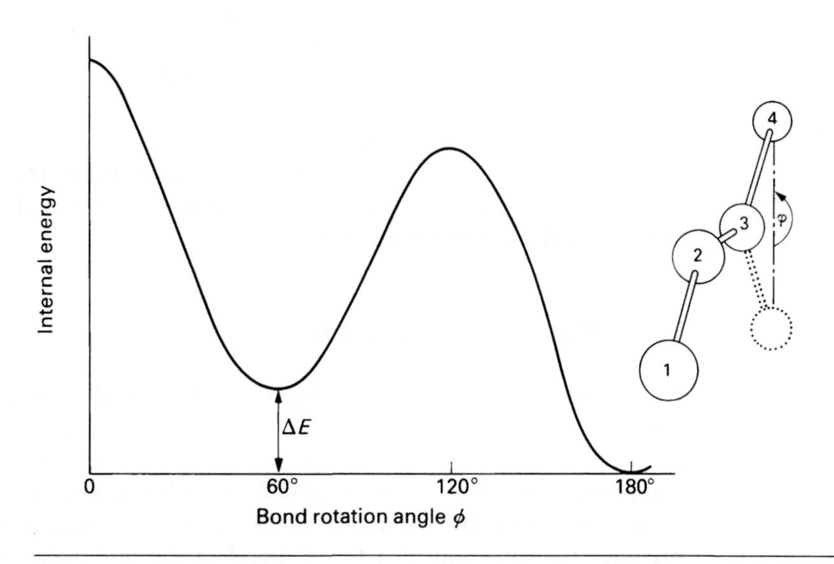

Figure 3.2 Internal energy variation with rotation of a C—C bond in polyethylene.

If $\phi = 0$, atoms that are covalently bonded to C_1 and to C_4 come into close contact; consequently the potential energy of that shape or *conformation* is high. However, if $\phi = 180°$, atom C_4 is at a maximum distance from C_1, and none of the atoms bonded to C_4 comes close to an atom bonded to C_1. Hence, the potential energy has a minimum value. This is referred to as the *trans* (*t*) conformation of the bond C_2—C_3. Further sub-minima at $\phi = +60°$ are referred to as *gauche*$^+$ (g^+) and *gauche*$^-$ (g^-) conformations. In the *rotational isomeric approximation* each C—C bond is assumed to be in a *gauche* or a *trans* conformation, and the shape of the polymer chain is defined by these conformations.

A chain of four carbon atoms has three rotational isomers, those of the bond C_2—C_3. A chain of five carbon atoms has 3^2 rotational isomers—the three possibilities of bond C_2—C_3 being combined with the three possibilities of bond C_3—C_4. Consequently, a polymer chain of n carbon atoms has 3^{n-3} rotational isomers. Only very few of these contain regularly repeating sets of bond rotations, such as the all-*trans* conformation that occurs in the polyethylene crystal, and the repeating sequence $tg^+ \, tg^+ \, tg^+ \ldots$ that occurs in the monoclinic crystalline form of isotactic polypropylene (Fig. 3.19). The overwhelming majority of conformations are irregular, and we now consider how to generate typical members of this set.

In the liquid state the C—C bonds transform from one rotational isomeric state to another, and the lifetime of a rotational isomer is about 10^{-10} s. For the polyethylene chain the potential energy ΔE of the *gauche* isomers is about $2\,\mathrm{kJ\,mol^{-1}}$ higher than that of the *trans* isomers. The relative numbers $n(g^+)$ and $n(t)$ of bonds in the isomeric states can be calculated using

$$n(g^+) = n(t)\exp\left(\frac{-\Delta E}{RT}\right) \tag{3.1}$$

where R is the gas constant and T the absolute temperature. Hence, at $T = 410\,\text{K}$, $n(g^+) = 0.26\ n(t)$, so the proportions of three isomers are roughly 1:1:4. These proportions can then be used in calculating the shape and size of the polyethylene chain.

3.2.2 Walks on a diamond lattice

The shape of a polyethylene molecule can be modelled as a random walk on the diamond crystal lattice (Fig. 3.3). In this face centred cubic crystal, each lattice vector has length a. Each step in the walk has Cartesian components $\pm\ a/4$, $\pm\ a/4$, $\pm\ a/4$. It is called a random walk because choices are made by a random number generator. Each step of the walk becomes the direction of a C—C bond in the molecule. In terms of the Miller index notation, the bond directions are of the <111> type. The < > brackets indicate that positive or negative components are possible. If the position of a C atom in the diamond lattice is approached along a [1 1 1] direction, the possible next steps in the walk are $[\bar{1}\,1\,1]$, $[1\,\bar{1}\,1]$ and $[1\,1\,\bar{1}]$, i.e. the sign of one component of the prior step is changed. In the computer programme that generates the walk, the choice of which component to change, is made at random. The current position (in units of $a/4$) is the sum of the last position and the Miller indices of the last step. Closed walks occur frequently; the smallest one with six steps has the shape of the

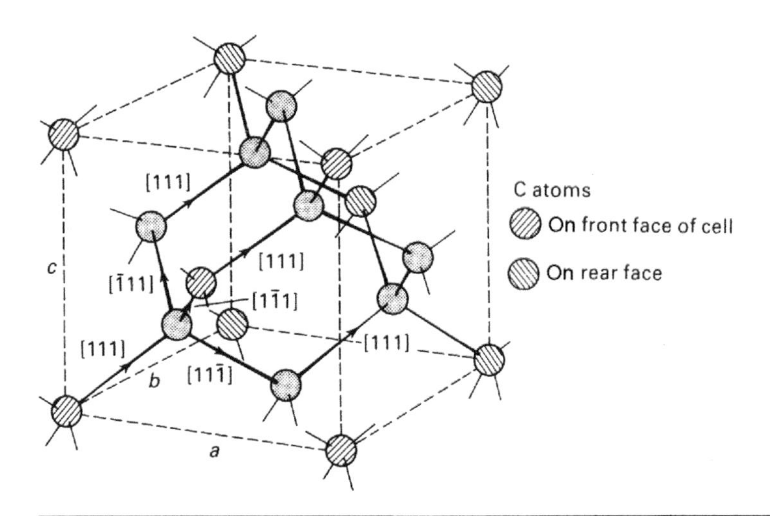

Figure 3.3 Miller indices of some covalent bond directions between C atoms in the diamond crystal lattice.

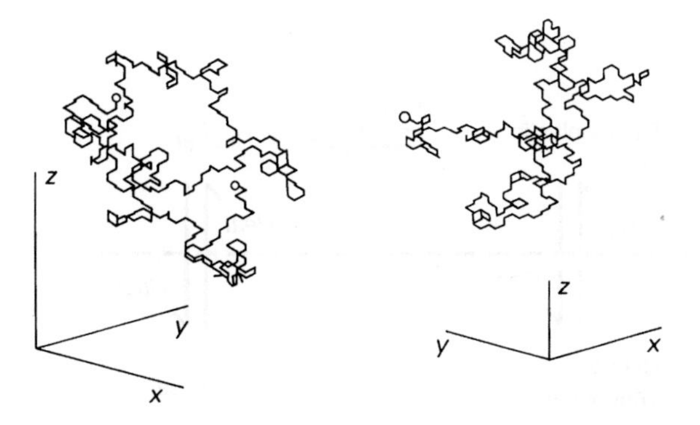

Figure 3.4 Two projections of a 400-step self-avoiding random walk on a diamond lattice.

cyclohexane molecule. Since we are modelling a linear polymer molecule, the walk should be self-avoiding. The current position must be compared with all previous positions to check that there is no duplication. If duplication occurs, part of the walk is erased and a further attempt is made to extend the walk. Figure 3.4 shows the coiled nature of a typical simulation of a 400 carbon atom polyethylene chain. When the computer simulation is repeated a large number of times (the Monte Carlo method!), a range of shapes and end-to-end lengths is found. A histogram can be generated of the frequency of different end-to-end lengths (Fig. 3.7) and the mean value computed.

The diamond lattice can also be used to generate the shapes of polymer chains in crystals, which involves regularly repeating sequences of *trans* and/or *gauche* bond conformations. Generating a *trans* isomer requires two successive sign changes of the same Miller index, for example successive bond directions $[\bar{1}\,1\,1]$, $[1\,1\,1]$ and $[\bar{1}\,1\,1]$; the first and third bonds are parallel. Generating a g^+ isomer requires changing the sign of progressively greater Miller indices, in the sequence 1st, 2nd, 3rd, 1st, 2nd, 3rd... (this is adding 1 in modulo 3 arithmetic); for example, the steps $[\bar{1}\,1\,1]$, $[1\,1\,1]$ and $[\bar{1}\,\bar{1}\,1]$. Generating a g^- isomer requires changing the sign of progressively smaller Miller indices in the sequence 3rd, 2nd, 1st, 3rd, 2nd, 1st; for example, the steps $[\bar{1}\,1\,1]$, $[1\,1\,1]$ and $[\bar{1}\,1\,\bar{1}]$.

3.3 Non-crystalline forms

3.3.1 The four main forms

A non-crystalline polymer can be transformed reversibly into another form by a temperature change (Fig. 3.5). Alternatively an irreversible crosslinking reaction can convert a linear polymer into a network polymer. This reaction can only occur when polymer chains can change shape, i.e. the polymer is

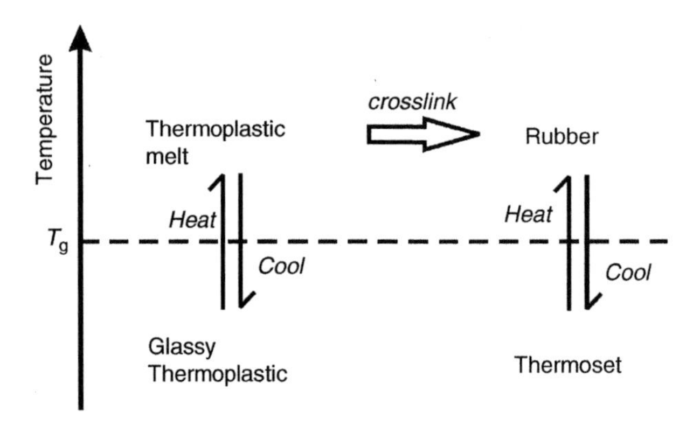

Figure 3.5 Reversible and irreversible conversion between four forms of non-crystalline polymer.

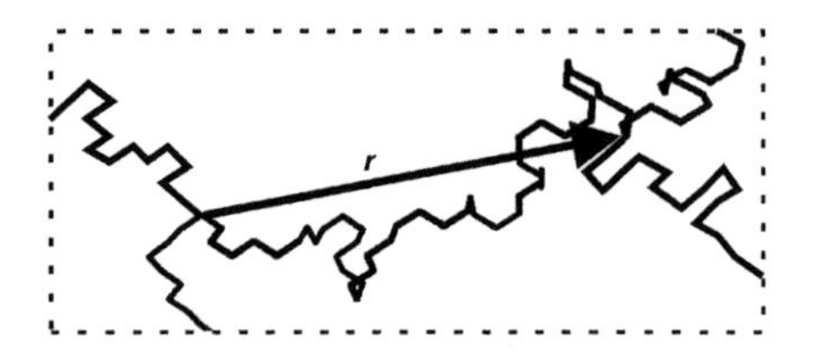

Figure 3.6 Planar projection of a network chain in a three-dimensional rubber network.

above T_g. Crosslinked polymers are called *thermosets* if T_g is above room temperature, or *rubbers* if T_g is below room temperature. The degree of crosslinking in most thermosets is higher than that in most rubbers. Thermosets are created by simultaneous polymerisation and crosslinking at temperature where the polymer chains are mobile. However, if the crosslinking process increases the T_g above the ambient temperature, the reaction stops.

In order to deal with the four non-crystalline forms in a unified way, we define a *network chain*, in a crosslinked system, as the section of network between neighbouring crosslinks (Fig. 3.6). The shape of both a network chain in a rubber, and a molecule in a polymer melt, can be changed dramatically by stress, and both can respond elastically. However, when the polymer is cooled below T_g, the elastic strains are limited to a few per cent (unless a glassy polymer yields), so the molecular shape is effectively fixed. If the melt or rubber was under stress when cooled, the molecular shape in the glass is non-equilibrium. This *molecular orientation* may be deliberate, as in biaxially stretched polymethylmethacrylate used in aircraft windows, or a by-product of processing, as the oriented skin on a polystyrene injection moulding. Details are discussed in Chapter 5.

3.3.2 Effect of molecular weight on molecular size

One way to specify molecular size is by the length of the *end-to-end vector* **r** (Fig. 3.6) that runs from the start to the end of an isolated chain, or between the crosslinks at the end of a network chain. The molecule is assumed to be unstressed, so its equilibrium shape has end-to-end vector $\mathbf{r}_0$. The *average length* of $\mathbf{r}_0$ will be calculated for a polymer with a C—C single bond backbone of a particular molecular weight. The average is over a large number of similar chains, and over a period of time.

The model used is a walk on a diamond lattice, in which *trans*, *gauche*$^-$ and *gauche*$^+$ conformations have equal probabilities of 1/3. To obtain a simple result, the self-avoiding condition is relaxed, so the walks can intersect with themselves. The end-to-end vector **r** is the sum of the step vectors $\mathbf{l}_i$. The mean square length of the walk is calculated from r_0^2, the scalar product of $\mathbf{r}_0$ with itself. The terms in the expansion are grouped according to the distance between the pairs of steps

$$
\begin{aligned}
r_0^2 &= \mathbf{r}_0 \bullet \mathbf{r}_0 = (\mathbf{l}_1 + \mathbf{l}_2 + \ldots + \mathbf{l}_n) \bullet (\mathbf{l}_1 + \mathbf{l}_2 + \ldots + \mathbf{l}_n) \\
&= \sum \mathbf{l}_i \bullet \mathbf{l}_i + 2\sum \mathbf{l}_i \bullet \mathbf{l}_{i+1} + 2\sum \mathbf{l}_i \bullet \mathbf{l}_{i+2} + \ldots \\
&= nl^2 + 2l^2 \sum \cos\theta_{i,i+1} + 2l^2 \sum \cos\theta_{i,i+2} +
\end{aligned}
\tag{3.2}
$$

where the summations are, for $i = 1$ to n, and $\theta_{i,j}$ is the angle between step i and step j. In the polyethylene chain (and diamond lattice) $\cos\theta_{i,i+1} = 1/3$. The correlation in direction between two steps decreases as the number of intervening steps increases. It can be shown that the average value $\overline{\cos\theta_{i,i+n}}$ is equal to $(1/3)^n$. Consequently the series for the mean square length can be summed to give

$$
\overline{r_0^2} = 2nl^2 \tag{3.3}
$$

Using the Monte Carlo method, for walks that can intercept, confirms this result. The root mean square length of the 100 step walks in Fig. 3.7 is 14.135. The theoretical distribution in the figure is derived in Section 3.4. It is the product of a $4\pi r^2$ term (the surface area of a sphere of radius r on which the chain end lies) and the Gaussian distribution of Eq. (3.13).

To experimentally verify Eq. (3.3), it needs to be expressed in terms of measurable quantities. Light or neutron scattering experiments can measure the radius of gyration r_g of isolated polymer molecules. r_g is defined (as in mechanics) in terms of the distribution of masses $m(R)$ of volume elements dV at radial distances R from the centre of mass, using

$$
r_g^2 = \frac{\int R^2 m(R)\,dV}{\int m(R)\,dV} \tag{3.4}
$$

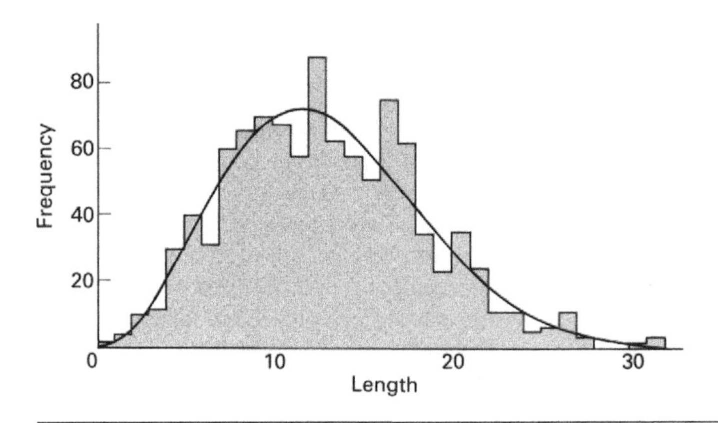

Figure 3.7 Histogram of end-to-end vector lengths for 1000 random walks of 100 steps, on a diamond lattice. The curve represents the theory of Eq. (3.13).

Schultz (1974) showed that

$$\overline{r_g^2} = \frac{1}{6}\overline{r_0^2} \tag{3.5}$$

so r_g is related to the mean square end-to-end length. The number n of C—C bonds in the backbone of an addition polymer chain is equal to twice the molecular weight M divided by the repeat unit mass M_r. Hence, the radius of gyration of monodisperse polymers, that have $n(\mathrm{t}) = n(\mathrm{g}^+) = 1/3$, should be related to M by

$$\overline{r_g^2} = \frac{2M}{3M_r}l^2 \tag{3.6}$$

The molecular weight dependence of r_g for polymethylmethacrylate (PMMA), both in the glassy state and in dilute solution, agrees with the form of Eq. (3.6) (Fig. 3.8), but the experimental values are about twice the theoretical values. The difference is due to a greater population of *trans* rotational isomers than that assumed. The solution values were obtained by light scattering measurements, but for the glassy state measurements, the PMMA was deuterated (replacing the hydrogen atoms by deuterium atoms) so that it had a greater neutron scattering cross section. About 1% of the ordinary polymer was dispersed in the deuterated glass so that the ordinary molecules were separated from each other. The angular distribution of neutron scattering was then analysed to find r_g.

The mean square end-to-end distance divided by nl^2 is used to quantify the relative size of polymer molecules of a given molecular weight (Table 3.1). The high value for polystyrene reflects the expansion in the chain size necessary to accommodate the large phenyl side groups. The

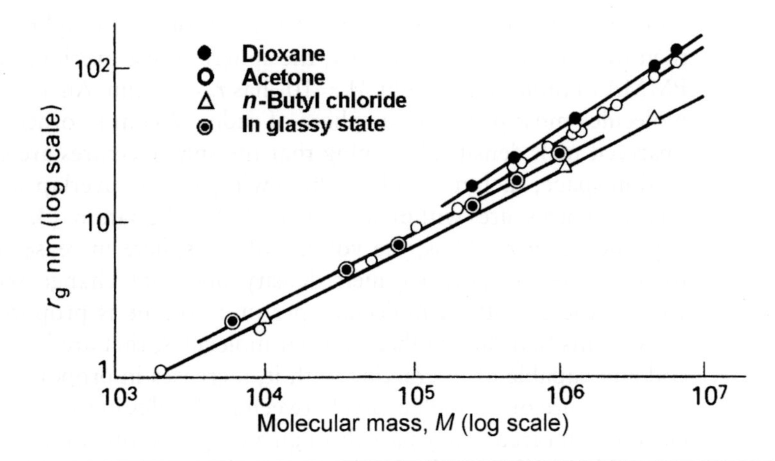

Figure 3.8 Comparison for PMMA in the glassy state and in dilute solution, of the radius of gyration vs. the weight average molecular weight (from Frischat, GH *The Physics of Non-Crystalline Solids*, Transtech, Switzerland, 1977).

Table 3.1 Chain size and entanglement data

Polymer	r_0^2/nl^2	*Entanglement molecular weight* ($g\,mol^{-1}$)	*Entanglement density* ($10^{-5}\,mol\,m^{-3}$)
PS	10.8	18 700	6
SAN	10.6	11 600	9
PMMA	8.2	9200	13
PVC	7.6	5560	25
POM	7.5	2550	49
PE	6.8	1390	61
PC	2.4	1790	67
PET	4.2	1630	81

values for PC and PET, which have benzene rings in the main chain, are not calculated on the same basis; rather the step length l is taken as the length of the in-chain rigid unit.

3.3.3 Entanglements in polymer melts

The molecular weight variation of the radius of gyration of polyethylene molecules, measured in the melt, was found to agree with Eq. (3.6), suggesting that individual molecules still have the random coil shape. If each molecule is assumed to reside inside a sphere of radius slightly larger than r_g, then to achieve the melt density, the spheres must overlap. For example, a PMMA of molecular weight $M = 10^6$ has $r_g = 25$ nm. An average of 40 such molecules must pack into a sphere of radius 25 nm in order to achieve the observed melt density. Assuming that the sphere centres are evenly distributed in space, each molecular sphere will partially overlap with 320 others whose centres are within a radius of $2r_g$. Equation (3.6) gives the M dependence of r_g. Thus, the volume of the sphere increases in proportion to $M^{1.5}$. As the polymer melt density does not change with molecular weight, the density of molecules per unit volume is proportional to M^{-1}. This means that the number of other molecules, that are inside a molecular coil and capable of interacting with it, increases in proportion to $M^{0.5}$.

The nature of entanglements between molecules in the melt is not exactly clear; one molecule may pass through a loop in another molecule; real knots are unlikely. There will also be a contribution from van der Waals forces between molecules. To quantify the entanglement effect, the concept of an entanglement molecular weight M_e is used. For this length of polymer chain, the inter-molecular forces can be replaced by one temporary crosslink (see Section 3.3.5 for the effects of permanent chemical crosslinks in a rubber). For polyethylene, $M_e \approx 1400$ (Table 3.1); so for a polyethylene with number average molecular weight $M_N > 2800$, the average molecule will be entangled with at least two neighbours and will transmit forces between them. Figure 3.9 shows some molecules in a flowing melt. If the forces transmitted by the entanglements become significant, then the sections of the molecule between entanglements will become both elongated and oriented in the direction of flow.

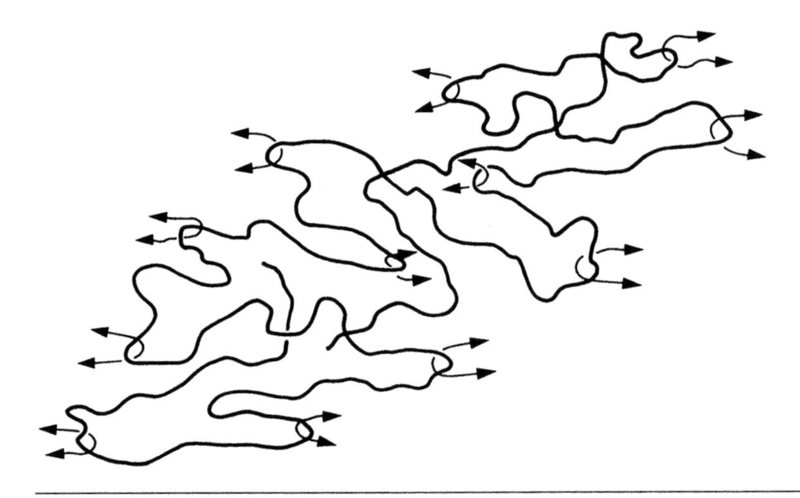

Figure 3.9 Sketch of the molecular orientation, of segments of a polymer molecule between entanglements, that develop in melt flow.

Appendix B explains how polymer melt flow curves can be derived, and defines apparent (shear) viscosity. It is difficult to correlate the apparent viscosity with a single molecular weight average, because it depends on the width of the molecular weight distribution. However, in the limit of very low shear strain rates $\dot{\gamma}$, when the entanglements between polymer chains produce negligible molecular extension, the apparent viscosity approaches a limiting value

$$\eta_0 = \lim_{\dot{\gamma}\to 0} \left(\frac{\tau}{\dot{\gamma}}\right) \tag{3.7}$$

This limiting viscosity is found to depend on M_W (Fig. 3.10). When $M_W < 2M_e$, the limiting viscosity is proportional to M_W, but when $M_W > 2M_e$, the effect of entanglements between molecules makes

$$\eta_0 = AM_W^{3.5} \tag{3.8}$$

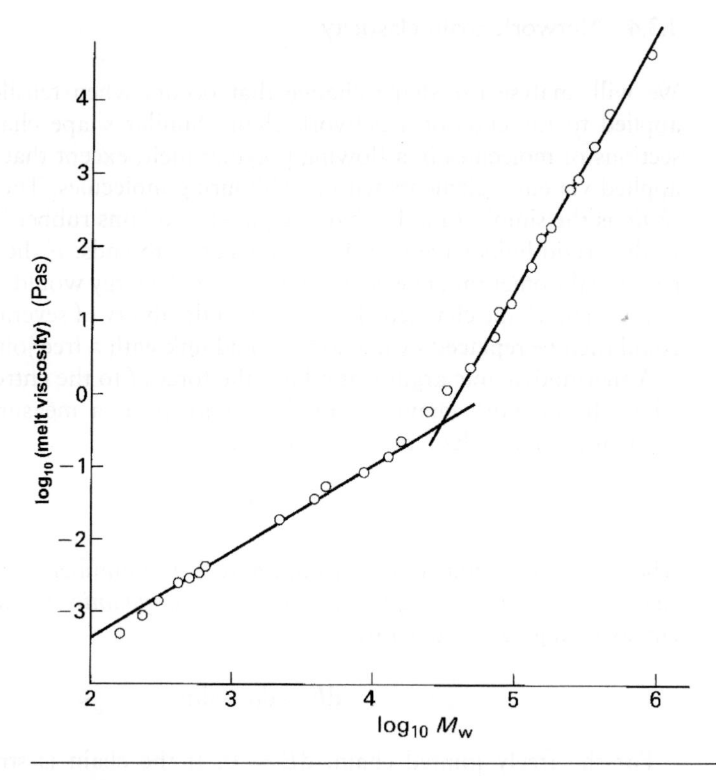

Figure 3.10 Log melt viscosity of polydimethylsiloxane at 20 °C vs. log weight average molecular weight (from Mills, N. J., *Eur. Poly. J.*, **5**, 675, 1969).

Values of the entanglement molecular weights (Table 3.1) have been derived from transient melt modulus measurements, using the equivalent of Eq. 3.20. There is some disagreement about the constant to be used in the formula, so the values should be taken as relative rather than absolute. The last column in Table 3.1 gives the entanglement density, calculated by dividing the amorphous density at 20 °C by the entanglement molecular weight. This assumes that the number of entanglements does not change on cooling from the melt. The entanglement densities, which vary considerably, will be used later to explain some mechanical properties of polymers.

De Gennes (1971) postulated that polymer molecules were constrained to move along a tube formed by neighbouring molecules. In a deformed melt, the ends of the molecules could escape from the tube by a reciprocating motion (reptation), whereas the centre of the molecule was trapped in the tube. When the chain end advanced, it chose from a number of different paths in the melt. This theory predicts that the zero-shear rate viscosity depends on the cube of the molecular weight. However, in the absence of techniques to image the motion of single polymer molecules in a melt, it is hard to confirm the theory.

3.3.4 Network chain elasticity

We will analyse the shape change that occurs when tensile forces f are applied to the ends of a network chain. Similar shape changes occur to sections of molecules in a flowing polymer melt, except that the forces are applied via entanglements with neighbouring molecules. The *freely jointed chain* is the simplest model that adequately explains rubber-like behaviour. In this, rigid links have free (ball) joints at both ends. If the more realistic rotational isomer model were used, its internal energy would change slightly as the chain shape changed. The restricted flexibility of several C—C bonds could then be replaced by one longer rigid link with a free joint at each end.

A thermodynamic argument relates the forces f to the entropy change $\mathrm{d}S$ when the network chain is stretched. Entropy is a measure of disorder, evaluated using Boltzmann's postulate that

$$S = k \ln W \tag{3.9}$$

where k is the Boltzmann's constant and W the number of distinguishable shapes of a chain of length r. For any *thermodynamic system*, the internal energy change $\mathrm{d}E$ is given by

$$\mathrm{d}E = \mathrm{d}q + \mathrm{d}w \tag{3.10}$$

For the freely jointed chain, $\mathrm{d}E = 0$. If the chain is stretched slowly enough for equilibrium to be maintained, the heat input $\mathrm{d}q$ is equal to the product of the absolute temperature T and the entropy change. When the forces f stretch the end-to-end vector by an amount dr, the work input dw is fdr, so Eq. (3.10) becomes

$$0 = T\mathrm{d}S + f\mathrm{d}r \tag{3.11}$$

Consequently, the force is given by

$$f = \frac{-T\mathrm{d}S}{\mathrm{d}r} \tag{3.12}$$

suggesting that the network chain acts as an *entropy spring*. This contrasts with the elasticity of glassy and semi-crystalline polymers, which is due to internal energy changes (Eq. 3.25).

To evaluate the chain entropy, the number of distinguishable chain shapes W must be counted, then Eq. (3.9) used. The general shape of the relationship between W and r is first calculated for a short one-dimensional chain (Fig. 3.11), for which enumeration is easy. In this chain each link, of unit length, is either in the positive or negative x direction. If there are five links, there is

1 Way to achieve an r value of 5   +++++

5 Ways to achieve an r value of 3   ++++−, +++−+, ++−++, +−+++, −++++

10 Ways to achieve an r value of 1   +++−−, ++−+−, +−++−, −+++−, ++−−+
+−+−+, −++−+ , +−−++, −.+.−.++, −−+++

By symmetry, there are 10, 5 and 1 ways to achieve r values of −1, −3 and −5, respectively. The histogram of W against chain length is already roughly similar to the Gaussian distribution. Next, the calculation is repeated for longer one-dimensional chains in which each link has length l. For a chain end-to-end length of ml, containing p positive and q negative links, the number of distinguishable chain shapes is

$$W = \frac{n!}{p!q!}$$

where the ! sign denotes factorial. If the number of links $n \gg 10$ and the chain is not fully extended, so $m \ll n$ and p and q are both large numbers, Stirling's approximation

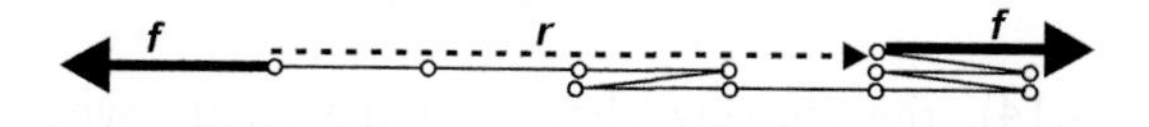

Figure 3.11 One-dimensional network chain, with end-to-end vector **r**.

$$\ln n! = n\ln n - n + \frac{1}{2}\ln(2\pi n)$$

can be used for n, p and q. After some algebric calculations, it can be shown that

$$\ln W = C - \frac{m^2}{2n}$$

where C is a constant. Since $m = r/l$, W can be written in terms of the end-to-end length r as

$$W = Ae^{-\frac{r^2}{2nl^2}} \tag{3.13}$$

where A is a constant. The entropy of the chain, from Eq. (3.9), is

$$S = kC - \frac{kr^2}{2nl^2} \tag{3.14}$$

The variation in entropy with chain length is an inverted parabola; therefore, using Eq. (3.12), the force is

$$f = \frac{kTr}{nl^2} \tag{3.15}$$

Equation (3.15) reminds us that rubber-like behaviour requires the thermal energy kT of rotating C—C bonds. The network chain acts as a linear spring, which is stiffer for short chains than for long chains.

The extension from one-dimensional to three-dimensional network chains is simple. The end-to-end vector **r** has components (r_x, r_y, r_z), while the force **f** has components (f_x, f_y, f_z). Equations like (3.15) link the x, y and z components.

3.3.5 Rubbers

Figure 3.12 shows a shear strain γ imposed on a rubber block containing a network chain with an end-to-end vector **r**. It is assumed that the crosslink deformation is *affine* with the rubber deformation, meaning that the components of **r** change in proportion to the rubber block dimensions. The components r_y and r_z are unchanged, but r_x becomes

$$r'_x = r_x + \gamma r_y$$

From Eq. (3.14), the entropy change of the chain, when the rubber is sheared, is

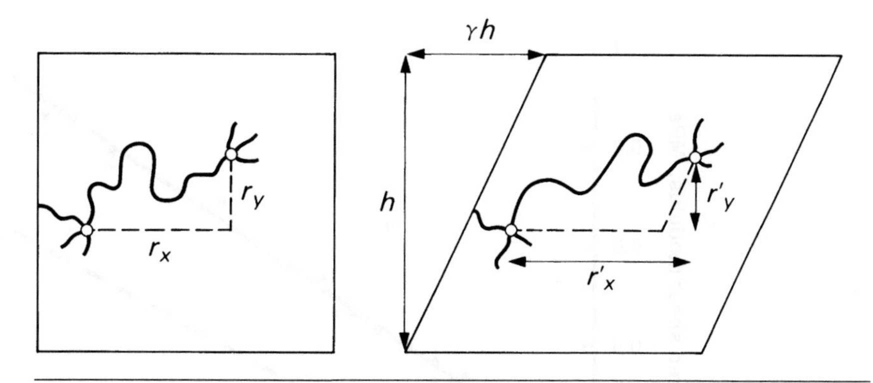

Figure 3.12 Affine deformation of the ends of a network chain when a shear strain γ is imposed on a rubber.

$$\Delta S = -\frac{k}{2nl^2}\left(r_x'^2 - r_x^2\right) = -\frac{k}{2nl^2}\left(\gamma^2 r_y^2 + 2\gamma r_x r_y\right) \tag{3.16}$$

On summing up for the N network chains in unit volume, equal numbers have positive and negative r_y values, so the second term on the right-hand side of Eq. (3.16) will cancel, leaving

$$\sum \Delta S = -\frac{Nk}{2nl^2}\gamma^2\overline{r_y^2} \tag{3.17}$$

The average value of r_y^2 in an unstressed chain is given by the equivalent of Eq. (3.2) for a freely jointed chain as

$$\overline{r_x^2} = \overline{r_y^2} = \overline{r_z^2} = nl^2 \tag{3.18}$$

Consequently, the shear stress is given by the equivalent of Eq. (3.15), as

$$\tau = -T\frac{\mathrm{d}\sum S}{\mathrm{d}\gamma} = NkT\gamma$$

which means that the shear modulus of the rubber is

$$G = NkT \tag{3.19}$$

This remarkably simple result has been checked experimentally by preparing rubbers in which each crosslinking molecule produces exactly two network chains. The positive deviation (Fig. 3.13) from the prediction of Eq. (3.19) is attributed to physical entanglements, which exist in the polymer before crosslinking. Therefore, rubbers are one of the few solids for

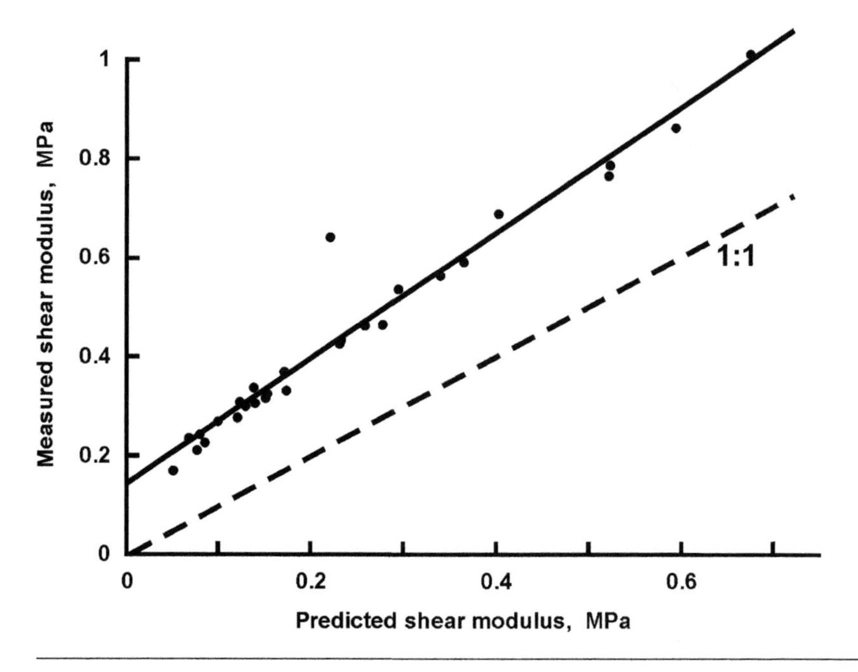

Figure 3.13 Comparison of measured rubber shear modulus with that calculated from the network chain density (from Treloar, L. G. R., *Introduction to Polymer Science*, Wykeham, 1970).

which the elastic modulus can be predicted, to within a small error, from the molecular structure.

The elastic moduli can be calculated from the network chain molecular weight M_c, and vice versa. Equation (2.13) can be substituted into Eq. (3.19) to give

$$G = 1000\frac{N_A k\rho T}{M_c} = 8310\frac{\rho T}{M_c} \tag{3.20}$$

where ρ is the density in kg m^{-3} and G is the shear modulus in Pa. The minimum shear modulus is determined by entanglements, whereas the maximum shear modulus occurs when the network chains are so short that rubber-like extensions are no longer possible. The approximation used to obtain Eq. (3.13) breaks down if the freely jointed chain has less than 10 links. For natural rubber, each isoprene repeat unit, of molecular weight 68, is approximately equivalent to a freely jointed link, so the minimum M_c of 680 corresponds with a maximum G of 3.2 MPa. The shear moduli of rubbers are usually in the range 0.3–3 MPa, among the lowest values observed for solid materials. Without the small restraining influence of the network chains, the shear modulus would be zero. If a rubber is swollen with a liquid (or uncrosslinked polymer) to produce a gel, the shear moduli can be less than 0.1 MPa.

The extension ratio λ of rubber loaded in tension is defined by

$$\lambda = \frac{\text{Deformed length}}{\text{Original length}} \tag{3.21}$$

As rubbers deform without change of volume, the extension ratios in the two lateral directions are $1/\sqrt{\lambda}$. Rubber elasticity theory leads to the prediction that the true stress (the force divided by the deformed cross-sectional area) is given by

$$\sigma_t = G(\lambda^2 - \lambda^{-1}) \tag{3.22}$$

an example of a non-linear elastic relationship. Natural rubber crystallises at extension ratios >6, which causes the stress to rise rapidly above that predicted by Eq. (3.22). This explains why it is impossible to stretch rubber bands to extension ratios more than 7.

3.3.6 Glass transition temperature

It is easier to describe the changes that occur at T_g, than to produce a molecular theory of the phenomenon. Chapter 7 will show that there is maximum mechanical damping at T_g. A graph of specific volume versus temperature, changes slope at T_g (Fig. 3.14). This contrasts with

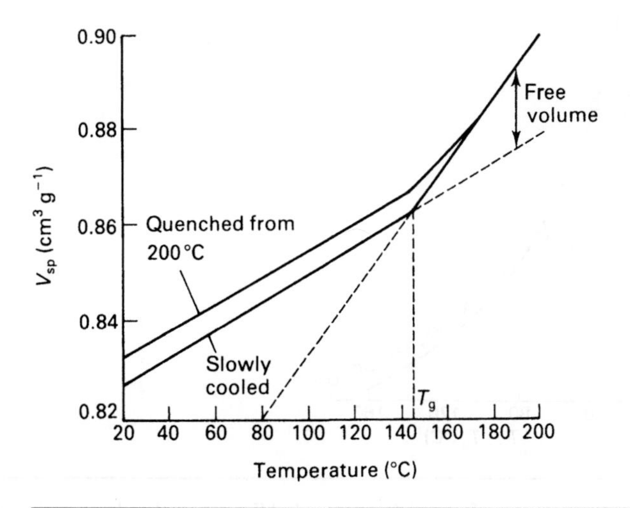

Figure 3.14 Specific volume vs. temperature for heating polycarbonate of two thermal histories (data from Hachisuka H *et al.*, *Polymer*, **32**, 2383, 1991), and the definition of free volume.

the step increase in specific volume when a crystalline phase melts (Fig. 5.4), with a change from a crystal lattice to random molecular coils with less compact packing. Consequently, there is no change in molecular arrangement or packing at T_g. When the glassy volume–temperature line in Fig. 3.14 is extrapolated above T_g, the difference between this and the melt volume is referred to as *free volume*. The increased molecular mobility in the melt or rubber state is attributed to free volume. If a glass is annealed at a temperature just below T_g, the specific volume slowly decreases to a limiting value, which lies on the extension of the melt volume–temperature line. These glass transition phenomena also occur for silicate and other inorganic glasses.

Clues about changes in molecular mobility, as the polymer approaches the glassy state, can be obtained from the increase in melt viscosity on cooling (Fig. 3.15), to a level beyond which further measurement is impossible. However, there is no quantitative model that relates the melt viscosity to the free volume. The line width in a nuclear magnetic resonance trace broadens on cooling below T_g, showing that protons (H atoms) experience different environments. It is narrow above T_g where rapid molecular motion gives every proton the same average environment. Experiments on polystyrene, diluted with a small amount of carbon tetrachloride, show that main chain rotation ceases when the polymer is cooled below the T_g of 90 °C. It is still possible for side groups to rotate in certain polymers, but the overall molecular shape becomes frozen.

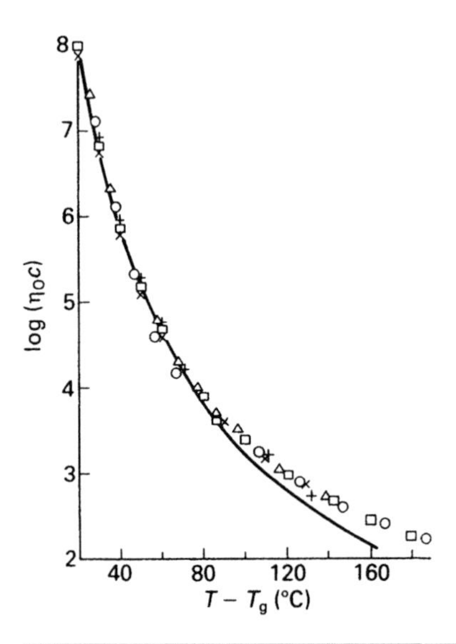

Figure 3.15 Log melt viscosity of polysulphones of different molecular weights vs. temperature; data shifted vertically for superimposition. The curve is $\log \eta = 2000/(T + 40 - T_g)$. (Mills NJ and Nevin A, J. *Macromol. Sci.-Phys.*, **4**, 1970).

Thermal vibration of the atoms, hence the *thermal energy* of the polymer, increases with absolute temperature T (for gas molecules the thermal energy is $3/2RT$ where R is the gas constant $8.3\,\mathrm{J\,K^{-1}\,mol^{-1}}$). When the thermal energy exceeds a critical value (at T_g), free volume is available for molecular motion. It is likely that the free volume is non-uniformly distributed in the melt, and that a number of lower density regions move rapidly through the melt (rather as dislocations move through a crystal lattice to allow plastic deformation).

There are rival theories of the glass transition; the Gibbs Dimarzio theory assumes that the configurational entropy of the chains approaches zero at T_g. Other researchers prefer a mode coupling theory (MCT), based on the dynamics of density fluctuations. However, it is difficult to extract a simple physical meaning from the complex equations that describe correlations between density fluctuations. Neither theory, at its current state of development, is particularly useful in understanding the properties of glassy polymers.

The value of T_g is expected to be high for polymers with strong forces between the chains (H bonds, or ionic forces). Polymers with stiff in-chain groups, such as single or multiple benzene rings, have high T_g values. There has been moderate success in calculating T_g by adding values assigned to the constituent chemical groups. In general, the processibility of the polymer decreases, and the risk of degradation during processing increases, as T_g increases. For most products it is not necessary to have a T_g in excess of 150 °C.

3.3.7 Glass microstructure

There is no experimental method of imaging the shape of polymer molecules in a glass, so we rely on simulations. Usually only a small region of glass is considered, with relatively short chains. The rotational isomeric model of Section 3.2.1 is used to generate the initial molecular shapes. For polypropylene, the probability of neighbouring bond pairs having particular rotational states was calculated from the steric interference of methyl side groups. The bond rotation angles were then varied somewhat from the 0° and ±120° positions to reduce the potential energy of the molecules. Figure 3.16 shows the molecular arrangement in a cube of side 1.82 nm; the periodic boundary conditions cause the pattern to repeat in neighbouring cubes. The experimental density of $890\,\mathrm{kg\,m^{-3}}$ at −40 °C was reproduced. The molecular radius of gyration was almost identical to that in a dilute solution, in agreement with experimental data of the type shown in Fig. 3.8. The model does not contain bundles of near-parallel chains. Neighbouring chains tend to be perpendicular to each other, but for neighbouring chains separated by more than 1 nm, there is no correlation of the direction of main chain C—C bonds. This supports indirect experimental evidence that glasses contain *frozen random coils*—molecules with shapes

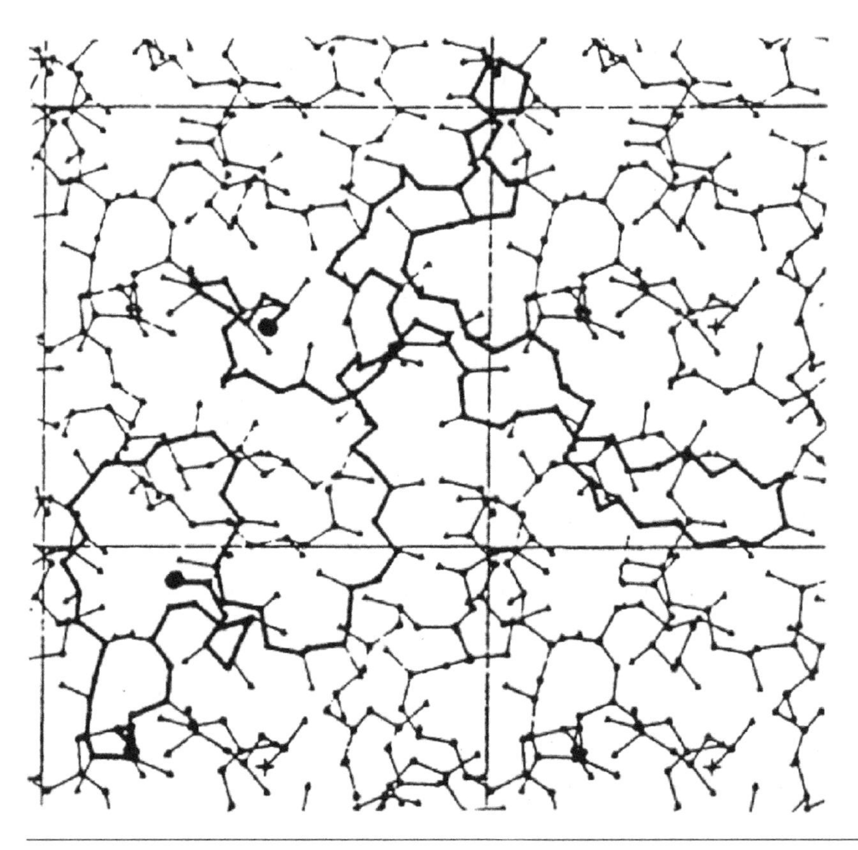

Figure 3.16 Projection of a 76 monomer polypropylene chain shape in the glassy state, with the parent chain shown in bold and the image chains shown as thin lines. The pattern in the cube of side 1.82 nm repeats in the neighbouring cubes (from Theodorou DN and Suter VW, *Macromolecules*, **18**, 1467, 1985).

described in Section 3.2. Consequently, the glassy state is a 'still photograph' of the structure of a polymer melt. The entanglement network, described in Section 3.3.3 will also exist in the glassy state.

For glassy polymers, the angular variation of intensity in an X-ray diffraction pattern has a single broad peak. This peak can be interpreted in terms of a *radial distribution function* (RDF), a graph of atomic density $\rho(r)$ as a function of the radial distance r from a reference atom. The RDF does not distinguish between the directions in which other atoms occur. Thus, it differs from the Bragg model for crystals (Section 3.4.9), which emphasises the spacing between close-packed planes. In the RDF for polycarbonate (Fig. 3.17), the initial sharp peaks between 0.1 and 0.25 nm are distances in the polymer chain between neighbouring atoms, whereas the broad peak at 0.55 nm reflects the range of nearest-neighbour distances between chains. Atoms further away than 1.5 nm appear to occur at random distances.

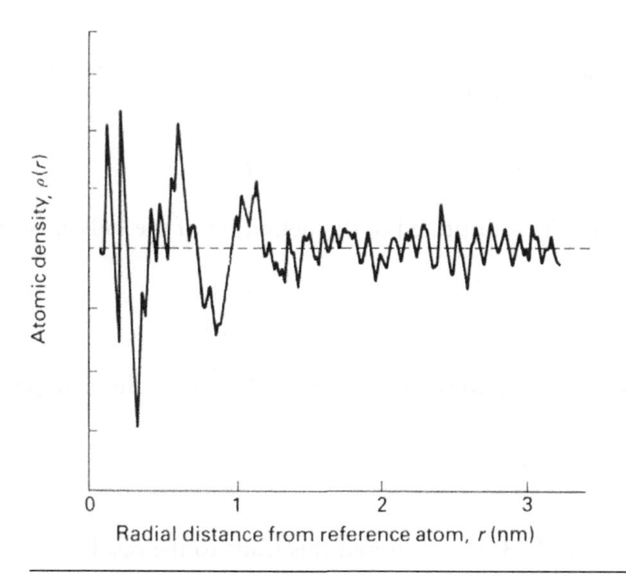

Figure 3.17 Radial distribution function for glassy polycarbonate (from Frischat, G *The Physics of Non-Crystalline Solids*, Trans Tech Publ., Switzerland, 1977).

The packing efficiency of polymer molecules in the glassy state can be estimated by dividing the glass density by the density of the crystalline form of the same polymer. The results range from 0.80 to 1.02; the average value for polymers with only H and F side groups is 0.88, smaller than the 0.96 value for polymers with bulky side groups. Consequently, the packing efficiency in the glassy state is relatively high.

3.3.8 Elastic moduli of glasses

The Young's moduli of glassy polymers at 20 °C range from 2 to 3.5 GPa, for polymers with T_g in the range 80–225 °C. This suggests that the weak van der Waals forces between the chains determine the magnitude of the modulus. This postulate was tested by analysing the pressure dependence of the bulk modulus of solid methane, which is also held together by van der Waals forces. Above 20 K methane has a cubic close packed structure in which every molecule has 12 neighbours at a distance of 0.41 nm. Hence, Eq. (2.1) when expressed in terms of the molar volume V, becomes

$$\frac{E}{E_m} = \left(\frac{V_0}{V}\right)^4 - 2\left(\frac{V_0}{V}\right)^2 \tag{3.23}$$

where E_m is now the molar constant and V_0 is the equilibrium molar volume. When the solid is compressed by a pressure p, the molecules move closer with a volume change dV, but since there is no change in the molecular disorder, hence no entropy change occurs. Equation (3.10) for internal energy change therefore, reduces to

$$dE = p dV$$

This can be substituted in the definition of the bulk modulus

$$K = -V \frac{dp}{dV} \tag{3.24}$$

to produce a relationship between the bulk modulus and the internal energy

$$K = -V \frac{d^2E}{dV^2} \tag{3.25}$$

When Eq. (3.23) is substituted this leads to the result

$$K = \frac{4E_m}{V_0} \left[5\left(\frac{V_0}{V}\right)^5 - 3\left(\frac{V_0}{V}\right)^3 \right] \tag{3.26}$$

for the variation of bulk modulus with applied pressure. The bulk modulus of solid methane at atmospheric pressure, when $V = V_0$, is given by Eq. (3.26) as

$$K = \frac{8E_m}{V_0} = \frac{N_A E_0}{V_0} \tag{3.27}$$

As solid methane has density $547\,\mathrm{kg\,m^{-3}}$ and molecular weight 16, the molar volume V_0 is $29 \times 10^{-6}\,\mathrm{m^3\,mol^{-1}}$, so Eq. (3.27) predicts a bulk modulus of 4.0 GPa.

It is more difficult to predict the bulk modulus of a glassy polymer because van der Waals forces occur between a variety of atoms, and the exact inter-molecular distances are not known. We assume that polymer chains are effectively incompressible along their lengths, and that densification only occurs in the plane perpendicular to the chains, resisted by the van der Waals forces. Therefore, the bulk modulus of a polymer should be 50% higher than that of solid methane, if the van der Waals forces have the same strength. This is confirmed by the atmospheric pressure bulk modulus of ~6 GPa for PMMA in Fig. 3.18. Although the value of E_0 is taken as an adjustable constant, the variation of the bulk modulus with the applied pressure agrees with the theory.

For isotropic materials, the Young's modulus E is smaller than the bulk modulus, given by

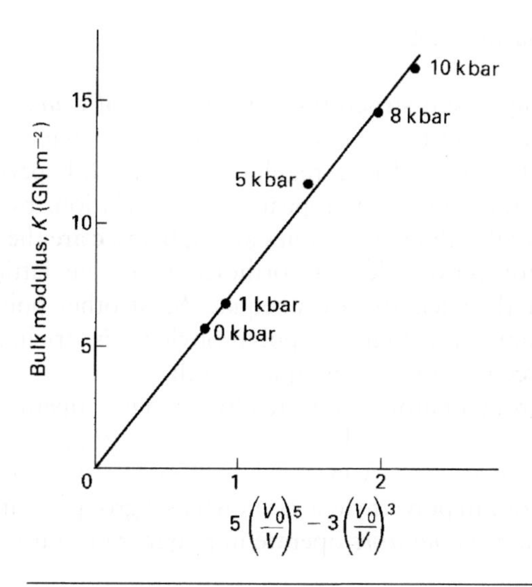

Figure 3.18 Bulk modulus of PMMA at different pressures vs. a function of molar volume. The straight line is the prediction of Eq. (3.26) for a van der Waals solid with $E_m = 137\,\text{kJ mol}^{-1}$.

$$E = 3K(1 - 2\nu) \tag{3.28}$$

where the Poisson's ratio ν for glassy plastics ranges from 0.3 to 0.4.

3.4 Semi-crystalline polymers

3.4.1 Introduction

In spite of much research, some details of the microstructure of semi-crystalline polymers are still unknown. Polymer development has proceeded empirically, with microstructural knowledge being acquired later, and then used to explain mechanical and physical properties. The order of presentation is that of increasing size scale: Bonding in the crystal unit cell, the shape of lamellar crystals, the microstructure of spherulites, the overall crystallinity and the processes of crystallisation. Details of polymer crystal structures and microstructures can be found in literatures listed in 'Further Reading'.

Although polymer crystal structures are known, and some slip mechanisms (slip plane and slip direction) determined, these are less important than for metals. Firstly, the amorphous phase plays an important part in the mechanical properties. Secondly, polymer yield strengths are not determined by obstacles to dislocation movement. However, it is possible to fabricate highly anisotropic forms of semi-crystalline polymers, so crystal characterization and orientation are important.

3.4.2 Crystal lattice and unit cell

Before discussing crystal structures, the concepts of a *unit cell*, *motif* and *symmetry operator* must be defined. A *unit cell* is a building block of the crystal lattice; it is stacked in a regular pattern to fill space. In crystallographic jargon, the unit cell is repeated by translation, by $h\mathbf{a} + k\mathbf{b} + l\mathbf{c}$ where h, k and l take all integer values and **a**, **b** and **c** are the lattice vectors. The unit cell for polyethylene is orthorhombic; the lattice vectors are orthogonal, but their lengths are unequal. Most other polymer unit cells have lower symmetry; that of polypropylene is triclinic, with non-orthogonal lattice vectors of non-equal length.

A *motif* is a group of atoms, repeated by symmetry operators to make the unit cell. The motif for polyethylene is the CH_2 group, whereas for polypropylene it is the atoms $—CH_2—CH(—CH_3)—$ from the monomer. In Fig. 3.19, the motif in polyethylene is shown as a group of three dark atoms.

Various *symmetry operators* operate in polymer crystals:

Translation moves the motif by a fraction of a lattice vector.

Mirror planes reflect the motif.

A *glide plane* combines a translation along an axis, and a reflection about a plane containing that axis.

A *screw axis* combines a translation along an axis, and a rotation about that axis. A 3_1 screw axis means that the rotation is $1 \times 360/3$ degrees.

These symmetry operators either act on the motif to construct the unit cell, or act on the whole crystal structure. In the polyethylene crystal (Fig. 3.19), the glide translation is **a**/2, and the reflection is about a plane perpendicular to **c**. There are 2_1 screw axes, parallel to **c**, at the positions shown—if two OHP transparencies of the structure are made, and one rotated by 180° about the 2_1 axis, the white atoms in one OHP lie over the dark atoms in the other, indicating a shift of **c**/2 would bring them into coincidence.

Three main principles explain the structures of polymer unit cells:

(a) *The polymer chain shape is a helix.* It is a regular winding along an axis (a 'spiral staircase' is strictly a helical staircase, as spirals are planar figures). In polyethylene (Fig. 3.19), the CH_2 motif is repeated by a glide operation, or by a 2_1 helix, along **c** to form a planar zigzag chain. For polypropylene, the motif is repeated by a 3_1 screw axis to form a helical chain (Fig. 3.20).

(b) *The internal energy of the molecules is a minimum.* The internal energy contributions are from the C—C bond rotational isomers and the van der Waals forces between atoms. In polyethylene, all the rotational isomers are *trans* (t), the lowest energy form (Fig. 3.2). In polypropylene, t and g^+ rotational isomers alternate to form left-handed helices (or t and g^- for right-handed helices). The resulting separation of the CH_3

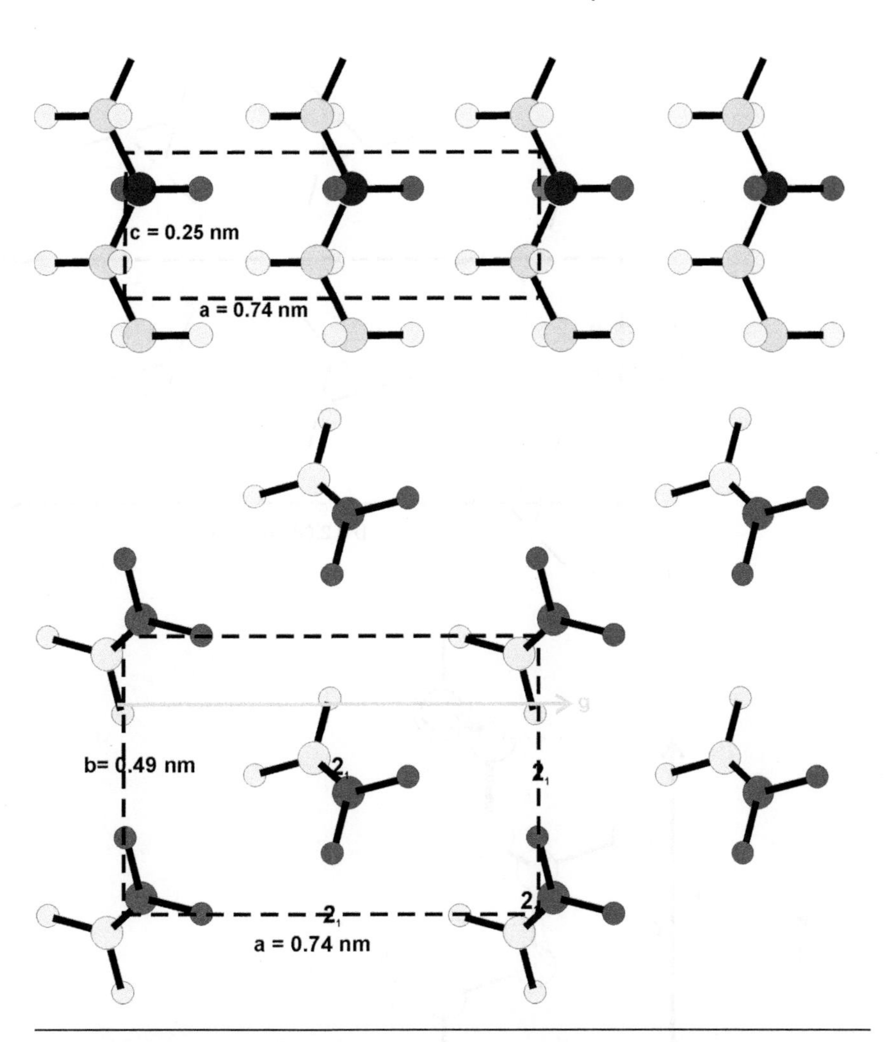

Figure 3.19 ac and ab projections of part of a polyethylene crystal, with the unit cell outlined by dashed lines. The motif of three darker atoms, at a height 3c/4 in the unit cell, are repeated by the glide plane symmetry operator.

side groups on the outside of the helix saves more energy than that used to create *gauche* isomers.

(c) *The chains pack together to maximise the (unit cell) density.* This is achieved by having the helix axes parallel to the **c** (chain) axis of the unit cell. In polyethylene, there are two angular settings of the C—C planar zigzags; each zigzag chain is surrounded by four chains with the opposite setting (Fig. 3.19), and hydrogen atoms on one chain fit into indentations in the surrounding chains. In isotactic polypropylene, the helices are displaced by units of **c**/12. If hydrogen bonds can form between chains, as in nylon 6,6, their number will be

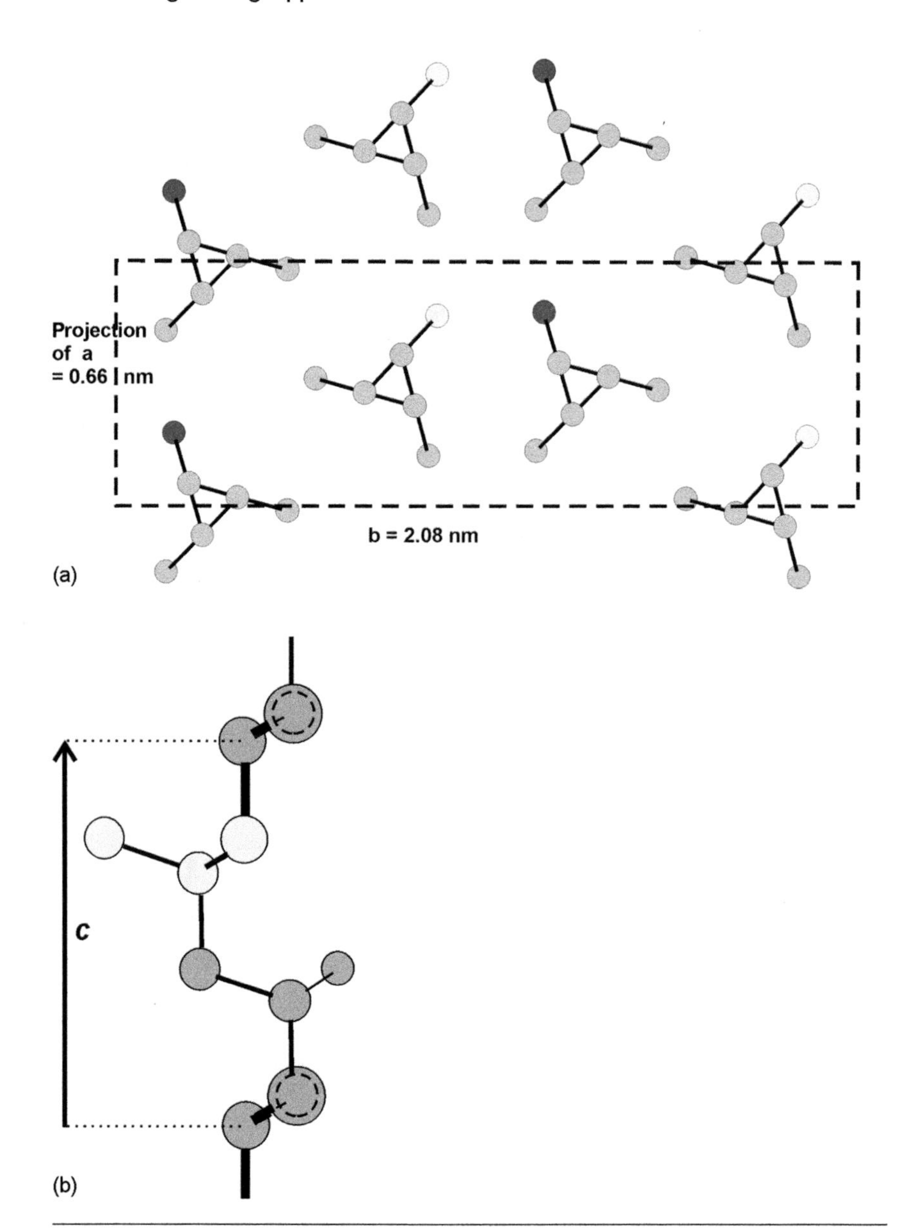

Figure 3.20 Polypropylene unit cell: (A) Projection on plane normal to **c** (dark atoms c/4, pale atoms 3c/4 above the ab plane. The unit cell is outlined by dashed lines. (B) One chain seen in **ac** projection—the motif of three carbon atoms is shaded differently as it moves up the helix axis.

maximised; the chains displace in the **c** direction and rotate, to allow hydrogen-bonded planes to form. This stabilises the crystals, increasing the melting temperature to 260 °C, compared with 135 °C for polyethylene.

3.4.3 Crystal elastic moduli

Polymer crystal unit cells have highly *anisotropic bonding*, in contrast with those of metals and ceramics. Covalent bonding is only continuous along the **c** axis of the cell; the much weaker van der Waals forces act in the **a** and **b** axis directions (Figs 3.19 and 3.20). The *ab* projection of the PE unit cell, with an area of 0.49 × 0.74 nm, contains an average of two chains (one at the centre and four shared ones at the corners). Hence, the density of chains crossing the **ab** plane is 5.5 chains/nm^2. The PP unit cell has an average of four chains (two internally and four shared on the sides) in an area of 2.08 × 0.66 nm, so the density of chains crossing the plane perpendicular to **c** is 2.9 chains/nm^2. The crystal Young's modulus in the **c** direction depends both on the chain density in the plane perpendicular to **c**, and on the deformation mechanisms. The PE planar zigzag can only stretch axially if the C—C atom distance is extended, or if the angle between two C—C bonds increases from 109.5°. The PP helix can also stretch if the C—C bonds rotate slightly away from *gauche* rotational isomer minima, so it is more compliant. The Young's modulus values E_c are 250 GPa for PE and 80 GPa for PP, at 20 °C. Most of the difference is due to the variation in chain packing density, but some is due to the additional deformation mechanism in PP. Therefore, polymer crystals can be divided into two main categories: those without large side groups having planar chains and E_c values in the range 250–350 GPa (PE, PET), and those with large side groups having helical chains and low E_c values (80 GPa for PP, 10 GPa for isotactic PS). The elastic moduli for stretching in the **a** or **b** directions, and for all the shear modes, are dominated by the high compliance of the van der Waals forces. There is anisotropy, with $E_a = 8$ and $E_b = 5$ GPa for polyethylene, because the inter-chain distances differ in the **a** and **b** directions. The hydrogen bonding direction in polyamide crystals has a higher Young's modulus.

3.4.4 Crystal shape

The shape and connectedness of the crystals is important. The usual form is *lamellar* (in the form of a thin plate or sheet), whereas a *fibrous* form is possible when a melt crystallises under a high tensile stress. Lamellar crystals were originally observed when dilute polymer solutions were crystallised slowly, but melt crystallised polymers also contain lamellar crystals. They are typically 10–20 nm thick, and of the order of 1 μm long and wide. They can be flat, or their surfaces may be curved (Fig. 3.21a). The fully extended length of a polyethylene molecule of $M = 10\,000$ is 90 nm, which is many times the lamella thickness. The crystal **c** axes lie within 40° of the normal to the lamella surface, so there must be regular chain folding at the upper and lower surfaces of single crystals grown from solution. The situation at the surfaces of melt crystallised lamella is not so simple. A number of nearly parallel lamellae grow together into the melt with layers

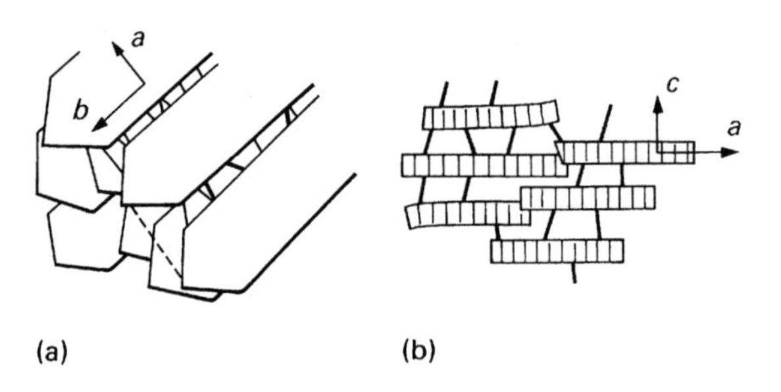

Figure 3.21 Sketch of inter-crystalline links near the surface of a growing spherulite. (a) Plan view of lamellar faces; (b) edge view of the lamellae.

of amorphous material between them. The geometry is similar to that of lamellar eutectics in metallic alloys, such as pearlite in steel. The time necessary for a randomly coiled molecule in the melt to change its shape into a regularly folded form, so that it can join a lamella, is much longer than that available during crystal growth. Consequently, sections of the molecule are incorporated into neighbouring lamellae, and the intervening lengths form part of the amorphous interlayers. In particular, sections containing short chain branches are rejected by the lamellae. The amorphous sections of molecules behave like network chains in rubber, since their ends are trapped in crystals. The crystallisation process induces molecular orientation and some bundles of elongated molecules coalesce to form *inter-crystalline links*. These fibrous extended chain crystals, of the order of 1 μm in length and less than 10 nm in diameter, connect the lamellae (Fig. 3.21b). They are revealed when mixtures of polyethylene and $C_{32}H_{66}$ paraffin are crystallised, then the paraffin removed.

3.4.5 Variety of crystal organisation

Recent advances in catalysis have allowed the production of polyolefins with low crystallinity. Spherulitic structures (see next section) only occur in propylene–ethylene copolymers when the crystallinity exceeds 45% (Fig. 3.22). Sheaf-like structures occur when the crystallinity is between 30 and 45%, whereas axialites and isolated lamellae occur between 15 and 30% crystallinity. Axialites are multi-layer aggregates of lamellar crystals which splay out from a common edge. Embryonic axialites occur for crystallinity from 5 to 15%. Therefore, as the crystallinity is reduced, the microstructures become simpler.

The tensile stress–strain curves, for the four microstructural types, cover the range from elastomers to typical semi-crystalline thermoplastics (Fig. 3.23). The lowest crystallinity material is a competitor with 'thermoplastic elastomers'.

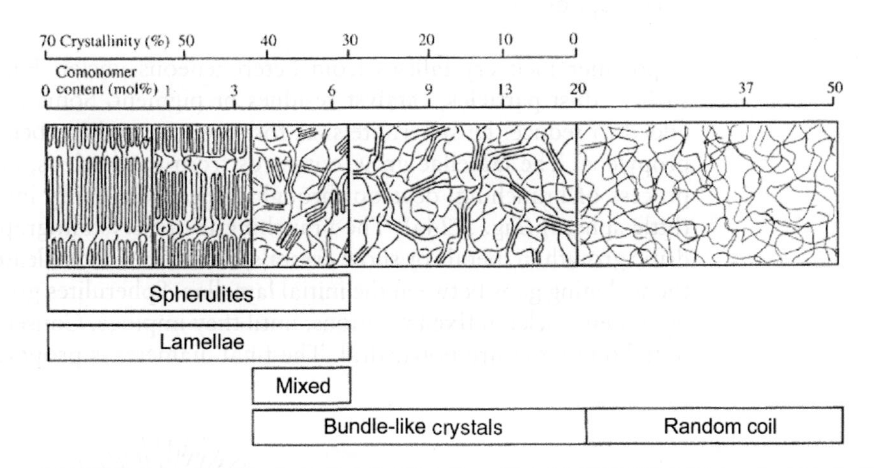

Figure 3.22 Range of microstructures in EP copolymers as a function of the overall crystallinity (Chen, H. Y. *et al.*, J. *Polym. Sci. B*, **39**, 1578, 2001) © John Wiley and Sons Inc. reprinted with permission.

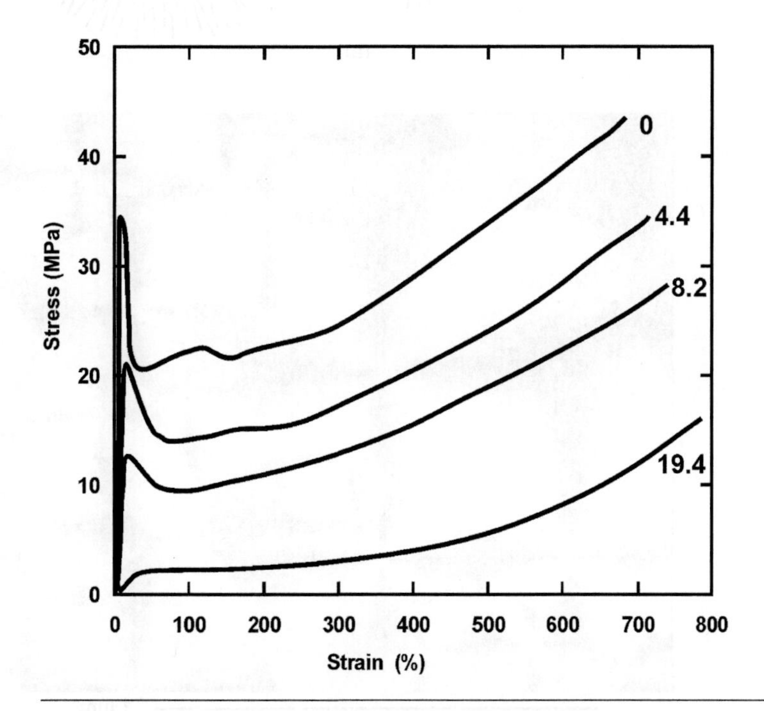

Figure 3.23 Tensile stress–strain curves for the four types of EP copolymers, labelled with the per cent octene, at 21 °C strain rate 100% min^{-1} (redrawn from Chum, S. *et al.*, ANTEC, 1775, 2003). © Society of Plastics Engineers Inc.

3.4.6 Spherulites

A polymer melt crystallises from heterogeneous nuclei: Foreign particles such as dust particles, catalyst residues or pigment. Some are deliberately added to reduce the spherulite size; for example, sodium benzoate in polypropylene. The lamellae grow outwards from the nucleus, and, through a process of branching, develop through a sheaf-like entity into a spherulite (little sphere) (Fig. 3.24a). The branching is not crystallographic as is dendritic growth in metals or snow crystals; new lamellae nucleate and grow in the widening gaps between the initial lamellae. Spherulites grow in the melt, with their nuclei in fixed positions, until they impinge. Consequently, spherulite boundaries are polyhedral. The final diameter is proportional to $d^{1/3}$,

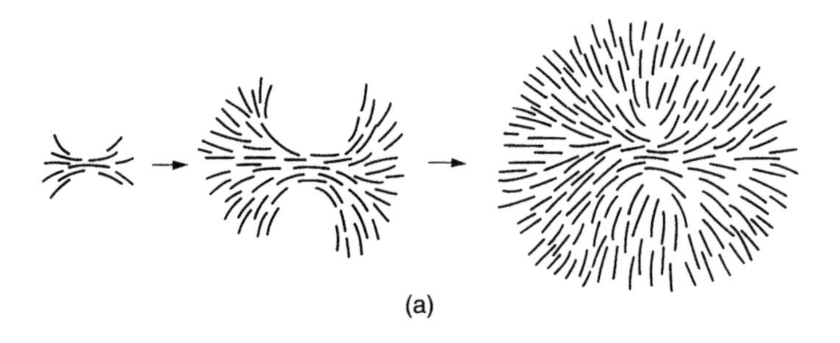

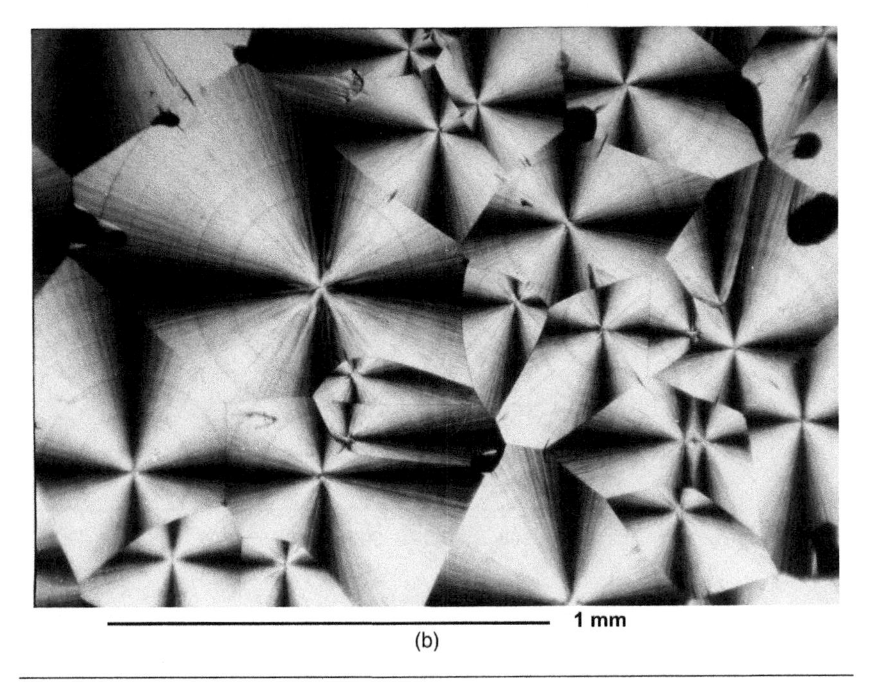

Figure 3.24 (a) Stages in the formation of a spherulite from a stack of lamellae; (b) Polarised light micrograph of two-dimensional spherulites grown in a thin film of polyethylene oxide. The polariser and analyser filters are vertical & horizontal.

where d is the number of nuclei per m^3. The range of d is considerable, with polymers such as PE and POM having spherulites of diameter $<10\,\mu m$, whereas low molecular weight polyethylene oxide can have spherulites 1 mm in diameter.

It is easy to observe spherulite growth in a thin film of low molecular weight polyethylene oxide, melt between a microscope slide and a cover slip, using polarised light microscopy. The spherulites grow as discs once their diameter exceeds the film thickness of about 0.1 mm. The discs have a radiating fibrous appearance and a Maltese cross pattern with arms parallel to the crossed polarising filters below and above the specimen (Fig. 3.24b). However, these two-dimensional spherulites are a rarity; in nearly all cases the spherulites are three-dimensional with polyhedral boundaries.

Isotropic materials (such as a non-oriented glassy polymer) have a single refractive index n which determines the speed C of light in the material

$$C = \frac{C_0}{n} \tag{3.29}$$

where $C_0 = 3.00 \times 10^8\,ms^{-1}$ is the speed of light in a vacuum. Anisotropic materials, such as polymer crystals, have orthogonal optic axes. The refractive indices n_1, n_2 and n_3 determine the speed of propagation of plane polarised light. A *polariser* filter, only allows the passage of light waves with their transverse electric field in a single direction. In Fig. 3.25, a light ray passes along the **b** axis of a polyethylene crystal; the polarisation direction (the electric vector **E**) is at an angle θ to the **a** axis. The light propagates as two components: One of magnitude $E\cos\theta$ and wavelength λ/n_a, polarised

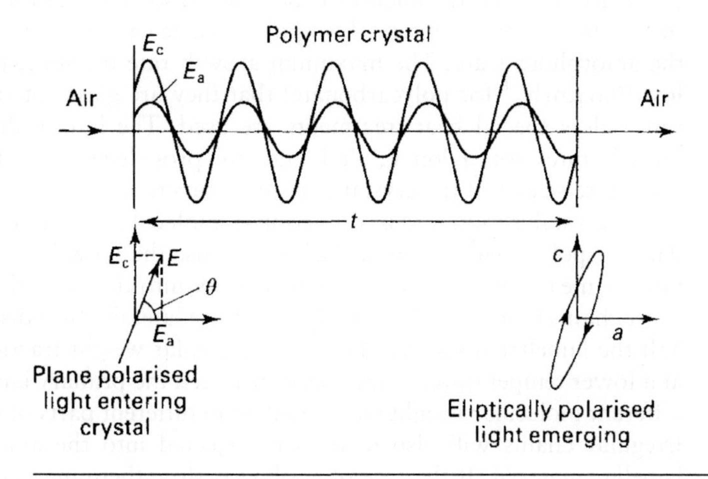

Figure 3.25 The effect of an anisotropic crystal on initially plane polarised light, producing elliptically polarised light. The number of wavelengths in the crystal has been reduced for clarity.

along the **a** axis, and one of magnitude $E \sin\theta$ and wavelength λ/n_c, polarised along the **c** axis. These components emerge from a thickness t of crystal with a phase difference $(n_c - n_a)t/\lambda$ wavelengths. Consequently, the emerging light is elliptically polarised, unless the crystal thickness just happens to make the phase difference an integral number of wavelengths. Thus, some light will pass through an *analyser* filter, set with its transmission direction at 90° to the polariser filter. However, if θ is 0° or 90°, the light will emerge from the crystal plane polarised, and is stopped by the analyser. The contrast pattern on the spherulite in Fig. 3.24b can now be explained. In the vertical arms of the dark crosses, the crystals have an optic axis along the spherulite radius, and in the horizontal arms they have an optic axis along the spherulite tangent direction; all these optic axes are parallel to the polariser. At other positions in the spherulite, θ is neither 0° nor 90°, so light is transmitted by the analyser. The cross remains stationary relative to the filters, if the polymer is rotated on the microscope stage, proving that the optic axes are tangential and radial in the spherulite. Electron diffraction shows that the crystal **c** axes are tangential in spherulites.

3.4.7 Crystallisation rate

The rate of crystallisation can be measured using a bulk specimen in a dilatometer. Alternatively, the radius of an individual spherulite can be measured during its growth between glass slides on a hot stage microscope. The crystallisation rate increases with supercooling (the amount by which the temperature is below T_m), until a maximum occurs, then decreases to zero at T_g (Fig. 3.26). For polyethylene, the peak spherulite growth rate ($\sim$100 $\mu m\ s^{-1}$) and the nucleation density are so high that the left-hand side of the curve is never observed, so polyethylene can never be quenched into the amorphous state. The maximum growth rate for some polymers is so low (0.6 $\mu m\ h^{-1}$ for polycarbonate) that they are glassy at room temperature unless special heat treatments are used. The intermediate spherulite growth rates for nylon 6, PET and polypropylene mean that both the spherulite sizes and the crystallinity are a function of the cooling rate.

The crystallisation rate is a function of molecular weight (Fig. 3.27). The data for polyethylene stops at 125 °C because the growth rate becomes too fast for measurement. As the lamellae grow into the melt, there is time for short chains that crystallise slowly, to diffuse laterally by a distance equal to half the lamellar thickness. This low molecular weight fraction crystallises at a lower temperature, in the spaces between the primary lamellae. Hence, different molecular weight chains end up in different parts of the spherulite. Irregular chains will also tend to be rejected into the amorphous interlamellar material. If the rate of cooling is slow there may be time for low molecular weight material to diffuse away, ahead of the growing spherulites, and end up as weak regions at inter-spherulitic boundaries.

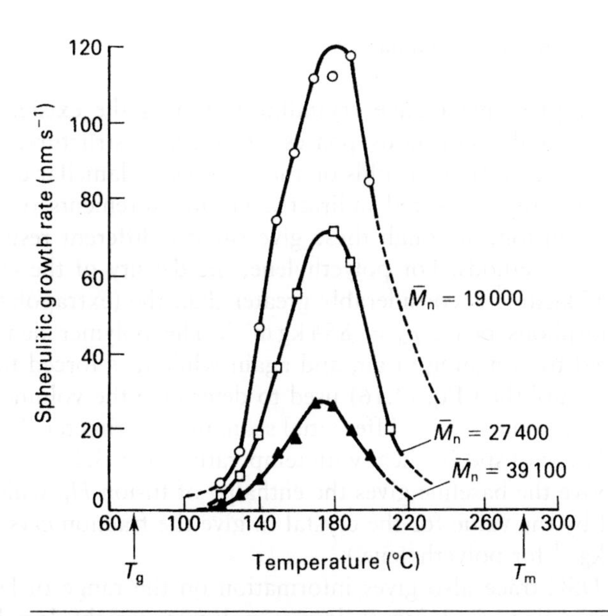

Figure 3.26 Growth rate of PET spherulites, of different molecular weights vs. crystallisation temperature (From *Polymer Engineering*, PT614, Unit 8 *Advanced Processing*, Open University Press, Milton Keynes, 1985).

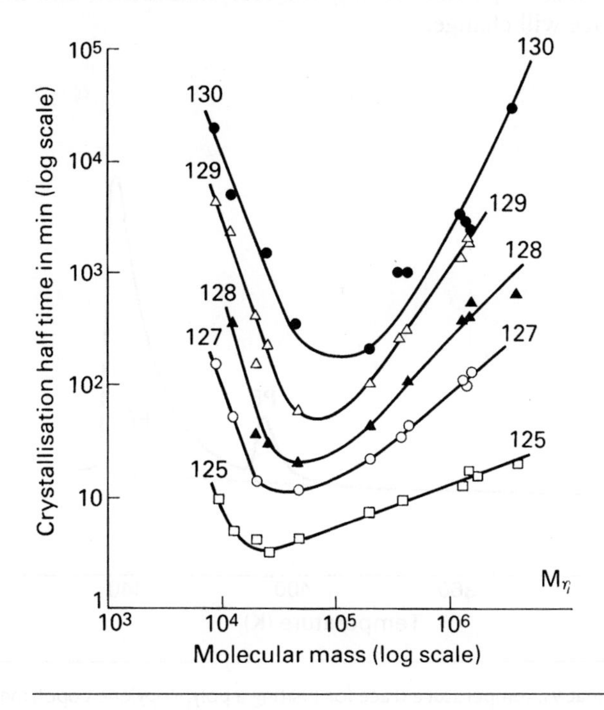

Figure 3.27 Logarithm of crystallisation half time vs. logarithm of molecular weight, for polyethylene crystallised isothermally at the temperatures indicated (from Mandelkern, L. J. *Mater. Sci.*, **6**, 615, 1968).

3.4.8 Percentage crystallinity

The concept of percentage crystallinity implies the existence of separate crystalline and amorphous phases of constant structure, whereas there may be defects inside, or folds on the surfaces of, lamellar crystals. Crystallinity is usually measured indirectly, via measurements of density or enthalpy of fusion, although these give slightly different results than X-ray diffraction methods. For polyethylene, the density of the crystal unit cell $\rho_c = 997\,\mathrm{kg\,m^{-3}}$ is considerably greater than the (extrapolated) density of the amorphous phase $\rho_a = 854\,\mathrm{kg\,m^{-3}}$. The polymer density ρ_p can be measured by weighing in air, and again while it is forced to be immersed in water, and then Eq. (2.16) used to determine the volume fraction crystallinity. Alternatively, a differential scanning calorimeter (DSC) can record the variation of specific heat with temperature (Fig. 3.28). Integration of the area above the baseline gives the enthalpy of fusion H_f, which can then be divided by the value for the crystal to give the fraction crystallinity. ($H_f = 295\,\mathrm{kJ\,kg^{-1}}$ for polyethylene.)

The DSC trace also gives information on the range of lamellar crystal perfection, since the thinnest, lowest molecular weight, lamellae melt some 30 °C below the final melting point. If a rapidly cooled polyethylene is subsequently annealed in this temperature range, the lamellae will thicken by a process of partial melting and recrystallisation, and the shape of the DSC trace will change.

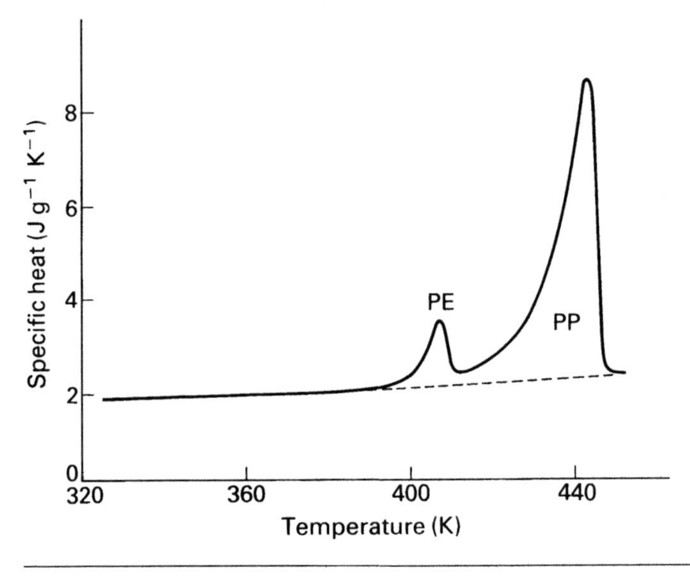

Figure 3.28 Specific heat vs. temperature trace for heating a polypropylene copolymer at $20^\circ\,\mathrm{min}^{-1}$ in a differential scanning calorimeter. The separate melting peaks indicate a polyethylene component, and a range of crystalline perfection.

3.4.9 X-ray diffraction

An X-ray beam, of a finite width, samples a small volume of the polymer structure. The diffraction pattern gives no information about the location of crystals within that volume, but it gives information about the range of crystal orientations in the volume; this can be used with optical microscopy to build up a picture of the microstructure. The crystal lattice model, used to interpret diffraction patterns, contains many sets of parallel planes. Polymer crystals often have lower lattice symmetry than metals, so the relationship between the interplanar spacing d and the Miller indices (hkl) of the plane are complex (Kelly and Groves, 1970). The Bragg condition

$$n\lambda = 2d \sin \theta \qquad (3.30)$$

gives the diffraction angle θ for the nth order diffraction peak from the (hkl) planes, where λ is the wavelength of the monochromatic X-ray beam. Figure 3.29 shows that the vectors **i** for the incident X-ray beam, **d** for the diffracted X-ray beam and **n** for the plane normal or *pole* are co-planar. The pole must have this specific orientation for the X-ray beam to be diffracted. The diffraction pattern for a spherulitic polymer consists of several complete concentric rings, each corresponding to a different set of planes. The angular intensity scan across such a pattern for polypropylene (Fig. 3.30) has four main peaks, one of which is composite. The inset diagram shows the positions of the (0 4 0) and (1 1 0) planes in the unit cell of polypropylene (shown in more detail in Fig. 3.20). The most useful information about crystal orientation would be from (0 0 1) planes, but the (0 0 1) diffraction is of negligible intensity. Consequently, information on the **c** axis orientation must be inferred from diffraction from other planes. For the spherulitic sample the (1 1 0), (0 4 0), (1 3 0) and (1 1 1) poles are randomly distributed in space, so it is reasonable to assume that the (0 0 1) poles are also randomly distributed. This fits with our model of the spherulite having radial symmetry.

The X-ray diffraction pattern from highly oriented polypropylene tape, used for parcel strapping, contains the remnants of the four diffraction rings

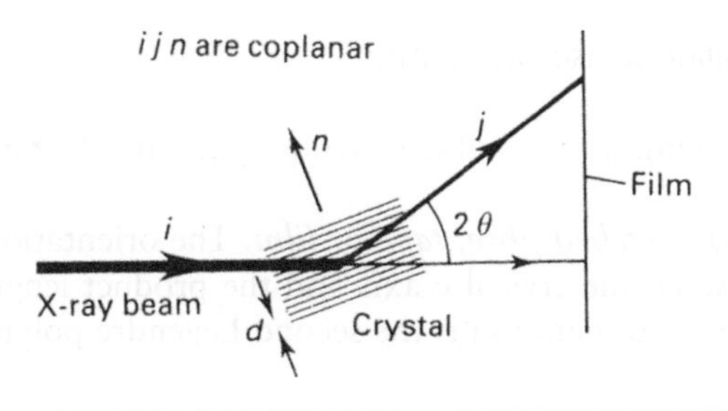

Figure 3.29 The Bragg diffraction condition for a set of planes in a crystal.

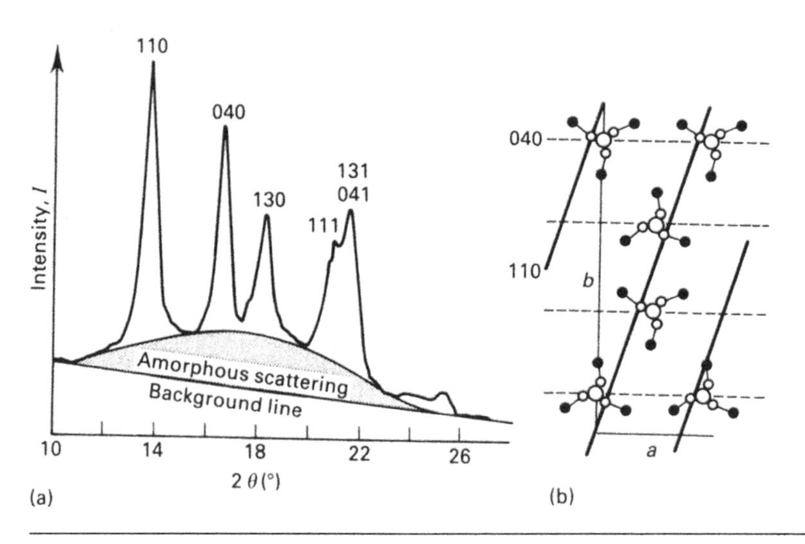

Figure 3.30 (a) Angular intensity scan of X-ray diffraction pattern for spherulitic polypropylene; (b) 1 1 0 and 0 4 0 planes in the unit cell, projected on to the **ab** plane.

(Fig. 3.31). The length axis L of the tape is vertical, so the (1 1 0), (0 4 0) and (1 3 0) diffraction peaks lie on the 'equator' of the figure. When semi-crystalline polymers are stretched, the crystal **c** axes tend to align with the tensile direction, here the L axis. The assumed crystal *orientation distribution* is of **c** axes perfectly aligned with the L axis, with **a** and **b** axes randomly distributed in the plane perpendicular to L. Therefore, the $(hk0)$ poles, at 90° to the **c** axis, should be perpendicular to the tape L axis. The diffraction peaks in Fig. 3.31 are consistent with this assumption. If in Fig. 3.29 the L axis is normal to the paper, the diffracting planes in the crystal shown are of the $(hk0)$ type. As the $(hk0)$ poles in the PP tape are randomly oriented in the plane of the diagram, many crystals will be positioned to produce diffraction spots on either side of the 'equator' of the pattern. To confirm the orientation distribution, further diffraction patterns should be taken as the sample is rotated around its L axis.

3.4.10 Crystalline phase orientation

The two most common types of oriented polymer product are:

(a) *Uniaxially stretched fibre, tape or film*. The orientation is defined by the angle θ between the crystal **c** axis and the product length axis L. One measure of average orientation is the second Legendre polynomial

$$P_2 = 0.5\left(\overline{3\cos^2\theta} - 1\right) \tag{3.31}$$

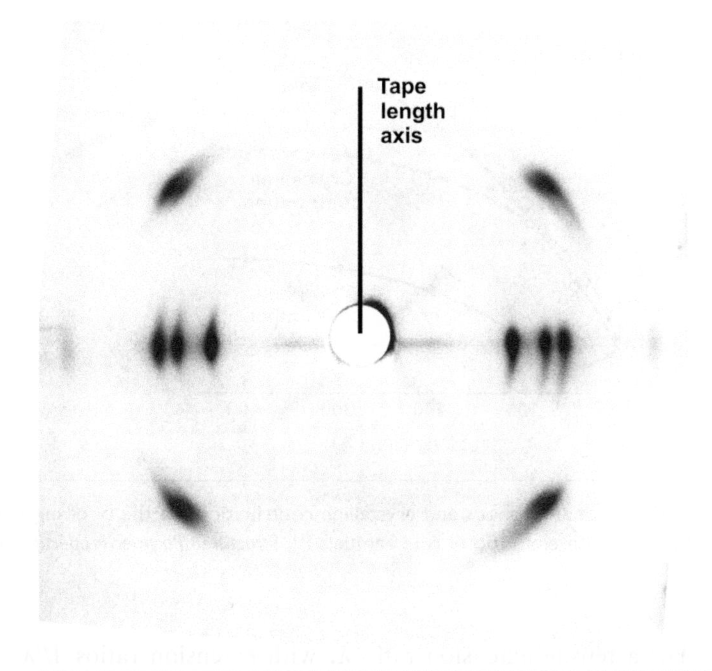

Figure 3.31 X-ray diffraction pattern for uniaxially oriented polypropylene strapping.

Since θ can only vary between 0° and 90°, P_2 can range from –0.5 to 1.0. P_2 can be calculated from the *birefringence* $n_L - n_T$, where n_L and n_T are the refractive indices for light polarised in the length (L) and transverse (T) directions of the product. If L polarised light passes through a polymer of thickness t and emerges with a phase difference of f wavelengths compared with T polarised light, the birefringence is calculated as

$$n_L - n_T = \frac{f\lambda}{t} \tag{3.32}$$

The crystalline phase birefringence can be divided by the refractive index difference $n_c - n_a$ for the crystal, to give P_2. However, there is also a contribution to the birefringence from the molecular orientation of the amorphous phase. Figure 3.32 shows the contributions to the overall bi-refringence of polypropylene films, hot stretched at 110°C by different amounts. The increase in the orientation with strain is non-linear and it differs between the phases.

A *pseudo-affine* model predicts the variation of P_2 with the deformation of a semi-crystalline polymer. It assumes that the distribution of crystal **c** axes is the same as the distribution of network chain end-to-end vectors **r**, in a rubber that has undergone the same macroscopic strain. Figure 3.12 showed the affine deformation of an **r** vector with that of a rubber block.

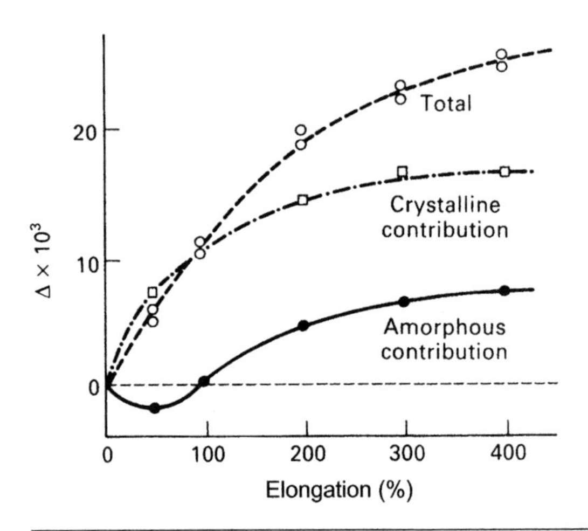

Figure 3.32 Variation of amorphous and crystalline contributions to the birefringence of polypropylene film with elongation (from Samuels RJ, *Structured Polymer Properties*, Wiley, 1974).

For a tensile extension ratio λ, with extension ratios $1/(\lambda)$ in the lateral directions, the angle θ in the deformed material is related to the value in the undeformed material by

$$\tan\theta = \frac{\lambda^{-0.5}}{\lambda}\tan\theta'$$

If the initial distribution of θ values is random, it can be shown that in the deformed material

$$\overline{\cos^2\theta} = \lambda^3(a^{-2} - a^{-3}\tan^{-1}a) \text{ where } a^2 = \lambda^3 - 1 \qquad (3.33)$$

This relationship is substituted into Eq. (3.31) and the theoretical curve plotted in Fig. 3.33. Given the simplicity of the model, the agreement is very good.

In a stretched rubber, the molecules elongate, and the **r** vectors move towards the tensile axis. Hence the variation of P_2 with extension ratio will differ from the pseudo-affine model. For moderate strains the increase of P_2 with extension ratio is linear, but at high extensions the approximation used in Eq. (3.12), that both q and q are large, breaks down. Treloar (1975) described models which consider the limited number of links in the network chains. Figure 3.33 shows that the orientation function abruptly approaches 1 as the extension ratio of the rubber exceeds $\sqrt{\lambda}$. Although the model is successful for rubbers, it fails for the amorphous phase in polypropylene (Fig. 3.32), presumably because the crystals deform and reduce the strain in the amorphous phase.

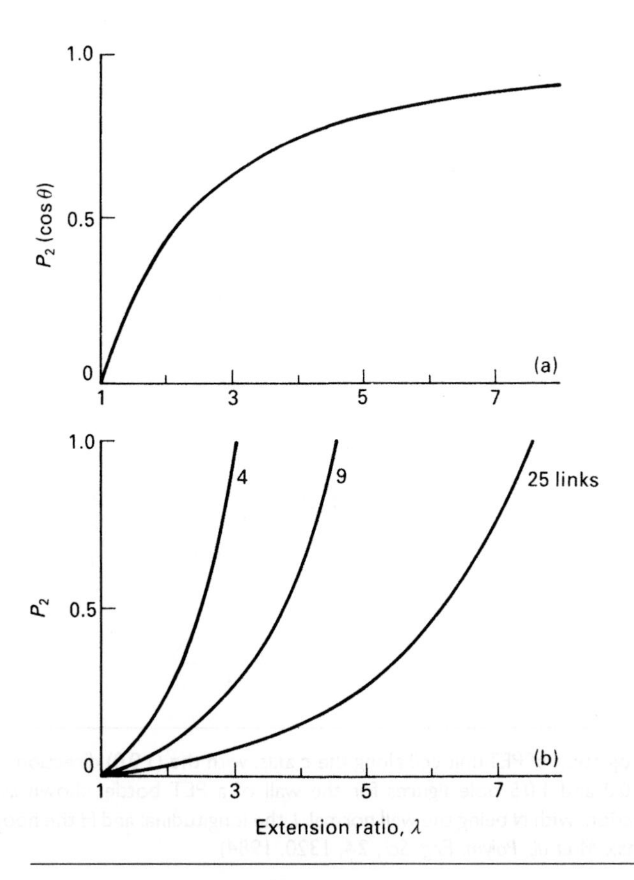

Figure 3.33 (a) Predicted crystal orientation function vs. tensile extension ratio, from the pseudo-affine model; (b) predictions for rubber networks of 4, 9 and 25 link chains.

(b) *Biaxially stretched products*. The optimum orientation has the crystal c axes in the plane of the product, with the majority in the direction of the highest stress. For cylindrical pressure vessels the hoop stress is twice the longitudinal stress (Appendix C), so there should be more crystal c axes in the hoop direction. In the stretch blow moulding of PET bottles (Section 4.5.2), the extension ratio in the hoop direction is greater than that in the longitudinal direction. X-ray diffraction *pole figures* are used to quantify the orientation distribution. Figure 3.34a shows the projection of the PET unit cell along the c axis. The unit cell is triclinic, with none of the cell angles equal to 90°. The (1 0 0) planes contain the in-chain benzene rings. In the (1 0 0) pole figure for a PET bottle, the intensity rises to five times random in the ND direction, normal to the bottle wall. Hence the benzene rings in the crystals tend to lie in the plane of the bottle wall. The (1 0 5) axis is close to the c axis; the intensity of (1 0 5) poles is a maximum of 1.8 times average in the hoop direction. The complete interpretation of the pole figures is difficult.

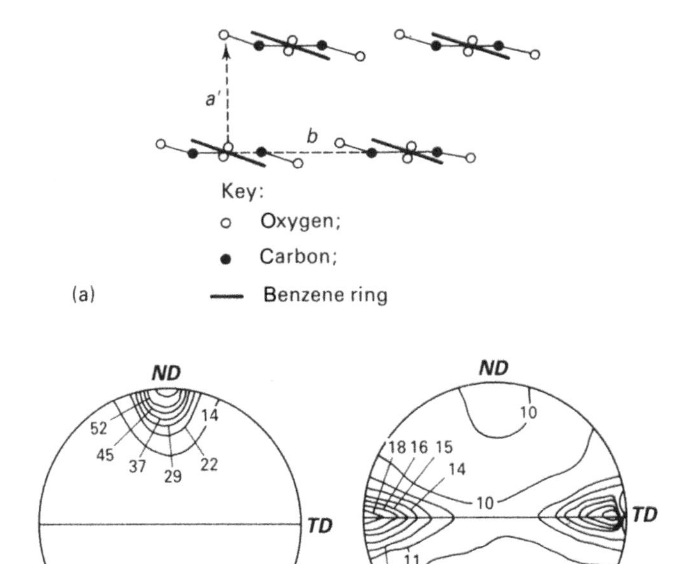

Figure 3.34 (a) Projection of PET unit cell along the **c** axis, with the (1 0 0) direction and **b** axis shown. (b) 1 0 0 and 1 0 5 pole figures for the wall of a PET bottle, shown as a stereographic projection, with *N* being the wall normal, *L* the longitudinal and *H* the hoop direction (from Cakmak *M et al.*, *Polym. Eng. Sci.*, **24**, 1320, 1984).

3.4.11 Summary of crystalline microstructure

The microstructural factors that have the greatest effect on mechanical properties are the per cent crystallinity and the preferred orientation of the crystals (if any). Composite mechanics concepts will be needed to explain the mechanical properties of spherulitic polymers; hence we return to them at the end of the next chapter.

Chapter 4

Polymeric composites

Chapter contents

4.1 Introduction

Many of the useful plastics are composites, obvious examples being rubber-toughened and fibre-reinforced grades. These additives are used to improve the mechanical properties of commodity plastics, allowing them to compete with engineering plastics. This chapter explores the mechanisms behind these improvements. Foamed plastics, which extend the mechanical property range to seating and protective packaging, will be included, since the gas phase is only an extreme case of a weak second phase.

Holliday defines a *complex material* as solid made by physically combining two or more existing materials, to produce a multi-phase system with different physical properties to the starting materials. If one material is a polymer, the other can be a glass, metal, air or another polymer. Polymeric composites can be classified into *macroscopic composites* (Sections 4.2 and 4.3), where the constituent materials can be distinguished with the naked eye, and *microscopic composites* (Sections 4.4–4.8), where they can be distinguished only with a microscope. Concepts, that explain the mechanical properties of macroscopic composites, are used later to explain those of microscopic composites. The examples will illustrate a variety of phase geometries, and methods of increasing stiffness, toughness and energy absorption. Some simple geometries are amenable to analytical solutions, but the majority of commercially important systems have such complex geometries that computer analysis is necessary.

To define a polymeric composite, it is necessary to specify, for each constituent material

(a) the geometry; the shape, size and orientation of the particles or phase, and the volume fraction V,
(b) the chemical composition.

The constituent materials are usually strongly bonded to each other; if not, the mechanical properties of the interface must also be specified. For a two-constituent system, such as polypropylene plus short glass fibres, if the volume fraction V_F of fibres is known, V_{PP} is also known, as $V_F + V_{PP} = 1$. The size of the glass fibres can be defined by their length distribution. The polypropylene phase is continuous, whereas the glass fibre phase is discontinuous. The chemical composition is also easy to specify. However, for rubber-modified polymer like ABS, the geometry of the rubber particles has to be determined experimentally. For both, glass reinforced PP and ABS, processing can affect the product microstructure; fibres become orientated, while rubber particles can become elliptical.

4.2 Elastic moduli

In order to model the mechanical properties of a composite, the relevant mechanical properties of the constituents must be known, e.g. elastic moduli for calculations of composite moduli. Two composites with simple geometries will be analysed for particular types of loading, to provide results of more general use. The macroscopic composite (laminated rubber and metal layers) and microscopic composite (parallel glass fibres in a polymer) are examples of useful composites.

4.2.1 Shear modulus of rubber/steel laminated springs

Chapter 3 explains how the very low shear modulus of rubbers is determined by the crosslink density. Consequently, a rubber block can replace metal leaf or coil springs, in which the bending or twisting of slender beams compensates for the high modulus of the metal. Laminated rubber/metal springs can replace multiple metal leaf springs in a heavy vehicle suspension, or roller expansion-bearings at the end of a bridge deck. Such laminated rubber designs (Fig. 4.1) require no maintenance, whereas the metal mechanisms must be lubricated.

The steel and rubber layers have much larger length and width than thickness. Consequently, any edge effects, where the rubber is close to a free surface, are insignificant. There is strong bonding at the rubber/steel interface. When a pair of shear forces F is applied to the top and bottom layers of the stack (Fig. 4.1), the force F is transmitted to the other layers, which are *loaded in series*. As the layers have equal areas A, the shear stress will be the same everywhere. This *uniform stress condition* is the essence of the analysis.

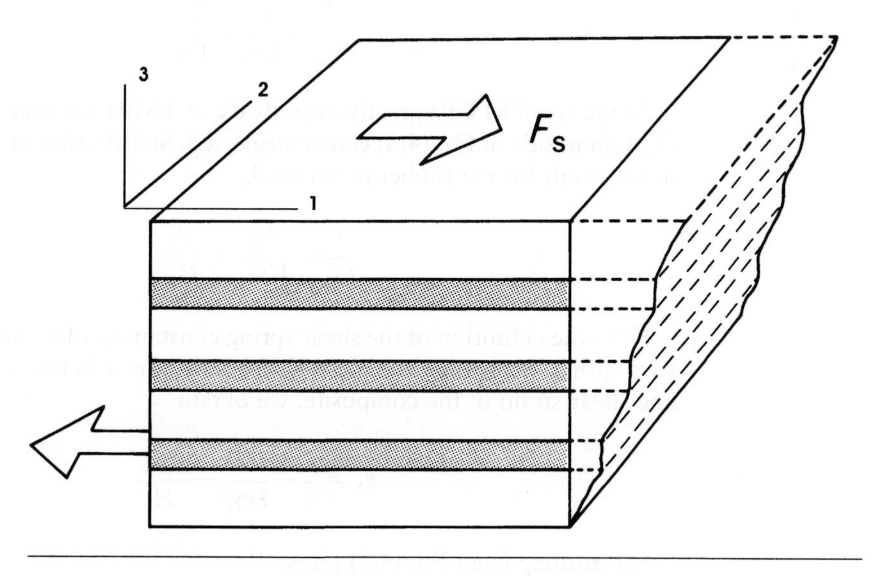

Figure 4.1 Steel and rubber spring, loaded in shear, showing coordinate system used later.

In shear, the volume of the materials remains constant, so the thickness of the layers remains constant. If the stack has total height H, then the volume fraction V_R of rubber is

$$V_R = \frac{h_R}{H} \tag{4.1}$$

where h_R is the total height of the rubber layers. The total shear deflection x_R in the rubber layers is

$$x_R = h_R \gamma_R \tag{4.2}$$

where γ_R is the shear strain in the rubber. There are similar expressions for the steel layers with subscripts S. Consequently, the total shear deflection x of the stack is

$$x = x_R + x_S = \gamma_R V_R H + \gamma_S V_S H$$

Dividing this equation by H, we obtain the average shear strain in the composite

$$\gamma_c = \gamma_R V_R + \gamma_S V_S$$

A further division, by the constant shear stress τ, gives the composite shear modulus G_c in terms of the rubber (G_R) and steel (G_S) shear moduli

$$\frac{1}{G_c} = \frac{V_R}{G_R} = \frac{V_S}{G_S} \tag{4.3}$$

As the G_S of 81 GPa greatly exceeds $G_R \cong 1$ MPa, the second term on the right-hand side of Eq. (4.3) can be neglected. Substitution of Eq. (4.1) leads to the result for the rubber/metal stack

$$\frac{1}{G_c} = \frac{h_R}{HG_R} = \frac{V_R}{G_R} \tag{4.4}$$

From the definition of the shear spring constant k_S of a laminated rubber/steel spring, expressing the force F and deflection x in terms of shear stress and shear strain of the composite, we obtain

$$k_S \equiv \frac{F}{x} = \frac{A\tau}{H\gamma_c} = \frac{AG_c}{H}$$

Substituting from Eq. (4.4) gives

$$k_S = \frac{AG_R}{h_R} \tag{4.5}$$

Hence, the spring shear stiffness only depends on the rubber shear modulus and its total thickness h_R, and is independent of the number of rubber layers.

4.2.2 Compressive stiffness of laminate springs

The laminated spring of the last section are often compressed in a direction perpendicular to the layers. The top and bottom surfaces of the rubber layers cannot expand sideways because they are bonded to steel plates. The effect of this restraint on the compressive response depends on the layer *shape factor* S (Fig. 4.3) defined as

$$S = \frac{\text{Top loaded area}}{\text{Bulge area}} \tag{4.6}$$

where the bulge area is the area of the four sides of the rubber layer. It is assumed that the width and length of the rubber block are comparable in magnitude. The compressive stiffness analysis follows the analysis for shear deformation in the last section. The equivalent of Eq. (4.5), for the compressive stiffness k_C is

$$k_C = \frac{AE(S)}{h_R} \tag{4.7}$$

where $E(S)$ is the *effective compressive modulus* of a rubber layer of shape factor S. When $S < 0.25$ (a cube), the bonded end surfaces hardly restrain the bulging of the sides of the rubber, so the conventional relationship with the rubber shear modulus holds

$$E(0.25) \cong 2G(1 + \nu) \tag{4.8}$$

As Poisson's ratio ν for a rubber $= 0.499$, this simplifies to $E = 3G$. Figure 4.2 shows a finite element simulation of the compression of a rubber cube by 25%. There is a relatively low compressive stress in the region that has bulged out at the sides of the block. The compressive stress distribution has a peak near the edges of the surface bonded to the metal plate, especially at the corner. There is a subsidiary stress maximum at the centre of the cube.

When $S \gg 1$, bulging can only occur near the sides of the block. Near the centre, the rubber compresses in volume with zero lateral strains, in uniform stress conditions. Therefore, the volume strain dV/V is approximately equal to the vertical compressive strain, except at the edges of the plates. From the definition of the bulk modulus K in Eq. (3.30), we find that, in the limit as S tends to infinity

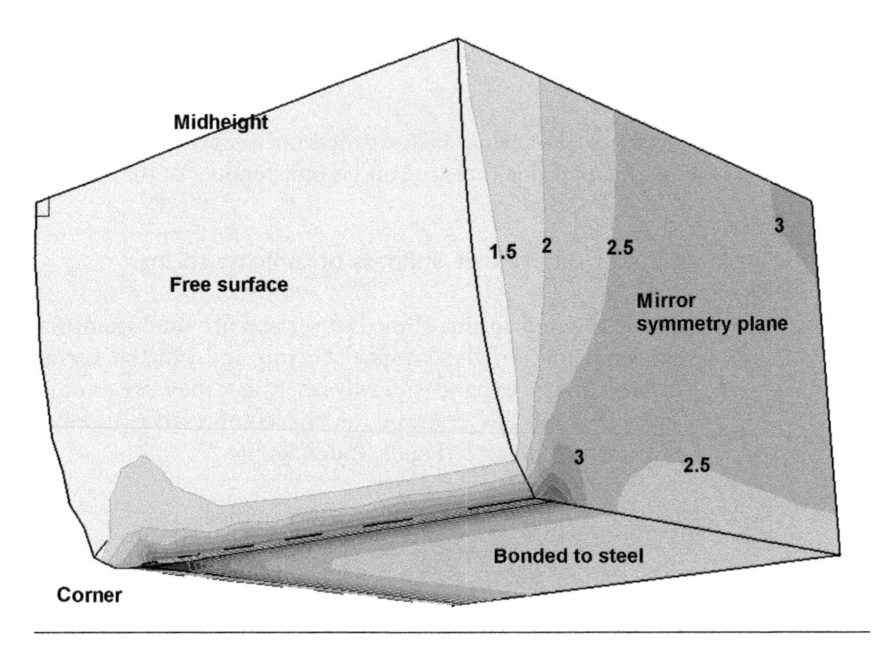

Figure 4.2 FEA of one eighth of a rubber cube, compressed in the vertical direction by 25%, showing vertical compressive stress contours (MPa). The surfaces, bonded to steel, are loaded.

$$E \equiv \frac{\sigma_x}{e_x} = \frac{-p}{\mathrm{d}V/V} = K \tag{4.9}$$

The bulk modulus of rubber, which depends on the strength of the van der Waals forces between the molecules, is 2 GPa. Therefore, the compressive modulus of a rubber layer increases by a factor of a thousand as the shape factor increases from 0.2 (Fig. 4.3). The responses are not shown for $S < 0.2$; such tall, thin rubber blocks would buckle elastically (Appendix C, Section C. 1.4), rather than deforming uniformly. When laminated rubber springs are designed, Eqs (4.5) and (4.7) allow the independent manipulation of the shear and compressive stiffness. The physical size of the bearing will be determined by factors such as the load bearing ability of the abutting concrete material, or a limit on the allowable rubber shear strain to $\gamma < 0.5$ and the compressive strain $e < -0.1$.

4.2.3 Young's modulus parallel to continuous aligned fibres

Figure 4.4 shows continuous fibres (or ribbons, or other constant cross section cylinders) aligned parallel to a tensile stress. The composite block is much longer than the fibre diameter, so end effects are negligible. The

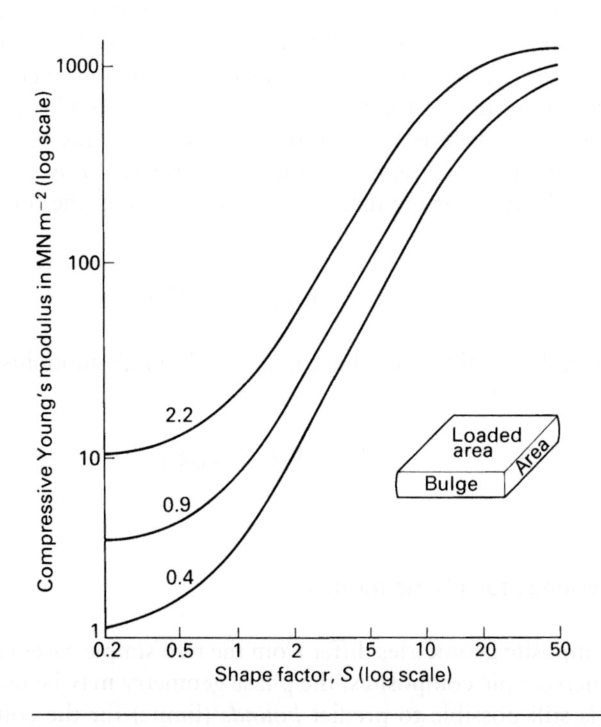

Figure 4.3 Compressive Young's modulus of a rubber spring vs. shape factor; curves labelled with the rubber shear modulus in MPa (from Lindley, P. B., *Engineering Design with Natural Rubber*, 4th Ed., Malayan Natural Rubber Producers Association, 1974).

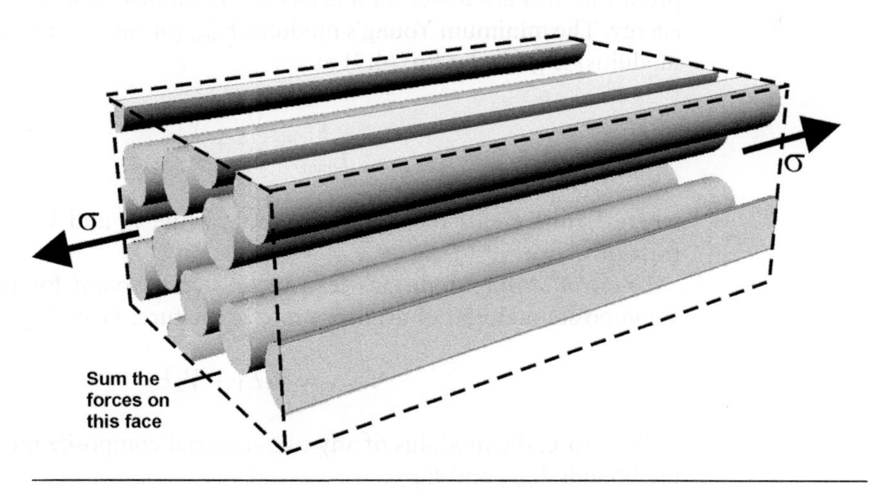

Figure 4.4 Model for the Young's modulus parallel to continuous fibres.

bonding at the fibre/matrix interface prevents the fibres from sliding in holes in the matrix, so the matrix and fibres experience *uniform strain e*. As they are *loaded in parallel*, the average tensile stress σ_c in the composite can be calculated by summing the forces on the ends of the fibres and matrix. In $1\,m^2$ of end face there is $V_f\,m^2$ of fibres (where V_f is the volume fraction of fibres), under a stress $E_f e$, hence, the total force is $V_f E_f e$. There is a similar expression $V_m E_m e$ for the matrix force, so summing the forces on the unit area gives

$$\sigma_c = V_f E_f\, e + V_m E_m e$$

Dividing by strain gives the composite Young's modulus $E_{\parallel}$ when the stress is parallel to the fibres

$$E_{\parallel} = V_f E_f + V_m E_m \tag{4.10}$$

4.2.4 Bounds for elastic moduli

Most composite geometries differ from the two simple cases considered. For many microscopic composites, the phase geometry may be unknown. However, it is still possible to predict *bounds* (limits) for the composite elastic moduli.

Uniform stress conditions produce a *lower bound* for the modulus of any two-material composite, independent of the material geometry. The materials are assumed to be isotropic, but neither needs to be continuous. The proof that this is a lower limit is based on minimisation of the stored elastic energy. The minimum Young's modulus E_{min} (or the shear G_{min} or bulk K_{min} modulus) is given by Eq. (4.3) as

$$\frac{1}{E_{min}} = \frac{V_1}{E_1} + \frac{V_2}{E_2} \tag{4.11}$$

where E_1 and E_2 are the material moduli, while V_1 and V_2 are their volume fractions.

Uniform strain conditions lead to an *upper bound* for the modulus of a composite material of arbitrary microstructure. From Eq. (4.10)

$$E_{max} = V_1 E_1 + V_2 E_2 \tag{4.12}$$

Therefore, the modulus of any two-material composite must fall between the bounds E_{min} and E_{max}.

4.3 Layered structures

4.3.1 Bending stiffness

A typical layered structure consists of two thin, glass-fibre-reinforced polymer skins bonded to a thick, lightweight honeycomb core (Fig. 4.5a). Such sandwich panels are used in railway carriages and aircraft; there are similar structures inside many skins. Other examples are less obvious; the space between the outer container and the toughened polystyrene liner of a refrigerator is filled with rigid polyurethane foam.

In these composites, the layers are bonded together. A sandwich panel beam is *symmetrical* if the skins have equal thickness, and are made of the same material. The neutral surface is at the mid-thickness, so the analysis of Appendix C can be used. Figure 4.6 shows the stress variation through a sandwich beam, calculated using Eq. (C.4) separately for the skins with high Young's modulus E_S, and the core with low modulus E_C.

Because the skin E_S and core E_C Young's moduli differ, the integral in Eq. (C.5) is separated into two contributions

$$MR = E_S \int w_S y^2 \mathrm{d}y + E_C \int w_C y^2 \mathrm{d}y \text{ or } MR = E_S L_S + E_C I_C \qquad (4.13)$$

where I_C is the second moment of area of the core section, and I_S that of the skins. A lightweight sandwich structure with a high bending stiffness can be constructed from thin, high modulus skins bonded to a thick, low density core of moderate modulus. The thick core moves the skins away from the neutral surface, so I_S is large. The core has secondary roles, of supporting the skins when surface compressive forces are applied, and transmitting shear forces from skin to skin. It is possible to reduce the beam mass, while maintaining the bending stiffness, by increasing the core thickness and decreasing the skin thickness. This process must not be taken too far because the skin tensile stresses, given by combining Eqs (C.4) and (4.13), increase in proportion to the total beam thickness. To avoid the risk of skin tensile failure or puncture by sharp objects, the skins must not be too thin.

4.3.2 Structural foam injection mouldings

The *structural foam* injection moulding process produces mouldings with solid outer layers and a foamed core (Fig. 4.5b). The average density is typically between 60 and 90% that of the solid polymer. A chemical blowing agent produces enough gas to foam the melt at atmospheric pressure, but the gas dissolves in the melt at a pressure of about 20 bar. The leading edge of the melt is at a low pressure as it enters the mould, so it foams. These foam bubbles are sheared against the mould surface and

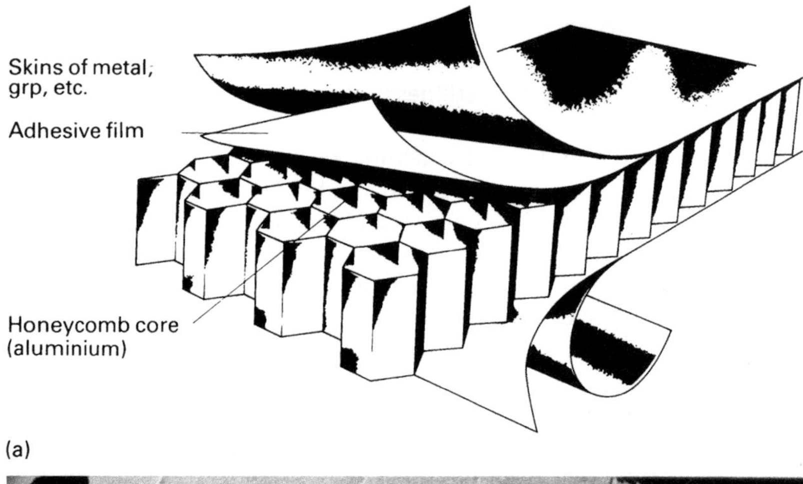

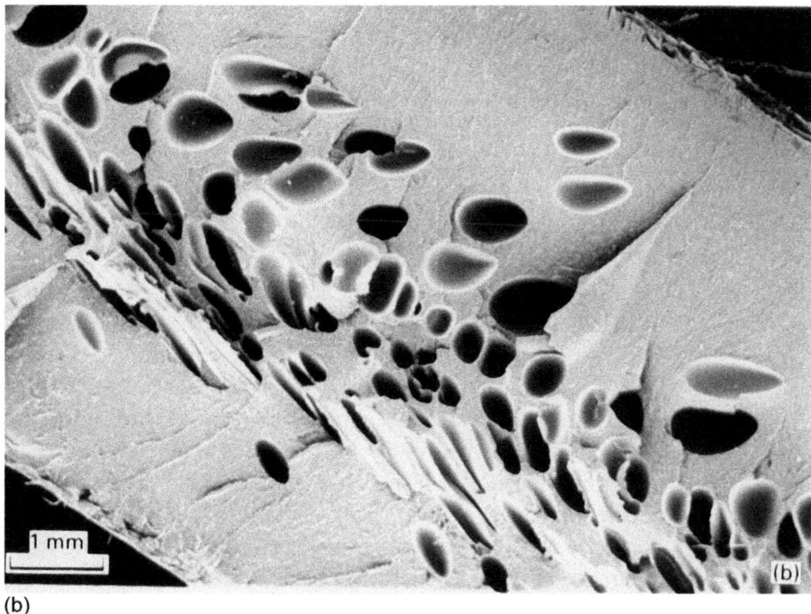

(b)

Figure 4.5 Layered structures with high bending stiffness. (a) Glass-fibre-reinforced skins on an aluminium honeycomb core; (b) structural foam injection moulding with maximum density at the skins.

solidify instantly, leaving a corrugated surface (visible in Fig. 4.5b). In the full mould, the pressure rises to about 40 bar. Consequently, the layers that solidify at that pressure contain no gas bubbles. The volume shrinkage of the cooling polymer (see Section 6.2) reduces the mould pressure; when it drops below 20 bar, the core of the moulding foams. In the final product,

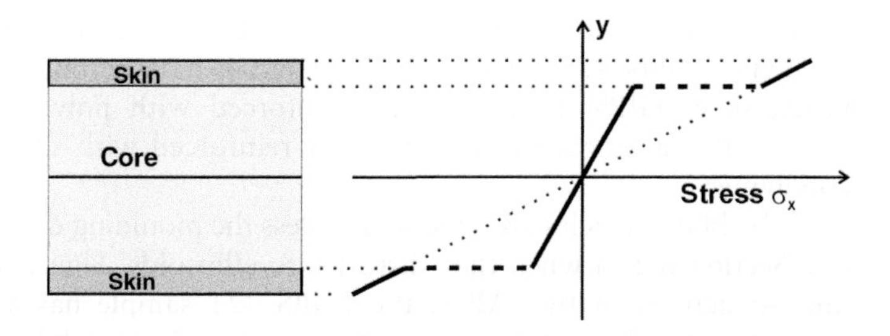

Figure 4.6 Variation of in-plane tensile stress through the thickness of a sandwich panel, loaded in bending.

there is a continuous increase in density from the core to close to the surface (Fig. 4.5b). The mouldings can vary in thickness from 3 to 8 mm without there being sink marks on the surface of the thicker sections—the foamed core is thicker in those regions. Morton-Jones ('Further Reading' for Chapter 13) gives more details of structural foam moulding used as washing machine tanks.

Section 4.7.3 describes how the Young's modulus of open-cell foams varies with the square of the foam density; a similar relationship applies to closed-cell foams. The bending stiffness of the structural foam panel can be calculated using a generalisation of Eq. (4.13), in which the Young's modulus is a function of the distance y from the neutral surface. The bending stiffness is typically 75% higher than that of a solid moulding of the same mass, partly due to the lower average density and partly because the highest modulus material is in the surface layers. However, structural foam mouldings are less tough than solid ones; the rough outer surface contains the equivalent of short cracks, which can propagate through the foamed core.

4.4 Rubber toughening

4.4.1 Toughening systems and their microstructure

Chapter 1 showed that glassy polystyrene is brittle in bending or tension, due to crazing. This response can only be suppressed under unusual conditions, such as testing at 85 °C, just below T_g, or imposing a hydrostatic pressure on top of the uniaxial tension. Therefore, for products subjected to impacts, rubber-toughened grades are used. The principles of rubber toughening are the same for all polymers, but the chemistry of specific polymers determines whether the dispersed rubber spheres are *well bonded* to the matrix. This rubber phase must be stable at high melt processing temperatures of the matrix. Polystyrene and styrene–acrylonitrile copolymer were among the first to be toughened, using polybutadiene or butadiene–styrene copolymer rubbers. Some toughened materials have acronyms; high-impact polystyrene (HIPS) is polystyrene reinforced with polybutadiene, while ABS is styrene–acrylonitrile copolymer reinforced with styrene–butadiene copolymer.

The rubber particles are spherical, unless the moulding orientation is high (see Section 6.2.3) when they distort into ellipsoids. Figure 4.7 shows the microstructures of two ABSs; the LABS 321 sample has a 12% rubber content with spherical particle diameters up to 3 μm, while the LABS 312 has 15% rubber and a bimodal particle size distribution. The matrix is of SAN copolymer, while the larger rubber particles have a salami-like structure, containing smaller spheres of SAN glass. *Phase separation* occurs because the two polymers are immiscible. The complex microstructure results from the polymerisation sequence. First, an emulsion of uniform sized butadiene droplets in water is polymerised; the resulting rubber latex, contains crosslinked rubber spheres less than 1 μm in diameter, which will become the rubber reinforcement in the ABS. Styrene and acrylonitrile, added to the emulsion, dissolve in the polybutadiene (PBD). When polymerisation is re-initiated, SAN copolymer grafts to double bonds in the PBD. Insoluble SAN copolymer separates inside the rubber spheres, and also forms a shell around them. The emulsion is precipitated and dried, then mixed with further SAN copolymer, before being extruded and pelletised. The volume fraction of rubber particles is usually between 10 and 30%, and the particle sizes in the range 0.1–5 μm. The ABS containing

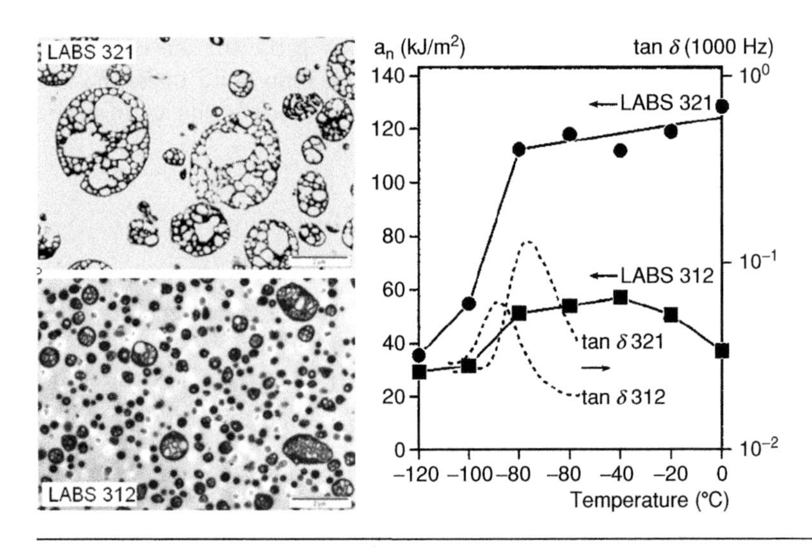

Figure 4.7 Transmission electron micrographs of ABS, with different particle size distributions, with the rubber phase stained dark, and the corresponding variation of impact strength with temperature (Heckmann *et al.*, *Macromol. Symp.*, **214**, 85, 2004) © Wiley-VCH.

larger rubber particles has a higher room temperature toughness, and a higher T_g of the rubbery phase, than that with small particles (Fig. 4.7b). The amount of grafted SAN is optimised to prevent rubber particles agglomerating during processing.

4.4.2 Elastic moduli and stress concentrations

Computer models are needed to calculate the modulus of rubber-toughened plastics. These are elastically isotropic on a macroscopic scale, while a simple cubic array of spheres is anisotropic, and the sphere separation varies with direction. Suitable models include spheres at random positions, within a cube having periodic boundary conditions (Fig. 4.8a). Another is a regular body centred cubic (BCC) array of spheres; Fig. 4.8b shows the stress distribution for a 0.19 volume fraction of rubber, for a 1% tensile strain along a cube axis ([0 0 1] direction). The stress in the polystyrene matrix is highest at the equators of the rubber spheres; the 'lines of force' travel through the glassy matrix, tending to avoid the low modulus rubber spheres. The lines of force bunch closest together in the matrix near the sphere equators, creating a stress concentration factor of 2.0. For comparison, the stress concentration factor is 1.9 for a single rubber sphere in an infinite block of glassy polymer, and 2.0 for an isolated spherical hole in an elastic matrix.

The modulus is nearly the same, whether the composite contains rubber spheres or spherical voids, since the rubber shear and Young's moduli are

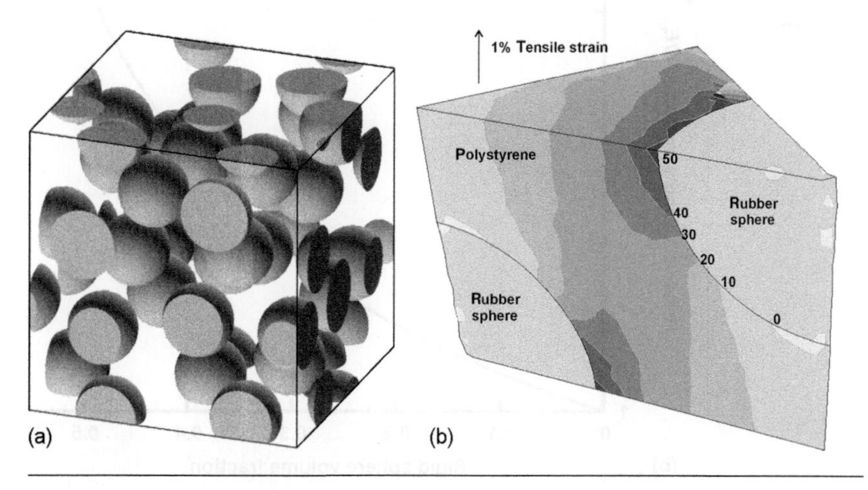

Figure 4.8 (a) Spheres at random positions within a cubic box (Segurado, J. and Llorca, J., *J. Mech. Phys. Solids*, **50**, 2107, 2002) © Elsevier. (b) Segment of a body centred cubic array of spheres (author's unpublished work) showing contours of tensile stress (MPa) in the vertical direction.

much lower than the matrix values. The variation of Young's modulus with void volume fraction, for the model in Fig. 4.8a, is slightly non-linear (Fig. 4.9a). A 25% void volume fraction causes a 40% reduction in Young's modulus compared with that of the pure matrix. However, this is acceptable, due to the gain in toughness. As the glassy matrix is continuous, the composite modulus is slightly below the upper bound of Eq. (4.12).

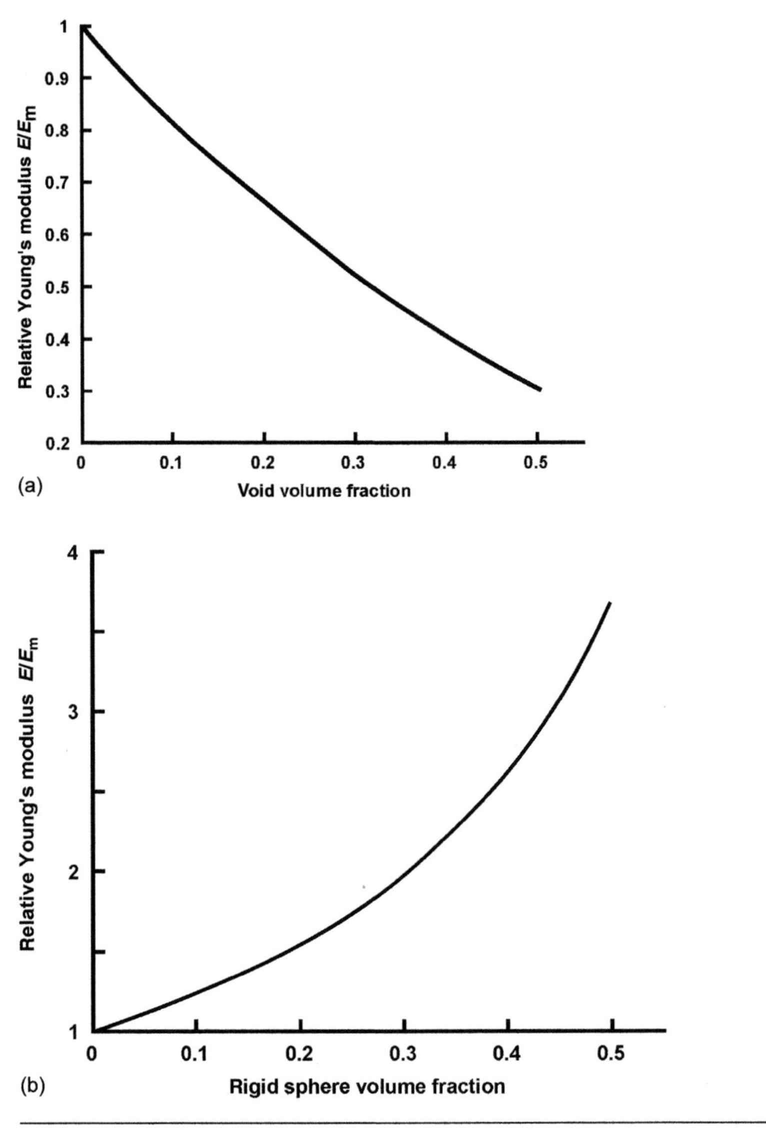

Figure 4.9 Predicted composite Young's modulus vs. particle volume fraction, for: (a) voids; (b) rigid spheres, for the random sphere model of Fig. 4.8a, with a matrix Poisson's ratio of 0.25 (Segurado, J., Llorca, J. *et al.*, *J. Mech. Phys. Solids*, **50**, 2107, 2002) © Elsevier.

A similar analysis predicts the stiffening effect of spherical inclusions with a much higher modulus than the polymer (Fig. 4.9b); the modulus increases by about 50% at a 0.2 volume fraction of spheres. However, Section 4.2.3 shows that it is more efficient to use continuous, aligned fibres to stiffen thermoplastics.

4.4.3 Initiation of crazes or yielding

When a high tensile stress is applied to a rubber-toughened plastic, the stress concentration around the equators of rubber spheres causes yielding or crazing (Chapter 8) to initiate and spread outwards. The presence of the graft copolymer across the glass–rubber interface prevents cracks occurring at the interface. TEM micrographs (Fig. 4.10) show that, when the overall plastic strain is high, multiple crazes have formed in the distorted rubber spheres. The complex internal structure of the rubber spheres is also visible.

Socrate *et al.* (2000) considered an axially symmetric problem, with a rubber sphere in the centre of a short cylinder of matrix; the spheres are in a row, aligned with the tensile stress axis. The potential positions of crazes were predetermined, initially running radially from the material interface, then becoming normal to the tensile stress along the cylinder. The initial stress concentration is greatest in the polymer near the equator of the sphere (Fig. 4.11a). The model, for a 20% volume fraction of rubber, predicts a yield point in the tensile stress–strain curve at an average strain of 1%, and 24 MPa stress, when the first craze propagates across the section. However, this relieves the stress in the polystyrene, and a tensile stress concentration

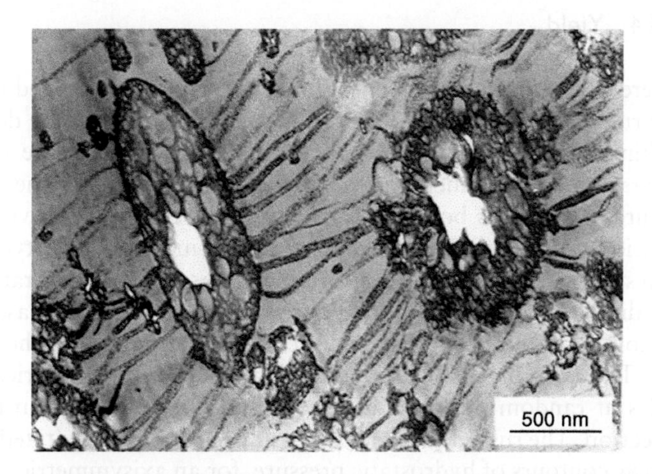

Figure 4.10 TEM of a highly deformed region of polystyrene, reinforced with natural rubber/polystyrene blend (Schneider et *al.*, *J. Mater. Sci.*, **32**, 5191, 1997) with kind permission of Springer Science and Business Media.

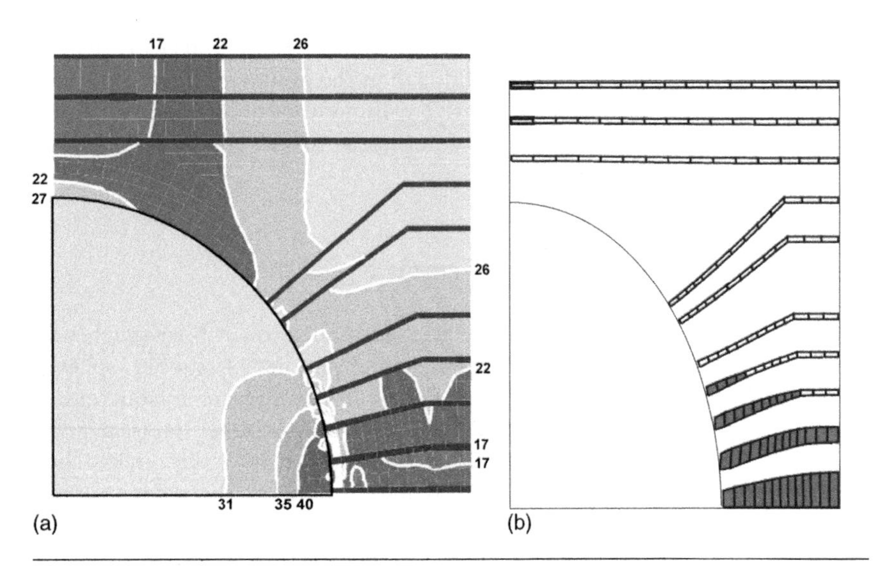

Figure 4.11 Modelling of 2% average tensile strain in a rubber-reinforced PS. (a) Contours of vertical tensile stress MPa; (b) predicted craze opening exaggerated by a factor of 10 (Socrate, S. *et al., Mech. Mater.*, **33**, 155, 2001) © Elsevier.

builds up in the rubber, near the equator of the sphere. Further crazes then initiate and propagate, at higher 'latitudes' on the sphere.

Figure. 4.11b shows, that at a mean strain of 2%, two crazes nearest to the particle equator have undergone through section yielding, whereas the two higher crazes are still growing in length.

4.4.4 Yield

There is experimental evidence, for many rubber-toughened polymers, that the rubber particles cavitate early in the deformation. The degree of cross-linking is kept relatively low in the polybutadiene phase of ABS to aid cavitation, and sometimes silicone oil is added for the same reason. Figure 4.12 shows both the conventional stress–strain curve and the volumetric strain versus tensile strain for rubber-modified polystyrene. When the polystyrene yields, the volume strain increases at a higher rate. Majority of the dilatational strain is due to cavitation in the rubber phase.

Computer modelling has explored the yield pattern in the polymer matrix. The two-dimensional model in Fig. 4.13a has cylindrical rubber particles at random positions in a box, under tensile strain in the horizontal direction. The rubber particles are assumed to be pre-cavitated. Figure 4.13b shows contours of hydrostatic pressure, for an axisymmetric approximation of a BCC array of rubber spheres. Cavitation relieves high hydrostatic tensions at the pole of the rubber particles. Consequently, well bonded, pre-cavitated inclusions stabilise the matrix yielding.

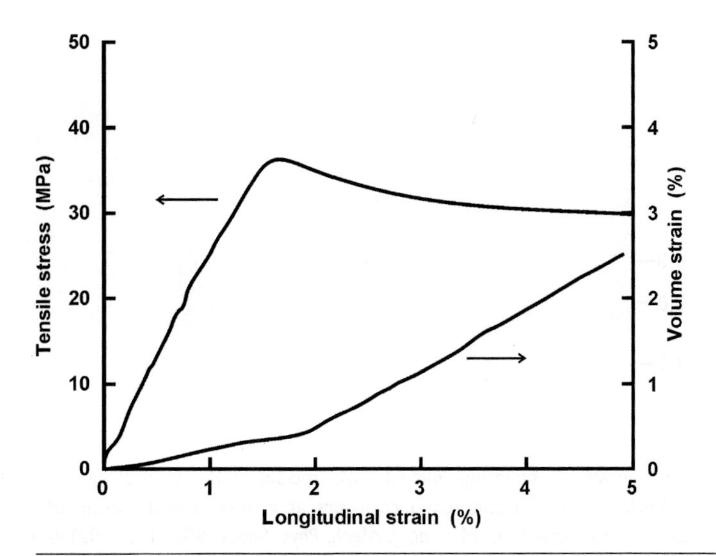

Figure 4.12 Variation of tensile stress and volumetric strain with tensile strain, for polystyrene reinforced with 10 and 25% of block copolymer rubber (reprinted with permission from Magalhaes, A. M. L. and Borggreve, R. J. M., *Macromol*, **28**, 5841, 1995) copyright American Chemical Society.

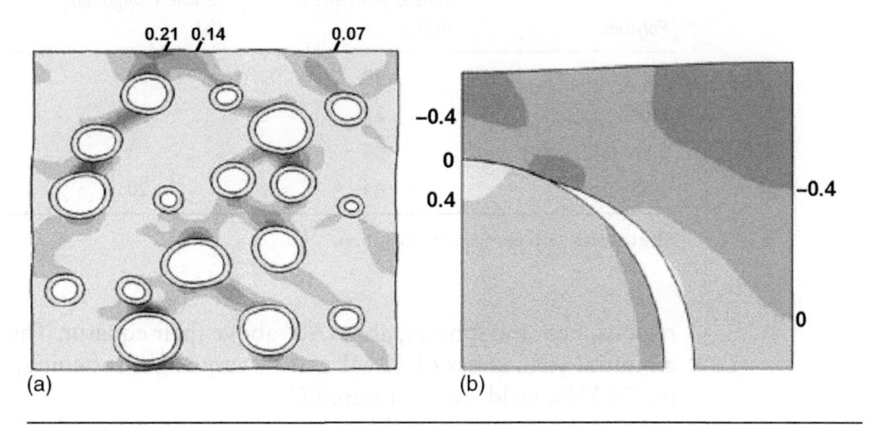

Figure 4.13 Models of pre-cavitated rubber-reinforced composites at 0.1 horizontal tensile strain (Dommelen, J. A. W. *et al.*, *Comput. Mater. Sci.*, **27**, 480, 2003) © Elsevier. (a) Plastic strain contours for cylinders at random positions; (b) hydrostatic stress contours (fraction of yield stress) for an axisymmetric approximation to a BCC array.

Danielsson *et al.* (2002) analysed a BCC lattice of rubber spheres, of volume fraction 0.25, in polycarbonate. The spheres were assumed to cavitate early in the process, so were replaced by voids. Figure 4.14 shows the plastic strain distribution in the repeating unit. In the left-hand view, a spherical surface is visible, whereas the right-hand figure is an isometric view of the other PC surfaces. The plastic strains are concentrated in matrix

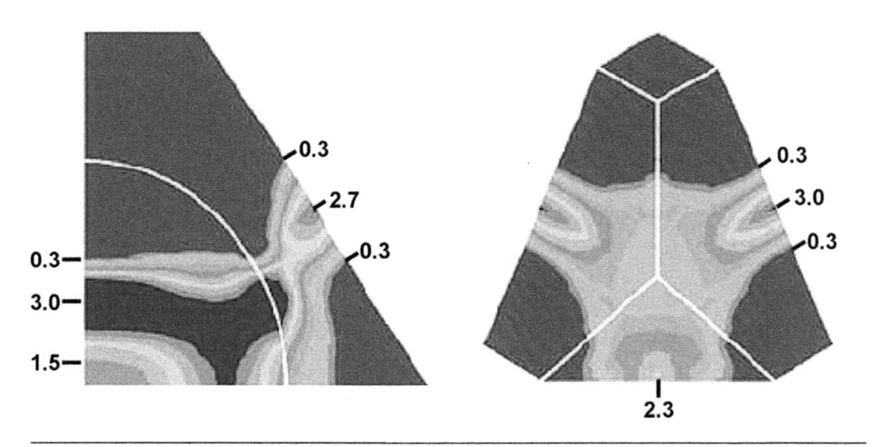

Figure 4.14 Two views of repeating unit of BCC model of PC containing 0.25 volume fraction of spherical voids, plastic strain contours at a mean tensile strain of 3.9% in the vertical direction (Danielsson M *et al.*, *J. Mech. Phys. Solids*, **50**, 315, 2002) © Elsevier.

Table 4.1 Yield (or fracture) stresses at 20 °C

Polymer	*Tensile yield stress (MPa)*	*Tensile elongation (%)*	*Compressive yield stress (MPa)*
PS	40	*	100
HIPS	22	>60	55
SAN (30% AN)	65	*	118
ABS	44	20	52

*Brittle with zero permanent elongation.

regions, near the spheres, about 40° above their equator. The model predicts an initial yield stress of 40 MPa, with very slight softening, compared with the 74 MPa yield stress of pure PC.

Table 4.1 compares yield stresses for matrix materials with those for the rubber-toughened versions of the polymer. Crazing can occur in the tensile tests, but not in the compressive tests. The rubber particles roughly halve the yield or fracture stress in the three cases where the deformation mechanisms do not change. The odd case is for SAN in tension, which fails by crazing. Adding rubber to produce ABS promotes shear yielding as the initial failure mechanism, and the tensile strength only falls by one third.

When a high tensile stress is applied, crazes or yielded regions grow from nearly every rubber inclusion. This contrasts with the untoughened polymer where crazes are separated by distances ~1 mm, and consequently, only a small fraction of the total material will craze; the average permanent elongation at fracture is less than 1%. In the rubber-toughened plastic, the multiple crazing and/or matrix yielding provide a high tensile elongation at break at the cost of losing half the tensile strength. Consequently, the energy absorbed before failure, proportional to the area under the stress–strain curve (Fig. 4.12), is much larger for the toughened polymer.

4.5 Phase-separated structures

Even a single polymer can have a composite structure. Here, the phase geometry and mechanical properties are considered for polymers that separate into two amorphous phases. Block copolymers usually have sufficient block lengths to allow micron-scale phase separation. Later on, we have considered smaller scale microstructures caused by the spinodal decomposition of polyurethanes. Semi-crystalline polymers will be considered in Section 4.6.

4.5.1 Block copolymers

In styrene–butadiene–styrene block copolymers, the butadiene is present as isolated spheres when the butadiene content is <20%, as cylinders when the butadiene content is 20–40% and as parallel lamellae when it is 40–60% of the total (Fig. 4.15b). The latter microstructure is geometrically similar to pearlite, a ferrous alloy containing lamellar stacks of cementite and ferrite. However, curved interfaces, with a greater energy, can sometimes occur when the phase ratio is unequal. Figure 4.15a shows a gyroid structure, in which both phases are continuous, in a styrene–isoprene–styrene tri-block copolymer, with a styrene volume fraction of 0.32. The mechanics of a lamellar stack, in which one phase is glassy and the other rubbery, can be treated using the model in Section 4.2.1. However, the lamellar stacks are randomly oriented, so the macroscopic Young's modulus can be calculated using a similar analysis as to that of spherulites in Section 4.6.

The microstructure depends on the details of the copolymer architecture. Figure 4.16 compares tensile stress–strain curves for styrene–butadiene–styrene tri-block copolymers; a 74 volume % polystyrene copolymer with a sharply defined central butadiene block contains rubber cylinders arranged in a hexagonal array; it has the highest yield stress, but very little elongation at break. With the same styrene content, but asymmetric length styrene blocks and a tapered butadiene to styrene transition, the lamellar structure has an intermediate mechanical response. Finally, with 65% styrene content, shorter end blocks, and a central random styrene butadiene central section, the gyroid structure has the lowest initial yield stress and the highest elongation at break.

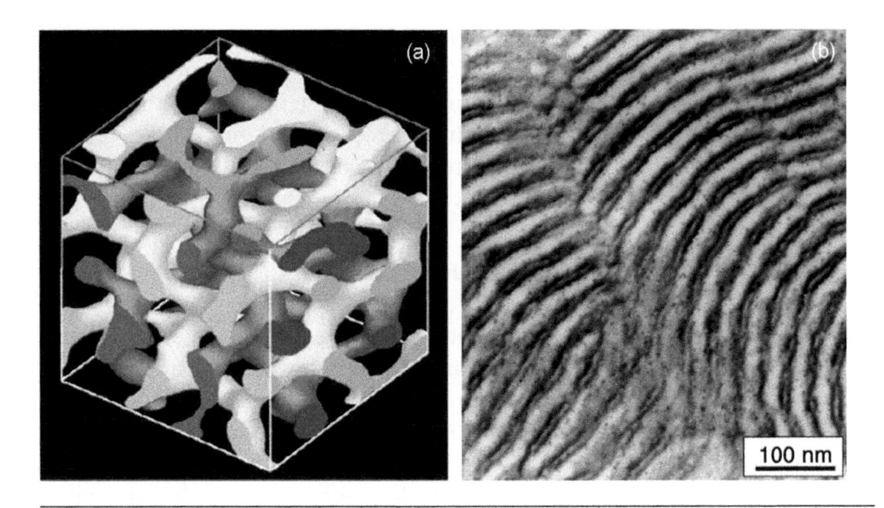

Figure 4.15 Tri-block copolymer morphologies: (a) Three-dimensional reconstruction of TEM images of gyroid structure in styrene isoprene styrene (Spontak, R. J. and Patel, N. P., *Curr. Opin. Coll. Interface Sci.*, **5**, 334, 2000) © Elsevier; (b) TEM of lamellar edges in styrene butadiene styrene (Huy, T. A. *et al.*, *Polymer*, **44**, 1237, 2000) © Elsevier.

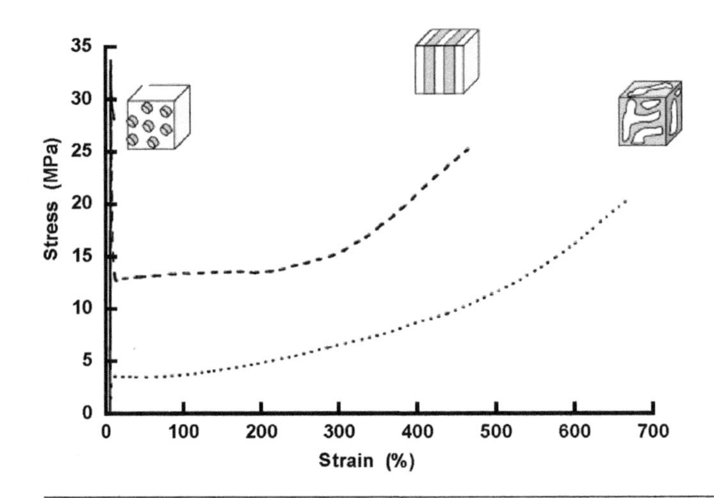

Figure 4.16 Tensile stress–strain curves for styrene butadiene styrene tri-block copolymers (Huy, T. A. *et al.*, *Polymer*, **44**, 1237, 2003) ©Elsevier.

4.5.2 RIM polyurethane

Reaction injection moulding (RIM) of polyurethane (Section 5.7) produces a block copolymer containing both hard and soft segments. The crystalline hard segments have melting points in excess of 150 °C, while the soft segments are polyether or polyester rubbers. A typical polyurea hard block is a step-growth polymer prepared by reacting a di-isocyanate (here 4,4′ diphenylmethane di-isocyanate—MDI) with an aromatic diamine extender (here the diamine of diethyl toluene—DETDA).

CH_3 NH_2 H_3C CH_3 C C H_2 H_2 NH_2

and MDI

O=C=N–⟨benzene⟩–CH_2–⟨benzene⟩–N=C=O

Typical polyurethanes use a diol extender such as 1,4 butane diol. The NH groups form hydrogen bonds with C=O groups on neighbouring chains, inside very small, hard blocks. The soft blocks are crosslinked polyethers or polyesters. A typical polyether is poly(propylene oxide) —[—O—CH_2—$C(CH_2)H$—]$_n$ pre-polymerised to a molecular mass M_N of about 6000, a degree of polymerisation $n = 140$. The PPO has a glass transition temperature of about $-60\,^\circ C$, so the crosslinked PPO is a rubber. This rubbery phase is connected, via the polyurethane molecules, to the crystalline phase.

For the crystal nucleation and growth mechanisms described in the last chapter, a graph of free energy of mixing versus composition has a positive curvature. However, when the graph has a negative curvature, a homogenous mixture can undergo *spinodal decomposition* into two phases with different compositions. It can be modelled in two dimensions (Fig. 4.17). An animation on the website www.physics.ndsu.nodak.edu/Wagner/spinodal/spinodal.html shows how the shapes stay similar, but the scale coarsens, with time. For an equilibrium 50:50 phase composition, both phases are continuous; in three dimensions, both phases should exist as

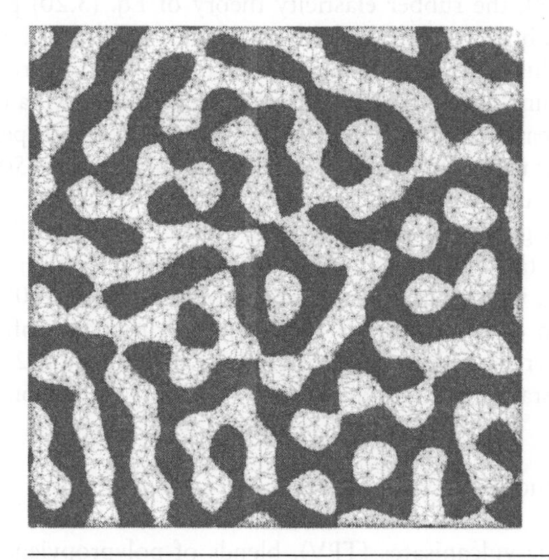

Figure 4.17 Two-dimensional model of spinodal decomposition of a 50:50 composition, taken to equilibrium (Read DJ *et al.*, *Eur. Phys. J.*, **E 8**, 15, 2002) © EDP Sciences.

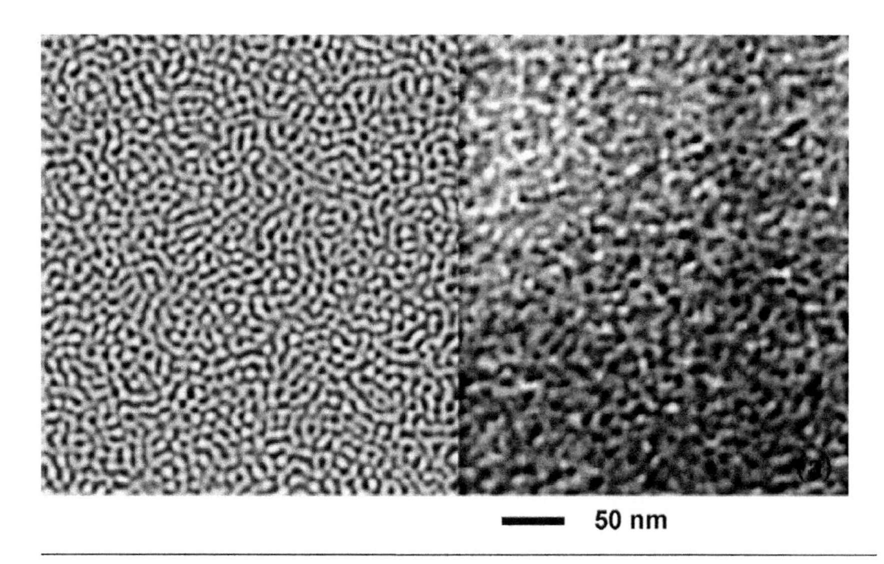

Figure 4.18 TEM of polyurethane microstructure (right), compared with intermediate stage of computer model (left). (Hamley, J. W. *et al.*, *Polymer,* **41**, 2569, 2000) © Elsevier.

convoluted finger-like regions, as in Fig. 4.15a. Cell dynamic simulations for an early stage of phase separation in a polyurethane (polyether–isocyanurate) are similar to the structure seen in the transmission electron microscope (Fig. 4.18). The inter-domain spacing is approximately 10 nm.

Neither the uniform strain model nor the uniform stress model is appropriate for this microstructure. Consequently, the elastic moduli of polyurethanes lie between the limits set by Eqs (4.11) and (4.12). For a network chain of $M_c = 6000$, the rubber elasticity theory of Eq. (3.20) predicts a shear modulus of about 0.4 MPa. The hard blocks will have the typical 3 GPa Young's modulus of glassy polymers. Increases in the hard block content cause the Young's modulus to increase from 30 to 500 MPa (Fig. 7.13). For automobile panel applications it is usual to have a high per cent of hard blocks so that the room temperature flexural modulus is $\sim$500 MPa.

When polyurethanes are stretched about 150%, the nearly-straight, short, soft segments crystallise. This increases the tensile strength and abrasion resistance of polyurethane rubbers. A similar strain-crystallisation phenomenon, which occurs in natural rubber at about 500% strain, limits the extension of rubber bands. Both the polyurethane soft segments and natural rubber have crystal melting points in the region 25–60 °C. In the unstretched state, the chain disorder prevents crystallisation.

4.5.3 Thermoplastic vulcanisates

Thermoplastic vulcanisates (TPV), blends of polypropylene with ethylene propylene diene (EPDM) copolymer, have replaced conventional rubbers in some applications, since they can be processed using thermoplastic machinery, yet have better fatigue response than many rubbers. They have a continuous matrix of thermoplastic PP, in which the EPDM phase exists as spheres of a few microns diameter (Fig. 4.19a). The process of blending, followed by injection moulding that causes the EPDM to crosslink, is referred to as *dynamic vulcanization*. The material behaves like a rubber with a Young's modulus of the order of 0.6 MPa. Different grades of *Santoprene* from Advanced Elastomeric Systems, with rubber contents from 40 to 90%, have different moduli. If these materials were related to the rubber-modified glassy thermoplastics considered in Section 4.4, the rubber phase should be continuous, with spherical PP inclusions. However, it appears that the shear flow of the EPDM during crosslinking creates a quite different microstructure.

Composite theory has been used to explain why TPVs behave like rubbers, with a low compression set, and low hysteresis when extended by 200%. Boyce *et al.* (2001) considered a two-dimensional array of identical sized rubber cylinders, with centres that were close to being on a body centred lattice. They considered the plane strain compression of this structure (Fig. 4.19c); there is zero strain in the direction perpendicular to the paper. Their FEA predicts that initially-thin PP regions yield and undergo high tensile strains, while other thicker PP regions remain undeformed. When the structure is unloaded, there is not much reverse yielding in the PP regions (white), so some rubber regions are also predicted to remain distorted. The predicted stress–strain curve, for a rubber volume fraction of 0.79, is close to the experimental data (Fig. 4.19b). However, a two-dimensional model, containing uniform sized rubber cylinders, is a poor approximation to the real structure. Wright *et al.* (2003) proposed that the PP phase has an open-cell foam structure, which is interpenetrated by a continuous EPDM phase. When etching was used to remove some of the elastomeric phase, the PP microstructure was found to be similar to a microcellular foam. However, they did not prove that the EPDM phase is continuous.

4.6 Modulus of spherulitic polyethylene

4.6.1 Deformation mechanisms in spherulites

In most spherulitic polymers, touching spherulites occupy whole of the space. Their microstructure is too complex to be completely modelled, especially if there is twisting of lamellar stacks about spherulite radii. Consequently, models simplify the structure, and use composite micromechanics concepts. A stack of parallel lamellar crystals with interleaved amorphous layers (Fig. 3.20) has a similar geometry to a laminated rubber/metal spring (Fig. 4.1). The crystals have different Young's moduli E_a, E_b and E_c (Section 3.4.3), and different shear moduli when the

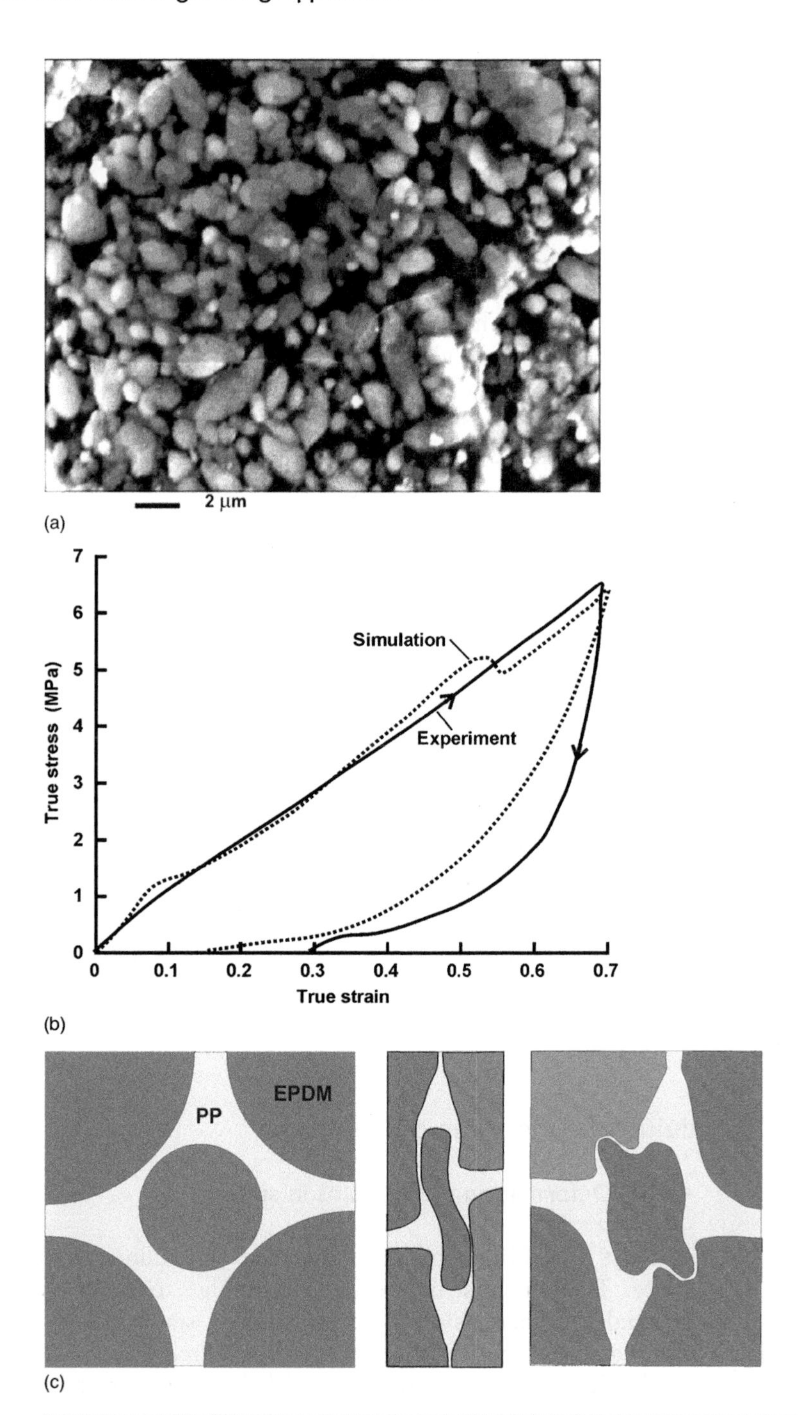

Figure 4.19 (a) Microstructure of a TPV, (Boyce MC *et al.*, *J. Mech. Phys. Solids*, **49**, 1073, 2001) © Elsevier; (b) predicted vs. experimental stress–strain curve (Boyce MC *et al.*, *J. Mech. Phys. Solids*, **49**, 1323, 2001) © Elsevier; (c) deformation—left to right: Undistorted; centre: Plane strain compression of 0.5; right: Unloaded (ibid).

shear stresses are in the *ab*, *bc* or *ac* planes. The amorphous layer is isotropic, with a shear modulus higher than that of a rubber because of the inter-crystalline links. If shear stresses are applied in the *ac* or *bc* planes, the lamellae move parallel to each other (Fig. 4.1). There are uniform stress conditions, so, by Eq. (4.3), the shear compliances of the phases add in proportion to their volume fractions. If a tensile stress is applied in the 3 direction (Fig. 4.1) normal to the lamellar surfaces, the uniform stress conditions mean that the tensile compliances in the 3 direction add in proportion to volume fractions of crystalline V_{cr} and amorphous material V_{am} giving

$$\frac{1}{E_3} = \frac{V_{cr}}{E_{cr}} + \frac{V_{am}}{E_{am}} \tag{4.14}$$

The high shape factor of the amorphous layers means that the amorphous Young's modulus E_{am} will be close to the amorphous bulk modulus of 2 GPa. The tensile compliance in the 3 direction will be dominated by the amorphous Contribution, because E_C of the polyethylene crystal is 250 GPa. Hence,

$$E_3 \cong \frac{E_{am}}{V_{am}} \tag{4.15}$$

Although E_3 is relatively high, the shear moduli G_{31} and G_{32} are very small, so the inter-lamellar layers will shear, if at all possible.

When tensile stresses or shear stresses act in the *12* plane, there are uniform strain conditions in the composite laminate. Consequently, the tensile moduli E_1 and E_2 can be added in proportion to their phase volume fractions, using

$$E_1 = V_{cr}E_{cr} + V_{am}E_{am} \tag{4.16}$$

or an equivalent equation for the shear moduli G_{12}. The crystal moduli are so high compared with the amorphous Young's modulus that the second term can be neglected, hence

$$E_1 \cong E_2 \cong V_{cr}E_{cr} \tag{4.17}$$

Fig. 4.20a shows a variety of deformation mechanisms at the equator of the spherulite, due to differing orientations of the lamellar stacks relative to the tensile stress axis. Computer models are needed to consider the variety of lamellar stack orientations, and calculate the macroscopic stresses. Using an axisymmetric model of a spherulite (in a regular array), the tensile yield stress was predicted to be a nearly linear function of the crystallinity (Fig. 4.20b), and in the same range as experimental data.

4.6.2 Elastic moduli of spherulitic polyethylene

The elastic moduli of lamellae stacks need to be averaged for all the orientations that occur in a spherulite. Where two spherulites meet at a boundary, the lamellar stacks have different orientations. When these spherulites deform, they remain in contact, so the deformation mechanisms must be coordinated locally. However, current models ignore interactions between spherulites. Some models that average the stiffness and the compliance of lamellar stacks provide upper and lower bounds, respectively for the polymer modulus. However, as the Young's modulus of the crystalline phase greatly exceeds that of the rubber-like amorphous phase, the upper bound greatly exceeds the lower bound, so the bounds are of little practical use. Moreover, as polyethylene is a viscoelastic material (Chapter 7), its time-dependent modulus can change by a factor of three or more according to the timescale of loading.

Figure 4.21 shows crystallinity as the main factor that determines the Young's modulus of polyethylene, with the aspect ratio (length/width) of lamellae, having a lesser effect. Lamellae with a high aspect ratio have a larger stiffening effect. The experimental data agrees with the predictions if the lamellar aspect ratio is in the range 20–40. Although aspect ratio may differ with polyethylene thermal history, the effect has not been confirmed experimentally.

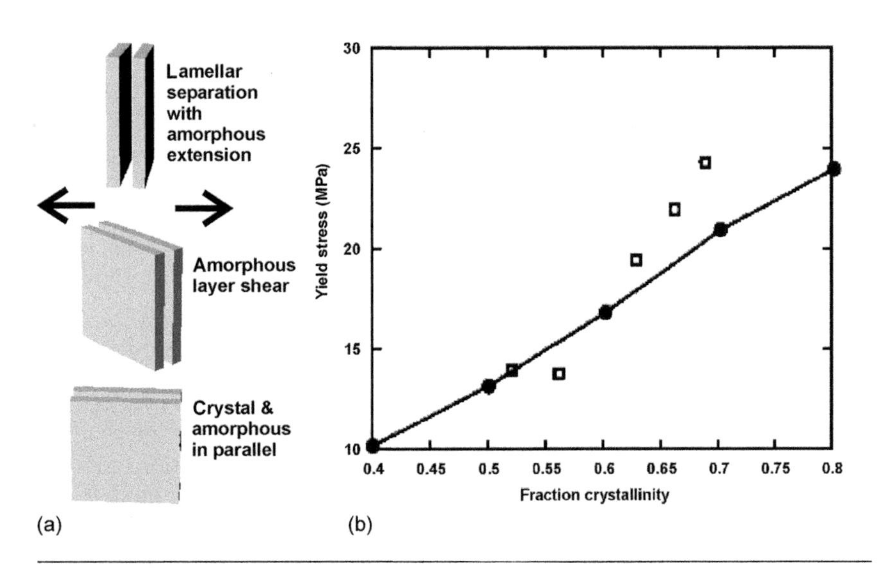

Figure 4.20 (a) Deformation modes in the equatorial region of a spherulite; (b) predicted yield stress vs. fraction crystallinity (redrawn from van Dommelen, J. A. W. *et al.*, *Polymer*, **44**, 6089, 2003) © Elsevier.

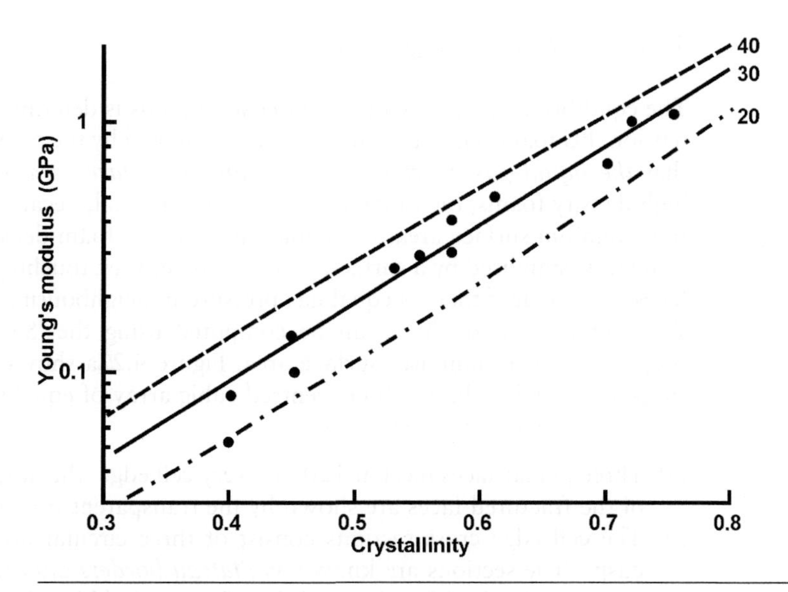

Figure 4.21 Predicted Young's modulus of spherulitic polyethylene vs. crystallinity, for lamella of various aspect ratios, compared with experimental data (Guan X and Pitchumani R., *Polym. Eng. Sci.*, **44**, 433, 2004) © John Wiley and Sons Inc. reprinted with permission

4.7 Foams

4.7.1 Polyurethane open-cell foam chemistry

The reactants for polyurethane flexible foams, of molecular weight <5000, are low viscosity liquids. CO_2 gas is generated by the reaction of the 1–4% water content with the di-isocyanate monomer

$$R{-}N{=}C{=}O + H_2O \rightarrow R{-}NH_2 + \uparrow CO_2$$

The amide groups, formed on both ends of the monomer, react with further di-isocyanate to form urea linkages. The polymer contains polyurea blocks, which phase separate later in the process, while proportion of tri-ol monomer present creates a network. Surfactants are used to modify the surface tension of the gas–liquid interfaces, hence, control the stability of the cell faces. The viscosity rises rapidly during polymerisation, and eventually a gel forms. At this stage, the cell faces, a fraction of a μm thick, are subjected to such large biaxial tensile strains that they fail. They retract into the surrounding edges, leaving an open-cell foam. However, sometimes thin cell faces survive, and some polyurethane foam cell faces have small central holes. The completion of the crosslinking reaction then stabilises the cell edges.

4.7.2 Open-cell foam geometry

The equilibrium geometry of low viscosity liquids is determined by surface tension. PU foam cell shapes are largely determined by the physical principle that *the liquid–gas interface has a minimum surface area (or energy).* In high-density foams, the isolated bubbles are spherical, because a sphere has the minimum surface area to volume ratio. As the foam density decreases, thin faces, stabilised by a surfactant, develop between touching bubbles; the faces are planar if there is equal gas pressure in neighbouring bubbles. The shape of the liquid phase can be computed using the 'Surface Evolver' program, which minimises surface area. Figure 4.22a shows the predicted shape of a single cell, in a body centred cubic array of equal sized bubbles, known as the *Kelvin foam*. In this

(a) Three planar faces meet at 120° at every cell edge (the original positions of the fractured faces are shown by the transparent triangular mesh).
(b) The cell edge cross sections consist of three circular arcs, meeting at cusps. The sections are known as *Plateau borders* after the nineteenth century scientist who observed them in soap bubbles. They are seen on the cut edges of the open-cell polyurethane (PU) foam in Fig. 4.22b.
(c) Four edges of equal length meet at each vertex, with inter-edge angles near to the 109.5° of tetrahedrally bonded carbon.
(d) Each cell has eight hexagonal and six square faces.

In PU foams, the cells have neither regular shapes, nor a uniform size. The foam microstructure can be specified in terms of the cell size distribution, and the cell shape anisotropy. PU slabstock foams rise while being supported on a moving belt, so the cells have a greater height than diameter.

4.7.3 Open-cell foam compressive response

Flexible PU foams need to have elastic moduli of the same order of magnitude (10 kPa) as human soft tissue, to be suitable for seating applications. Their density is usually below 40 kg m^{-3}. Low stresses can cause high elastic compressive strains because

(a) The >0.95 volume fraction of air provides space for edges to undergo large bending deflections without contact.
(b) The slender edges have low bending and torsional stiffness.
(c) The Young's modulus of the PU is low at approximately 50 MPa.
(d) The strains in the bent edges are <20%, so the PU can recover completely on unloading.

Micro-mechanics models for foam deformation are simplifications of the real structure. Figure 4.23 shows a repeating element of the Kelvin foam cell of Fig. 4.22, prior to deformation. The flat surface at the front is a mirror symmetry plane through the polymer structure, as is the hidden flat surface

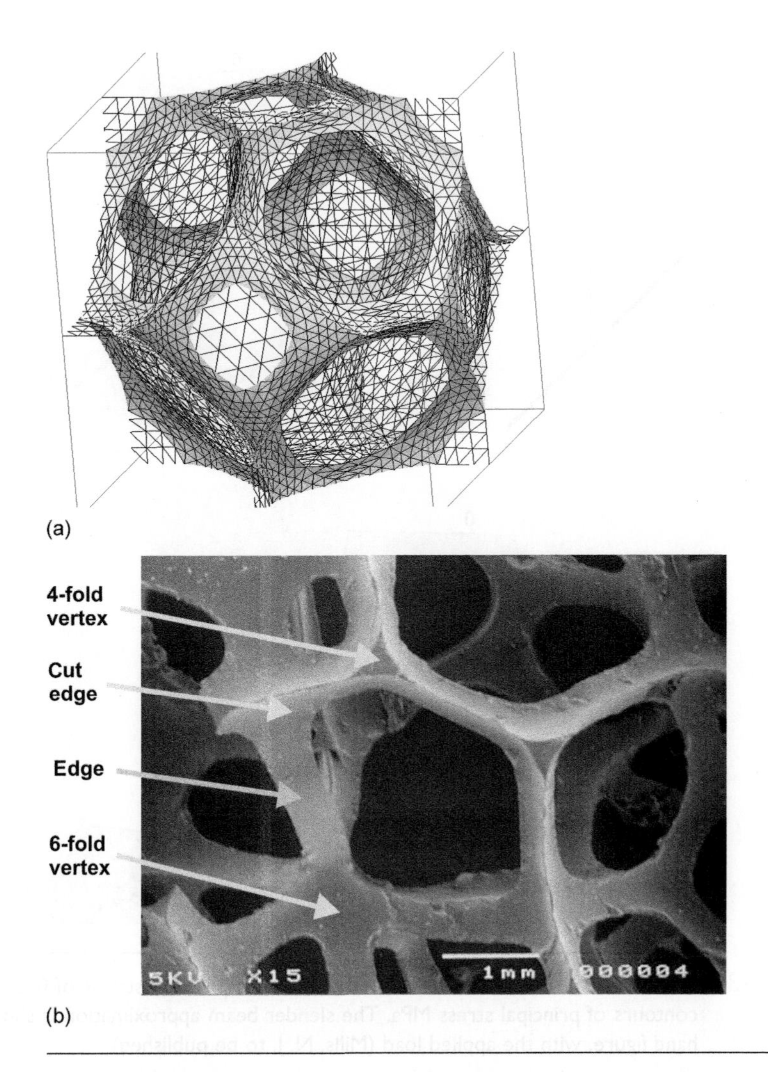

Figure 4.22 (a) Cubic repeat unit of a Kelvin open-cell foam structure, of relative density $R = 0.0276$, generated by Surface Evolver; (b) scanning electron micrograph of open-cell polyurethane foam with $R = 0.026$.

on the left. When the foam is compressed vertically (along a [0 0 1] direction), the cell edges bend progressively but do not twist. At 50% compressive strain, the edges have not touched.

There is a simple analytical solution, treating the cell edges as slender beams with a constant cross section. The deformed shape of the edge DC has a centre of symmetry at its midpoint O (Fig. 4.23). The half-edge shape is that of a cantilever beam of length $L/2$, built in at vertex C, loaded at its free end O by a force $F/2$. Large deformation beam theory relates the local curvature to the moment M by Eq. (C.15). The compressive stress σ_z in the

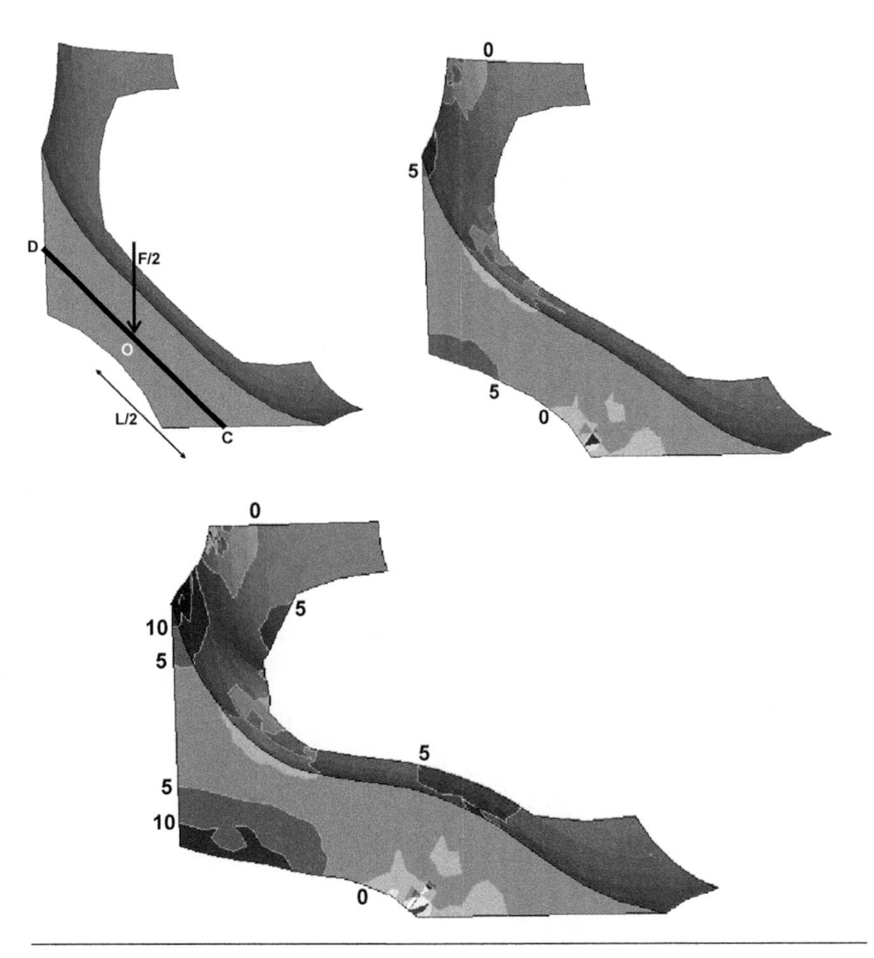

Figure 4.23 Section of Kelvin foam model, $R = 0.027$, at compressive strains of 0, 20 and 50%, with contours of principal stress MPa. The slender beam approximation is shown on the left-hand figure, with the applied load (Mills, N. J. to be published).

foam was shown (Zhu and Mills, 1997) to depend on the edge bending stiffness *EI*, being given by

$$\sigma_z = \frac{2EIF^2(\alpha)}{L^4} \tag{4.18}$$

where $F(\alpha)$ is an elliptic integral

$$F(\alpha) \equiv \int_{\delta}^{\pi/2} \frac{d\phi}{\sqrt{1 - p^2 \sin^2 \phi}} \tag{4.19}$$

$p \equiv \sin \alpha/2$, and ϕ is defined by $\sin \theta/2 = p \sin \phi$. $F(\alpha)$ is a function of the inclination α of the edge at O. The lower limit of integration is

$$\delta = \sin^{-1}\left(\sin {}^{\beta}\!/_{2} \Big/ \sin {}^{\alpha}\!/_{2}\right) \tag{4.20}$$

where β is the edge inclination at D. The second moment of area I, for an edge with a Plateau border cross section of area A, is

$$I = 0.1338\, A^2 \tag{4.21}$$

Consequently, the foam Young's modulus E_f in the [0 0 1] direction is related to the polymer Young's modulus E, and to the square of the foam relative density, by

$$E_f = 1.01\, ER^2 \tag{4.22}$$

The foam *relative density* R is defined in terms of the foam ρ_F and polymer ρ_P densities

$$R \equiv \frac{\rho_F}{\rho_P} \tag{4.23}$$

The foam model is nearly elastically isotropic, with a 6% lower modulus in the [1 1 1] direction, due to a contribution from edge torsion. If the Plateau border width varies realistically, being thinnest at mid-edge, the constant in Eq. (4.22) is approximately 2.3. Further corrections occur if the model has a distribution of cell sizes and shapes. The predicted compressive stress (Fig. 4.24) hardly increases for strains between 10 and 60%, in agreement with the experimental data. The non-linear response is due to the large change in the foam edge shape, rather than to the material non-linearity. However, the elastic model cannot predict the hysteresis observed on unloading.

4.7.4 Closed-cell foam geometry

When gas bubbles grow in a highly viscous thermoplastic melt, cell wall thinning is resisted by the 'hardening' of the biaxially extended melt. Figure 4.25 shows polystyrene foam of density $40\,\mathrm{kg\,m^{-3}}$, having closed-cells with 4, 5 and 6-sided faces. More than 90% of the polymer is in the cell faces, which are of near-uniform thickness of a few μm. Consequently, the cell edges play little part in the mechanics of the foam compression.

4.7.5 Closed-cell foam compressive response

When closed-cell foams are compressed, the stress is taken both by the compressed air in the cells, and by the bent and stretched cell faces. The

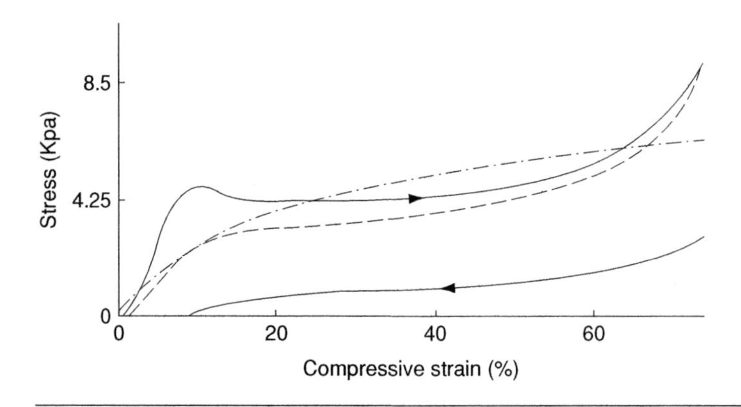

Figure 4.24 Compressive stress–strain curve for PU foam of density 31 $kg\,m^{-3}$, compressed in-plane —, and through thickness - - -, compared with the Kelvin foam prediction for compression along [1 1 1] direction –-,–,–,–,–.

Figure 4.25 SEM of section through closed-cell polystyrene foam with relative density $R = 0.025$.

macroscopic (average) strain is the same in the polymer and the air, and the loads are taken in parallel (Section 4.2.3). For analysis, the response of 1 m^3 of foam containing trillions of cells is the same as that of a single macro-cell (Fig. 4.26) containing the total gas volume. For most such foams, after the initial elastic stage, the cell walls under tension prevent further lateral expansion, so Poisson's ratio is zero. Thus, the compressive strain ε is also the volumetric strain, and the air is compressed as in a piston. As the polymer is effectively incompressible, the gas volume in the compressed foam is $1 - \varepsilon - R$. Using the ideal gas law for isothermal compression, the absolute gas pressure p in the foam is given by

$$p_a(1 - R) = p(1 - \varepsilon - R) \tag{4.24}$$

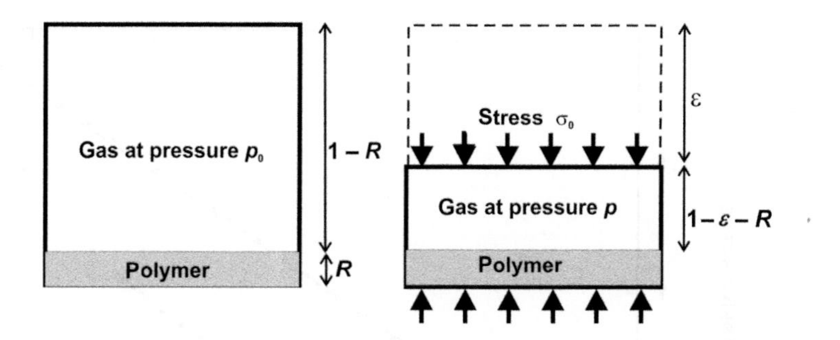

Figure 4.26 The gas volumes in a foam of zero Poisson's ratio and relative density R, before and after uniaxial compression. Heights of the phases in meters.

The applied compressive stress σ is the foam gas pressure minus atmospheric pressure

$$\sigma = p - p_a = p_a \frac{\varepsilon}{(1 - \varepsilon - R)} \tag{4.25}$$

It is assumed that the polymer contribution σ_0 to the foam stress is constant, so

$$\sigma = \sigma_0 + \frac{p_0 \varepsilon}{1 - \varepsilon - R} \tag{4.26}$$

For low-density foams σ_0 can be evaluated by fitting the loading part of a graph of stress against $\varepsilon/(1 - \varepsilon - R)$ with a straight line, and extrapolating to zero strain (Fig. 4.27). The compressive behaviour of closed-cell foams varies from elastic for ethylene vinyl acetate copolymer foams in trainer midsoles, to viscoelastic for polyethylene foams used in camping mats, through yielding with polystyrene foams, to brittle with methacrylate foams. Chapter 8 considers yielding in polystyrene foams.

4.8 Short fibre reinforcement

4.8.1 Fibres and their orientation

Glass fibres, with Young's moduli of 72 GPa, are much stiffer than polymers, and, if undamaged their tensile strengths of 1–2 GPa are much greater. Their temperature resistance, with a T_g exceeding 500 °C, is much higher than that of the polymer. However, glass is a brittle elastic solid, and the fibres are easily damaged in plastics processing. The total volume fraction of glass is restricted to $V_g \leq 0.2$ to prevent the melt viscosity becoming excessive. Manufacturers quote the percentage *by weight* of

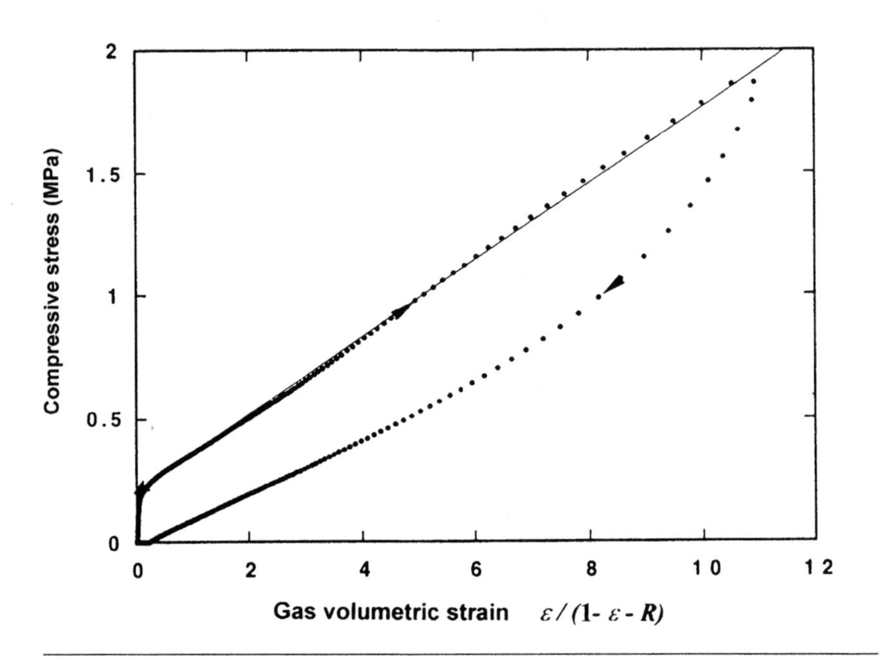

Figure 4.27 Stress–strain relation for PP foam of density 43 kg m^{-3}, plotted to fit Eq. (4.26).

glass fibres; as the glass has a density of 2540 kg m^{-3} compared with 900–1400 kg m^{-3} for the plastic matrix, 30% by weight of glass is equivalent to $V_g \cong 0.15$.

The microstructure of a glass-reinforced thermoplastic is characterized by the fibre volume fraction V_f, its length distribution, and its orientation distribution. The glass fibre length may average 3 mm before it is incorporated into the plastic, but high-flow stresses in the extruder barrel rapidly comminute the fibre length. A typical fibre length distribution in a moulded part is shown in Fig. 4.28a. The fibres are 10–12 μm in diameter; thus they can have aspect ratios (length/diameter) of up to 100:1.

The fibre orientation in sections of an injection-moulded bar can be determined by contact micro-radiography, which shows X-ray 'shadows' of the fibres, or by optical microscopy of polished surfaces. Image analysis of the latter can detect the axial ratio and major axis orientation of the elliptical sections of glass fibres. In modelling the composite mechanics, it does not matter if the angle between the fibre length and the flow direction is θ or $-\theta$; the sign of θ cannot be determined experimentally. In Fig. 4.28b the histogram of orientation angles is from a simulated *maximum entropy* distribution, with parameter $\overline{\cos^2\theta} = 0.7$. The orientation factor $\overline{\cos^2\theta}$ varies through the thickness of PP injection mouldings, typically being 0.65 at the surface, 0.75 at 0.5 mm below the surface, and 0.6 at the mid-plane.

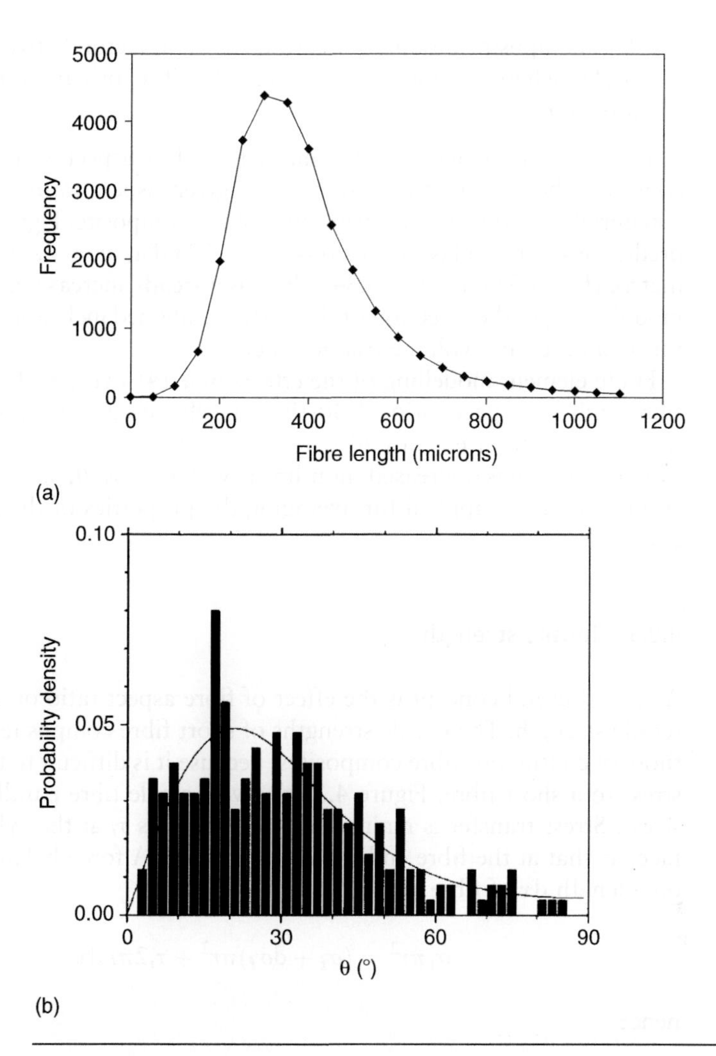

Figure 4.28 Characterising glass fibres in an injected moulded polypropylene plate: (a) Fibre length distribution (Hine, P. J. *et al.*, *Compos. Sci. Tech.*, **62**, 1445, 2002); (b) theoretical angular distribution (Hine, P. J. *et al.*, *Compos. Sci. Tech.*, **64**, 1081, 2004) © Elsevier.

4.8.2 Young's modulus

Two concepts from fibre-reinforcement theory are of general use in polymer mechanics. The first is the effect of inclusion shape on the anisotropic elastic moduli of a composite. Two rules control the effects of fibre reinforcement:

(a) A set of fibres, with a distribution of lengths and mean length $\overline{L}$, acts like a set of fibres of uniform length $\overline{L}$.

(b) The composite modulus determined by a factor between $\overline{\cos^2\theta}$ and $\overline{\cos^4\theta}$, where θ is the angle between the fibre axis and the tensile stress direction.

The mean fibre length determines the fibre aspect ratio. One theory considers the effects of inclusions, of a given aspect ratio, surrounded by a material with the average properties of the composite. Figure 4.29a shows predictions for E–glass inclusions ($E = 73$ GPa, $\nu = 0.22$) in an epoxy matrix ($E = 5.35$ GPa, $\nu = 0.34$); there is a steady increase in the composite modulus E_{11} in the direction of the perfectly aligned inclusions, as the aspect ratio and the fibre volume fraction increase.

Finite element modelling of the effects of a 14% volume fraction of glass fibre with an aspect ratio of 30 (Fig. 4.29b), used fibre orientations that fitted a maximum entropy distribution. It predicted that the longitudinal Young's modulus increased non-linearly with $\overline{\cos^2\theta}$, and that constant strain conditions applied for averaging the properties of the unidirectional composite.

4.8.3 Tensile strength

A second useful concept is the effect of fibre aspect ratio on the composite tensile strength. The tensile strengths of short fibre composites are less than those of continuous fibre composites, because it is difficult to transfer a high stress to a short fibre. Figure 4.30 shows a single fibre parallel to a tensile stress. Stress transfer is mainly via shear stresses τ_i at the cylindrical interface, as that at the fibre ends can be neglected. A force balance calculation on a length dx of fibre gives

$$\sigma_f \pi r^2 = (\sigma_f + d\sigma_f)\pi r^2 + \tau_i 2\pi r\,dx$$

hence

$$\frac{d\sigma_f}{dx} = \frac{2\tau_i}{r} \tag{4.27}$$

Assumptions must be made about the matrix and interface behaviour before Eq. (4.27) is integrated. If the matrix remains elastic and the interface does not fail, the shear stress rises to a maximum at the fibre ends, where the tensile strains in the fibre e_f and the matrix e_m differ the most. However, a ductile polypropylene matrix is assumed to yield in shear at a stress $\tau_y \cong 20$ MPa. This will occur near the fibre ends, so the interface shear stress is

$$\tau_i = \pm\tau_y \text{ when } e_f < e_m$$

$$\tau_i = 0 \text{ when } e_f = e_m$$

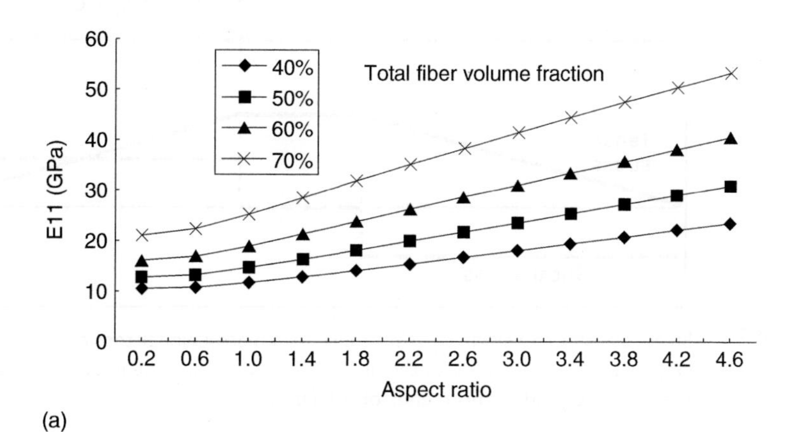

(a)

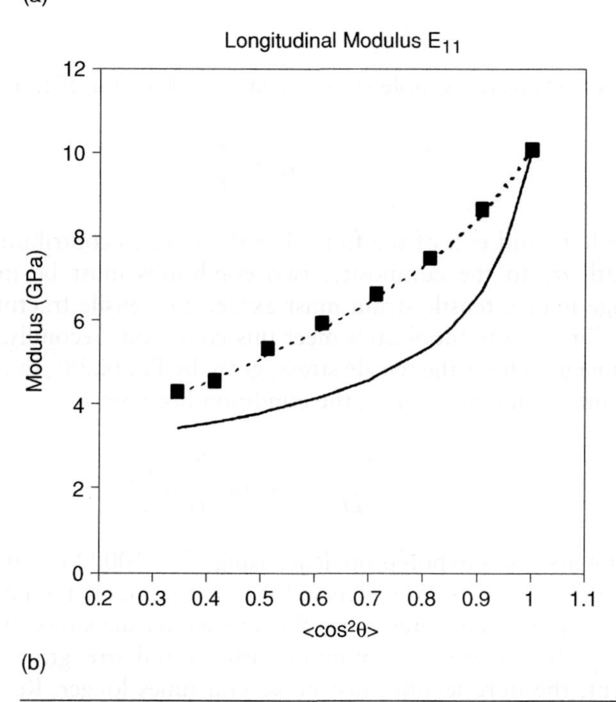

(b)

Figure 4.29 Variation of longitudinal Young's modulus of glass-fibre-reinforced PP with: (a) Fibre aspect ratio and volume fraction (Chao, L.-P. and Wang, Y.-S., *Polym. Compos.*, **21**, 20 2000) © John Wiley and Sons Inc. reprinted with permission; (b) orientation average $\overline{\cos^2\theta}$ (Hine, P. H. *et al.*, *Compos. Sci. Tech.*, **64**, 1081, 2004) © Elsevier. Dashed line, constant strain; solid line, constant stress averaging.

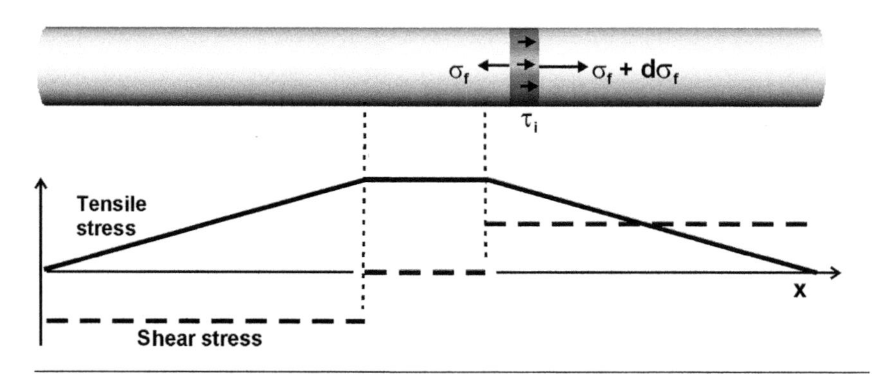

Figure 4.30 Stress transfer to a single fibre in a thermoset matrix, under tension in the *x* direction. The graph shows the variation of the fibre tensile stress and the interfacial shear stress, when the interface yields at both ends of the fibre.

This leads to the simple stress variations shown in Fig. 4.30, with

$$\sigma_f = \frac{2\tau x}{r} \tag{4.28}$$

at the left-hand end of the fibre. For the fibre to contribute its full tensile strength σ_f^* to the composite, two conditions must be met. Firstly, the average matrix tensile strain must exceed the tensile fracture strain of the glass fibre e_f^*; thermoplastics meet this condition. Secondly, the fibre must be long enough for the tensile stress, given by Eq. (4.28), to reach σ_f^*. As the maximum value of x is $L/2$, the condition becomes

$$\frac{2\tau_L x}{2r} > \sigma_f^* \text{ or } \frac{L}{D} > \frac{\sigma_f^*}{2\tau} \tag{4.29}$$

For glass fibres in polypropylene, using $\sigma_f^* = 1000$ MPa and $\tau_y = 20$ MPa, the condition is $L > 25D$. Hence, for $D = 10$ μm, the minimum fibre length $L = 250$ μm. Such fibres cause the average tensile stress in the fibre to be half σ_f^*. To achieve the optimum stiffness and strength for a given glass content, the fibre length must be several times longer. Re-examination of Fig. 4.28a reveals that the fibres in a typical glass reinforced thermoplastic are insufficiently long for optimum reinforcement. Consequently, process development has aimed at increasing the fibre length to approximately 5 mm. The fibre surfaces need chemical treatment to achieve adequate shear strength at interfaces with non-polar polyethylene and polypropylene. Treatment with silane coupling agents achieves optimum strength and toughness in the composite.

Chapter 5

Processing

Chapter contents

5.1 Introduction

The main polymer processes are described, to explain the shapes that can be made, and to indicate the order of capital cost. For details of individual processes, refer to the specialised texts listed in 'Further Reading'. The majority of processes for thermoplastics contain three stages (Fig. 5.1). They are first heated into the melt state, at a relatively low temperature; the methods and rates of heat transfer are studied, and the efficiency of the screw extruder is explained. In the second stage the shape of the melt changes. Polymer melts differ from those of metals, in being highly viscous, yet non-Newtonian. This allows the stable inflation of bubbles. Melt flows are analysed to relate the output to the required pressure. In the third stage, cooling fixes the product shape; the low thermal conductivity has repercussions both on product design and process productivity. The next chapter will discuss the effects of processing on product microstructure. Some secondary processes, such as the warm stretching of solid fibres and the machining of ultra high molecular weight polyethylene (UHMWPE), are considered in Chapter 15.

Thermoset processes involve an additional stage of polymerisation and/or chemical crosslinking. The starting materials for reaction injection moulding are of low viscosity, so are easy to mix and pump into the mould. This process is examined at the end of the chapter.

5.2 Heat transfer mechanisms

The main heat transfer processes that are come across in plastics processing are conduction, convection and viscous heating, with radiation only playing a role in thermoforming. Most products are much thinner than they are wide, so only one-dimensional heat flow (Fig. 5.2) will be considered. The heat flow direction is along the x axis, perpendicular to the surface of the product; there are planar isotherms perpendicular to the x axis. The heat flow Q is considered across an area A of the isothermal surface.

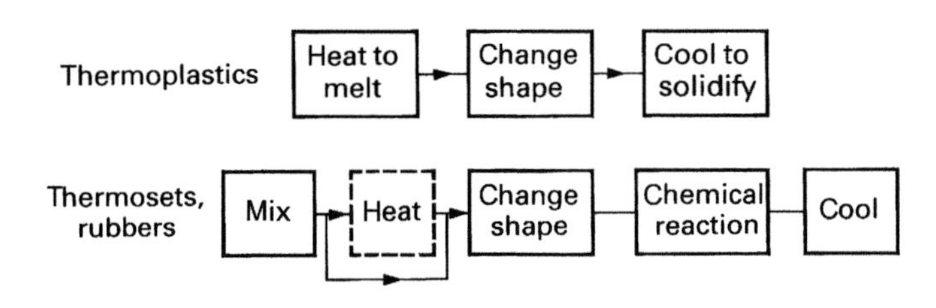

Figure 5.1 Block diagrams of the stages in processing thermoplastics, and thermosets or rubbers.

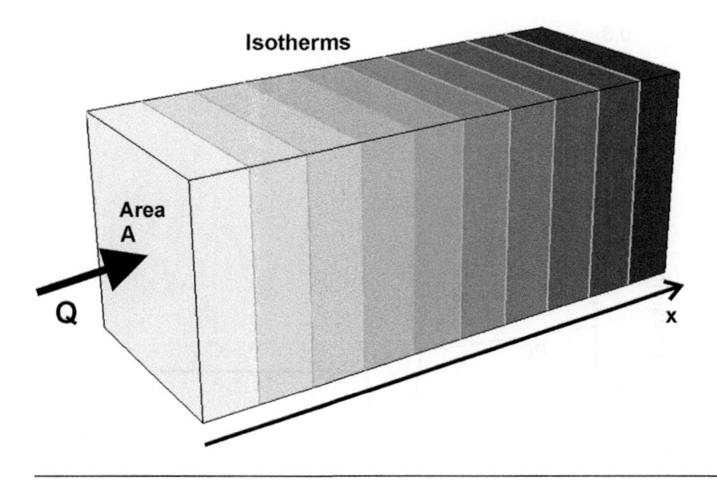

Figure 5.2 Isotherms for steady one-dimensional heat flow through a plastic, due to a heat flow Q onto the surface area A, from either radiation, convection or conduction from a metal.

5.2.1 Conduction

The thermal conductivity k of a material is defined by *steady-state conduction*. The heat flux Q (W) is parallel to the negative temperature gradient $-\mathrm{d}T/\mathrm{d}x$

$$Q = -kA\frac{\mathrm{d}T}{\mathrm{d}x} \tag{5.1}$$

For polymers k is of the order of $0.2\,\mathrm{W\,m^{-1}\,K^{-1}}$. This is much smaller than the $50\,\mathrm{W\,m^{-1}\,K^{-1}}$ for steel, due to the lack of free conduction electrons, and the weak forces between polymer chains. *Steady-state conduction* occurs through the foam-insulated wall of a domestic refrigerator; the temperature at any point in the foam remains constant, however.

Transient conduction conditions occur in polymer processing. Appendix A derives Eq. (A.14) for one-dimensional transient heat flow, which contains the *thermal diffusivity* α. This is the combination $k/\rho c_\mathrm{p}$ of the thermal conductivity k, density ρ and specific heat c_p. For most polymer melts α is approximately equal to $0.1\,\mathrm{mm^2\,s^{-1}}$ (Fig. 5.3). For the melting of low-density polyethylene in an extruder, typical conditions are: a barrel temperature of $T_0 = 220\,°\mathrm{C}$, an initial polymer temperature $T_\mathrm{P} = 20\,°\mathrm{C}$, and a melting process complete at $T = 120\,°\mathrm{C}$. Consequently, using Eq. (C.19), after a contact time t, the melt front is at a distance x_m from the barrel given by

$$x_\mathrm{m} \cong \sqrt{\alpha t} \tag{5.2}$$

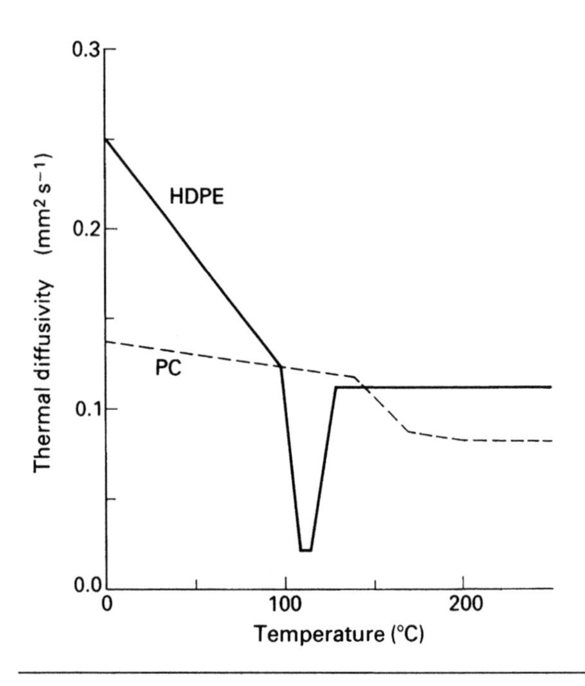

Figure 5.3 Variation of thermal diffusivity with temperature, for amorphous polycarbonate and semi-crystalline polyethylene.

Hence, as $\alpha = 0.1\,\mathrm{mm^2\,s^{-1}}$, the thickness of melt x_m in millimetres is related to t in seconds by

$$x_m \cong 0.3\sqrt{t} \tag{5.3}$$

This result governs melting, both in an extruder and in hot-plate welding. It only applies while x_m is much less than the sheet thickness. It shows that conduction alone is too slow to melt thick layers of plastic; it would take 100 s to melt a 3 mm surface layer.

In the cooling stage of processing, the product must completely solidify before the mould can be opened. Figure A.4 in Appendix A shows the dimensionless temperature profiles at different times since the start of cooling. Products cooled from both sides, such as injection mouldings, cool faster than products of the same thickness (extruded pipe, blow moulded and thermoformed) where one side is effectively thermally insulated. The temperature profiles are given in terms of the Fourier number, a dimensionless time defined by Eq. (A.26) as

$$Fo \equiv \frac{\alpha t}{L^2}$$

where t is the cooling time, L is the half thickness of an injection moulding or the full thickness of a blow moulding. The product is completely solidified when $Fo \cong 0.3$, so, if $\alpha = 0.1$, the solidification time t_s in seconds is given by

$$t_s \cong 0.3\frac{L^2}{\alpha} \cong 3L^2 \tag{5.4}$$

where L is measured in mm. Hence, it is necessary to keep the product thickness to a minimum if the productivity of the mould is to be high.

In reality α is not constant. Figure 5.3 shows that for amorphous polycarbonate, α increases on passing from the melt to the glassy state. For semicrystalline polyethylene on cooling, α is very low through the crystallisation temperature range because the specific heat is high. Computer methods are needed to obtain solutions to transient conduction problems using such a temperature-dependent diffusivity.

5.2.2 Convection

Convection cooling occurs in water and in air; their density decreases on heating and the buoyancy of the hot 'liquid' is sufficient to cause a moderate flow rate in the low viscosity medium, so long as it extends for a sufficient vertical distance. However, convection currents do not occur in polymer melts, which are far too viscous. Convection cooling can occur in the liquid or gaseous environments at polymer surfaces. The heat flow across the polymer/liquid interface is

$$Q = hA(T_s - T_0) \tag{5.5}$$

where h is the heat transfer coefficient, T_s is the polymer surface temperature and T_0 the environmental temperature. The heat transfer coefficient depends to some extent on the size, shape and orientation of the object, but approximate values for different cooling media are given in Table 5.1.

We are familiar with the cooling effect of wind and rain on our own exposed skin; weather forecasters refer to the 'wind chill factor'. The data in Table 5.1 shows the effectiveness of water sprays in removing heat from the surface of extruded plastic pipes. Consequently, the pipe cooling rate is limited by the transient conduction within the plastic; the melt surface can be treated as if kept at the water temperature, and the analysis of the last section be used.

5.2.3 Biot's modulus

When a polymer slab of thickness L is cooled on one side by convection, we need to know if there is a significant temperature gradient inside the plastic

Table 5.1 Heat transfer coefficients

Medium	*Heat transfer coefficient* ($W m^{-2} K^{-1}$)
Still air	10
Air at velocity $5 \, ms^{-1}$	50
Water at 5 °C	1000
Water spray	1500

during cooling. This can be found by calculating Biot's modulus, a *dimensionless group* defined by

$$B \equiv \frac{hL}{k} \tag{5.6}$$

If $B \gg 1$, there is marked temperature gradient, but if $B \ll 1$, there is not. For a sheet of polymer of thickness >1 mm cooled by water or by contact with a steel mould, $B \gg 1$. When $B > 10$, it is a good approximation to say that the polymer surface temperature is immediately reduced to that of the cooling medium. For a blow-moulded parison (Section 5.5.2) of thickness $L = 0.002$ mm, cooled in still air on the outside, $B = 2$. If the lower end of the parison cools for 150 s before the mould closes, it is a reasonable approximation to ignore the temperature gradient through the wall, and calculate the average temperature drop at the end of the parison.

5.2.4 Radiation

In the thermoforming process, the temperature of the ceramic or metal radiant heaters is typically 400 °C (673 K). These can be treated as black bodies at an absolute temperature T, for which the radiation heat flux from an area A is

$$Q = A\sigma T^4 \tag{5.7}$$

As the constant $\sigma = 5.72 \times 10^{-8} \, W m^{-2} K^{-4}$, the heat flux from heater at 673 K is $12 \, kW m^{-2}$. The power spectrum of black-body radiation shifts to shorter wavelengths as the temperature of the body increases. For a heater at 673 K, most of the spectrum lies in the infrared region at wavelengths 2–5 μm. Although some polymers are transparent in the visible region, all polymers strongly absorb in the infrared, which excites vibration of the covalently bonded atoms. Consequently, the radiation is absorbed in the surface layer of the polymer. As the polymer surface temperature rarely exceeds 200 °C, Eq. (5.7) shows that the losses from re-radiation are small.

5.2.5 Viscous heating

The flow of a viscous fluid generates heat throughout the fluid. This should not be confused with frictional heating, which occurs at the interface between two solids in relative motion. The power dissipated in a small cube of melt in a shear flow, is the product of the shear force on the top and bottom surfaces and the velocity difference between these surfaces. When this quantity is divided by the volume of the cube, the power dissipated per unit volume W is found to be

$$W = \tau\dot{\gamma} \tag{5.8}$$

In a pressure flow, the greatest power is dissipated near the channel walls where the shear strain rate is highest. Hence, the viscous heating will lead to temperature differences between the core and the surface of the melt. When a melt falls in passes down a flow channel, under the influence of a pressure drop Δp, we can assume adiabatic conditions, so that no heat is transferred to the channel walls; the average temperature rise of the melt is

$$\Delta p = \rho C_{\mathrm{p}} \Delta T \tag{5.9}$$

For a pressure drop Δp of 50 MPa into an injection mould, the temperature rise $\Delta T = 40\,^{\circ}\mathrm{C}$.

5.3 Melt flow of thermoplastics

Flows can be classified into *streamline*, when particles in the fluid follow paths (streamlines) that remain constant with time, and *turbulent*, when vortices cause unpredictable changes in the flow pattern with time. The changeover occurs at a critical value of the Reynolds number, which is defined as the melt velocity, divided by the viscosity times the channel diameter. The high viscosity of thermoplastic melts causes velocities to be low. Hence, the Reynolds number is very low and the flows are streamline. We will consider steady flows, and ignore the start and end of injection and blow-moulding flows, when the melt accelerates and decelerates, respectively. However, in the RIM process (Section 5.6.5), turbulent flow of the low viscosity constituents in the mixing head achieves intimate mixing.

5.3.1 Shear flows

In a simple *shear flow*, the streamlines are parallel. The velocity along each streamline remains constant, with a velocity gradient at right angles to the streamline. If the x axis lies along the streamline and the y axis lies in the direction of the greatest velocity gradient, the shear strain rate is defined by

$$\dot{\gamma} \equiv \frac{\partial V_x}{\partial y} \quad (5.10)$$

where V_x is the velocity component along the streamline.

Polymer melts adhere to metals, so there is no slip at the metal/polymer interface. When one metal boundary of the melt moves parallel to another at a velocity V, the *drag flow* causes a shear flow with a constant shear rate (Fig. 5.4). At the interfaces, the polymer and metal velocities are equal (0 at the stationary surface and V_m at the moving surface. The shear strain rate $\dot{\gamma}$ is given by

$$\dot{\gamma} = \frac{V_m}{h} \quad (5.11)$$

where h is the gap between the surfaces. This result is valid whether or not the melt has a Newtonian flow law (see below). Drag flow occurs in an extruder barrel, as a result of the screw rotation.

Pressure flow is a shear flow between fixed metal boundaries, due to a pressure gradient in the melt. The pressure p falls down the streamlines, which are perpendicular to the isobars (Fig. 5.4). Appendix B derives the relationship between the pressure gradient, the channel dimensions and the flow law of the fluid. For rectangular, circular or annular cross sections, the shear stress τ varies linearly across the channel, and the velocity is maximum at the centre. The *Newtonian flow law* in Chapter 3 is

$$\tau = \eta \dot{\gamma} \quad (5.12)$$

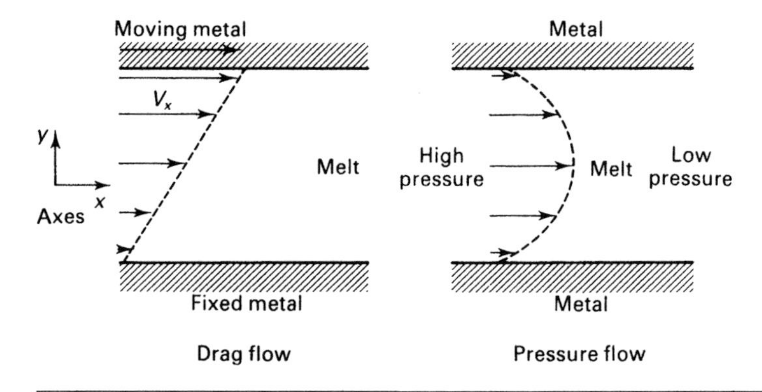

Figure 5.4 Velocity fields in shear flows caused by (a) drag flow and (b) pressure flow.

where the constant η is the (shear) viscosity. Polymer melts only obey this law at very low shear rates $\dot{\gamma}$ (section 3.3.3); a more realistic approximation is the *power law fluid* for which

$$\tau = k\dot{\gamma}_a^n \tag{5.13}$$

k and n are constants and $\dot{\gamma}_a$ is the apparent shear strain rate (the reason for use of the adjective 'apparent' is given in Appendix B). It is customary to define an apparent shear viscosity using

$$\eta_a \equiv \frac{\tau}{\dot{\gamma}_a} \tag{5.14}$$

even though this is a function of shear strain rate, rather than a constant. Section 2.3.3 explained that, as the shear rate is reduced, the viscosity tends to the zero-shear rate viscosity η_0.

Many shear flows are combinations of pressure and drag flows.

5.3.2 Extensional flows

In extensional flows, the velocity increases (fibre melt spinning) or decreases (radial flow from the sprue in an injection mould) along the streamlines, but there is no velocity gradient in the perpendicular direction. Figure 5.5 shows fibre melt spinning where the velocity V_x increases with distance x from the spinneret, as the result of a tensile stress σ_x along the fibre. The tensile strain rate $\dot{e}_x$ is defined by

$$\dot{e}_x \equiv \frac{\partial V_x}{\partial x} \tag{5.15}$$

and the tensile viscosity is defined by

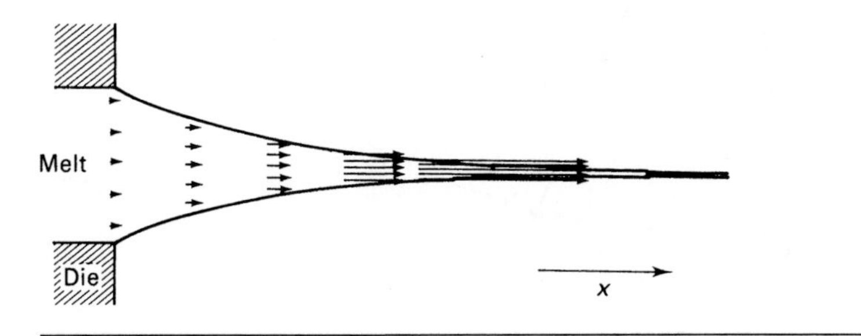

Figure 5.5 Velocity field in a tensile extensional flow for fibre spinning.

$$\eta_T \equiv \frac{\sigma_x}{\dot{e}_x} \tag{5.16}$$

The tensile stress increases with x, as a result of the fibre cross section decrease. This increase is exponential, if the melt does not cool and the tensile velocity η_T is independent of the strain rate. The velocity is given by

$$\frac{V_x}{V_0} = \exp\left(\frac{\sigma_{x0}x}{\eta_T V_0}\right) \tag{5.17}$$

where σ_{x0} is the tensile stress and V_0 the average velocity, at the die exit. However, as a result of heat transfer to the air, the melt temperature decrease causes the tensile viscosity to rise, which limits the increase in the fibre velocity. Finally, polymer crystallisation prevents any further flow.

When melt enters a constant-thickness injection mould cavity through a central sprue, there is radial flow (Fig. 5.6). The radial velocity component V_r is inversely proportional to the radial position, while the tangential velocity component V_θ is zero. The equivalent of Eq. (5.15) shows that the radial tensile strain rate is negative, decreasing in magnitude with radial distance. The melt has constant volume, so the sum of the strain rates in the r, θ and z directions is zero. As the strain rate is zero in the z direction, normal to the mould wall, the strain rate in the hoop direction is positive. For a purely viscous liquid, there is a constant negative pressure gradient in the r direction. However, elastic stresses in the melt add a non-constant term to the pressure gradient.

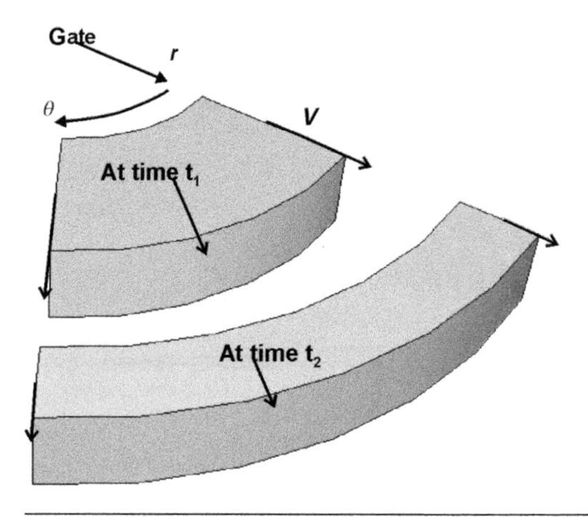

Figure 5.6 Radial flow from a gate, with the position of a block of melt at two times.

5.3.3 Molecular weight influences on flow

Much effort has gone in trying to predict melt rheological response from the molecular structure. Physical chemists have considered the modes of vibration of short-chain molecules (to explain the low molecular weight shear viscosity data in Fig. 3.10), applied mathematicians have attempted to explain the non-Newtonian and elastic properties of melts from the lifetimes of temporary entanglements between molecules, while physicists have used the snake-like motions of sections of polymer chains (reptation) for the same purpose. None of these approaches has been completely successful.

The qualitative explanation, for the variation of apparent viscosity with shear strain rate, is that the zero-shear rate viscosity is due to entanglements between the molecules. This viscosity increases with the polymer molecular weight, because each molecule is entangled with more of its neighbours. Under the high melt stresses of commercial processes, sections of molecules between entanglements become extended. Hence, the average value of the end-to-end vector r becomes larger than the equilibrium value r_0. In a shear flow, this molecular elongation causes the apparent viscosity to fall, either due to a decrease in the number of entanglements, or due to a reduction in the time for which they act, or both.

To explain the effect of the MWD on melt flow properties, we first consider the shear flow of a monodisperse polyethylene of molecular weight M. Figure 5.7 shows, on logarithmic scales, how the shear stress τ varies with the shear strain rate $\dot{\gamma}$. At low $\dot{\gamma}$ values, τ is proportional to $\dot{\gamma}$ and the

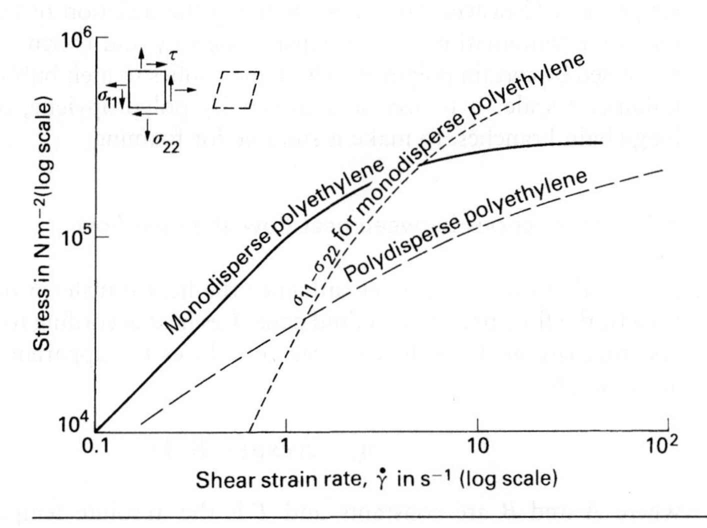

Figure 5.7 Shear stress vs. shear strain rate for a monodisperse polyethylene, and for a polydisperse PE of the same M_W value. The normal stress difference is also shown for the monodisperse PE.

elastic stresses are insignificant. In this Newtonian region, the shape of polymer molecules is still the equilibrium random coil of Fig. 3.4.

The *elastic effects* in polymer melts are associated with the molecular coil deformation shown in Fig. 3.9. The effects include die swell, a diameter increase when the melt exits from a die and flow instabilities such as melt fracture (causing a rough surface). One measure of the elastic effects is the tensile stress difference $\sigma_{xx} - \sigma_{yy}$ that occurs in shear flow in the xy axes. There can be a tensile stress σ_{xx} in the direction of flow, or a compressive stress σ_{yy} on the channel walls, or a combination of the two. Figure 5.7 shows that, as the shear rate increases, the value of $\sigma_{xx} - \sigma_{yy}$ increases with $\dot{\gamma}^2$ until it is of the same magnitude as τ. The sections of the molecules between entanglements are now elongated by the elastic stresses, and the increase in shear stress is no longer proportional to $\dot{\gamma}$. Elastic deformation of the melt is always associated with non-Newtonian viscous behaviour.

The flow curve of a broad MWD polyethylene is more non-Newtonian than that of a narrow MWD polyethylene (Fig. 5.7). These polymers have the same M_W, so, by Eq. (3.8), have the same zero-shear rate viscosity. The elastic stresses at low shear rates are influenced by the high molecular weight tail of the MWD. When the tensile stress difference is small, it can be described by

$$\sigma_{xx} - \sigma_{yy} = B\dot{\gamma}^2 (M_W M_Z)^{3.5} \tag{5.18}$$

where B is a constant. Therefore, increasing the breadth of the MWD, in particular the parameter M_Z/M_W, increases the melt elasticity, thereby decreasing the exponent n in the power law Eq. (5.13). The MWD of most addition polymers can be tailored so the melt flow properties suit a particular process. However, processes involving the inflation of bubbles of melt require a combination of high tensile viscosity and thermal stability, only possessed by certain polymers. The same applies to melt bubbles on the sub-millimetre scale: It is necessary to modify polypropylene, by introducing long-chain branches, to make it suitable for foaming.

5.3.4 Interactions between heat flow and melt flow

Mixing flows in an extruder influence the heat transfer process, and conversely, the flow of a viscous fluid generates heat according to Eq. (5.8). The resulting rise in the melt temperature reduces the apparent melt viscosity according to

$$\eta_a = A \exp(-B/T) \tag{5.19}$$

where A and B are constants and T is the absolute temperature. These interactive effects mean that any realistic calculations of melt flows and pressures must be computer based, with the temperature and viscosities of melt elements being updated at the end of every calculation step.

5.4 Extrusion

5.4.1 Melting and plasticisation

Extruded products, such as pipe, sheet or complex profiles for window frames, have a constant cross section. Figure 5.8 shows part of a pipe extrusion line; the continuous output must either be coiled, if it is sufficiently flexible, or cut into lengths and stacked for distribution. All parts of the line must be carefully controlled to keep the product dimensions within the acceptable limits. We will concentrate on the analysis of the extruder output.

The extruder screw (Fig. 5.9) has three main sections. Solid granules fall under gravity into the *feed section*. In some machines, the barrel wall has longitudinal grooves in the feed section to aid the forward conveying of granules. The flight of the screw is usually at an angle $\theta = 18°$ to the direction of rotation, so that the pitch of the screw is equal to its

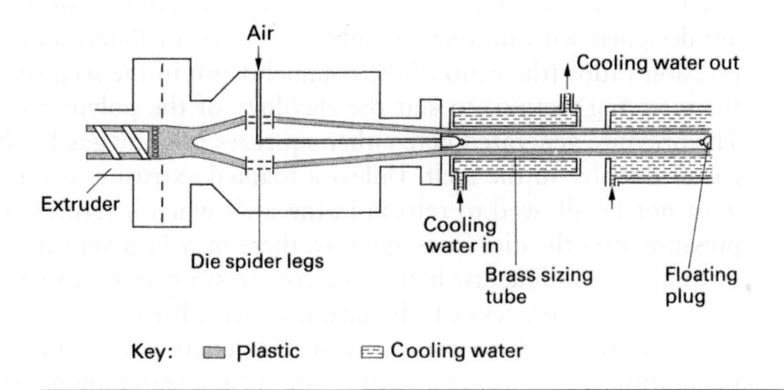

Figure 5.8 Die and cooling sections in an extrusion line for the manufacture of plastic pipe.

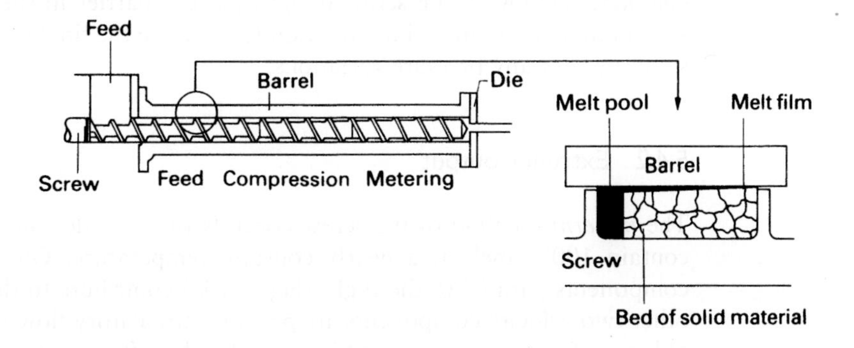

Figure 5.9 The three sections of an extruder screw. The detail shows the melting mechanism in the feed section.

diameter D. The main function of the feed section is to melt granules. This occurs by conduction from the electrically heated barrel, and by viscous heating of the melt from the mechanical work input of screw rotation. The insert in Fig. 5.9 shows that a thin film of melt develops in contact with the barrel. Since the screw rotates at about 60 rpm, Eq. (5.3) shows that a 0.3 mm thick layer of melt develops in the 1 s between passes of the screw land. This melt layer is scraped from the barrel wall, once a second, by the relative motion of the screw flight, to form a melt pool on the forward face of the land. After 10 revolutions, a volume equivalent to a 3 mm layer of melt on the barrel wall has accumulated in the melt pool. If the screw were stationary, it would take 100 s for a similar amount of melt to be plasticised. As the polymer progresses down the screw channel, the width of the melt pool increases, and the average temperature of the polymer increases.

Once the polymer is molten, viscous dissipation can occur. For typical polyethylene melts, the shear stress τ is of the order of 10^5 Pa when the shear strain rate $\dot{\gamma}$ is $100\ \mathrm{s}^{-1}$. Therefore, by Eq. (5.8), the power input is of the order of $10^7\ \mathrm{W\,m}^{-3}$. This power is dissipated in the molten layer, increasing its temperature and thickness.

The channel depth decreases in the extruder *compression section*. Screws are designed with different numbers of turns of flight, and different compression ratios (the ratio of the channel depth in the feed section to that in the metering section) to suit the rheology of the polymer being extruded. The pressure generated here either squeezes out any gas bubbles, or causes gas to dissolve in the melt. Unless a foamed extrusion is required, bubbles must not be allowed to reform in the melt when it returns to atmospheric pressure after the die. Consequently, there may be a vent to the atmosphere or to a vacuum line just before the compression section to aid degassing.

The melting process can be unstable, with breaks occurring in the solid bed continuity. These cause pressure fluctuations at the die, and hence fluctuations in the volume output rate, which cause the pipe wall thickness to vary. Such fluctuations are more likely when the screw speed is increased; the polymer residence time in the extruder can become insufficient for complete melting. Some screw designs place a barrier in the compression section that allows melt, but not granules, to pass; the final metering section smoothes out any pressure variations.

5.4.2 Extruder output

The *metering section* of the screw controls the extruder output. It should contain 100% melt at a nearly constant temperature. Only the velocity components parallel to the flight (Fig. 5.10a) contribute to the output; the other two velocity components are part of a circulatory flow that mixes the polymer. The output is a combination of a drag flow and a pressure flow. The drag flow is due to the motion of the screw surface with circumferential velocity

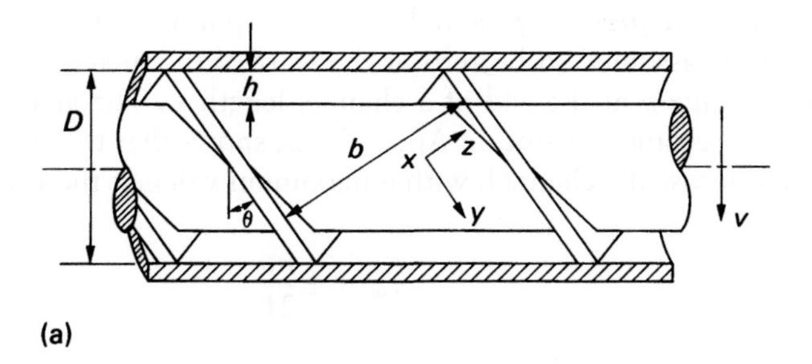

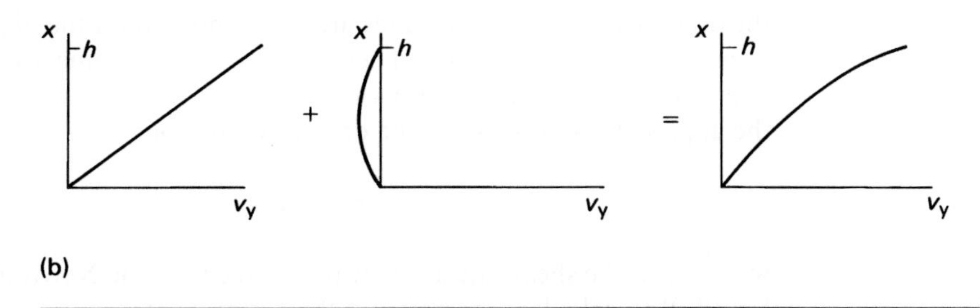

Figure 5.10 (a) Geometry of flow in the metering section of an extruder screw, with axes relative to the moving screw surface. (b) Velocity components V_y due to drag flow, pressure flow and a combination of the two.

$$V = \pi DN \tag{5.20}$$

where N is the rotation speed (rev s^{-1}). If we ignore the effects of channel curvature and channel edges (as the breadth b is much greater than the height h), the drag flow velocity components, relative to the screw surface, are shown in Fig. 5.10b. V_y increases linearly from zero at the screw surface to $V\cos\theta$ at the barrel surface. Consequently, the average value of V_y is 1/2 $V\cos\theta$, and the drag flow output is

$$Q_{\text{drag}} = \frac{V}{2}\cos\theta\, bh \tag{5.21}$$

The shear strain rate in the metering section is relatively low; for an extruder of diameter $D = 50$ mm and a channel depth of $h = 2$ mm rotating at 1.5 rev s^{-1} it is

$$\dot{\gamma} = \frac{\pi dN}{h} = 118\ \text{s}^{-1}$$

This is an order of magnitude smaller than the shear strain rates in injection moulding. Consequently, higher molecular weight polymers can be processed by extrusion.

A backward *pressure flow* reduces the output of the metering section. The pressure is assumed to be zero at the start of the metering section, and to reach a value p at the end of a channel length L. The analysis of pressure flow in a rectangular slot, in Appendix B, shows that the shear stress varies linearly across the channel, with a maximum value at the walls (Eq. B.5) of

$$\tau_W = \pm \frac{hp}{2L} \tag{5.22}$$

It appears that we should use the non-Newtonian shear flow curve to calculate the strain rates. However, the pressure flow is much smaller than the drag flow, and its shear rates are superimposed on the higher shear rate $V\cos\theta/h$ in the drag flow. Therefore, we can assume that the melt is approximately Newtonian for the pressure flow, with a viscosity equal to the apparent viscosity η_d in the drag flow, and put

$$\tau_W = \eta_d \dot{\gamma}_W \tag{5.23}$$

where $\dot{\gamma}_w$ is the shear rate at the wall. We then use the Newtonian version of Eq. (B.9) to calculate the pressure flow output rate

$$Q_P = -\frac{b\dot{\gamma}_w h^2}{6} = -\frac{bh^3 p}{12\eta_a L} \tag{5.24}$$

When the drag and pressure flow are added, the total metering section output is given by

$$Q_M = \frac{1}{2} bhV\cos\theta - \frac{bh^3 p}{12\eta_a L} \tag{5.25}$$

The metering section output must be the same as the flow rate through the die, as must the peak pressure p where the metering section meets the die. For a pipe die (Fig. 5.8), consisting of a channel of circumference b, height h and length L (or a set of such channels in series), the output, from Eqs (B.5) and (B.9) is

$$Q_D = \frac{bh^{2+1/n}}{2(2+1/n)}\left(\frac{p}{2kL}\right)^{1/n} \tag{5.26}$$

Equation (5.25) for the metering section and Eq. (5.26) for the die, can be solved graphically. The solution, lying at the intersection of the two curves (Fig. 5.11), is known as the extruder *operating point*. The performance of a real extruder at different screw speeds (varying V) and with different dies (varying the pressure flow component) can be used to construct the screw and the die characteristics, and confirm the analysis.

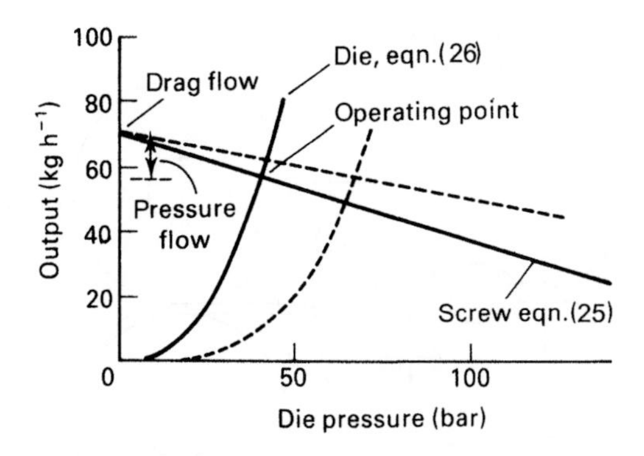

Figure 5.11 Extruder operating diagram: Pressure before the die vs. the output rate. The solid lines are for a lower viscosity melt than for the dashed lines.

Extruder/die combinations are designed so that the pressure flow is less than 10% of the drag flow, so the latter is a reasonable estimate of the total output. They are run as fast as possible, consistent with the melting process being complete and the output being stable. A pack of wire-mesh screens, supported on a perforated steel 'breaker plate' between the screw and the die, filters out any large foreign particles, and slightly increases the pressure.

5.4.3 Extrudate solidification

When the melt emerges from the extruder die, its shape must be fixed within a short distance. The haul-off mechanism pulls the solid extrudate forwards, and either air pressure or vacuum forces the outer melt surface into contact with a cooled metal calibrating section. Once the outer skin of the extrudate has solidified, cold-water baths or sprays complete the process. The cooling section is relatively long because conduction is the only mechanism of removing heat from the extrudate; calculation of its length is an important part of the process design.

Figure 5.12 shows another type of cooling. One surface of an extruded sheet is in contact with a cooled metal roll for a distance $\pi D/2$, then the other surface is cooled. It is possible to use an analytical solution (see Fig. A.4 in Appendix A) for the cooling of a sheet of thickness L on the first roll; the roll surface temperature T_0 is constant, and the sheet surface in contact with the air is effectively insulated. The temperature distribution after the cooling of the second surface on the lower roll can only be calculated numerically. Cooling of first one and then the other surface for certain total time, is more efficient than cooling on one side only for the same time.

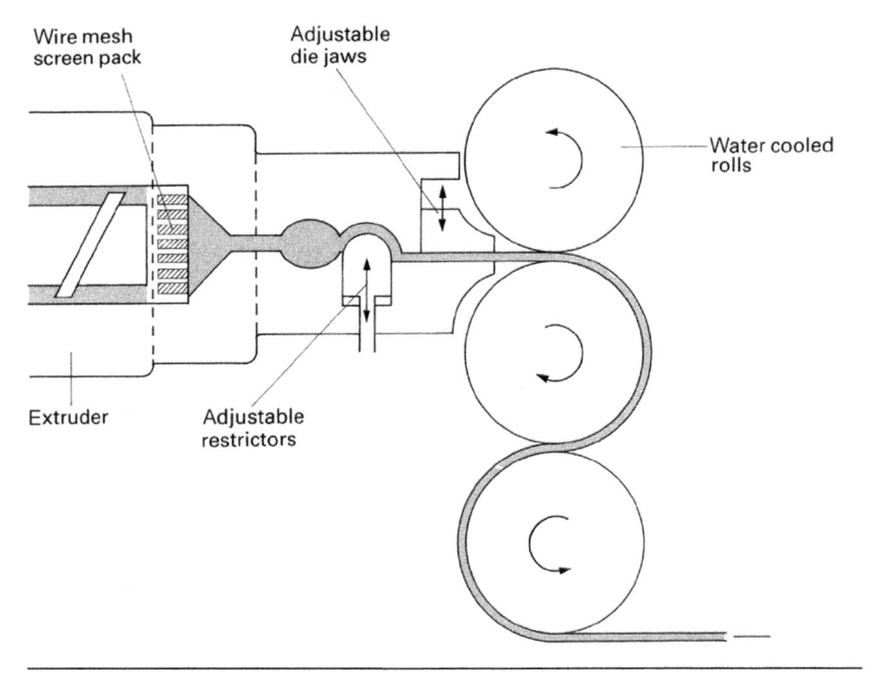

Figure 5.12 Solidification of an extruded sheet of plastic as a result of contact with steel rolls.

5.4.4 Ram extrusion of UHMWPE powder

Section 2.7.1.1 described ultra-high molecular weight polyethylene (UHMWPE). Its molecular weight causes the melt viscosity to be far too high for conventional melt processing. However, it can be Ram extruded into rods (Fig. 5.13). This process uses a reciprocating Ram to push small 'feeds' of powder through a long heated barrel. The back pressure from the high friction at the polymer/steel interface means that there is a high pressure in the melt. As the production rate is of the order of 1 mm min^{-1}, the PE spends several hours under pressures of 10–20 MPa and a temperature of about 200 °C. There is very little shear as the rod, of diameter 80–150 mm, passes through the barrel, so the main process is pressure sintering of the powder.

5.5 Processes involving melt inflation

In a number of processes, air pressure is applied to a bubble of polymer melt to change its shape. We will examine one continuous process—blown film production—and two cyclic processes—blow moulding and thermoforming. All of them involve some melt extensional flow with a resultant thinning of the bubble, and at least one side of the polymer solidifies without the constraint of contact with a mould.

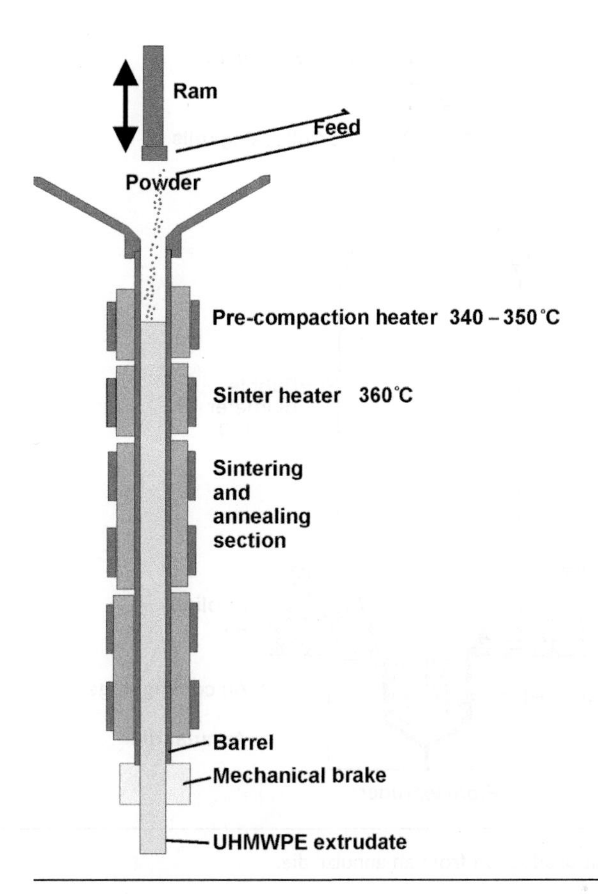

Figure 5.13 Ram extruder for producing UHMWPE rod.

5.5.1 Blown film

In blown film production, an annulus of melt rises vertically from a die attached to the end of an extruder (Fig. 5.14). The melt, during its passage through the die, must pass over two or more spider legs that support the core of the die against high melt pressures. The melt streams weld together above each spider leg, and these weld lines could be the regions of weakness in the melt bubble. To overcome this, dies incorporate one or more spiral mandrels, in which the main melt supply spirals upwards, gradually leaking the flow into a vertical motion.

The melt bubble is stretched vertically and circumferentially by a factor of 2 or more, so that an initial melt thickness of about 1 mm is reduced to between 250 and 100 μm. In the biaxial tensile flow, the melt stress in the hoop (H) direction can be calculated from the pressure p inside the bubble, the current bubble radius r and thickness t, using Eq. (C.22) of Section C.3.

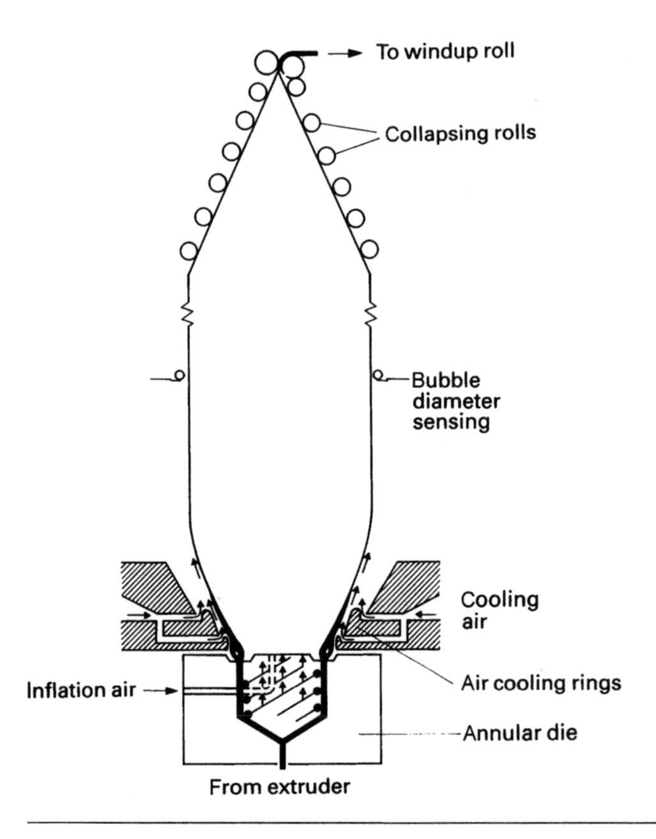

Figure 5.14 Blown film production from an annular die.

The longitudinal stress is also affected by tensile wind-up force F, therefore, Eq. (C.21) is modified to

$$\sigma_L = \frac{pr}{2t} + \frac{F}{2\pi rt} \tag{5.27}$$

The only way to generate data for this type of unsteady biaxial tensile flow is to instrument a blown film machine. The tensile viscosity, defined by Eq. (5.16), hardly changes with the tensile strain rate. Figure 5.15 shows data for the uniaxial stretching of an LDPE and an HDPE. The apparent tensile viscosity increases with strain rate for the more elastic LDPE, in contrast with the non-Newtonian reduction in viscosity in shear flows.

As there is no external control of the bubble shape, it is possible for shape instabilities to occur if excessive internal pressures are used, or if the polymer melt has an unsuitable extensional viscosity response.

Cooling is provided by an annular air jet which blows upwards on the outside of the bubble with an initial velocity of about $1\,\mathrm{m\,s^{-1}}$. The heat transfer coefficient becomes smaller as the air velocity falls but

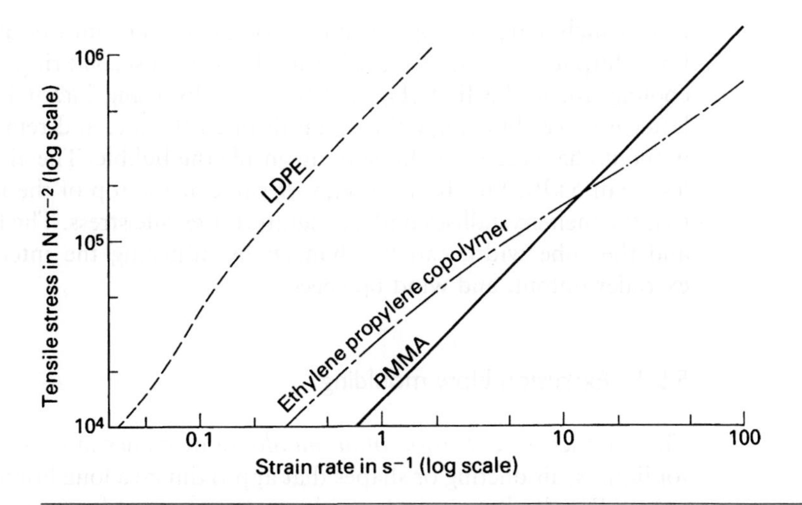

Figure 5.15 Tensile stress vs. tensile strain rate (log scales) for the uniaxial extensional flow of LDPE, ethylene propylene copolymer and PMMA.

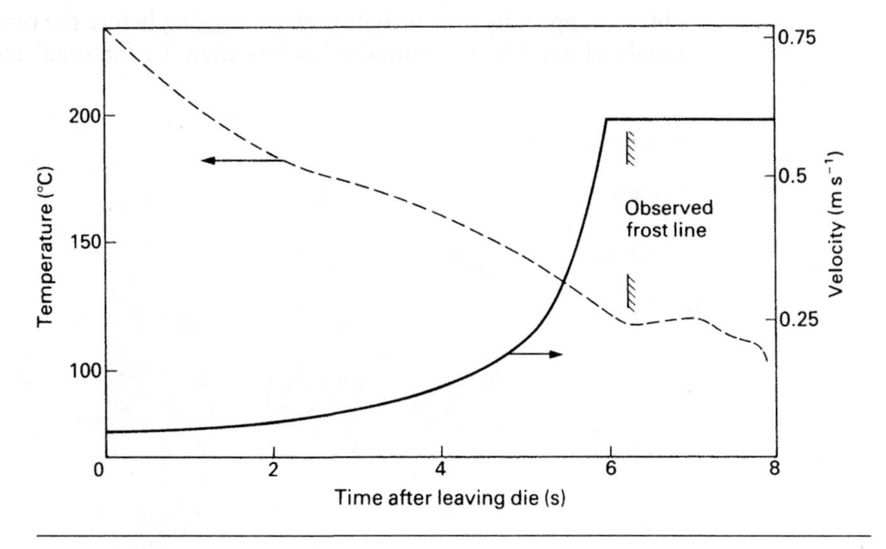

Figure 5.16 The temperature and the velocity of an LDPE film vs. the time after leaving the die (from Dowd LE, *Soc. Plast. Eng. J.*, **28**, 22, 1972).

the reduction in thickness means that the surface area to volume ratio increases. Figure 5.16 shows the temperature profile of an LDPE film with a blow-up ratio (= final bubble diameter/die diameter) of 3.5. The melt temperature falls nearly linearly with height until crystallisation occurs at about 120 °C, at the 'frost line' where the film becomes opaque. The melt accelerates in the longitudinal direction until just before it crystallises. The rolls, which flatten the bubble into a collapsed tube, must be above the frost line. Output has been increased either by using a second ring of external air cooling above the first (Fig. 5.14) and/or by using internal cooling. The latter involves blowing cold air up through the die, and removing hot air, without changing the volume of air inside the bubble. The air pressure p is less than 5 kPa, but the very high r/t value at the top of the bubble means that the melt crystallises under a significant tensile stress. The film thickness and the tube width can be changed by adjusting the internal pressure, extruder output, and wind-up speed.

5.5.2 Extrusion blow moulding

The products of *extrusion blow moulding* machines are hollow containers for liquids, air ducting or shapes that approximate a long hollow tube, e.g. a canoe. For the larger products the melt generated by an extruder is held temporarily in an accumulator chamber (preferably annular in shape) before being extruded rapidly by a piston. The parison emerges downwards from the die and hangs under the forces of gravity (Fig. 5.17). It must be able to support its own weight without sagging before the two halves of the mould close. The time involved is less than 1 s for small mouldings, and

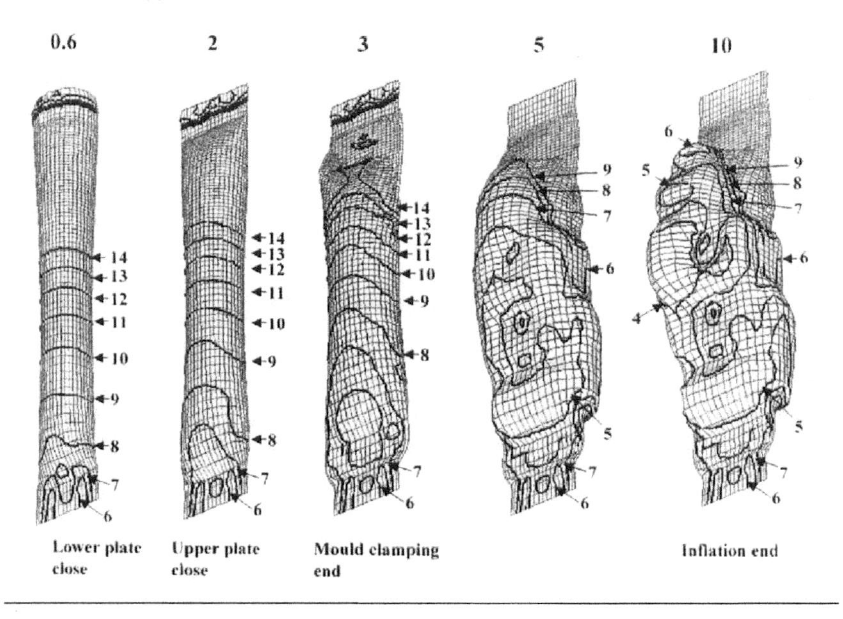

Figure 5.17 Predicted time sequence of the parison thickness during the inflation of a fuel tank with thickness contours(mm) (Tanifuji, S.-I. *et al.*, *Polym. Eng. Sci.*, **40**, 1878, 2000) © John Wiley and Sons Inc. reprinted with permission.

upto 2.5 min for large mouldings. The parison thickness profile is programmed; a hydraulic actuator moves the conical die interior vertically to initially restrict, and then open up the die width. This compensates for the stretching of the parison under gravity and during inflation, so the resulting container has a nearly uniform wall thickness.

When the melt emerges from a die gap of width h at a velocity V, the shear strain rate at the die wall is (Appendix B)

$$\dot{\gamma}_{aw} = \frac{6V}{h} \tag{5.28}$$

To avoid surface roughness with high molecular weight polyethylenes, the shear stress under these conditions must be less than 100 kPa. When the extrudate hangs under its own weight, the vertical tensile stress at the top is

$$\sigma = \rho g L \tag{5.29}$$

For a length $L = 0.2$ m, a density $\rho = 750\,\text{kg m}^{-3}$ and $g = 9.8\,\text{m s}^{-2}$, the stress $\sigma = 1.5$ kPa. To avoid significant stretching, the tensile strain rate must be less than $0.2\,\text{s}^{-1}$, which means that the tensile viscosity must exceed $7500\,\text{N s m}^{-2}$. When these two conditions are imposed on a typical shear flow curve in Fig. 5.18, it is clear that the melt must be highly non-Newtonian. A similar process cannot be used for Newtonian silicate glasses

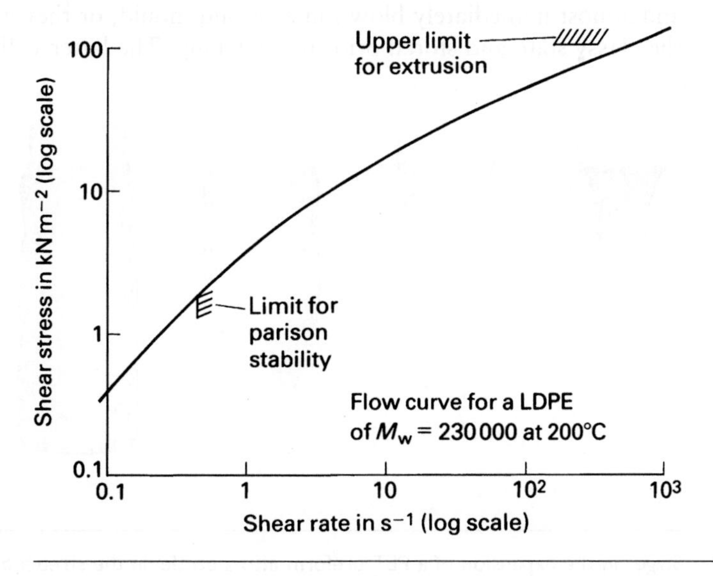

Figure 5.18 Shear stress vs. shear strain rate for a polyethylene used for blow moulding. The limits shown are for the stability of the hanging parison and for the parison to have a smooth surface.

to make glass bottles. The molecular weight must be $M_W > 150\,000$ for polyethylene to provide suitable parison stability.

During the slow extrusion of large mouldings, the lowest part of the parison may have cooled from 200 to 160 °C. When the mould halves close, they collapse the lower end of the tube, and then press the two layers of melt together, forming a weld. Inflation of the parison from the other end causes rapid extensional flow until the melt contacts the cooled aluminium mould. Figure 5.18 shows the thickness profile of the parison during the extrusion, and then the blowing of a car fuel tank. The final wall thickness varies from 4 to 9 mm.

The moulds need only to resist an air pressure of 10 bar. Consequently, aluminium moulds can be used, and the high thermal conductivity of this metal aids the cooling process. The conduction cooling from one side only is relatively slow for large containers. Cooling from both the sides would be four times as fast. Interior cooling by injecting liquid CO_2 has been attempted to increase the productivity of the machines.

5.5.3 Injection blow moulding

The *stretch blow-moulding* process involves the injection moulding of a preform, and its subsequent stretching. There are no weld lines in the preform (Fig. 5.19), which is gated at the base. Consequently, in the stretching operation, high pressures can be used without the risk of the melt splitting. Preforms are either removed from the mould at about 100 °C and almost immediately blown in a second mould, or they are cooled into the glassy state and stored prior to stretching. The latter will be described.

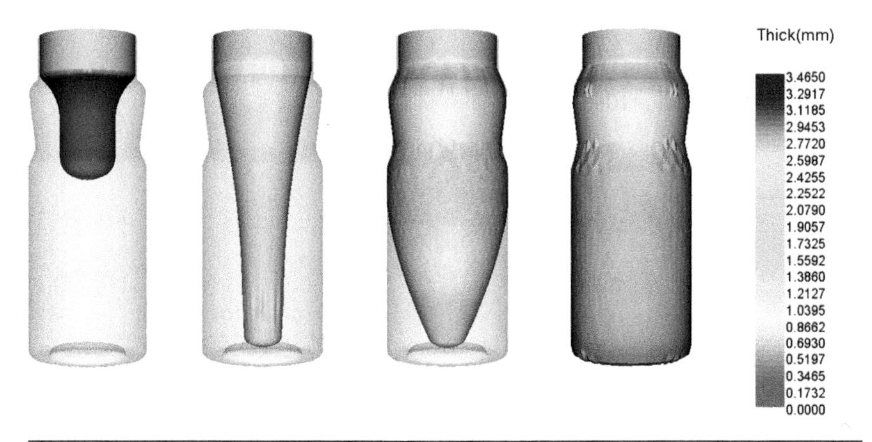

Figure 5.19 Stages in the expansion of a PET preform into a bottle, in the stretch blow moulding of PET. The stretch rod extends the length before the internal pressure expands the diameter (Pham X.T. *et al.*, *Polym. Eng. Sci.*, **44**, 1460, 2004) © John Wiley and Sons Inc. reprinted with permission.

The crystallisation kinetics of PET suit the process (Fig. 3.26). A glassy preform can be injection moulded with a wall thickness of up to 4 mm. If the mould is kept at 10 °C, the inner layers of the polymer cool fast enough for the crystallinity to be negligible. High molecular weight PET, with $M_N \cong$ 24 000, is used so that the rate of crystallisation is suitably low.

The preform, heated to 100 °C, is in the rubbery state. It is stretched in length by a rod inserted through the neck, which moves down at about 1 m s^{-1}. It is then inflated by air at about 4 bar to the dimensions of the mould (Fig. 5.19). The stretched rubbery PET crystallises when the extension ratio exceeds 2 (Fig. 8.16), before the preform hits the mould wall. This helps to stabilise its shape. The neck, usually with a screw thread, is not expanded, and there is a lower degree of expansion in the base. This can be revealed by placing a PET bottle in an oven at 120 °C for about 10 min. The neck and base, which were glassy, will crystallise in spherulitic form and be opaque, yet hardly shrink. However the wall, which had formed crystals smaller than the wavelength of light, will shrink noticeably while remaining transparent. Section 11.4.2 explains why light scattering occurs in semi-crystalline polymers when the crystal size is of the order of the wavelength of light. The high hoop stress just before the bubble reaches the mould wall (45 MPa for a radius $r = 45$ mm, wall thickness $t = 0.5$ mm and pressure $p = 5$ bar) can easily stretch the semi-crystalline polymer, so the crystals in the bottle wall are highly oriented (Section 2.4.7).

5.5.4 Thermoforming

The secondary process of thermoforming converts extruded sheet or film into curved parts with non-re-entrant shapes (e.g. margarine tubs, baths and curved panels). Thermoforming is possible for the majority of polymers. The first stage in the process is to heat an appropriate-sized sheet into the melt state (semi-crystalline polymers may be just below the final crystal melting temperature). Thick sheets, such as the 6 mm PMMA for baths, are preheated in ovens, but sheets of 2 mm and less are heated over the mould by electric radiant heaters. As explained in Section 5.2.4, the great majority of the radiation heat energy is in the infrared. All plastics strongly absorb in the infrared, even those transparent to light, and this heats the top surface of the sheet rapidly. However, heat conduction through the sheet thickness is relatively slow. For sheets less than 0.5 mm thick, there is hardly any temperature gradient through the thickness, but for thicker sheets the heating rate is limited to avoid top surface overheating before the lower surface reaches the forming temperature. Figure 5.20 shows the temperature profiles in 1 mm sheet, calculated for a radiant temperature of 500 °C. A heating time of 12 s is required. Cycling the heaters on and off every 5 s or so is a possible way of reducing the temperature differential across a thicker sheet. The time for deforming the melt bubble is low and the thin-walled products solidify almost immediately as they contact the mould surface.

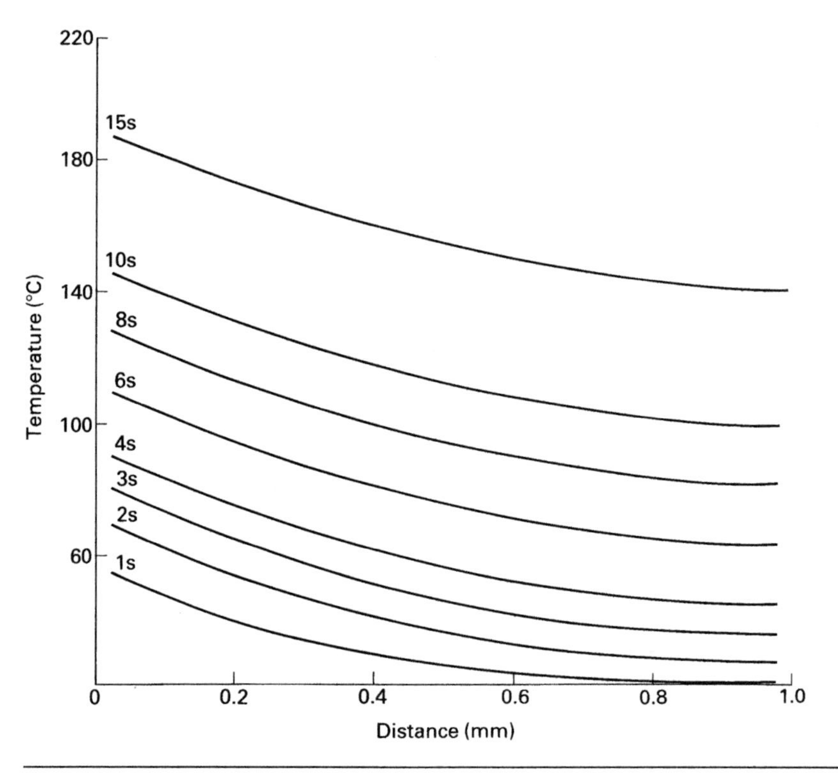

Figure 5.20 Calculated temperature profiles in a 1 mm plastic sheet, as a function of the exposure time to a radiant heater at 500 °C.

Hence, the overall cycle times is dominated by the heating time, if it is necessary to heat the sheet over the mould.

The edges of the plastic sheet are clamped to the frame of the moulding table by an air-tight rubber gasket. In the basic process, the melt bubble is sucked down into a 'female' (concave) mould by a partial vacuum, formed when the air in the mould is evacuated. Once the melt contacts the cold mould, its outer surface rapidly cools to a temperature at which it can no longer extend. Consequently, stretching is limited to the remaining part of the bubble, out of contact with the mould. If the sides of the mould are nearly vertical, the wall thickness of the moulding decreases exponentially, and the thinnest part will be at the bottom corner. The draw ratio of the mould cavity is defined by

$$\text{Draw ratio} = \frac{\text{Mould depth}}{\text{Mould width}}$$

Figure 5.21 shows one quarter of a mould with a draw ratio of 0.5. The ABS material had thinned from 1.53 to 0.2 mm at the corner. Such corners

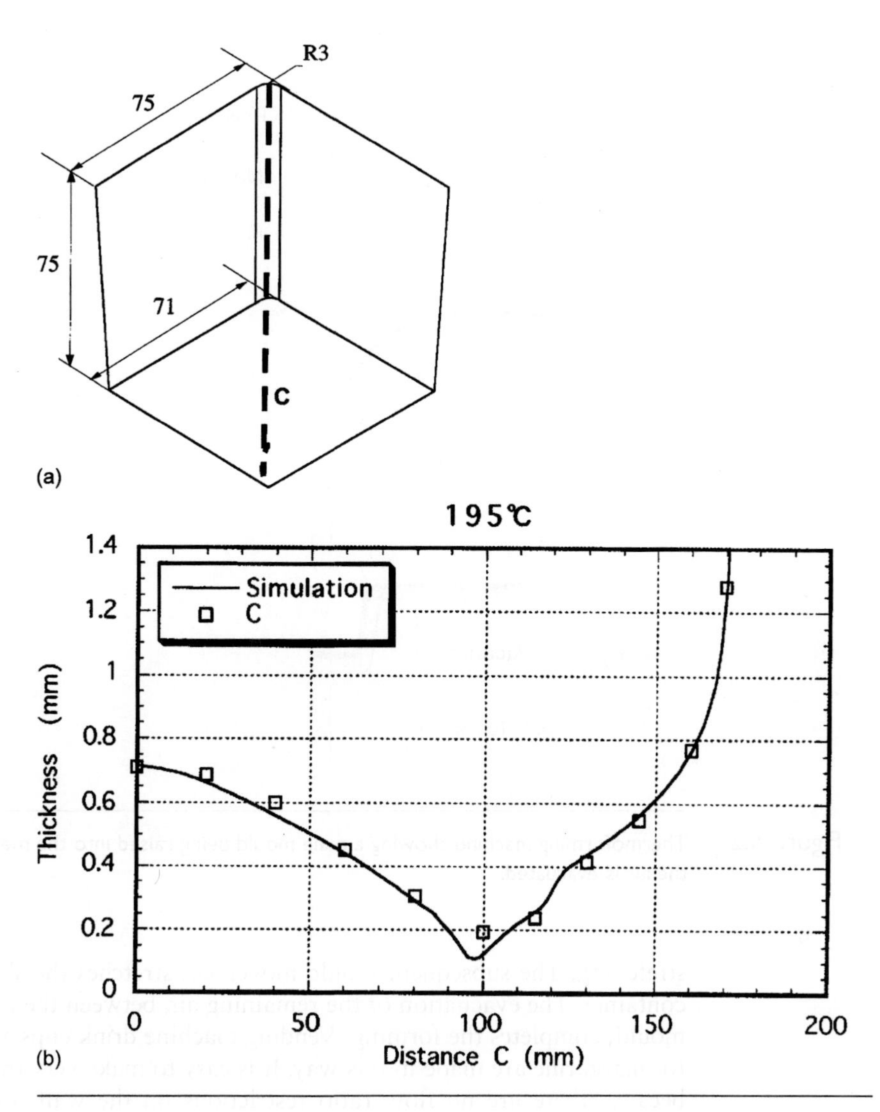

Figure 5.21 (a) Quarter of mould, (b) thickness profile along the axis C for ABS, compared with finite element simulation (Wang, A. *et al.*, *J. Mater. Proc. Technol.*, **91**, 219, 1999) © Elsevier.

would be vulnerable in use; boxes when dropped tend to fall on a corner, so corners should be radiused and as thick as elsewhere. Consequently, vacuum forming into female moulds is only used for products with a draw ratio much less than 1.

To form a container with a thicker base than sides (Fig. 5.22), a movable 'male' (convex) mould is used to determine the internal shape of the container. A positive air pressure may be used to inflate the sheet of melt upwards, before the mould is raised into the interior of the bubble. When its top surface contacts the sheet, the base of the moulding will stop

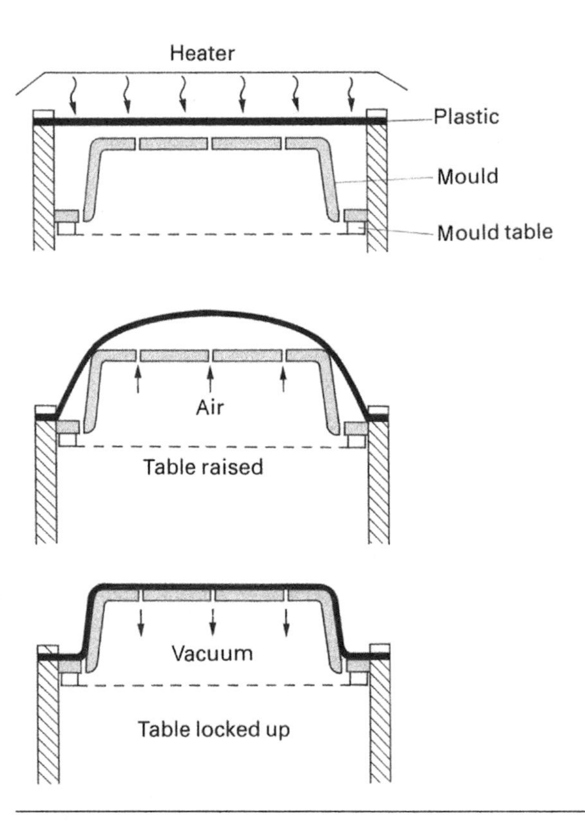

Figure 5.22 Thermoforming machine showing a male mould being raised into the melt bubble before the air is evacuated.

stretching. The subsequent mould movement stretches the sidewalls of the container. The evacuation of the remaining air, between the bubble and the mould, completes the forming. Vending machine drink cups and containers for margarine are made in this way. It is easy to make very thin containers, because there are no flow ratio restrictions on the wall thickness, as in injection moulding. However, to make a container with a height several times the width, blow moulding must be used.

The low pressures in thermoforming mean that cast aluminium moulds are adequately strong, and their high thermal conductivity is an advantage over the steel required for injection moulds. As the forces to move the mould are low, the thermoforming process has low capital costs compared to injection moulding. However, there is no independent control of the product thickness, and it is impossible to include reinforcing ribs. The thin product walls, if flat, will bend easily. Consequently, corrugations are used (Fig. 1.16) to increase the second moment of area of the cross section, and hence, the bending stiffness. This prevents hand pressure from distorting a disposable drink cup. Refrigerators liners are thermoformed, with a draw ratio of about 1. However, they are supported by rigid polyurethane foam, poured between them and the outer protective sheets. The bond between the polyurethane foam and the ABS liner creates a stiff sandwich structure (Section 4.3).

5.6 Injection moulding

5.6.1 Mould design

The injection-moulding process can make complex shaped parts, with exact replication of both mould surfaces. However, the product must be extracted when the mould cavity opens. For two-plate moulds with no moving cores, the part cannot have undercuts, and the axis of any holes must be parallel with the mould opening direction. Internal threads can be made using unscrewing cores, attached to the main mould. Holes, with their axis at an angle to the mould opening direction, can be made using sliding cores. Products with a re-entrant shape, such as the shell of a full-face motorcycle helmet, can be made if the male mould half splits; a wedge-shaped central piece is retracted, allowing the side pieces to move closer, releasing the helmet. Hollow or re-entrant parts can be made by assembling or welding together several injection mouldings. Moulds are normally designed using computer aided design (CAD). The shape files are converted into instructions for numerically controlled machines that mill the cavities from alloy steel blocks.

5.6.2 Cycle of operations

Figure 5.23 shows the main parts of an injection-moulding machine. To minimise the cycle time, several operations in the cycle are carried out simultaneously. Thus, a new batch of melt is prepared during the solidification of the moulding. The method of melting is the same as in extrusion, and the extruder operating diagram (Fig. 5.11) can be applied to plasticisation in injection moulding. However, as the screw rotates, it moves backwards. The back pressure determines the rate of melt accumulation at the front of the screw; a high back pressure slows down the rate of melt accumulation, but gives better mixing and generates more viscous heating. At the beginning of a cycle, the mould closes hydraulically and the extruder moves forward until the nozzle contacts the fixed mould half. The non-return valve at the screw tip prevents the melt from drooling during plasticisation, and prevents air admission when the nozzle retreats; it opens fully when the screw moves forward to inject the melt. The injection pressure is controlled so that it is high during the rapid mould filling but low during the feeding of the full mould. The rates of melt flow

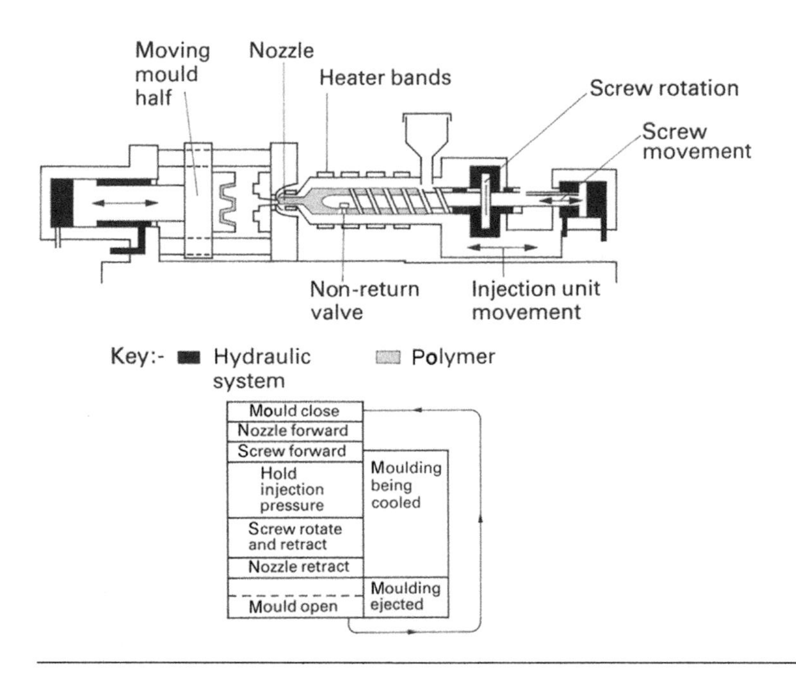

Figure 5.23 An injection-moulding machine with hydraulic mould closing, showing the cycle of operations.

can be set, and there are various control options for the mould filling (see Section 5.6.3).

The two main parameters for machine size are, the maximum *shot size* that can be injected in a single forward movement of the screw, and the maximum mould *clamping force* F_{clamp}. The shot size is usually quoted in terms of the mass of polystyrene that can be injected. The clamping force, typically in the range from 100 kN to 100 MN, restricts the maximum projected area A of a moulding onto the mould parting plane (Fig. 5.24). A typical average melt pressure p in the mould is 20 MPa, and the condition

$$pA < F_{\text{champ}} \tag{5.30}$$

must be obeyed if the mould is not to open and flash be formed at the edges of the moulding.

Moulds must resist high pressures without distortion, and resist wear over 10^5 cycles or more; they are usually made from forged blocks of low-alloy steel, air-hardened after machining. Moulds act as thick-walled pressure vessels, with high tensile stresses in the walls. For a melt pressure of 50 MPa and a cavity diameter that is four times the wall thickness, the average hoop stress in the wall is 100 MPa from Eq. (C.21). Concave corners in the mould cavity act as stress concentrating features, so to

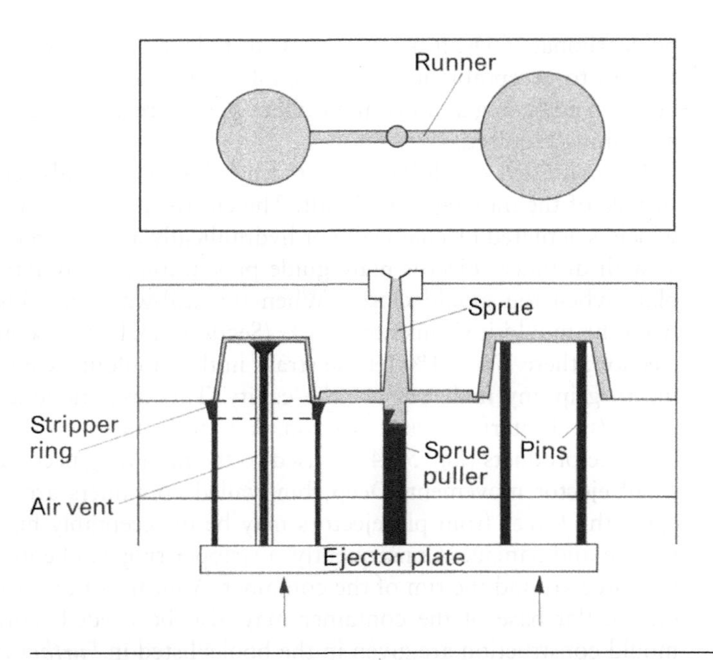

Figure 5.24 **Plan and elevation of a two-plate mould with twin cavities. The plan shows the moulding projected area. Alternative methods for ejecting a cup moulding are shown; the stripper ring and air vent are for very thin moulding.**

avoid localised yielding at such locations, the mould must be made from a steel of yield stress exceeding 300 MPa.

The mould contains features that add to its complexity. Four guide pins, at the corners of one side of the mould, engage with four sleeve inserts in the other mould half; these ensure an exact alignment as the mould closes. A tapered sprue (in a separate steel cylinder) leads through the fixed mould half to a series of runners that distribute the melt. Gates, at the entries to the cavities, restrict and control the flow into the mould; they assist the removal of the sprue and runners. Gate locations are selected so that the flow orientation in the mould is favourable, and the flow paths to the farthest points in the mould are approximately equal. Thus, parts with axial symmetry are preferentially gated in the centre. The gate cross-sectional area influences the time for which more melt can be packed under pressure into the cavity; the gate will usually be the first part of the moulding to completely solidify. Provision must be made for air to vent from the cavity—a gap of 25 μm between the mould blocks is sufficient.

Moulds are kept at a constant temperature by re-circulating water or oil through cooling channels. Polyethylene when injected has a heat content of about 700 J g^{-1}, of which only half is left in the ejected warm moulding. The mould temperature is set to suit the product and polymer. For some products it is low, to maximise the cooling rate. For other products, such as polycarbonate CDs, it is set at 90 °C to reduce the orientation and residual stresses to acceptable levels. The cooling channel system is a compromise between achieving a uniform mould temperature and the complexity of the machining required.

The solidified moulding is ejected by pins, with heads set flush into the surface of the moving mould half. The ejector pins are mounted in a plate which is actuated mechanically or hydraulically after the mould has opened a small distance. Ejector plate guide pins automatically retract the ejector plate when the mould closes. When the still-warm moulding is released from the mould it shrinks by $\sim 1\%$ (Section 6.3.1). Consequently, prior to ejection, there is a $\sim 1\%$ tensile strain in the moulding, causing the moulding to grip any male part of the cavity. This and frictional forces at the plastic/steel interface resist ejection. A 1° to 2° taper on the inner walls of box-like products (Fig. 5.24), is used so the moulding becomes loose after a small ejector movement. Deep thin-walled containers are troublesome to eject; the forces from pin ejectors may be unacceptably high, causing distortion and damage. Consequently, a stripper ring can be used to distribute the force around the rim of the container. A means of breaking the vacuum on the flat base of the container may also be needed. Further details of mould construction are given in the books listed in *Further reading*.

As moulds have a high capital cost, injection moulding is usually economic, only if more than 10 000 products are made. It is preferable to reduce the cycle time than to invest in a second mould to achieve the required production rate. The mould only has a high productivity if the cooling time, which dominates the cycle time, is minimised. Consequently, the product wall thickness is minimised (Eq. 5.4 predicts the effect on the cooling time). Sections 3 and 4 of Chapter 13 consider further suitable product shapes for injection moulding.

5.6.3 Control of mould filling

Melt flow and heat transfer interact strongly during mould filling in the typically 1–3 s that it takes for the highly viscous melt to fill the mould. During this time a skin of solid polymer builds up at the mould walls; this oriented skin is considered further in the next chapter. The temperature profile through the moulding has an intermediate peak due to the viscous heating that occurs in the high shear stress regions.

The mould cavity must be filled completely without any flash occurring. The simplest control for mould filling uses limit switches on the screw travel. These actuate after the correct volume of melt has been injected, reducing the injection pressure to a lower holding value. However, any melt leakage past the screw non-return valve, or at the nozzle, causes the part mass to vary, with a consequent variation in the product dimensions. A better, more direct, control method uses a melt pressure transducer in the mould, which detects the rapid pressure rise, once the mould is full; the

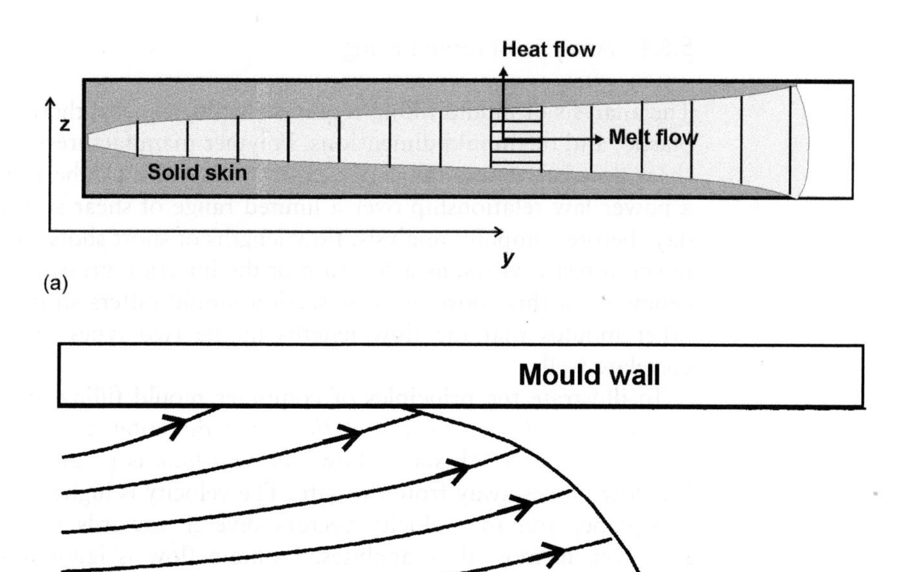

Figure 5.25 (a) One-dimensional flow along a mould (or runner) of constant cross section. (b) Fountain flow predictions: Change in the melt front with time (Chang R.Y. and Yang W.H., *Int. J. Numer. Methods Fluids*, **37**, 125, 2001) © John Wiley and Sons Inc. reprinted with permission .

injection pressure is switched to the holding level once the mould pressure reaches a set level. The cavity pressure reaches an initial peak (Fig. 5.25) just after the mould is full. Subsequent feeding of the moulding occurs at a nearly constant holding pressure until the gate freezes; thereafter cooling and contraction of the moulding, causes the mould pressure to fall to zero.

Computer controlled injection-moulding machines store the optimum process parameters for each mould/polymer combination. However, a number of machine settings affect each polymer variable; the melt temperature prior to injection is influenced by the screw speed, the back pressure on the screw, and the barrel and nozzle heater temperatures. There is no simple way of measuring the melt temperature at the nozzle; any thermocouple projecting into the melt would be sheared off by the high flow stresses. Therefore, there is open-loop control on the melt temperature, with the value fluctuating both during and between cycles.

5.6.4 Analysis of mould filling

The analysis of mould filling requires rheological and thermal data for the plastic, and the mould dimensions. Polymer manufacturers usually provide shear flow curves at a range of temperatures; these can be approximated by a power law relationship over a limited range of shear strain rates. In the days before computer analysis, flow lengths of short shots were determined in spiral test cavities, as a function of the injection pressure. However, the geometry of this constant cross section mould differs so much from most other moulds that the flow lengths in the two types of mould do not correlate well.

To illustrate the principles of computer mould filling analysis, we first consider *one-dimensional heat flow* in a direction normal to the mould surface. Figure 5.25b shows how *fountain flow* is predicted to develop as the flow moves away from the gate. The velocity is highest near the mould mid-plane, and the velocity vectors diverge towards the melt surfaces. However, in most flow analyses, fountain flow is ignored, and the melt front assumed to be normal to the mould surface. Figure 5.25a shows the build up of a solid layer on the surface of a mould, of rectangular cross section, gated at one end. The heat flow is determined by the differential equation

$$\rho C_p \left(\frac{\partial T}{\partial t} + v \frac{\partial T}{\partial y} \right) = k \frac{\partial^2 T}{\partial z^2} + \eta \dot{\gamma}^2 \tag{5.31}$$

The V term on the left hand side represents the movement of hot melt into an element, while the right hand side is a combination of thermal diffusion and viscous heat generation. If the computation shows that an element is solidified, the channel is narrowed at that position for the subsequent time step.

For such a one-dimensional flow, the volume flow rate at each segment is equal to that at the gate Q. Because of material continuity, the average melt velocities V are known at each segment of the cavity. The apparent shear rate $\dot{\gamma}_{aw}$ at the segment wall is calculated using Eq. (B.11) in Appendix B. The shear flow curve at the appropriate melt temperature is then used to find the shear stress for that $\dot{\gamma}_{aw}$ value. Finally, the pressure difference Δp across the segment of length L and thickness h is given by

$$\Delta p = 2L \frac{\tau_w}{h} \tag{5.32}$$

The sum of the Δps for all the full mould cavity segments gives the injection pressure.

Next, we consider a *two-dimensional, isothermal, viscous flow analysis.* Most mould cavities are much longer and wider than they are thick, so the flow is approximately two-dimensional. In the Hele-Shaw flow approximation, both fountain flow and melt inertia are ignored. Consequently, the flow velocity component normal to the mould surface is ignored. For isothermal flow in a constant-thickness mould, the pressure p variation as a function of the x,y coordinates of the mould is determined by a form of Laplace's equation

$$\frac{\partial^2 p}{\partial x^2} + \frac{\partial^2 p}{\partial y^2} = 0 \tag{5.33}$$

This equation, related to Poisson's equation of Section 13.6.1, states that the mean curvature of the pressure surface is zero. Figure 5.26 shows the predicted isobars for a flow into a mould with a cut-out (computed using the steady-state heat flow analogue). The velocity vectors are perpendicular to the isobars; the circular arc isobars near the gate show there is radial flow, but in sections with parallel side walls, the velocity is parallel to the walls.

Commercial mould filling programs, such as *Moldflow*, combine the one-dimensional heat flow calculations with two-dimensional viscous flow calculations. Keitxmann *et al.* (1998) gave the governing differential equation as

$$\frac{\partial}{\partial x}\left(S\frac{\partial p}{\partial x}\right) + \frac{\partial}{\partial y}\left(S\frac{\partial p}{\partial y}\right) = 0 \tag{5.34}$$

where $S = \int_0^b z^2 \mathrm{d}z/\eta$ is an integral in the thickness direction from the mid-plane to the mould wall. This equation reduces to Eq. (5.33) when the flow is isothermal.

In the finite element solution, the melt viscosity η is a function of position, due to its pressure and temperature dependence. Heat transfer as a result of material transport (the melt velocity has components u, v), diffusion in the thickness z direction, and viscous generation, is described by

Figure 5.26 Isobars for two-dimensional isothermal flow into a mould, computed using a thermal heat flow analogy.

$$\rho C_p\left(\frac{\partial T}{\partial t}+u\frac{\partial T}{\partial x}+v\frac{\partial T}{\partial y}\right)=k\frac{\partial^2 T}{\partial z^2}+\eta\dot{\gamma}^2 \tag{5.35}$$

This equation is used alternately with the flow equation, to update the melt temperature distribution. If the injection pressure is below the limit of the machine, the new position of the melt front is computed. The programmes output the melt front shape and melt pressure isobars at various times, hence predict whether a mould can be filled satisfactorily. If not, modifications can be made to the CAD file for the mould geometry, and the analysis repeated, before the mould cavity is machined. Figure 5.27 shows the predicted flow fronts for an instrument panel moulding.

5.6.5 Reaction injection moulding

In reaction injection moulding (RIM), two monomers are injected into a mould, where polymerisation and crosslinking occur. It is used mainly with polyurethanes to make large automotive panels. The chemistry of polyurethanes was described in Chapter 4. Other systems used include a block copolymer between a crystalline polyamide (nylon 6) and a rubbery polyether (polypropylene oxide). In principle, any polymerisation reaction that can be substantially completed after about 30 s in the mould is a candidate for RIM.

The two components are kept in temperature-controlled tanks, with pumped re-circulation when injection is not taking place. For the polyurethane system, one tank contains an isocyanate (usually MDI) and the other a mixture of polyol, chain extenders, catalyst and mould release (and possibly blowing agent or reinforcing additives). An amine catalyst accelerates the initiation of the polymerisation, while an organotin catalyst

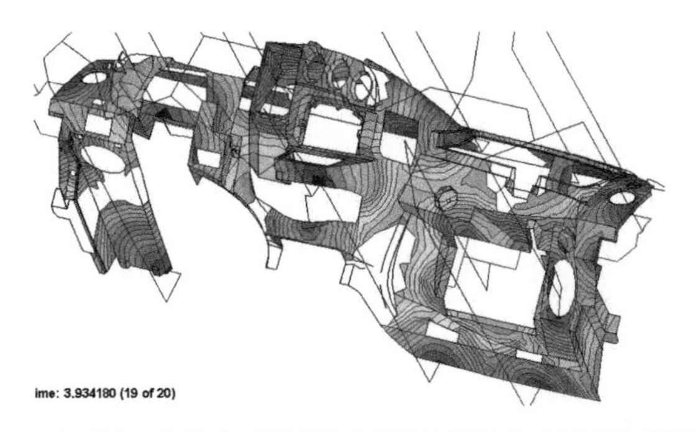

Figure 5.27 Melt front and pressure contours during the filling of an instrument panel support moulding (Bayer Material Science AG Leverkusen, Germany).

accelerates the gelling (when the liquid gels, its viscosity becomes infinite and it acts as a rubbery solid). A surfactant enables the polymer to wet the mould surface to obtain a better surface finish. The two components are low viscosity liquids with viscosities of the order of $1\,\mathrm{N\,s\,m^{-2}}$. They are pumped in accurately metered amounts to the mixing head (Fig. 5.28a); when the valve is opened, the 100–200 bar pressure causes the two streams of liquid to meet head-on at a $100\,\mathrm{m\,s^{-1}}$ velocity in a small ($<5\,\mathrm{cm^3}$) chamber. The *Reynolds number* of the flow exceeds 200, so efficient turbulent mixing occurs on a scale less than 0.1 mm. The consequent low diffusion distances for the chemicals allow a rapid reaction. The liquids passes through an after-mixer, further improves the mixing, before entering the mould. The pressure drop in mixing generates heat so that the mixture is at about 50 °C as it enters the mould. At this temperature the components are highly reactive.

The gate design and the mould filling flow are more akin to those in gravity casting of metals than they are to the injection moulding of thermoplastics. The moulds are fed at the lowest point, and a laminar flow into the mould is required so that no air bubbles become entrapped. Consequently, long film gates are often used; the liquid flowing through the gate as a 1–2 mm thick film at about $1\,\mathrm{m\,s^{-1}}$. The mould must be filled in 1–2 s because the gel time is approximately 10 s. The strongly exothermic reaction (Fig. 5.29) can cause gas bubbles to be generated. The aluminium or steel mould is controlled at about 60 °C so the surface of the moulding never heats above this. The poor thermal conductivity, however, means that the centre can reach 150 °C. Consequently, the polyurethane cures first in the interior. Foaming of the core can be used to compensate for the high polymerisation shrinkage. Air or nitrogen, dissolved into the holding tanks under pressure, causes the liquids to froth as they enter the mould, but this gas re-dissolves when the mould is full and the pressure rises to 5 bar. Later, polymerisation shrinkage causes the pressure to drop and the bubbles to reappear. The final microstructure has a solid skin of density $1100\,\mathrm{kg\,m^{-3}}$ and a foamed core of density $800–950\,\mathrm{kg\,m^{-3}}$. The crosslinked products are form-stable, so cooling may be completed after the removal of the product from the mould.

The overall cycle times are currently 30–90 s. Internal mould release agents such as 1–2% of zinc stearate plus a fatty acid are incorporated into the constituents, but it is still necessary to spray a layer of mould release agent into the mould approximately every 30 mouldings. The advantage vis-a-vis conventional injection moulding is that the liquid pressure in the mould, hence the mould clamping force for a given moulding projected area, is reduced by more than 95%. The capital costs of both the machine and the mould are thus considerably smaller than for injection moulding. There is an overall energy saving as the polymerisation is carried out in the mould, cutting out all the extrusion, granulation and processing operations needed with thermoplastics.

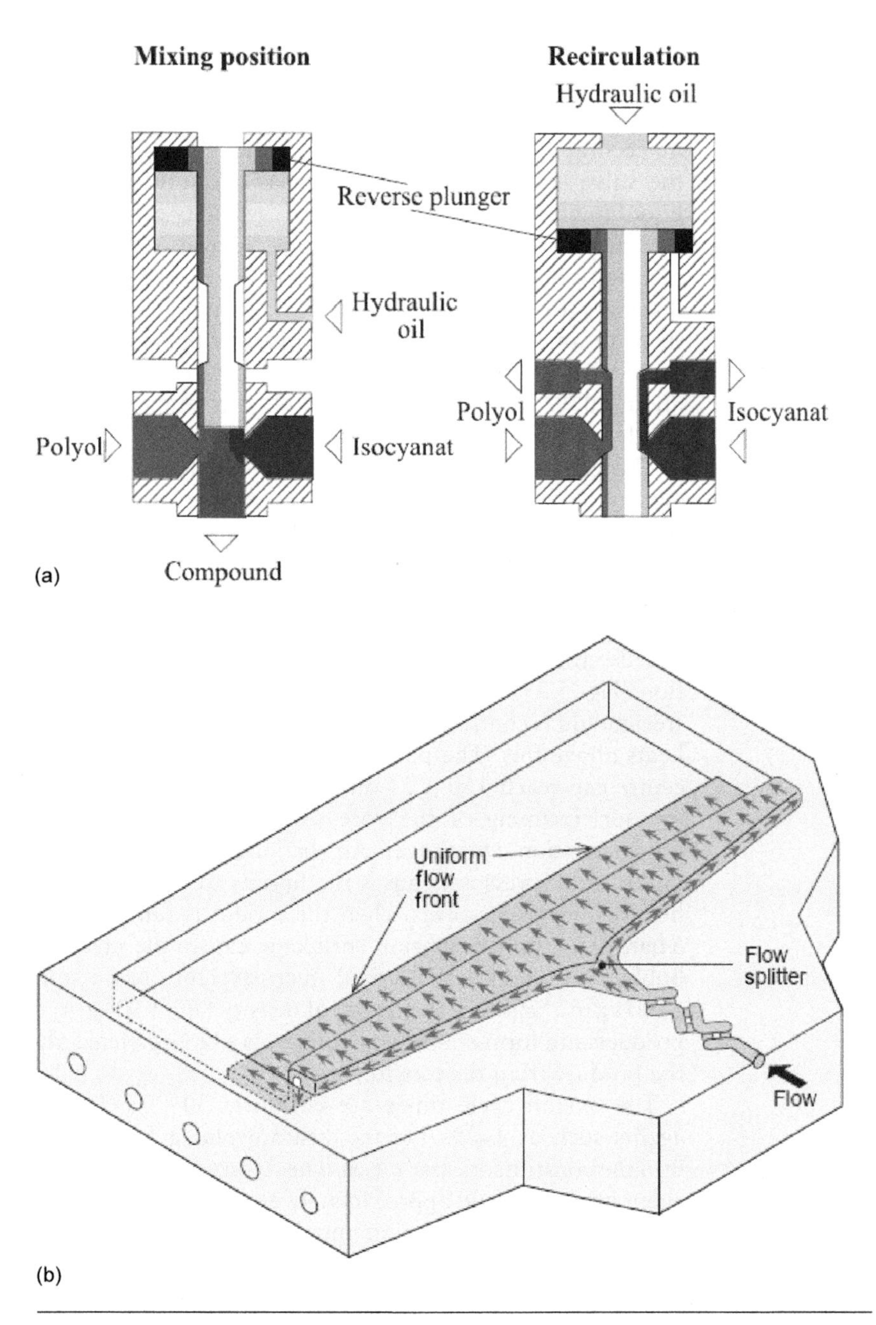

Figure 5.28 (a) Impingement mix head (Trautmann, P. and Piesche, M., *Chem. Eng. Tech.*, **24**, 1193, 2001) © Wiley-VCH. (b) Mould gating used in the reaction injection-moulding process (Bayer booklet Palmosina, MF, Gating for the RIM process, © Bayer USA).

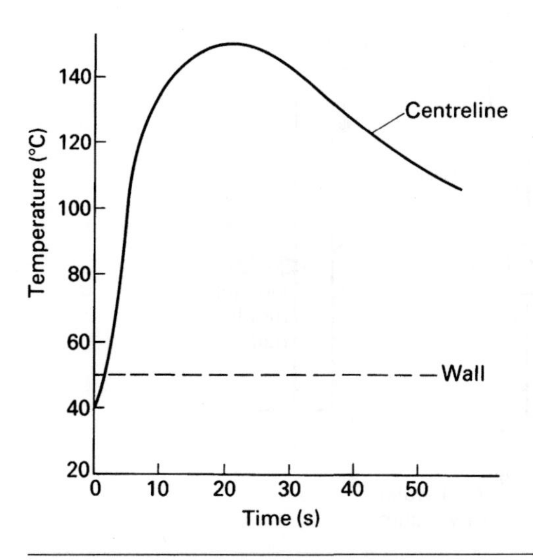

Figure 5.29 Temperature history at different positions in a polyurethane RIM moulding 5 mm thick, injected at 40 °C into a mould at 50 °C.

The applications of RIM have been widened by introducing 20% by weight of chopped glass fibres into the polyol component. Apart from changes in the pump type, to cope with the higher viscosity liquid and the abrasive wear, there is little modification in the process. The flexural modulus is increased by a factor of 3 or so, to be comparable with that of thermoplastics.

5.6.6 Bead foam moulding

Figure 5.30 shows a machine for moulding expanded polystyrene or polypropylene beads. The pentane containing beads are pre-expanded with steam, and then matured for the order of a day to allow the ingress of air. The closed mould is evacuated, and a fixed mass of beads blown in. Steam is passed into the mould cavity, via a number of 12 mm diameter inserts each containing about 20 sub-millimetre diameter holes, out of the other side of the mould. The superheated steam can diffuse inside the beads; it both melts the polystyrene and expands the beads. For a short time the steam pressure is allowed to rise, causing the beads to deform into polyhedral shapes, and the inter-bead surfaces to fuse. Since the steam pressures are less than 10 bar, the moulds can be made of aluminium, and the clamping forces are low. Marks on the exterior of the mouldings show the locations of the steam entry points. The process is relatively slow, with cycle times of a few minutes for products 30 mm thick. The

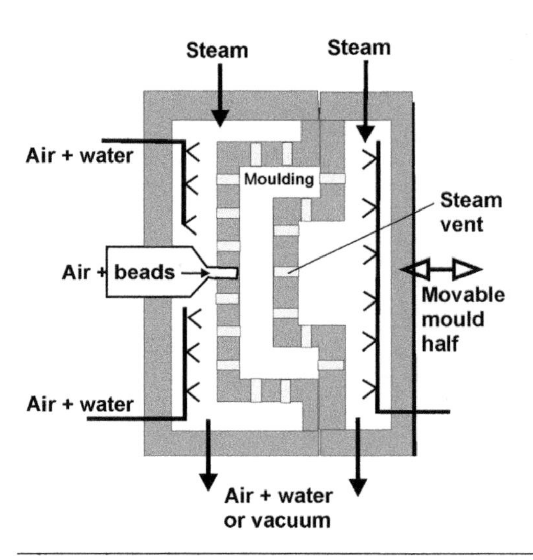

Figure 5.30 Mould for EPS beads, with inputs and outputs shown.

process has been optimised by keeping inter-bead channels open (they have the same geometry as the open-cell foam shown in Fig. 4.22b). They reduce the distance for steam diffusion to a maximum of one bead radius; steam flow from the moulding is a more efficient cooling mechanism than that of thermal conduction.

5.7 Rapid prototyping

When products are designed by CAD, it is relatively easy to produce prototypes by converting the files into input instructions for a number of techniques, either based on three-dimensional milling machines, or the assembly of two-dimensional layers. The latter will be described, since it can apply to the photo-curing of a vat of polymer, or the sintering of layers of polymer powder, or ink jet based methods. This is the automation of a traditional process of making complex shapes by assembling a stack of thin wood layers. The design is converted into slice cross sections. In the resin-based stereolithography process, an ultraviolet laser scans the surface of a tank of resin (typically an epoxy resin with a reactive diluent). This locally heats the resin to 65 °C, and crosslinks it, with a 10% volume shrinkage as it converts into a solid. The table holding the prototype then sinks in the resin by one increment and the process is repeated. When the complete product has been made, it is raised from the tank and post-cured at an elevated temperature. The product consists of layers (Fig. 5.31), but the dimensional accuracy is good.

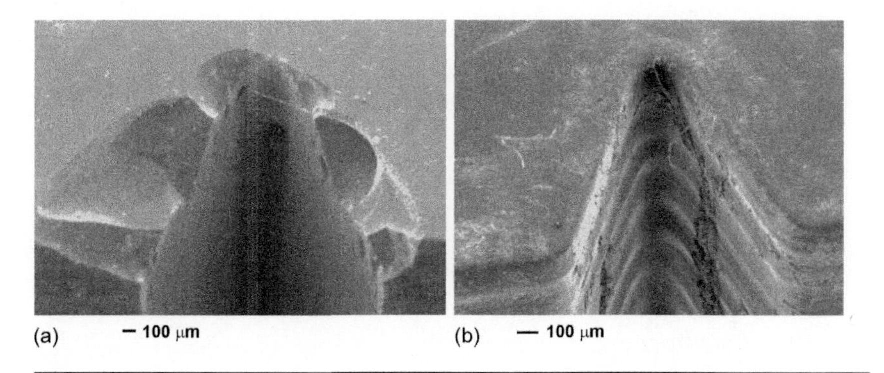

Figure 5.31 Acura SI 40 epoxy resin. (a) Machined notch, with brittle fractures. (b) Stereolithography notch (Hague, R. *et al.*, *J. Mater. Sci.*, **39**, 2457, 2004) with kind permission of Springer Science and Business Media.

Rapid prototypes do not have the mechanical durability of the thermoplastic product, but are adequate to assess the product appearance, assembly and some mechanical properties. The CAD instructions, used for making the prototype, can also be used to machine the alloy steel blocks for injection moulding, reducing the production time.

Chapter 6

Effects of melt processing

Chapter contents

6.1 Introduction

Melt processing has a number of effects on plastic products; microstructural effects, such as crystallinity changes, and macroscopic effects, such as product shrinkage. These effects are permanent, since plastics are never annealed to cause re-crystallisation, as is possible for cold-worked metal sheet. The cooling rate of the polymer in the later stages of processing, and the stress in the melt as solidification occurs, are the main process parameters. Non-uniform shrinkage can cause residual stresses or the product to warp. The effects of processing on the particle microstructure of PVC and ultra high molecular weight polyethylene (UHMWPE) will be considered at the end of the chapter. Chapter 13 will consider how choice of process affects the product design.

6.2 Microstructural changes

6.2.1 Effects of cooling rate on crystallinity and density

The cooling rate in all plastics processes is relatively fast. It varies with position and can be orders of magnitude larger in the outer layers than in the centre of a product. Figure 6.1 shows the variation in cooling rate through the glass transition temperature, in a sheet cooled from both sides by a medium with an infinite heat transfer coefficient. If the maximum possible crystal growth rate in a semi-crystalline polymer (Section 3.4.7) is moderate, it may be possible to cool thin mouldings sufficiently fast enough to avoid any significant crystallisation. Thus, injection-moulded PET bottle preforms (Section 5.5.3) are cooled rapidly enough to be glassy. In PBT, crystallisation rates are higher, so it is not possible to significantly change the average crystallinity of a 3 mm thick moulding, although the crystallinity of a 100 μm thick surface layer can be reduced.

Metastable crystals are formed in some polymers at high cooling rates. In polypropylene, a hexagonal form of crystal is formed just below the rapidly cooled skins of injection mouldings. This metastable form will re-crystallise as the stable triclinic form if the moulding is annealed at a sufficiently high temperature. For most polymers, the degree of crystallinity increases with the time spent in the crystallisation temperature range. Figure 6.2 shows the density variation through the 22 mm thick wall of an extruded HDPE pipe; the crystallinity is 5% higher at the pipe bore than at the outer surface, because of the lower cooling rate.

The density of glass-forming polymers increases slightly as the cooling rate into the glassy state decreases. Slower cooling gives the polymer more time to relax towards an equilibrium glassy state; this may be associated with changes in the local conformation of the polymer chains. This slow approach towards an equilibrium state continues if the polymer is held at a

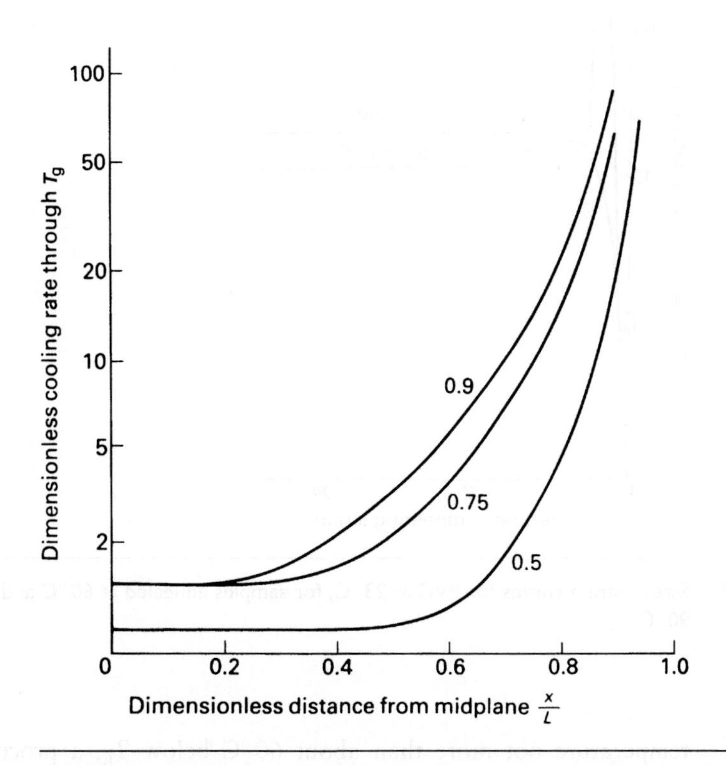

Figure 6.1 Cooling rate through T_g in a sheet of plastic cooled from T_m to T_b on both sides, as a function of the distance from the midplane, for different values of $(T_g - T_b)/(T_m - T_b)$.

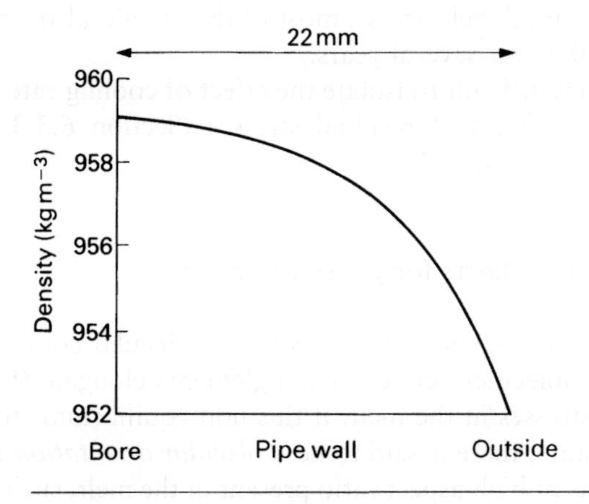

Figure 6.2 Variation of density, through the thickness of a 22 mm wall HDPE pipe.

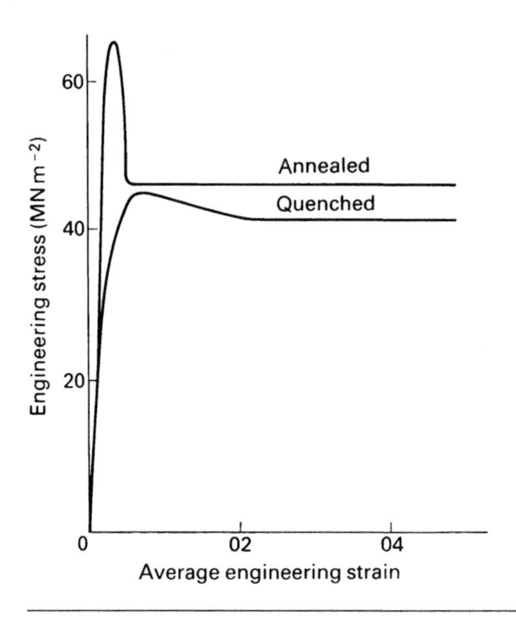

Figure 6.3 Stress–strain curves for PVC at 23 °C, for samples annealed at 60 °C and quenched from 90 °C.

temperature not more than about 60 °C below T_g, a process known as *ageing*. Ageing can occur at 20 °C for polymers like PVC with a low T_g. Figure 6.3 compares the stress–strain curve of a 1.7 mm thick PVC specimen immediately after quenching from 90 °C into cold water, with that of a specimen annealed at 65 °C then at 40 °C. Annealing causes a considerable increase in the initial yield stress; most of this would also occur if the PVC were kept at 20 °C for several years.

It is sometimes difficult to isolate the effect of cooling rate on mechanical properties from effects of residual stresses (Section 6.3.3) or molecular orientation (next section).

6.2.2 Melt stress effects for glassy polymers

Both shear and extensional flows affect the molecular conformations in the melt. Parts of molecules between entanglements elongate (Fig. 2.6), giving rise to tensile stresses in the melt. If this non-equilibrium structure persists into the solid state, there is said to be *molecular orientation* in the product. Rigid inclusions of high aspect ratio present in the melt, such as glass fibres, can align relative to the flow direction. Alternatively, initially isotropic but deformable inclusions, such as rubber spheres, can both elongate and align with the flow direction. The product is said to contain *oriented inclusions*.

Orientation is greatest where the polymer melt solidifies under high stress. Thus, biaxially stretched film is likely to have strong molecular orientation, as is the *skin* of an injection moulding, which solidifies while the mould is being filled. The *core* of an injection moulding is defined as the region that solidifies after the mould cavity is full. Due to the pressure gradient down the mould while it is being filled (Fig. 6.4a), the shear stress in the melt is maximum at the mould wall. The molecular orientation in polystyrene, measured using the optical birefringence technique (Fig. 6.4b),

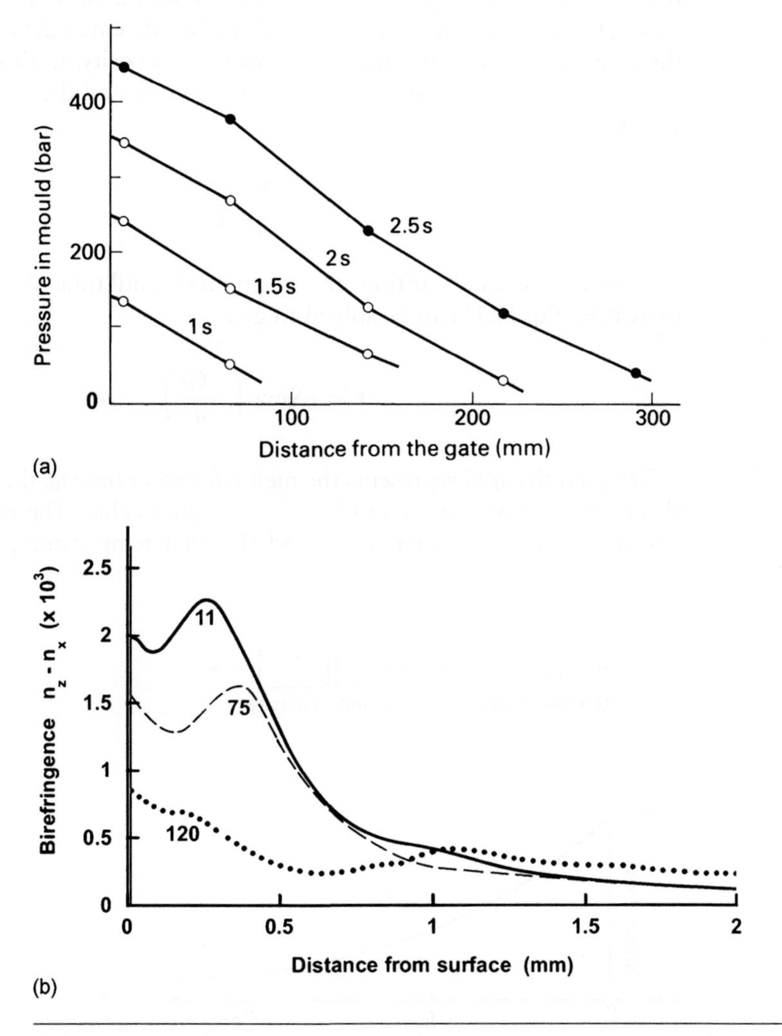

Figure 6.4 (a) Pressure distribution in a large injection mould at various times during filling (from Wales J.L.S., *Polym. Eng. Sci.*, **12**, 360, 1972). (b) Refractive index difference vs. position through the thickness of a polystyrene moulding at different distances from the gate, in a mould 127 mm long. The *x* and *z* axes used are defined in Figure 6.11.

is highest in the skin, and the skin is thickest near the gate of the moulding where the flow continues for the longest time. The lower orientation at the surface is due to fountain flow (Fig. 5.25b), in which a relatively unstressed melt from near the centre of the channel comes into contact with the mould wall under a low shear stress (Section 5.6.4).

In the packing stage of injection moulding, the only significant flow is near the gate. In liquid metals, the stresses become zero immediately after the flow ceases, but in polymer melts the stresses decay over a measurable time period. Figure 6.5 shows the Maxwell viscoelastic model (Chapter 7) for the relaxation of molecular orientation once the flow ceases. The spring represents the temporary shear modulus G of the entanglement network in the melt, and the damper represents the melt viscosity. In this series model, the shear stress τ is constant in both the elements, and the shear strain rates can be added giving

$$\dot{\gamma} = \frac{\tau}{\eta} + \frac{\dot{\tau}}{G} \tag{6.1}$$

If there is steady shear flow at a strain rate S until time $t = 0$, and no flow thereafter, Eq. (6.1) can be solved to give

$$\tau = \eta S \exp\left(\frac{-Gt}{\eta}\right) \tag{6.2}$$

The quantity η/G represents the melt *relaxation time* t_0, the time in which shear stress decays to $1/e$ (34%) of its original value. The relaxation time depends on the molecular mass and the melt temperature, factors which

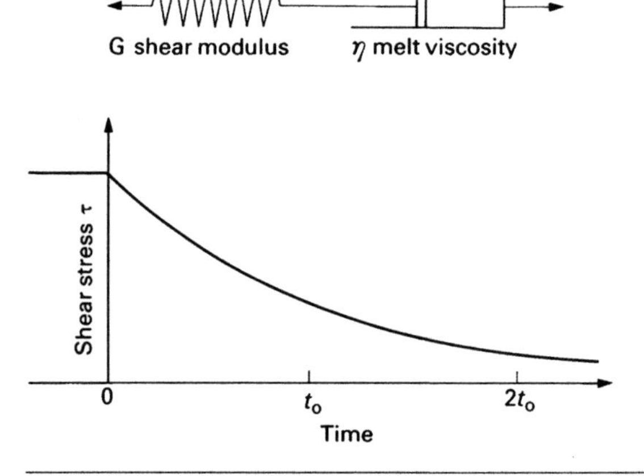

Figure 6.5 Maxwell viscoelastic model and its prediction of stress relaxation in a melt after the cessation of steady shear flow. t_o is the relaxation time.

affect the viscosity. Stress relaxation data, with a more complex time-dependence than Eq. (6.2), can be modelled by a number of Maxwell units in parallel, having a range of relaxation times (see Section 7.2). Solidification takes longest at the centre line of an injection moulding, so there is more time for relaxation after the flow. Consequently, the molecular orientation is the least (Fig. 6.4b).

In the blow moulding and thermoforming processes, solidification commences only when the melt contacts the cold mould wall, and flow stops. The time available for melt stress relaxation depend on the thickness of the product, and how close the melt is to the cold mould. There is likely to be high orientation in the thin side walls of a thermoformed disposable cup, made from a highly viscous glassy polymer that has solidified rapidly (hence the heat reversion shown in Fig. 1.16). Hence, if an empty cup is squeezed flat, vertical cracks tend to occur in the side walls, in the direction of orientation.

High levels of uniaxial molecular orientation are usually detrimental, since strengthening in the orientation direction is offset by weakening in the perpendicular direction. Impacts on the product surface tend to cause cracking in the weakest direction. If a small spherical indenter presses on the surface, the stress field is symmetrical about an axis normal to the surface, and tensile surface stresses occur just outside the contact area. The cracks that develop in brittle glassy plastics can be used to map out the flow pattern in a large moulding (Fig. 6.6), because they align with the flow direction. There is a particular weakness near the gate of the moulding where the orientation is largest. If feeding is allowed to occur under a high holding pressure, the problem will be exacerbated.

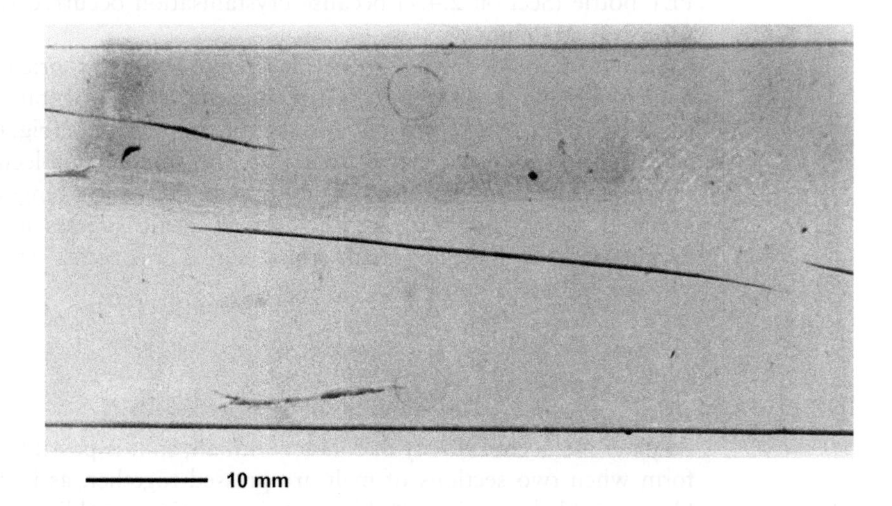

Figure 6.6 Cracks in a polystyrene injection moulding, due to surface indentations, lie along the flow direction.

6.2.3 Melt stress effects for semi-crystalline polymers

If crystallisation occurs in an oriented melt, then non-spherulitic microstructures can form, with *preferred orientation of the crystals* (Section 3.4.10). Fibrous nuclei, believed to contain fully extended polymer chains, can form in an oriented melt. Figure 6.7a shows several fibrous nuclei, in a polyethylene injection moulding, aligned with the flow direction. On either side of these dark nuclei is a bright layer, where lamellar crystals have grown from the nucleus. The **c** axes of the lamellar crystals are parallel to the fibrous nucleus; the microstructure of platelet crystals skewered by a rod-like nucleus has been described as a shish kebab. The rest of the microstructure consists of small spherulites.

The skin layers of injection mouldings can have a preferred crystalline orientation. Figure 6.7b shows a section through a polypropylene moulding; the 0.3 mm thick skins have a different microstructure than the core. This oriented microstructure is used in hinges between the two halves of a box (Fig. 1.15). The mould is gated so that the melt flows through a constriction that is about 0.4 mm thick. This ensures that the whole of the hinge has the highly oriented microstructure, hence it is strong in bending. Splitting of the hinge parallel to the orientation when it is flexed has no effect on its strength; it causes it to whiten, and it reduces the stiffness for subsequent flexure.

In an extrusion blow-moulded polyethylene container with a 1 mm wall, about 5 s elapse before crystallisation takes place at the inner surface (Eq. 5.2). During this time, the melt tensile stresses relax, so the microstructure will be spherulitic, even though the spherulites may be somewhat distorted. In contrast there is high orientation in the wall of a stretch blow-moulded PET bottle (Section 2.4.7) because crystallisation occurred while the preform was stretching.

Semi-crystalline polymers can also have molecular orientation in the amorphous phase. One method of assessing such orientation is to heat the product into melt state and observe the shape change (Fig. 6.8). Heating activates the entropic elastic forces in the oriented molecules. As they retract to their equilibrium coiled shapes, the melt changes shape. The shape change is not evidence of residual elastic stresses in the product, because the average residual stress across the cross section is zero (Section 6.3.4).

6.2.4 Weld lines

Polymer melts cool slowly in contact with air, so it is possible for a weld to form when two sections of melt are pressed together, as in the base of a blow-moulded container. Welds occur in injection mouldings when the melt from neighbouring gates meet (Fig. 5.27), or when the melt stream parts to flow past a hole in the product, and then recombines. Extruded products

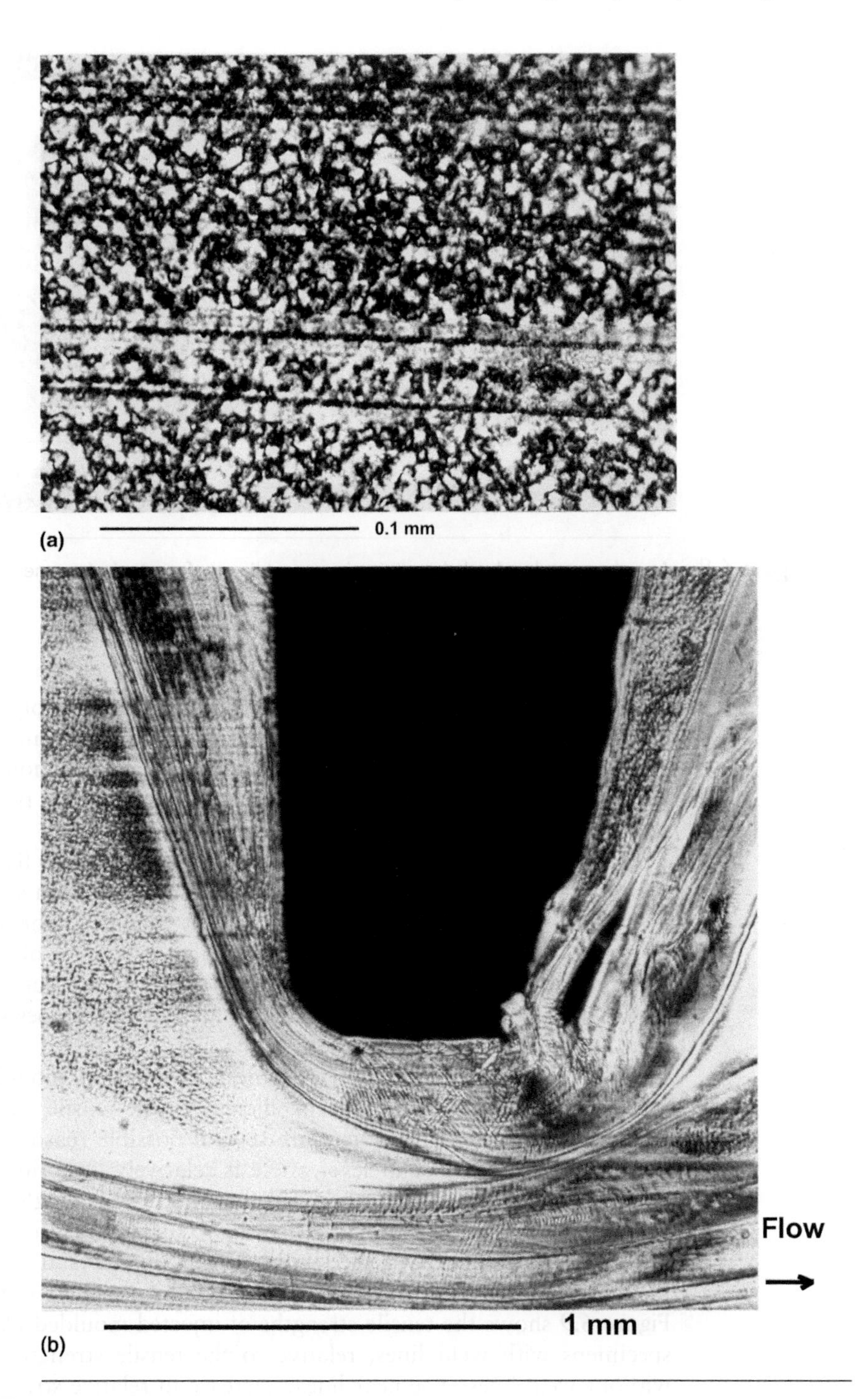

Figure 6.7 Polarised light micrographs of: (a) Fibrous nuclei in a polyethylene injection moulding surrounded by parallel lamellar crystals, then by spherulites. (b) Oriented skin of a polypropylene moulding, forming the hinge of a box where there is a thickness restriction.

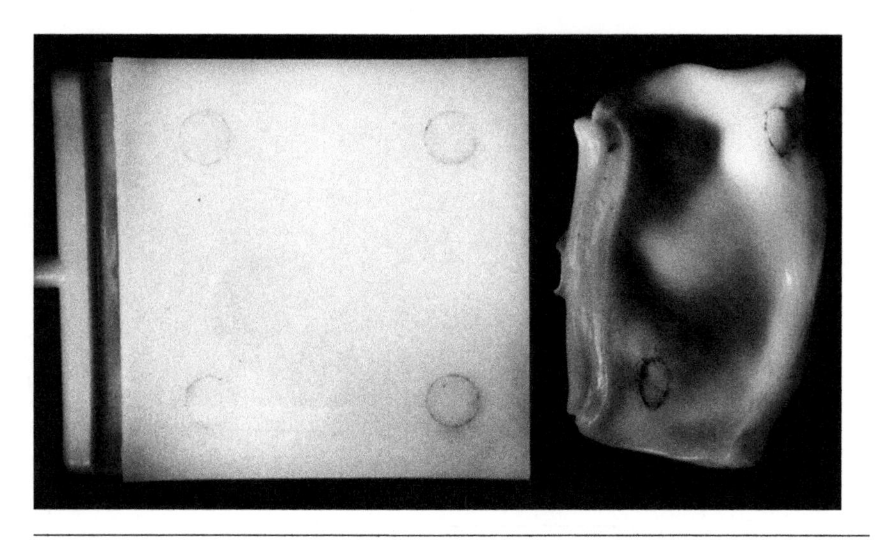

Figure 6.8 An edge gated polyethylene injection moulding, before and after melting, in a bath of silicone oil.

can be joined together by hotplate welding, for example polyethylene pipes (Section 14.2.5) or PVC window frame sections. The flat end surface of the extrusion is pressed against a metal heating plate with a non-stick coating for a couple of minutes to prepare a layer of melt 2–3 mm thick. When the two extrusions are pressed together, a weld is formed.

The common feature of these processes is that two flat, or slightly convex, melt surfaces come together under pressure, with some outwards flow. The orientation produced is at right angles to the original flow direction. In the welding of extrusions, there is a layer of soft semisolid polymer behind the melt layer. When the melt flows outwards to form a bead at the free surfaces, this semisolid material undergoes shear deformation.

Weld lines can be points of weakness in injection mouldings. Tensile specimens, with a weld line perpendicular to the tensile stress, often fail just after the yield point. There are several possible reasons for this phenomenon. On a molecular level, there is relatively little time for polymer molecules to diffuse across the weld plane, so the entanglement network across the weld line is insufficiently strong to survive the process of necking. On a microstructural level, the region of transverse crystal orientation near the weld line is weaker than the microstructure elsewhere in the specimen. Figure 6.9 shows the tensile strengths of injected moulded PS and PMMA specimens with weld lines, relative to the tensile strength of specimens without weld lines. The near linear increase in relative strength with melt temperature was predicted by a complex model which considered diffusion processes.

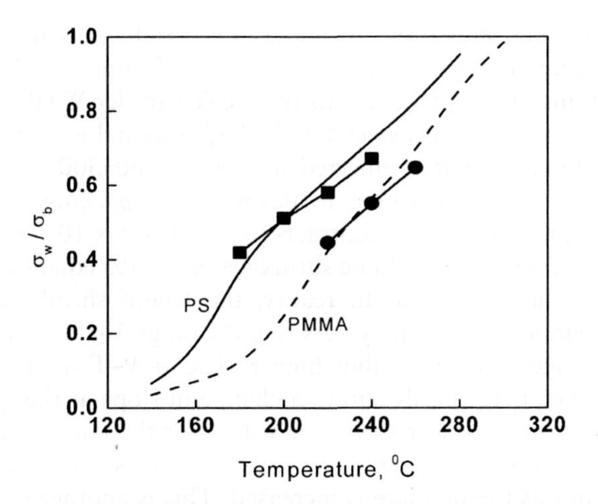

Figure 6.9 Weld line strength of PS and PMMA vs. melt temperature, with predictions from a diffusion model (Gao, S. *et al., Polymer,* **45**, 2911, 2004) © Elsevier.

6.3 Macroscopic effects

6.3.1 Shrinkage and distortion

The dimensional reproducibility, between a number of injection mouldings from a mould, is usually extremely good. However, if two parts are to be assembled without the application of excessive force, the separation of the assembly points must fall between close limits. The linear shrinkage S of a moulding is defined by

$$S = \left(\frac{100(M - L)}{M}\right)\% \tag{6.3}$$

where L is the length of the moulding at 23 °C, and M the equivalent length in the mould. The shrinkage must be considered when the mould is machined. Shrinkage depends both on the polymer microstructure and the processes that occur in moulding. Polymers have high thermal expansion coefficients, because the increased molecular vibrations at higher temperatures are only weakly resisted by the van der Waals forces between the polymer chains. Thermal expansion is anisotropic if molecular orientation is present, with a smaller expansion coefficient in the direction of the majority of covalent bonds; however, in the following, it is assumed that the polymer is isotropic.

There is a near-step increase in specific volume when the crystals in a semicrystalline polymer melt. In the pressure–volume–temperature (p–V–T) data for polyethylene (Fig. 6.10a), the crystalline phase finishes melting at 130 °C at 1 bar pressure. Pressure is included as a variable, because large pressure changes occur during the solidification of injection mouldings. The low bulk modulus of the melt, due to the weak van der Waals forces between the chains, partly compensates for the high thermal expansion coefficient. If a polyethylene melt is injected at 200 °C and 300 bar ($V = 1.286 \times 10^{-3}\,m^3\,kg^{-1}$ according to Fig. 6.10a) without a packing stage, the specific volume of this mass of melt contracts to $V = 1.035 \times 10^{-3}\,m^3\,kg^{-1}$ at 20 °C and 1 bar pressure. The volume shrinkage of 19.5% equates to an isotropic linear shrinkage of 7.0%. In reality, the length shrinkage of injection-moulded semi-crystalline polymers is in the range 1–3%, whereas the thickness shrinkage is considerably higher. The p–V–T data for amorphous polymers (Fig. 6.10b) only shows a change in slope at the glass transition, so we expect much lower shrinkage values for these materials. Figure 6.10 shows how the T_m of polyethylene and the T_g of polystyrene shift to higher temperatures as the pressure is increased. This is another reason for injecting the melt at a temperature well above these values.

Figure 5.25a showed a flat section of a moulding, with a thickness (in the z direction) much smaller than the other dimensions. Solid skins develop adjoining the mould surfaces, forming a closed box around the molten core. Most of this low modulus skin is easily sucked inwards by the contracting core, once the pressure on the mould surface has fallen to zero. Apart from near the ends of the mould, the solid skins do not resist the contraction of the core in the z direction. However, the skins and core are thin layers, connected in parallel in the x and y directions, having the same length L. Therefore, the skins resist the overall shrinkage in the x and y directions. Consequently, there are residual stresses in these directions in the cold moulding (Section 6.3.3).

The shrinkage of a moulding can be reduced, either by increasing the holding pressure or time, or by enlarging the gate, to increase feeding. Figure 6.11a shows the effect of changing the holding time on the pressure in the mould. The pressure rises rapidly once the mould is full, then decreases during solidification, as the melt channel between the injection unit and the mould constricts in diameter. If the holding pressure is removed before the gate freezes, there is a rapid pressure drop as some melt flows out of the mould. However, once the gate solidifies, no further feeding can occur. Figure 6.11b shows how the mass of the moulding reaches an asymptotic value as the holding time is increased. Shrinkage measurements show that most of the extra material, injected in the holding phase, has been used in reducing the thickness shrinkage. The length and width shrinkage decrease less, because of the constraining influence of solid skins.

The shrinkage of semi-crystalline injection mouldings can be anisotropic, due to the crystal orientation in the direction of flow. The shrinkage in the flow direction tends to be higher than the shrinkage in the perpendicular direction in the plane of the product. If a thin disc (the lid of a box) is gated at the centre, its radial shrinkage will be larger than the circumferential

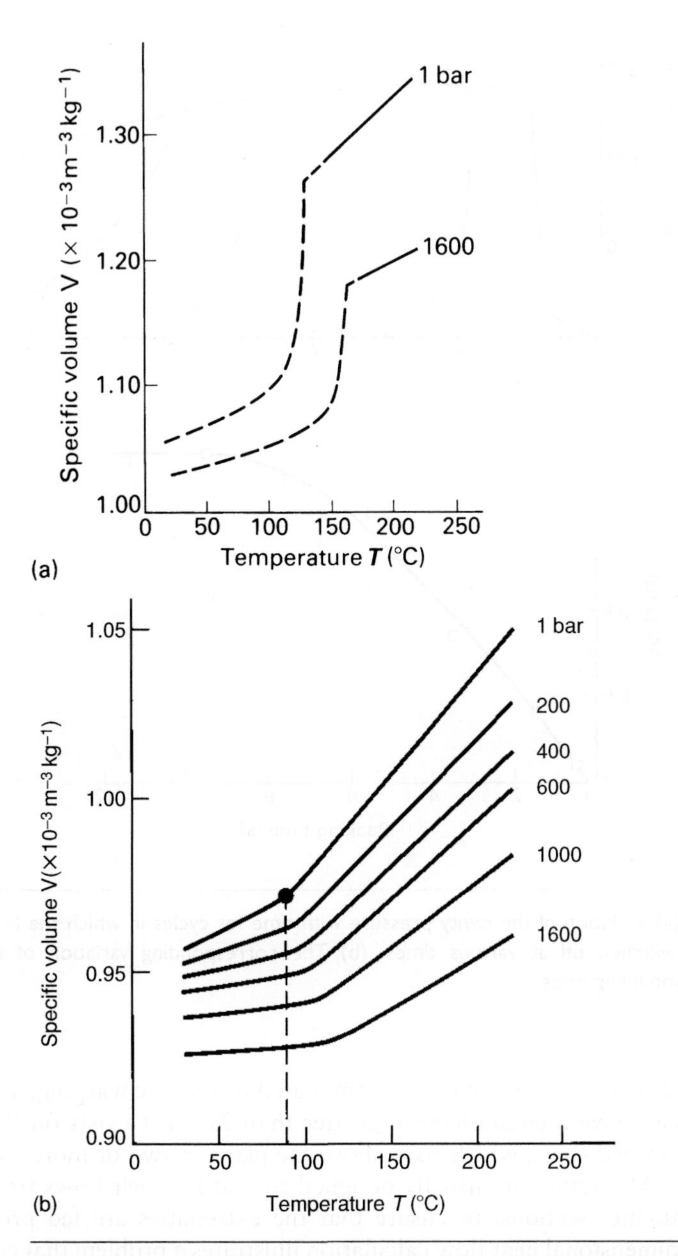

Figure 6.10 Pressure–volume–temperature relationships for: (a) Polyethylene and (b) polystyrene (from Wang K.K., *Polym. Plast. Technol. Eng.*, **14**, 88, 1980, Marcel Dekker Inc., NY and Menges G., *Polym. Eng. Sci.*, **17**, 760, 1977, Soc. Plastics Eng. Inc.).

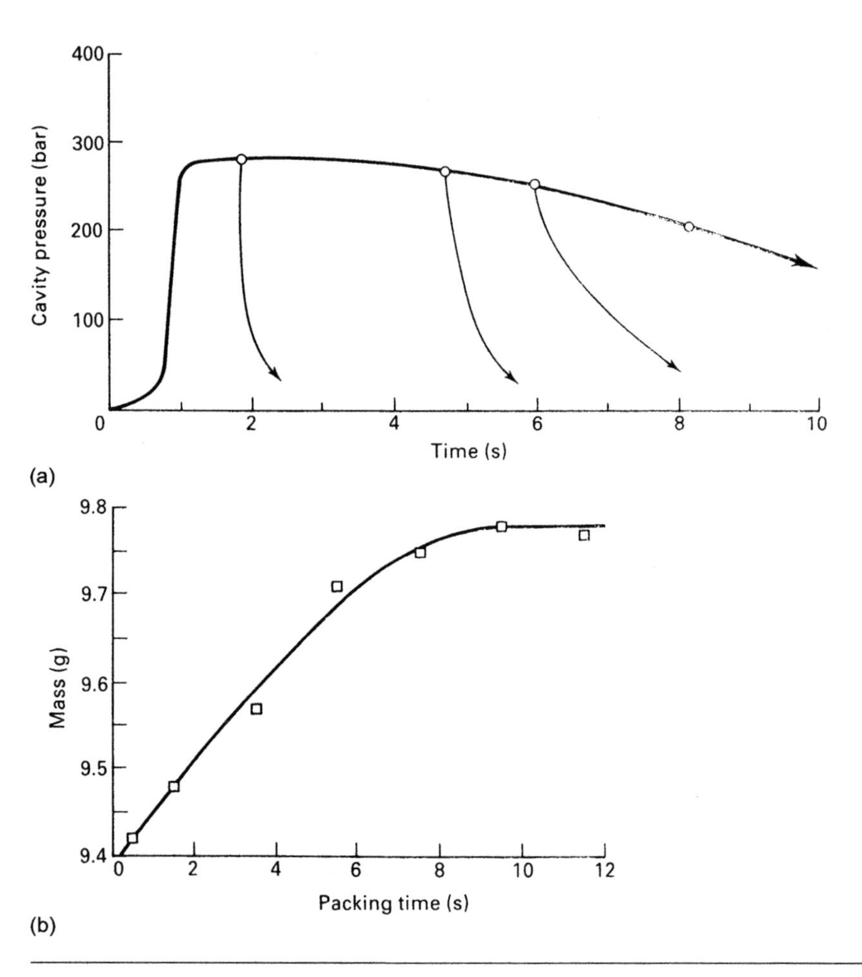

Figure 6.11 (a) Variation of the cavity pressure with time for cycles in which the holding pressure is switched off at various times. (b) The corresponding variation of the polyethylene moulding mass.

shrinkage. This can be accommodated by the disc warping; a disc of radius r can have a circumference greater than $2\pi r$ if it exists on the surface of a warped sheet, which rises above the plane in two or more sectors.

Mouldings are usually designed so that the melt flows from thicker into thinner sections, to ensure that the extremities are fed properly. A two-dimensional heat flow calculation illustrates a problem that can arise during solidification. Figure 6.12a shows isotherms for the cooling of a rib, with a heat transfer coefficient of $1000\,\mathrm{W\,m^{-2}\,K^{-1}}$ at the mould interface, calculated by finite element analysis. The thinner rib solidifies first, and an isolated island of melt, cut off from the melt supply, is left at the intersection of the rib and the plate. Contraction of this melt either causes sink marks in the plate surface, if the moulding is relatively thin and the skin pulls in, or it leads to a central shrinkage cavity in thicker mouldings. A sudden increase in cross section, in the direction of the flow, increases the probability of a shrinkage cavity forming. Figure 6.12b shows a section through an injection-moulded gear, with a shrinkage cavity where the flange intersects the rim that supports the gear teeth. The two-dimensional heat flow computation in Fig. 6.12a, which uses the same dimensions, predicts the position of the cavity.

6.3.2 Surface roughness

A number of types of surface roughness can occur on plastics products. These are especially noticeable on surfaces that have solidified in contact with air. Thus, if you look down the bore of an extruded pipe towards a light source, it is often possible to see three or four regularly spaced surface grooves, marking the positions where the melt passed the spider legs of the die. There is a slight difference in the elastic recovery of the melt where the two melt streams welded together. A more significant roughness can occur on the interior of blow-moulded containers, known as *shark skin*. This is due to flow instability as the polymer melt rapidly extrudes from the die. A periodic slipping of the melt at the die wall produces ridges at right angles to the flow direction (Fig. 6.13a). In addition to being unsightly, these could be sites for crack initiation. The roughness becomes more severe as the molecular mass of the polymer increases, or the extrudate velocity increases.

The surface of injection mouldings ought to be a replica of the surface of the steel mould. However, waves, running at right angles to the flow direction (Fig. 6.13b), may be visible on parts of the surface that are intended to be flat. These waves are prominent if the mould is filled slowly. They form on the skin of the polymer during mould filling, due to the tensile stresses in the core of the melt, acting parallel to the flow direction (Section 5.3.3). As there is no net tension on the advancing melt front, these stresses put the soft skin of the melt into compression, causing it to buckle.

6.3.3 Residual stresses in extrudates

Residual stresses occur both in extruded products and in injection mouldings. Extruded products are easier to analyse because the pressure is constant and the extrudate length is constrained during solidification. Consider a flat area in an extruded product, similar to that shown in Fig. 6.14. For modelling purposes the polymer is divided into a number of thin parallel layers. It is assumed that the flow stresses in the molten polymer are negligible, so the polymer layers are stress-free as they solidify. The solidification temperature T_s is taken as T_g for a glassy polymer, or

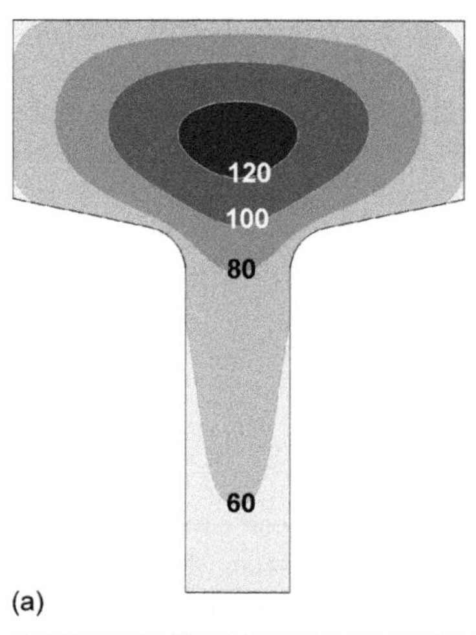

(a)

(b)

Figure 6.12 (a) Isotherms (°C) for 2-D heat flow, after 7.9s, for the gear section of a rib thickness 2 mm with a plate of thickness 3 mm, during cooling. (b) section of a polypropylene gear showing a shrinkage cavity at the intersection of the flange and the gear.

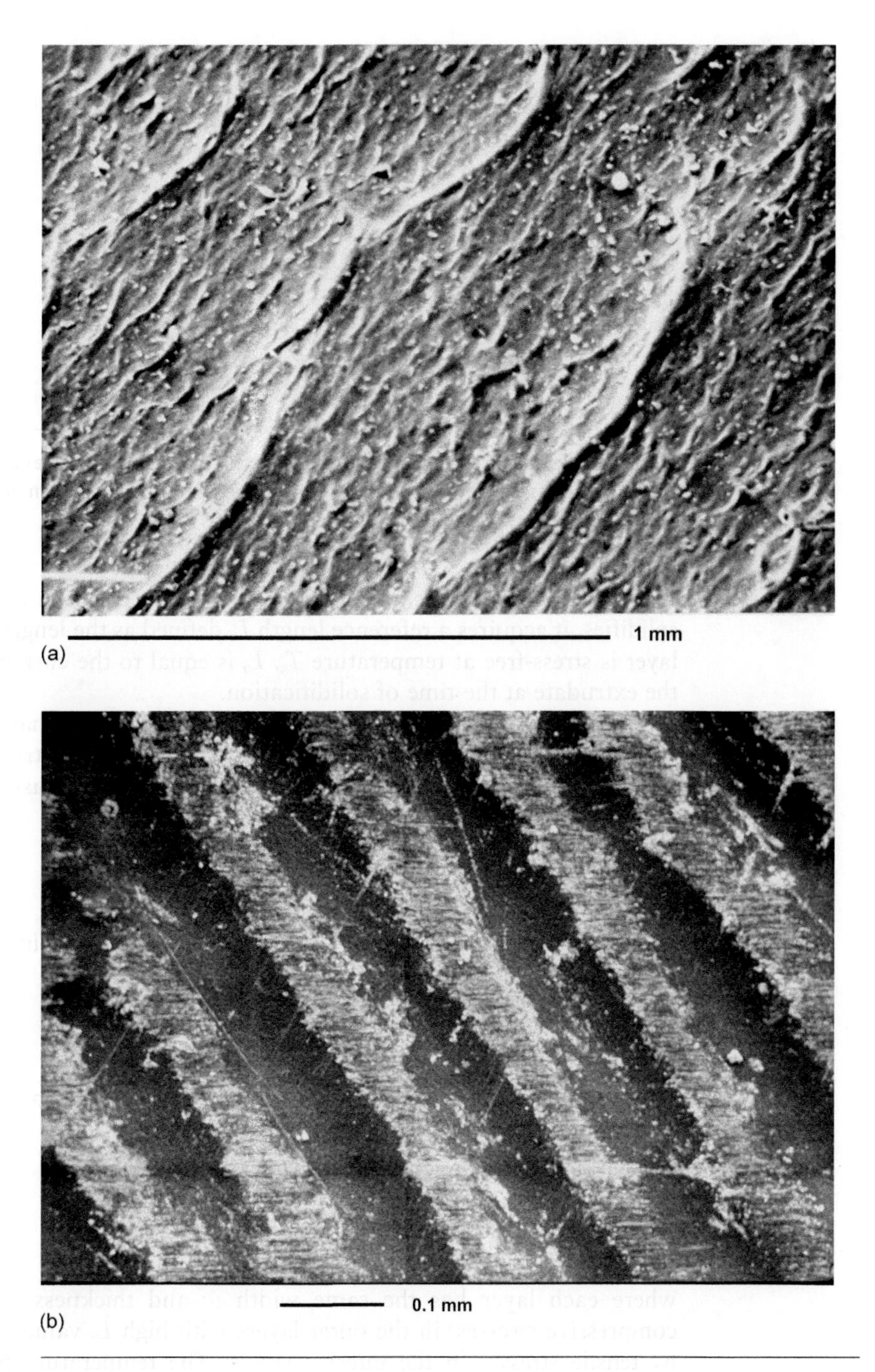

Figure 6.13 Scanning electron micrographs of surface defects on: (a) The inner surface of a blow-moulded HDPE bottle; (b) a polyethylene injection moulding. In both micrographs the ridges run at 90° to the flow direction.

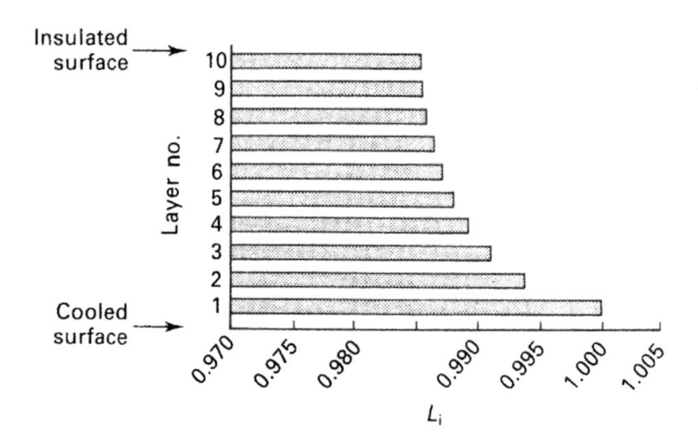

Figure 6.14 Six-mm thick MDPE extrudate split into layers, each with reference length at which it would be stress-free at the solidification temperature T_s, cooled on one surface. The extrudate has been cooled on one side only.

the temperature at which crystallisation is 75% complete. As the ith layer solidifies, it acquires a reference length L_i defined as the length at which the layer is stress-free at temperature T_s. L_i is equal to the current length L of the extrudate at the time of solidification.

The total strain, in any solid layer, is the sum of the thermal strain and the elastic strain. If the plate is at a temperature T, its thermal strain is $\alpha(T - T_s)$ where α is the linear thermal expansion coefficient. The elastic strain in the x direction, due to a biaxial stress system

$$\sigma_{xx} = \sigma_{yy} = \sigma_i$$

in the ith layer, is $\sigma_i(1 - \nu)/E$. Consequently, the total strain is

$$e_i = \alpha(T - T_s) + \frac{\sigma_i(1 - \nu)}{E} \tag{6.4}$$

Since there is no external force on the plate in either the x or y directions, the total internal force on a cross section

$$w\Delta z \sum_{i=1}^{n} \sigma_i = 0 \tag{6.5}$$

where each layer has the same width w and thickness Δz. Therefore, compressive stresses, in the outer layers with high L_i values, are balanced by tensile stresses in the interior layers. The temperature profile during cooling can be calculated using the finite difference methods of Appendix A. Equations (6.4) and (6.5) then give the equilibrium length of the extrudate. The value of L_i varies from a maximum, at the surfaces which solidified first, to a minimum at the centre which solidified last. Figure. 6.15a shows how the surface and the centre stresses vary with Biot's modulus (Section 4.2.3). The stresses are proportional to a quantity σ^* defined by

$$\sigma^* = E\alpha(T_s - T_b) \tag{6.6}$$

Their magnitude can be reduced by using a higher cooling bath temperature T_b or reducing the value of Biot's modulus. Pipe extrudates are only cooled from the outside. Consequently, the residual stresses are compressive at the outer surface and tensile near the bore (Fig. 6.15b). If a sheet is cooled unequally from two sides, it bends until internal stresses have a zero net bending moment. Therefore, if a moulded part is bent or warped, this suggests that the cooling has been uneven.

6.3.4 Residual stresses in injection mouldings

Variations in the cavity pressure p during solidification lead to variations in the reference lengths L_i of layers in the moulding. The latter are defined as

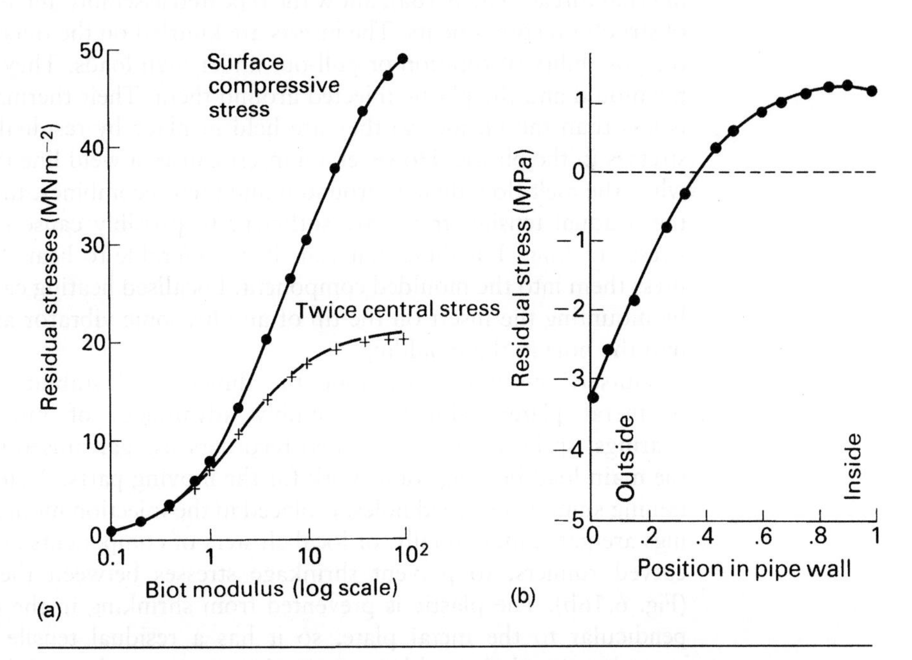

Figure 6.15 (a) Predicted residual stresses in a sheet of polycarbonate vs. Biot's modulus (from Mills, N. J., *J. Mater. Sci.*, **17**, 558, 1982). (b) Measured residual stresses in an extruded polyethylene pipe wall (from Plastics Pipes V, 1982, Plastics and Rubber Inst. Conference).

the lengths at which the layer is stress-free at atmospheric pressure and temperature T_s. If a layer has solidified at a pressure p_i, an additional term $(p - p_i)/3K$ must be added to the right-hand side of Eq. (6.4) for the total strain, when the moulding is at pressure p. K is the bulk modulus of the melt. The complex shape of most moulds means that the solidifying moulding must have the length L of the mould. Hence, the left-hand side of Eq. (6.4) is zero while the moulding is in the mould. After the moulding has cooled to room temperature, the stress distribution in the mould is

$$\sigma_i = \frac{E}{1-\nu}\left(\alpha(T_s - T) - \frac{p_i}{3K}\right) \tag{6.7}$$

Once the moulding is ejected from the mould, the average residual stress on the cross section falls to zero according to Eq. (6.5). There are many possible residual stress distributions across the thickness of injection mouldings, because of the varied pressure histories in the mould. If the cooling of one side of the mould is more effective than the other, parts of the moulding may bow.

The shrinkage of thermoplastics can be turned to advantage if metal components are to be attached to a moulding. Metal inserts, with an internal thread (Fig. 6.16a), allow the repeated assembly and dismantlement of structural components. The inserts are knurled on the outside to prevent the possibility of rotation or pull-out under high loads. They are placed in the mould and the plastic injected around them. Their thermal contraction is less than the plastic, so they are held in place by residual tensile hoop stresses in the plastic. However, an insert causes a weld line (Section 6.2.4) when the melt flow divides around it and then recombines. In some plastics the residual tensile strains are sufficient to possibly cause environmental stress cracking. For these materials it is preferable to heat the inserts and press them into the moulded component. Localised heating can be provided by mounting the insert on the tip of an ultrasonic vibrator and pressing it into the hole in the moulding.

'Outsert' mouldings combine the dimensional stability and stiffness of metal plates with the assembly advantages of moulded plastics bearings, springs, etc. Some video recorders use galvanised steel plates as the main load-bearing framework for the moving parts. A steel plate, containing suitable punched holes, is placed in the injection mould. The mouldings are gated individually, or local clusters of components are fed through curved runners, to prevent shrinkage stresses between the components (Fig. 6.16b). The plastic is prevented from shrinking in the direction perpendicular to the metal plate, so it has a residual tensile stress in the central region, balanced by compressive stresses at the periphery that grips the plate.

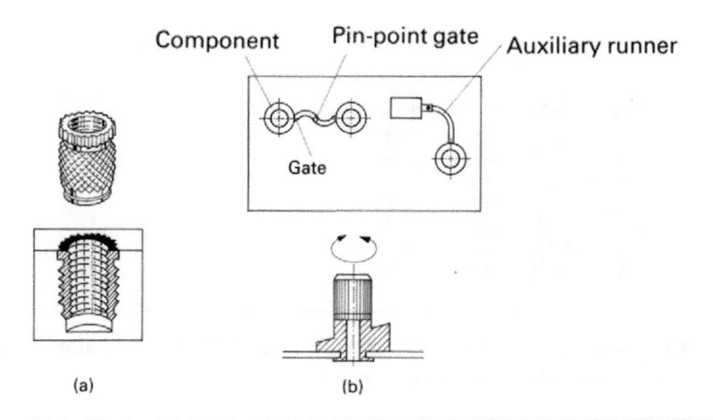

Figure 6.16 (a) A metal insert before and after insertion into a plastic part. (b) Injection-moulded outserts on a steel plate and a cross section of an outsert knob (from Lexan Noryl Valox booklet, General Electric Plastics, and Outsert Moulding with Hostaform booklet, Hoechst, 1978).

6.4 Fusion of particle and bead polymers

6.4.1 Mixing

Often the reason for a failure in a moulding will be found to be the poor dispersion of pigments and other additives, associated with high orientation. Polyolefins in particular are weakly bonded to pigments. Elongated strings of pigment act in the same way as bands of inclusions in wrought steel products, so a fracture can initiate from one of these strings of pigment when a bending impact occurs.

The two main types of mixing relate to the nature of the phase to be dispersed. *Distributive mixing* distributes fine solid particles or liquids evenly throughout the melt, whereas *dispersive mixing* breaks up agglomerates of particles. In a shear flow, as in the barrel of an extruder, the distributive mixing increases with the total shear strain. Figure 6.17 shows how the layer thickness S, of regions of initially pigmented and natural polymer, decreases according to

$$S = \frac{S_0}{\gamma} \tag{6.8}$$

Equation (6.8) fails when the layer thickness approaches the diameter of any solid particles in the melt. The total shear strain received by polymer, passing down an extruder screw of length L equal to 40 times the diameter,

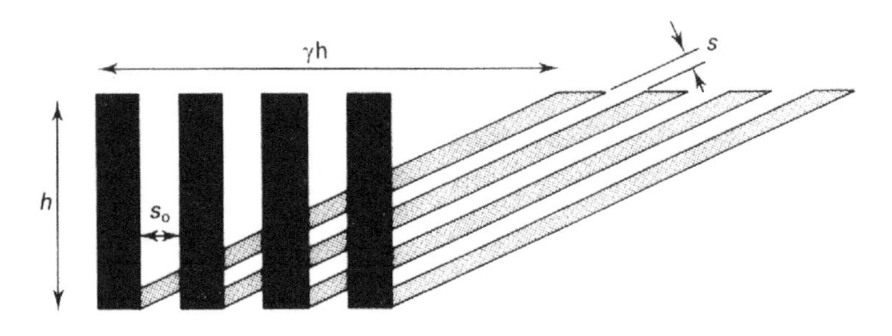

Figure 6.17 Spatial disposition of layers of coloured polymer, before and after a shear strain γ in a distributive mixing shear flow.

exceeds 500. Hence, granules of coloured masterbatch polymer are adequately dispersed by the time they leave the extruder.

Extrusion is not adequate to disperse agglomerated powders such as carbon black of a high 'structure factor'. Higher stress mixing of a batch of plastic can be achieved using an internal mixer with intermeshing blades. An alternative is the cavity transfer mixer, that fits onto the front of an extruder screw, which has a cutting and folding action on the melt.

6.4.2 PVC powder processing

The effects of processing on PVC are more complex than those of most other polymers because the suspension polymerisation particles have an internal structure. This structure would be irrelevant to the product properties if the particles melted into a homogeneous melt on heating. However, PVC has about 10% crystallinity and the crystalline regions bind the particles together. Transmission electron microscopy reveals a micro-domain substructure inside the primary particles, with crystals smaller than 10 nm. The crystals melt in the range of 200–240 °C. Lower temperatures are used in melt processing to avoid thermal degradation.

PVC powder blends are typically processed through a twin-screw extruder (Fig. 6.18) before extrusion. The counter-rotation of the two intermeshing screws, inside a barrel with a figure of eight cross section, effectively pumps a powdery melt that can slip against the barrel wall. Its action is close to that of a positive displacement pump, with C-shaped segments of material being passed from one screw to the other. The initial stage of processing occurs once the PVC is heated to above its T_g of 80 °C. The grains are compacted together to increase the bulk density from about 500 to 1200 kg m^{-3} (Fig. 6.19). At this stage, the various solid additives are at the particle boundaries. Particle densification then occurs. The increase in density starts at the surface and spreads to the interior as the porosity is eliminated. The particles are deformable and elongate in the direction of the

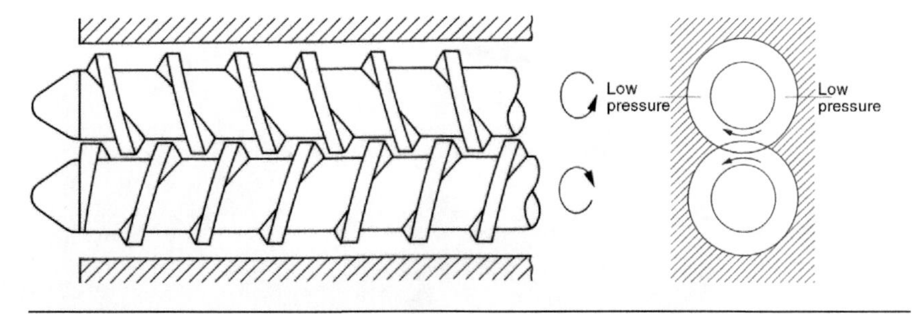

Figure 6.18 Intermeshing counter-rotating screws in a twin-screw extruder.

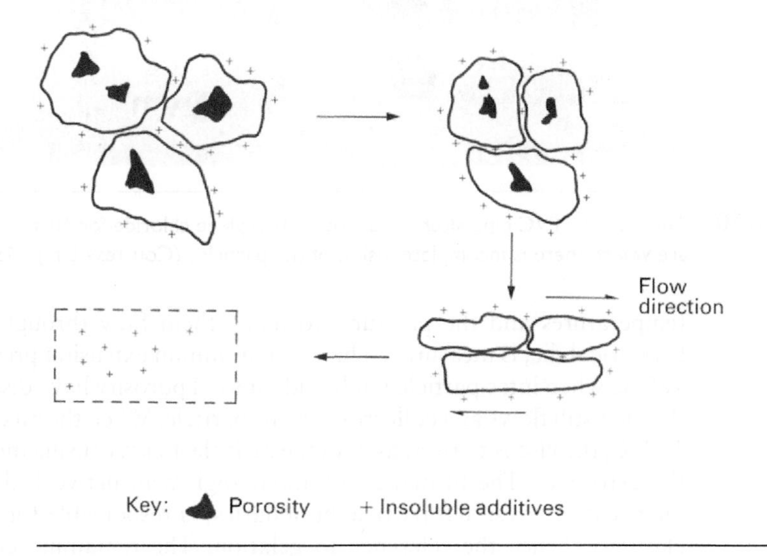

Figure 6.19 Mechanism of particle fusion during extrusion of PVC.

flow. Finally particle boundaries fuse, leaving a melt containing only primary particles. However, if the powder is compounded in an internal mixer, the kneading blades impart a much higher level of shear at an early stage in the process, breaking the particles into fragments.

There are various methods of assessing the state of fusion of PVC. Intact particle boundaries, coated with stabilisers, etc. are very weak. Consequently, immersing the PVC in a liquid such as methylene dichloride causes the particles to separate, giving rise to a white powdery surface (Fig. 6.20). The wall of the pipe has been chamfered to expose the different layers. Fusion is only complete at the inner and outer surfaces, that experienced the highest levels of shear in passing through the extrusion die. A pipe with this level of fusion would have inferior mechanical properties. The fusion of PVC particles is referred to as *gelation*. As PVC is mixed in an internal mixer, viscous dissipation causes a steady rise in temperature. Samples are removed at various

Figure 6.20 The wall of a PVC pipe after exposure to methylene chloride for 10 min. The white areas are where there is incomplete fusion of the particles (Courtesy Dr. J. Marshall).

temperatures and the pressure, to make them flow through a very short cylindrical die, is measured. There is a minimum extrusion pressure at about 160 °C when inter-particle voids and internal porosity have disappeared, but the melt still flows as a collection of sub-particle. When the mixing is taken to 180 °C, the viscosity rises, as does the melt elasticity, causing melt fractures in the extrudate. The formation of an entanglement network throughout the melt, with no residual particle boundaries, is responsible for the high melt elasticity, hence the reference to gelation. The remaining crystals at this temperature physically bind the molecules into an entanglement network.

PVC pipes used for water distribution, are subjected to an internal pressure, which can vary according to the water demand. The resulting fluctuating, stresses the pipe wall causing fatigue loading. Fatigue stresses, if high enough, may eventually initiate small cracks at weak particle boundaries in the PVC. Subsequent fatigue cycles cause the crack to grow slowly until failure occurs. Section 6.3.3 explained how pipes, cooled on the outside, contain residual tensile hoop stresses at the bore. Consequently, cracks tend to start near the bore of the pipe from a point of weakness.

6.4.3 UHMWPE powder processing

UHMWPE powder has a size of about 200 μm (Fig. 6.21) with sub-particles of size 1 μm visible on their surfaces.

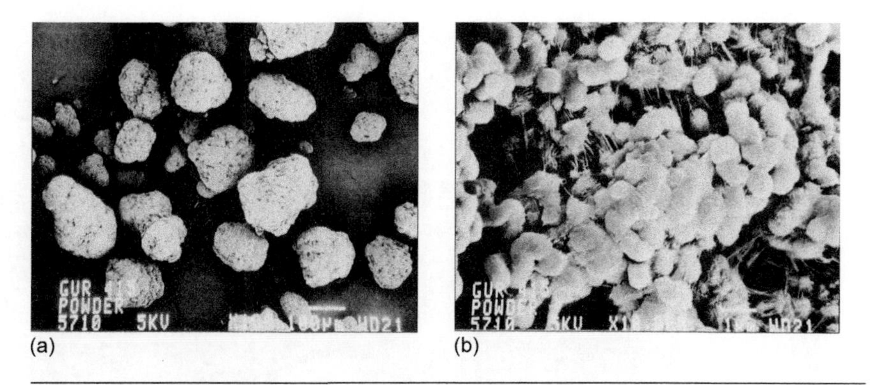

Figure 6.21 SEM of UHMWPE powder: (a) Low; (b) higher magnification (Sedel, L. and Cabanela, M. E., Eds., *Hip Surgery—Materials and Developments*, Martin Dunitz, London, 1998).

During ram extrusion (Section 5.4.4), the molten particles are forced into contact with each other. Polyethylene chains diffuse across the particle boundaries, and, with time, a strong bond develops. The UHMWPE however retains a memory of its particle structure, especially when calcium stearate was used prior to the late 1990s—the stearate coats the particles before processing, and it makes the interfaces more visible when microtomed slices are examined in an optical microscope (Fig. 6.22a). There are parallels with PVC particle fusion, but it is not possible to use shear flow to disrupt the UHMWPE boundaries.

6.4.4 Polystyrene foam bead processing

The steam moulding process for EPS beads occurs on a cycle time of 1–10 min. Compared to the sintering of UHMWPE, there is far less time for polymer diffusion across the bead boundaries. Such diffusion is essential for the product to have a reasonable bending or tensile strength. Figure 6.23a shows the fracture surface of an EPS bead moulding, in which some bead boundaries are intact—a sign of lack of fusion—but the majority of the fracture has run across beads, exposing the cell structure. Much EPS packaging for consumer products will fracture in such a manner if bent. This is not critical if the packaging is protected by a strong cardboard box, but in application such as bicycle helmets (Chapter 14) it is more important for the bead boundaries to be strong. The tensile failure stress of foam mouldings of density 22 kg m^{-3} increases almost linearly with the degree of boundary fusion (Fig. 6.23b). Coatings, such as calcium stearate, are used on the beads to aid the bonding process.

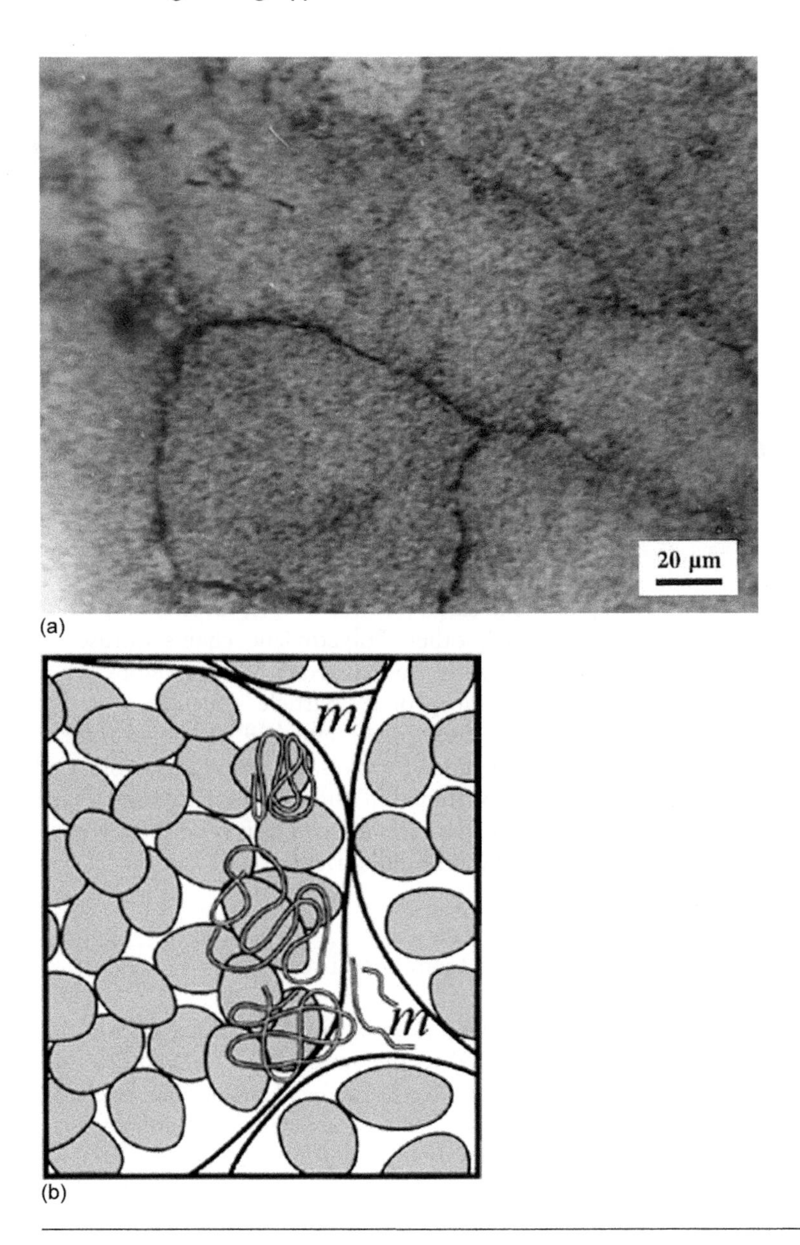

Figure 6.22 (a) Optical micrograph of UHMWPE section, dark at particle boundaries. (b) Schematic of polyethylene chain diffusion at particle boundaries (Olley R.H. *et al.*, *Biomaterials*, **20**, 2037, 1999) © Elsevier.

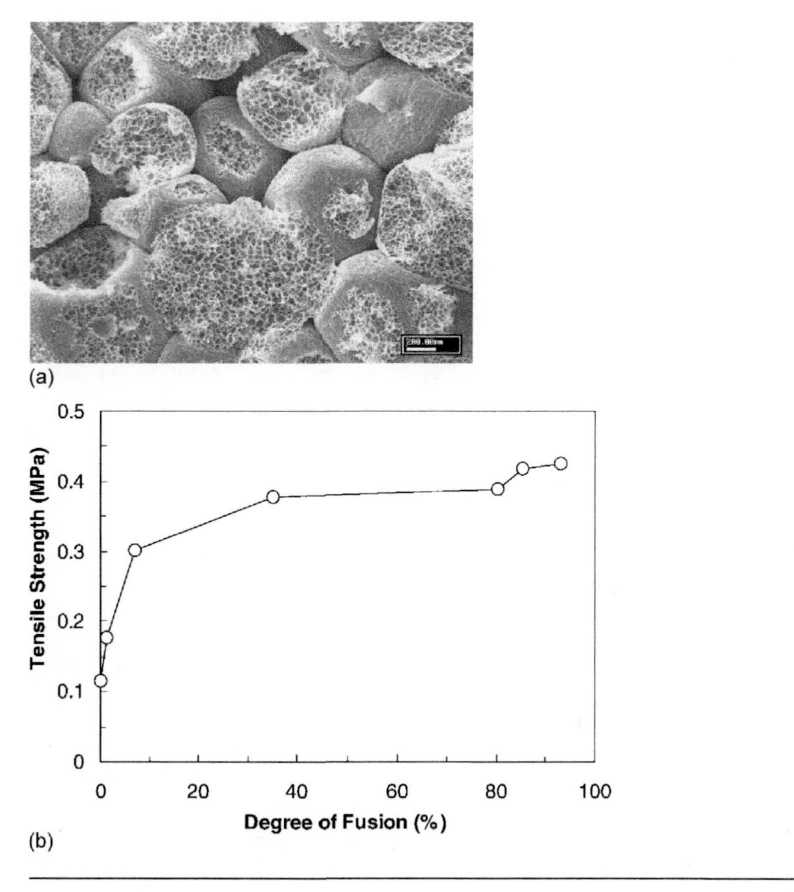

Figure 6.23 EPS of density 22 kg m^{-3}. (a) Fracture surface and (b) variation of tensile strength with degree of fusion (Rossacci, J. and Shivkumar, S., *J. Mater. Sci.*, **38**, 201, 2003) with kind permission of Springer Science and Business Media.

Chapter 7
Viscoelastic behaviour

Chapter contents

7.1 Introduction

The mechanical behaviour of plastics is time-dependent, and often non-linear. This contrasts with metals and silicate glasses which behave as linear elastic materials at ambient temperature. Hence, the concept of Young's modulus, defined for linear elastic materials, strictly does not apply to polymers. Nevertheless, a Young's modulus with an associated timescale is still useful for comparing the responses of polymers. Most solid polymers can be treated as being viscoelastic, when the stresses are moderate. At high stresses, yield (Chapter 8) or fracture (Chapter 9) can occur, so the polymer never recovers its original shape on unloading.

The mechanical behaviour of viscoelastic materials combines elements of both viscous fluid and elastic solid responses. Viscoelastic effects are strongest when the polymer is in transition between the glassy and rubbery liquid state. For instance, slow-recovery foams for mattresses retain the indentation of a hand for many seconds after the pressure is removed, because their T_g is approximately 20 °C. Table 7.1 lists a number of simple experiments that demonstrate viscoelastic behaviour. There are everyday examples of each: Sitting on a plastic chair for an hour subjects it to creep loading, whereas a plastic spring deflected by a fixed amount will undergo stress relaxation. When plastic gears rotate, the teeth are loaded and unloaded once per revolution, a form of cyclic loading. When a tensile test is performed, the machine cross-head speed affects the result. More complex types of loading can be a combination of these simple cases. Time-dependent phenomena also occur at high stresses, as is shown in the subsequent chapters on yielding and fracture phenomena.

The chapter starts by analysing viscoelastic models, to show how the different phenomena are connected. This is followed by engineering design calculations of the creep deflections of viscoelastic structures, using modifications of methods for elastic materials. Finally, vibration damping and energy losses in cyclic loading are considered.

Table 7.1 Simple cases of viscoelastic behaviour

	Variation with time of	
Test	*Strain*	*Stress*
Creep	Increases	Constant
Stress relaxation	Constant	Increases
Cyclic	Sinusoidal	Sinusoidal but phase shift
Tensile test	Constant rate of increase	Usually increases

7.2 Linear viscoelastic models

7.2.1 The Voigt model for creep

Simple viscoelastic models can mimic the phenomena mentioned in Table 7.1. Although the models are inadequate at high stress levels, they aid understanding, and are the basis for more complex treatments. They are **mechanical analogues** of viscoelastic behaviour, constructed using the linear mechanical elements shown in Table 7.2. They are **linear** because the equations relating the force f and the extension x only involve the first power of both the variables.

The elements can be combined in series or parallel as shown in Figs 7.1 and 5.5. The convention for these models is that **elements in parallel undergo the same extension.** It is obvious that elements in series experience the same force. Thus, in the *Maxwell model*, the spring and dashpot in series experience the same force, while in *Voigt model* the spring and dashpot in parallel experience the same extension x. The total force f across the Voigt model can be written as the differential equation

$$f = kx + c\frac{\mathrm{d}x}{\mathrm{d}t} \tag{7.1}$$

where t is the time. Such a combination of mechanical elements occurs on the rear suspension of many cars, where a helical spring surrounds a shock absorber (an oil-filled cylinder with a small orifice to restrict the oil flow), and both connect the wheel hub to the car body.

The viscoelastic analogue of the Voigt model is produced by imagining the elements to be contained in a unit cube; the tensile stress σ equals the force on the unit end face area, and the tensile strain e equals the extension per unit length. The analogy is not perfect, as the Voigt mechanical model only operates along one axis, whereas viscoelastic materials are often isotropic. Hence, the mechanical model, inside a $1\,\mathrm{m}^3$ black box, must be aligned with the stress direction. The differential equation of the Voigt viscoelastic model is

$$\sigma = Ee + \eta\frac{\mathrm{d}e}{\mathrm{d}t} \tag{7.2}$$

Table 7.2 Elements in linear viscoelastic models

Element	*Symbol*	*Equation*	*Constant*
Spring	/\/\/\/\/\/\	$f = kx$	k
Dashpot	(dashpot symbol)	$f = c\,\mathrm{d}x/\mathrm{d}t$	c

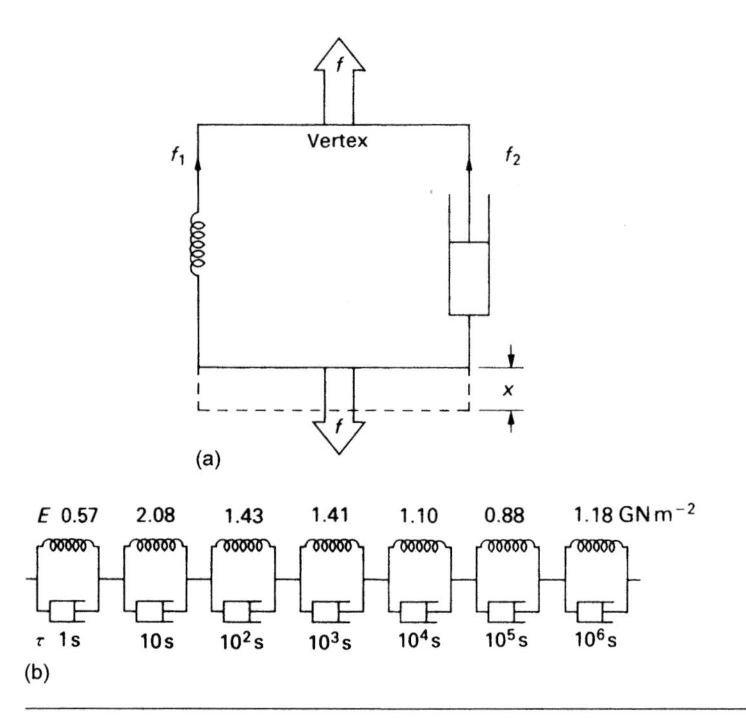

Figure 7.1 Viscoelastic models employing: (a) a single Voigt element; (b) multiple Voigt elements connected in series. The values of the moduli and retardation times are used to model the creep of HDPE in Figure 7.3 also.

This has the same form as Eq. (7.1); the renamed constants are a Young's modulus E, and a viscosity η. It is not possible to directly link these constants to the modulus of the crystalline phase and the viscosity of the amorphous inter-layers in a semi-crystalline polymer. Hence, the Voigt model is an aid to understanding creep, and relating it to other viscoelastic responses, rather than a model of microstructural deformation.

Creep loading means that the stress is given by

$$\sigma = 0 \quad \text{for} \quad t \leq 0$$

$$\sigma = \sigma_0 \quad \text{for} \quad t > 0$$

For the Voigt model of Eq. (7.2), substituting the constant stress σ_0 for $t > 0$, and dividing by η, gives

$$\frac{\sigma_0}{\eta} = \frac{Ee}{\eta} + \frac{de}{dt}$$

If both sides of this equation are multiplied by exp (Et/η), it can be integrated to give

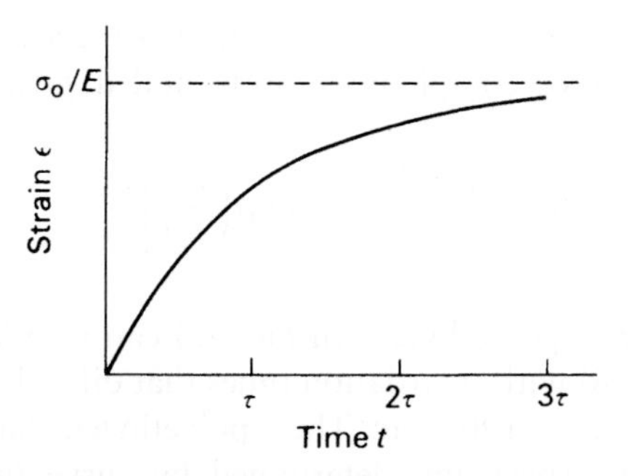

Figure 7.2 Response of a single Voigt element to a creep stress σ_0 applied at time $t = 0$, shown on linear scales. τ is the retardation time.

$$\frac{\sigma_0}{E} \exp\left(\frac{Et}{\eta}\right) = e \exp\left(\frac{Et}{\eta}\right) + A$$

Substituting the initial condition, that $e = 0$ when $t = 0$, reveals that the constant of integration $A = \sigma_0/E$. Therefore, the solution is

$$e = \frac{\sigma_0}{E}\left[1 - \exp\left(\frac{Et}{\eta}\right)\right] \tag{7.3}$$

The physical response is clarified by introducing the *retardation time* $\tau = \eta/E$. Figure 7.2 shows that 63% of the creep strain occurs in the first τ seconds and 86% within twice the retardation time.

7.2.2 Creep compliance of the generalised Voigt model

In order to compare the prediction of the model with experimental creep data we need to define the *creep compliance* of the model or of a polymer by

$$J(t) \equiv \frac{e(t)}{\sigma_0} \tag{7.4}$$

where the parentheses indicate that both J and e are functions of the time t since the creep stress σ_0 was applied. For the Voigt model the creep compliance

$$J(t) = \frac{1}{E}\left[1 - \exp\left(-\frac{t}{\tau}\right)\right] \tag{7.5}$$

Figure 7.3 shows that the predicted creep of the Voigt model is a poor representation of the creep of a polyethylene. Better predictions can be obtained by combining, in series, Voigt models with different retardation times, as in Fig. 7.1b. The series connection applies the same stress to each Voigt model, so their creep compliances can be added, giving

$$J(t) = \sum_{i=1}^{n} \frac{1}{E_i} \left[1 - \exp\left(\frac{t_i}{\tau_i} \right) \right] \tag{7.6}$$

The creep response of polyethylene in Fig. 7.3 can be adequately reproduced by using Eq. (7.6) with retardation times that differ by powers of 10, i.e. $\tau_1 = 1$ s, $\tau_2 = 10$ s, $\tau_3 = 100$ s, etc. Thus, polyethylene has a *spectrum of retardation times*. The spectrum, determined by curve fitting the creep response, can be used to predict other forms of viscoelastic behaviour.

7.2.3 Prediction of stress relaxation

The generalised Voigt model (Fig. 7.1b) will be used to predict stress relaxation. The Maxwell model (Fig. 6.5), or a combination of Maxwell models in parallel, could be used to model stress relaxation, but the purpose is to show the predictive power of a known viscoelastic model. A single Voigt element cannot exhibit stress relaxation. A constant extension causes a constant force in the spring, and a zero force in the stationary dashpot. However, when the model in Fig. 7.1b is used, the distribution of the total extension among the Voigt elements changes with time, allowing stress relaxation.

First, we need a rule to predict the effect of time-varying loads on a viscoelastic model. When a combination of loads is applied to an elastic material, the stress (and strain) components caused by each load in turn can be added. This addition concept is extended to linear viscoelastic materials. The *Boltzmann superposition principle* states that if a creep stress σ_1 is

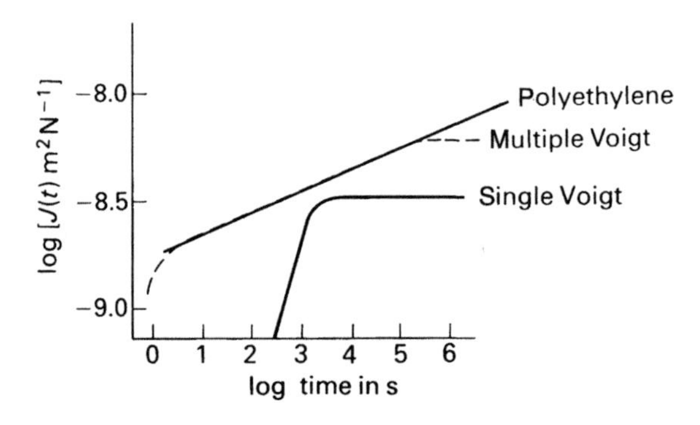

Figure 7.3 The shear creep compliance of an HDPE at 29 °C vs. time on logarithmic scales. The dashed curve is for the multiple Voigt element model of Figure 7.1. The response of a single Voigt element having $E = 300$ MPa and $t = 1000$ s is also shown.

applied at time t_1, and the further creep stress σ_2 is applied at time t_2, the total creep strain at times $t > t_2$ is

$$e = \sigma_1 J(t - t_1) + \sigma_2 J(t - t_2) \tag{7.7}$$

This is shown schematically in Fig. 7.4 for the case when σ_2 is negative. The creep strain is equal to e_0 at two different times. This suggests a method for predicting stress relaxation. Further negative stress increments

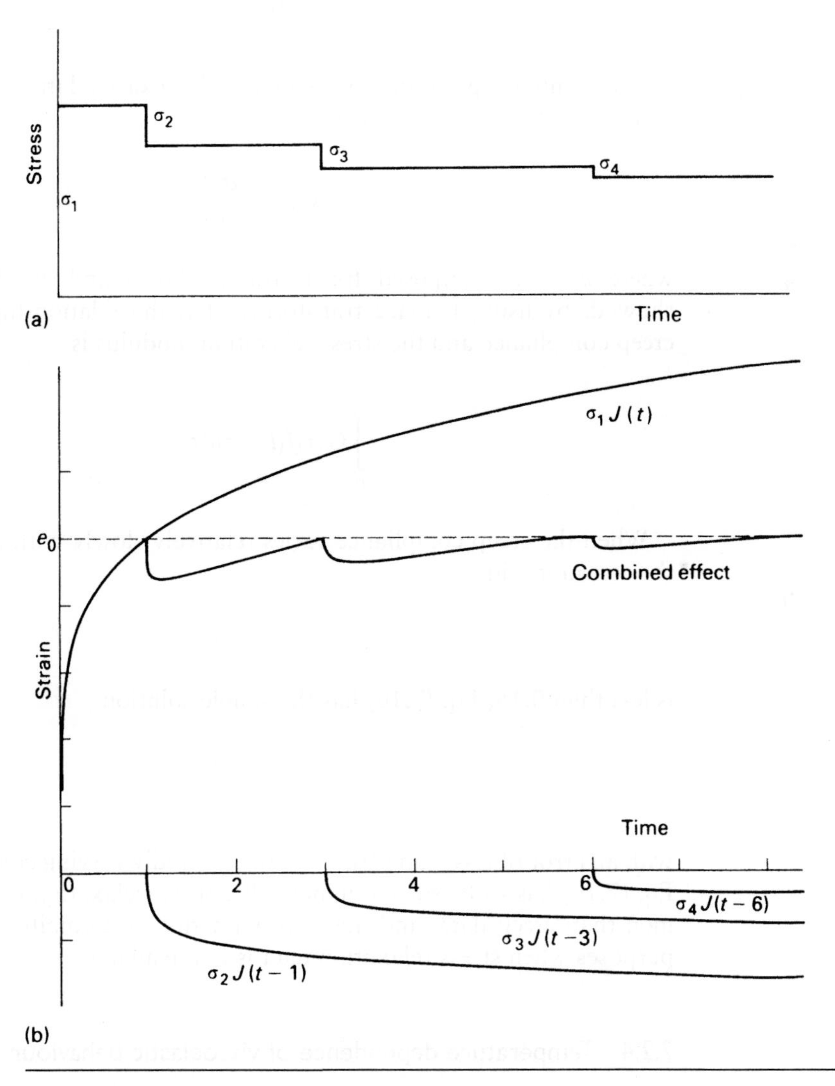

Figure 7.4 (a) Creep loading that consists of step changes in stress σ_1, σ_2, σ_3 at times t_1, t_2, t_3. (b) The creep strains due to the separate application of the stresses, and their combined effect according to Boltzmann's superposition principle.

$\sigma_3, \sigma_4, \ldots$, made at times $t_3, t_4, \ldots$, are chosen so that the strains, calculated from Eqs (7.6) and (7.7), are $e(t_4) = e_0$, $e(t_5) = e_0$, etc.

A computer program can calculate the necessary stress increments at sufficiently short time intervals, and hence predict the stress relaxation curve. In the limit that the applied stress changes continuously at a rate $d\sigma/d\tau$, Eq. (7.7) becomes the convolution integral

$$e = \int_0^t \frac{d\sigma}{d\tau} J(t - \tau) d\tau \tag{7.8}$$

The computer program approximates this exact relationship. The *stress relaxation modulus* $G(t)$ is defined by

$$G(t) \equiv \frac{\sigma(t)}{e_0} \tag{7.9}$$

where e_0 is the imposed fixed strain. Akonis and MacKnight (1983) showed, by using Laplace transforms, that the relationship between the creep compliance and the stress relaxation modulus is

$$\int_0^t G(\tau) J(t - \tau) d\tau = t \tag{7.10}$$

When the creep compliance varies relatively slowly with time, i.e. when the constant n in

$$J(t) = J_0 t^n$$

is less than 0.15, Eq. (7.10) has the simple solution

$$G(t) \cong \frac{1}{J(t)} \tag{7.11}$$

with an error of less than 4%. For a more rapidly varying creep compliance, Eq. (7.10) has to be solved numerically. Stress relaxation data is less common than creep data, and this approximation is normally used for design purposes when stress relaxation data is not available.

7.2.4 Temperature dependence of viscoelastic behaviour

Viscoelastic measurements can be performed at a series of absolute temperatures T. The temperatures intervals need to be small when near the glass transition temperature of an amorphous polymer. The data is taken at a low stress level, in the linear viscoelastic region. If graphs of the viscoelastic function versus log time or log frequency are compared, the shapes are similar for measurements made at neighbouring temperatures. Figure 7.5a shows shear creep compliance graphs for polystyrene, plotted on logarithmic scales. It was found that the curves superimpose to form a *mastercurve*, if each curve was shifted horizontally by an empirical factor $a(T)$. This is equivalent to multiplying all the creep times by a constant factor. A small vertical shift factor $\rho T/\rho_0 T_0$, where ρ is the density and the subscripts refer to the reference temperature T_0, is sometimes used to conform to the predictions of rubber elasticity theory.

This *time–temperature superposition* of linear viscoelastic data means that all the retardation times τ_i of the linear viscoelastic model have a common temperature shift factor $a(T)$

$$\tau_i = \tau_{i0} a(T) \tag{7.12}$$

where τ_{i0} are the reference values at temperature T_0. The elastic moduli E_i in the model are temperature independent, so the temperature dependence of τ_i is provided by the viscosity η_i. Now the temperature dependence of melt viscosity is independent of molecular weight (Chapter 3). Consequently, the width of the retardation time spectrum partly depends on the width of the molecular weight distribution, with different molecular weight entities in the amorphous phase having different viscosities. However, a retardation time spectrum is required to describe the response of a monodisperse polymer.

The shape of the mastercurve is related to the polymer microstructure. That for polystyrene at 100 °C (Fig. 7.5b) shows a transition from a glassy compliance at 1 s to a rubbery one at times exceeding 10^7 s. It continues to 10^{10} s, so it can be used for extrapolation to times longer than those accessible by experiment. Time–temperature superposition for semi-crystalline polymers, such as polyethylene, may be successful for a limited temperature range, i.e. 20 °C–80 °C. As polyethylene starts to re-crystallise if heated within 50 °C of T_m, and residual stresses may start to relax, data for higher temperatures will not superimpose.

7.3 Creep design

7.3.1 Creep data

Products are used at different temperatures, and each has a unique stress distribution. Consequently, design requires creep data under a wide variety of conditions. However the test programme to generate such data is excessively long. For most plastics there will be tensile creep data, for times up to

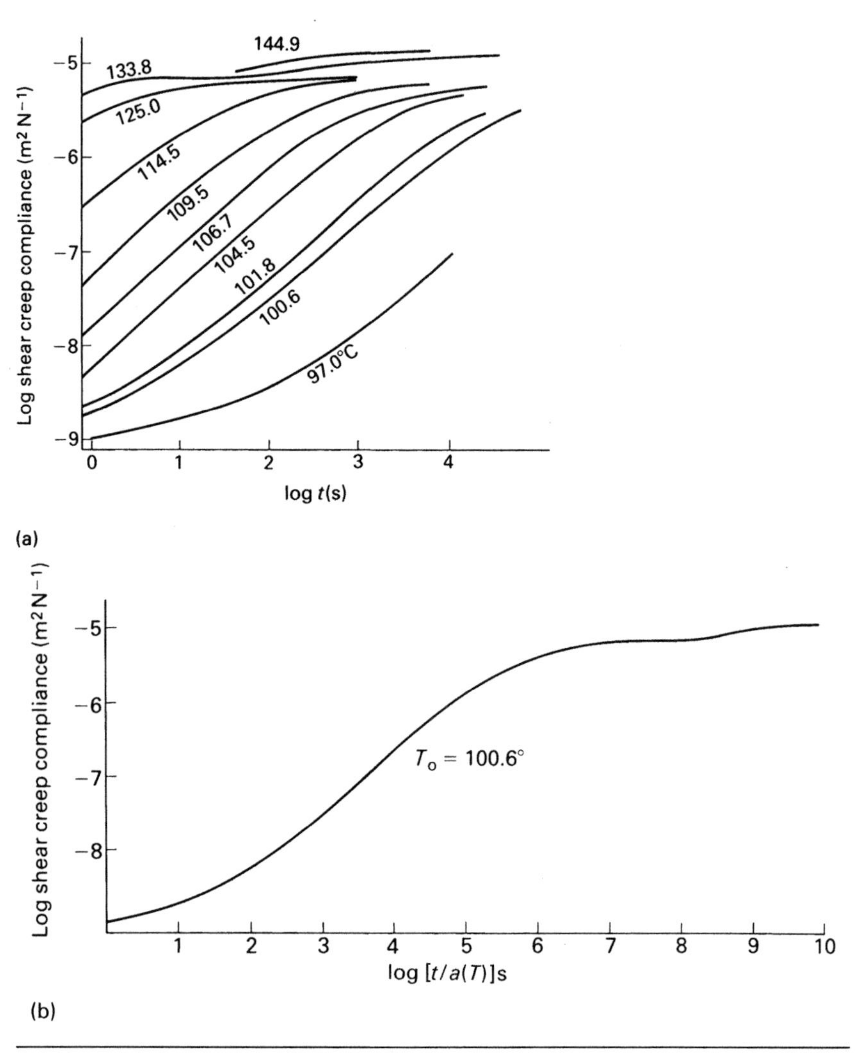

Figure 7.5 (a) Shear creep compliance of polystyrene vs. the creep time at various temperatures. (b) Master curve at 100 °C constructed by shifting the curves horizontally until they superimpose.

perhaps 1 year, at 5–10 different stress levels (Fig. 7.6). The lowest stress will cause a strain of a fraction of 1% at long times, while the highest stress will cause yielding within less than 1 h. There may also be creep data at selected elevated temperatures. It is normal to plot log strain versus log time, so that the creep data is approximately linear, and extrapolation to longer times is straightforward.

A cross-plot can be made of the tensile creep strains for a specific time versus the creep stress. When a smooth curve is fitted to the discrete

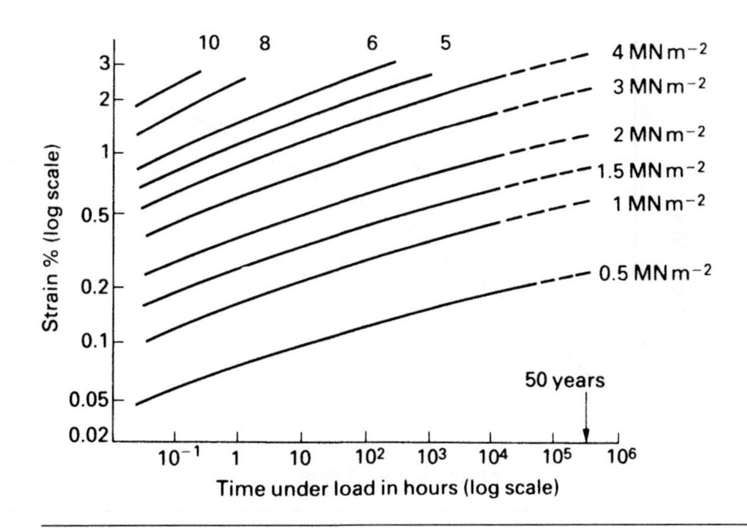

Figure 7.6 Tensile creep data for HDPE at a range of creep stresses (from Pipes: Hostalen GM5010 T2, GM7040G, 1980, Courtesy Hoechst AG).

set of points, it is known as an *isochronous stress–strain curve*. It has the same appearance as a tensile stress–strain curve, but each point refers to the same time. Isochronous stress–strain curves can be used for interpolation, to estimate the creep strain for stresses that lie between the data points.

There is *linear viscoelastic behaviour* in the stress region where the isochronous stress–strain curve is linear (to within 5%). The creep compliance $J(t)$, defined by Eq. (7.4), is independent of stress. However, above this stress region (stresses >1 MPa for the data in Fig. 7.7 for a time of 1 year) there is *non-linear viscoelastic behaviour* and the creep compliance becomes stress dependent

$$J(\sigma, t) \equiv \frac{e(\sigma, t)}{\sigma} \tag{7.13}$$

7.3.2 Linear viscoelastic design

A simple product that undergoes creep is shown in Fig. 7.8. A cantilever arm, with a float at one end, operates a water valve in a domestic cold-water tank. The load at the free end, due to the buoyancy of the float, is 5 N. The deflection of the free end of the arm should not exceed 30 mm after 1 year. The arm is to be injection moulded from polyethylene, and the cross section of the arm must be determined.

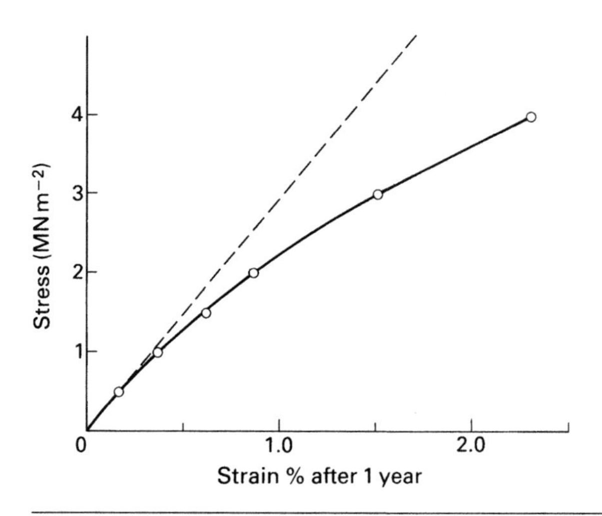

Figure 7.7 Isochronous stress–strain curve at a time of 1 year constructed from the creep data in Figure 7.6. The broken line represents linear viscoelastic behaviour.

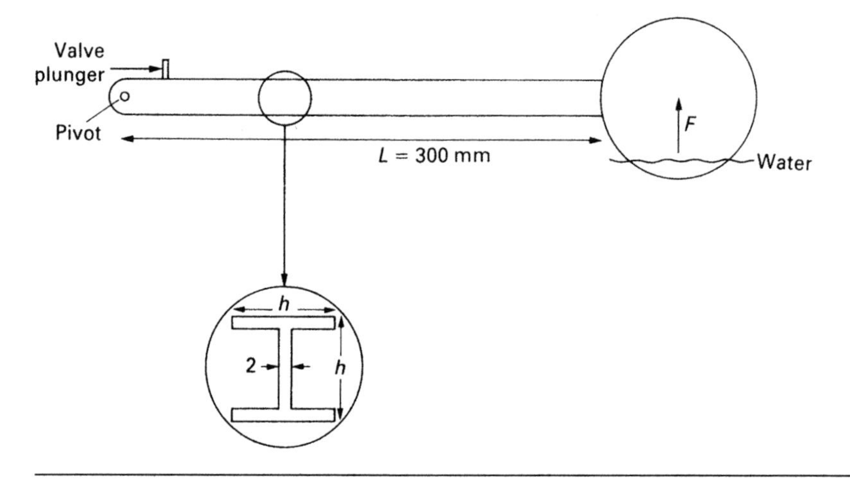

Figure 7.8 Cantilever beam used with a spherical float to operate a valve in a cold water tank. The insert shows a possible cross section for the beam.

We start with analysis for the deflection of an elastic arm. For a point force F on the end of cantilever beam of length L, Appendix C shows that the deflection δ at the free end is

$$\delta = \frac{FL^3}{3EI}$$

where E is the Young's modulus, and I the second moment of area of the beam cross section.

The rule for linearly viscoelastic design is: *Replace 1/E in the elastic formula by the creep compliance J(t), to obtain the time-dependent deflection.* Applying this rule, the cantilever arm deflection is given by

$$\delta(t) = \frac{FL^3 J(t)}{3I} \tag{7.14}$$

To use this equation, the polymer response must be in the linear viscoelastic region.

Figure 7.8 shows a possible cross section for the beam. The 2 mm thick section was chosen so the injection moulding cycle time is short, while the I beam is efficient in bending (Chapter 13). From the linear portion of the isochronous stress–strain curve, the linear viscoelastic compliance is $J(1\ \text{year}) = 3.3 \times 10^{-9}\ \text{m}^2\ \text{N}^{-1}$. Substituting this and the deflection limit in Eq. (7.14), the required second moment of area $I = 5.0 \times 10^{-9}\ \text{m}^4$. This can be provided if the beam height is $h = 18$ mm.

7.3.3 Pseudo-elastic creep design

It is wasteful of material to design a product to be in the linear viscoelastic region. The *pseudo-elastic* design method, for non-linear viscoelastic materials, gives a more reasonable design. The process requires an initial design, which is taken from the previous section. The stages are:

(1) Make an elastic stress analysis of the product, which is an approximation. We calculate the maximum stress σ_m in the beam using Eq. (C.8) of Appendix C

$$\sigma_m = \frac{y_m M_m}{I}$$

where y_m is the maximum distance from the neutral axis and M_m is the maximum bending moment. Substituting for I, $y_m = 9$ mm and $M_m = 5 \times 0.3$ Nm at the left-hand end of the beam; the stress $\sigma_m = 2.65$ MPa. The longitudinal strain e_z varies linearly with the distance y from the neutral surface (Appendix C), whether the material is elastic or viscoelastic. For linearly viscoelastic materials the stress variation is also linear, so the concept of second moment of area remains valid. However, for non-linear viscoelastic materials, the stress variation with y has the same shape as the isochronous stress–strain curve, and the stresses at the upper and lower surfaces are smaller than those calculated by Eq. (C.8).

(2) In the elastic deflection formula replace $1/E$ by the stress-dependent creep compliance to obtain

$$\delta(t) = \frac{FL^3 J(\sigma_m, t)}{3I} \tag{7.15}$$

(3) Calculate the creep compliance, for the design time and σ_m, from the isochronous curve (Fig. 7.7). It is $5.8 \times 10^{-9}\,m^2\,N^{-1}$, 76% higher than the linear viscoelastic value. When the compliance is substituted in Eq. (7.15), the beam deflection of 51 mm is found to be too large. The cross-sectional dimensions are then increased slightly to $h = 20$ mm to meet the new I target and the calculations repeated. As this gives a slightly too low deflection, the design has a slight safety margin.

The pseudo-elastic approach overestimates the real deflection, so leads to a conservative design. It assumes that the product geometry does not change sufficiently under load to invalidate the moment calculations, and that the compressive loads are not high enough to cause buckling. However, creep strains of 1% or 2% in comparatively thin products can lead to large bending or torsional deflections. Consequently, it is common to construct and test product prototypes, rather than rely entirely on a mechanics analysis. Simple shapes can be made by welding flat parts together, while more complex shapes can be made by rapid prototyping (Chapter 5).

7.3.4 Recovery and intermittent creep

When plastics are unloaded, the creep strain is recoverable. This contrasts with metals, where creep strains are permanent. The Voigt linear viscoelastic model predicts that creep strains are 100% recoverable. The fractional recovered strain is defined as $1 - e/e_{max}$, where e is the strain during recovery and e_{max} is the strain at the end of the creep period. It exceeds 0.8 when the recovery time is equal to the creep time. Figure 7.9 shows that recovery is quicker for low e_{max} and short creep times, i.e. when the creep approaches linear viscoelastic behaviour.

Many products are only loaded intermittently. It is impossible to produce data for all loading histories, but experiments have been made with regular loading/unloading cycles. Figure 7.10 compares creep strains, as a function of the cumulative creep time, for continuous loading and a 6 h per day loading cycle. The polypropylene (PP) recovers significantly in each 18 h recovery period. The peak strain at the end of each loading period increases only slightly with the cumulative loading time. In continuous loading, the creep strain increases more rapidly with time, and the creep time per day accumulates four times as fast. Consequently, the intermittently loaded sample has half the creep strain after 1 year that the continuously loaded polymer has after 3 months. In general intermittently loaded products are unlikely to fail from excessive creep strain.

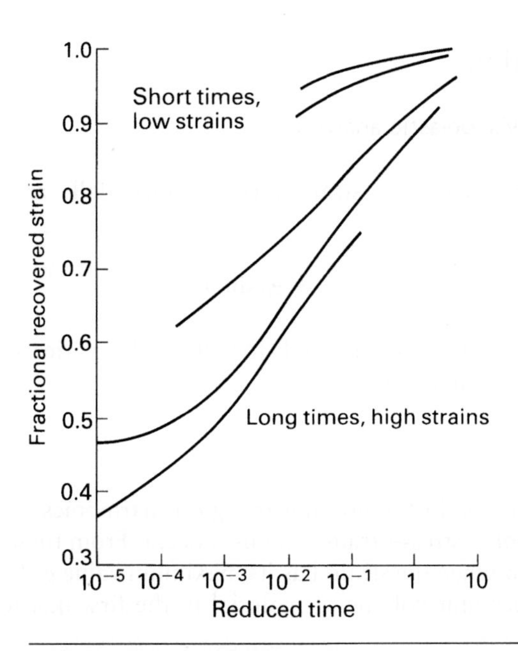

Figure 7.9 Fractional recovered strain vs. reduced time (= recovery time/creep time) for acetal copolymer at 20 °C and 65% humidity. The data from different creep times and stresses do not superimpose (from *Thermoplastics and Mechanical Engineering Design*, ICI Plastics Division, booklet G117).

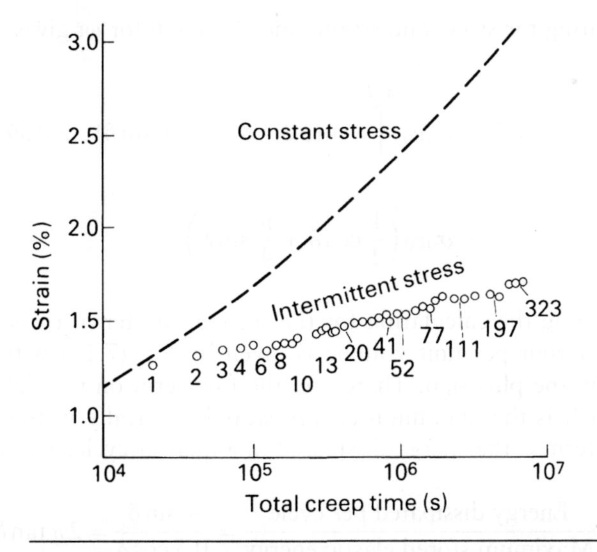

Figure 7.10 Comparison of continuous and intermittent creep (6 h per day) of polypropylene at 20 °C and 10 MPa stress. The maximum strain at the end of each intermittent creep period is plotted against the cumulative creep time (from *Plastics and Mechanical Engineering Design*, ICI Plastics Division, booklet G117).

7.4 Cyclic deformation

7.4.1 Linear viscoelastic analysis

We analyse the effect of a strain varying sinusoidally at an angular frequency ω

$$e = e_0 \sin \omega t \tag{7.16}$$

For a linear viscoelastic material, the stress also varies sinusoidally, but leads in phase by an angle δ

$$\sigma = \sigma_0 \sin(\omega t + \delta) \tag{7.17}$$

At higher strains, in the non-linear region, harmonics can be generated and more complex stress–strain responses occur. From these equations, the stress is $\sigma_0 \cos \delta$ when the strain has its maximum value e_0 (Fig. 7.11a). The energy input per unit volume of material in the first quarter of the strain cycle is

$$W = \int_0^{e_0} \sigma \, de$$

Substituting for stress and strain, and writing θ for ωt gives

$$W = \sigma_0 e_0 \int_0^{\pi/2} (\sin\theta \cos\delta + \cos\theta \sin\delta) \cos\theta \, d\theta$$

$$= \sigma_0 e_0 \left(\frac{1}{2} \cos\delta + \frac{\pi}{4} \sin\delta \right) \tag{7.18}$$

Repeating the calculation for the second quarter of the strain cycle, the energy output per unit volume is given by Eq. (7.18) with a minus sign replacing the plus sign. Therefore, the first term on the right-hand side of Eq. (7.18) is the maximum energy stored elastically in the cycle, and the second term is the energy dissipated in a quarter cycle. We can write

$$\frac{\text{Energy dissipated per cycle}}{\text{Maximum stored elastic energy}} = \frac{\pi \sin\delta}{0.5 \cos\delta} = 2\pi \tan\delta \tag{7.19}$$

Tan δ is a useful parameter of energy dissipation.

Dynamic mechanical thermal analysers (DMTA) are semi-automated machines for determining the Young's or shear modulus of polymers as a

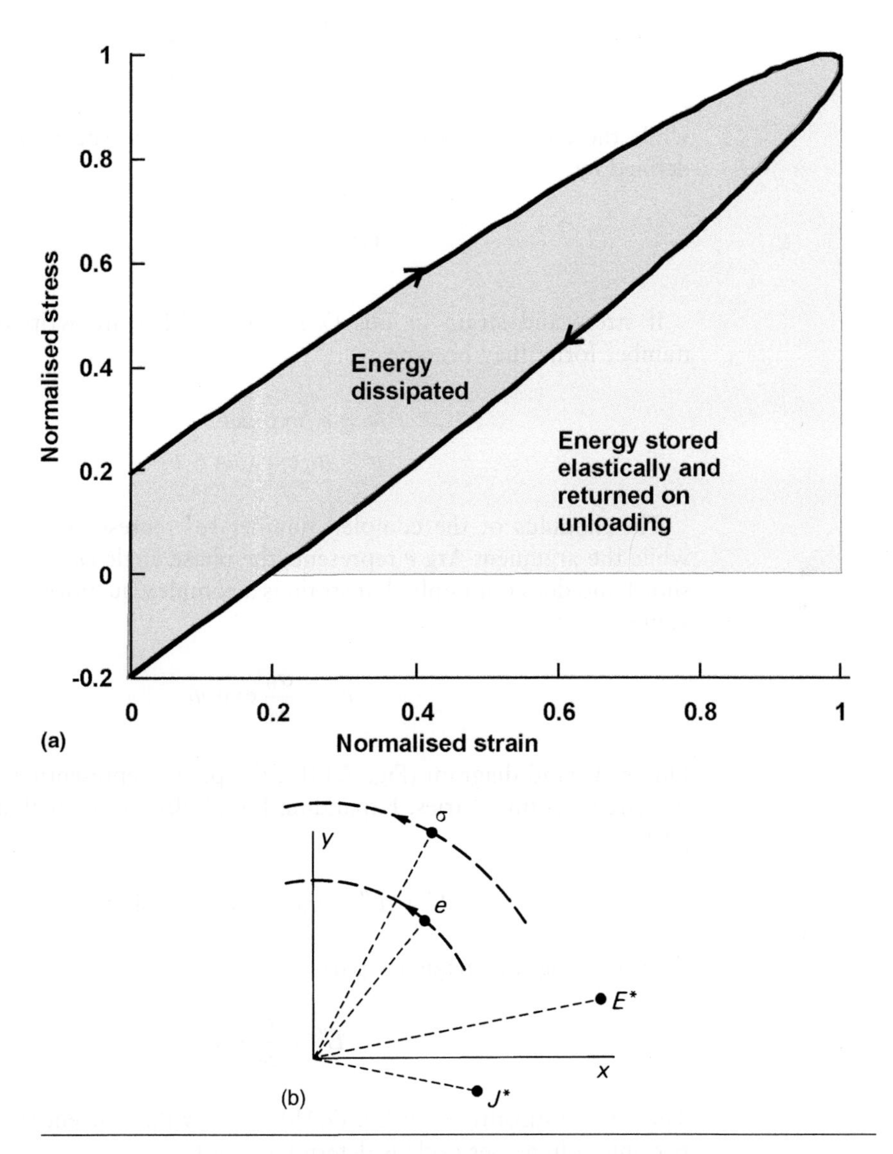

Figure 7.11 (a) Stress vs. strain for the first half cycle, showing the stored elastic energy, and the energy dissipated in the first half cycle. Drawn for $\tan\delta = 0.2$. (b) Positions of the stress, E; strain, e; Young's modulus, E^* and compliance, J^* in an Argand diagram, for one point on the ellipse in part (a).

function of temperature, or a limited range of frequencies, usually 0.1–100 Hz. The sample is loaded in compression, bending or shear to a low strain level, and the sinusoidal load signal is compared with the displacement signal. In such tests, the components E' and E'' of the complex Young's modulus E^* are defined by

$$E^* = E' + iE'' = \frac{\sigma}{e} \tag{7.20}$$

while the components G' and G'' of the complex shear modulus G^* are defined by

$$G^* = G' + iG'' = \frac{\tau}{\gamma}$$

If stress and strain in Eqs (7.16) and (7.17) are written in complex number form, they become

$$\begin{aligned} e &= e_0 \exp i\omega t \\ \sigma &= \sigma_0 \exp (i\omega t + \delta) \end{aligned} \tag{7.21}$$

The modulus of the complex number $|e|$ represents the amplitude e_0, while the argument Arg e represents the phase angle ωt. The mathematical shorthand does not imply that strain is a complex quantity. E^* has the fixed values

$$E^* = \frac{\sigma_0}{e_0} \exp i\delta \tag{7.22}$$

On an Argand diagram (Fig. 7.11b), the points representing e and σ trace out circles as time varies. Expanding Eq. (7.20) into its real and imaginary parts

$$E' + iE'' = \sigma_0/e_0(\cos\delta + i\sin\delta) \tag{7.23}$$

then equating the imaginary parts, gives

$$E'' = \frac{\sigma_0}{e_0} \sin\delta \tag{7.24}$$

This can be substituted in Eq. (7.18) to show that the energy W dissipated per unit volume per cycle is determined by E''

$$W_{\text{dis}} = \pi e_0^2 E'' \tag{7.25}$$

Figure 2.18 showed how G' varies with temperature for plasticised PVCs. For an HDPE (Fig. 7.12), there is a general decrease in the storage Young's modulus E' over the temperature range, but peaks occur in E'' near the polymer T_g at $-120\,^\circ$C and in the melting range from $50\,^\circ$C upwards. The results are a function of the sinusoidal frequency, as a result of the time–temperature interdependence.

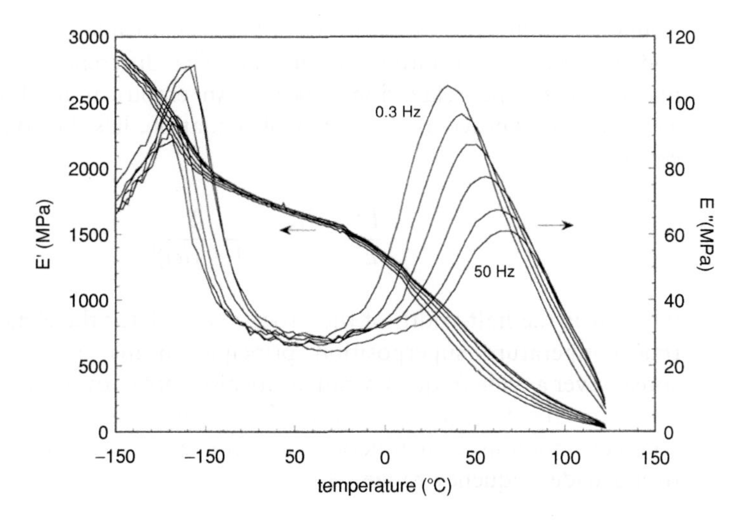

Figure 7.12 Temperature variation of the Young's modulus of HDPE (Pegoretti, A. *et al.*, *Comp. Sci. Tech.*, **60**, 1181, 2000) © Elsevier.

It is useful to see how the Voigt linear viscoelastic models of Section 7.2 behave with a sinusoidal strain input. When the strain variation equation (Eq. 7.21) is substituted in the model constitutive equation (Eq. 7.2), the stress is given as

$$\sigma = Ee + \eta i\omega e \tag{7.26}$$

Therefore, the complex compliance J^* (the reciprocal of E^*) of the Voigt model is

$$J^* = \frac{e}{Ee + \eta i\omega e} = \frac{1}{E(1 + i\omega\tau)} \tag{7.27}$$

so

$$J' - iJ'' = \frac{1 - i\omega\tau}{E(1 + \omega^2\tau^2)} \tag{7.28}$$

where τ is the retardation time η/E. The single Voigt model is a poor representation of polymer behaviour because $\tan\delta$ is predicted to increase linearly with frequency. However, the model of Fig. 7.1b, with a number of Voigt elements in series, is better. The complex compliances of the elements can be added, so a summation sign is placed in front of the right-hand side of Eq. (7.27), for a range of τ_i values.

The Zener viscoelastic model is a modification of the Voigt model, in which a spring is placed in series with the dashpot. This causes the

Young's modulus relaxation to occur between a value E_0 at zero frequency, and a value E_∞ at infinite frequency. The difference $E_\infty - E_0$ is the magnitude of the relaxation process. An empirical modification of the Zener model, known as the Cole-Cole equation, has the complex modulus equation

$$\frac{E^* - E_\infty}{E_0 - E_\infty} = \frac{1}{1 + (i\omega)^\alpha} \tag{7.29}$$

where α is the half-width of the process ($\alpha = 1$ for the Zener model). The time–temperature superposition principle means that damping peaks, spread over a wide frequency range, are also spread over a wide temperature range for fixed frequency testing. It is more convenient to measure the complex modulus as a function of temperature, than it is to measure it over a wide frequency range.

7.4.2 Isolation of machine vibration

Vibration is caused by out-of-balance rotating or reciprocating machinery, or by impulse loading (the Fourier transform of a step loading contains a wide range of frequencies). Resonance will occur if the natural frequency of the structure is excited. We consider the vertical vibration of a machine of mass m, supported on a number of mountings of total spring constant k. It is treated as a mass supported on a damped spring. The single degree of freedom is the displacement x of the mass (Fig. 7.3a). When the vibration applies a force F_0 to the mass, the transmissibility T is defined in terms of the amplitude ratio of F_0 and the force F_T transmitted to the support surface. Alternatively, the applied and transmitted displacements x_0 and x_T can be used

$$T = \frac{|F_T|}{|F_0|} = \frac{|x_T|}{|x_0|} \tag{7.30}$$

The differential equations for the forces, in terms of the spring constant k, the dashpot constant c and the time derivatives $\dot{x}$ and $\ddot{x}$ of the displacement, are

$$\frac{F_T}{F_0} = \frac{kx + c\dot{x}}{kx + c\dot{x} + m\ddot{x}}$$

Substituting the sinusoidal displacement $x = \exp(i\omega t)$ gives

$$\frac{F_T}{F_0} = \frac{k + ic\omega}{k + ic\omega - m\omega^2}$$

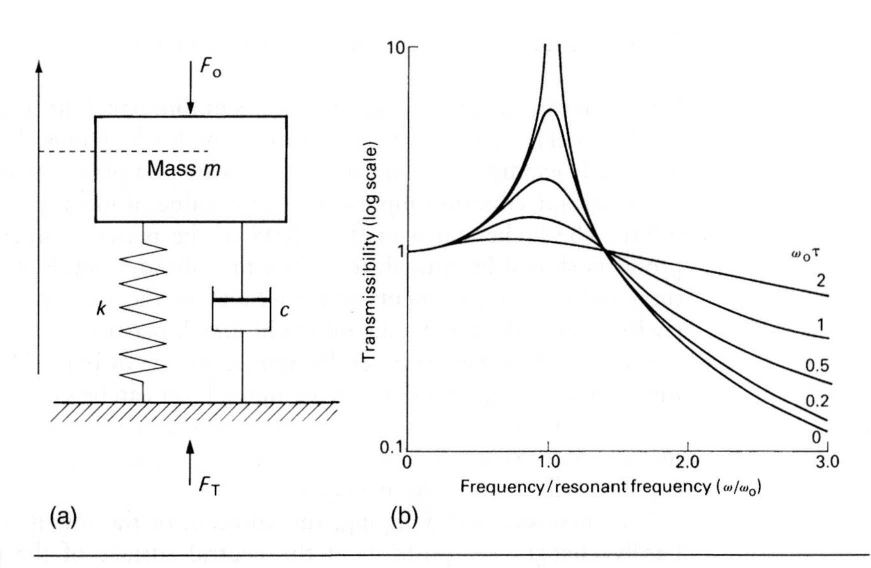

Figure 7.13 (a) Isolation of a vertical vibration of a mass m by a damped spring. (b) Transmissibility as a function of the normalised vibration frequency, for various values of the parameter $\omega_0 t$.

The modulus of the complex quantities is taken, yielding

$$T^2 = \frac{k^2 + c^2\omega^2}{(k - m\omega^2)^2 + c^2\omega^2}$$

The result is re-expressed in terms of the resonant frequency $\omega_0{}^2 = k/m$ of the undamped system, and the characteristic damping time $\tau = c/k$ of the spring–dashpot combination, giving

$$T^2 = \frac{1 + \omega^2\tau^2}{(1 - \omega^2/\omega_0^2)^2 + \omega^2\tau^2} \tag{7.31}$$

Figure 7.13b shows the transmissibility versus frequency, for various values of $\omega_0\tau = c/(mk)^{0.5}$. Strong damping of the resonance requires $\omega_0\tau > 1$. However, to isolate the vibration of a machine from the floor, the resonant frequency must be less than 40% of the vibration frequency, and the best performance occurs with little or no damping. Rubber and steel laminated bearings (Section 4.2.1) have low resonant frequencies, and the moderate $\tan\delta$ value of the rubber (0.07–0.30) prevents severe resonance on start up.

7.4.3 Constrained layer damping of metal panels

Vibration can excite bending resonances in thin panels at frequencies in the 100 Hz–5 kHz range that generate unacceptably high noise levels. The most noticeable examples are metal panels in cars and planes, which are subject to significant vibration inputs. The $\tan\delta$ value of metal panels is typically 0.001. The higher $\tan\delta$ of 0.01–0.05 of the plastic casings of consumer products should be enough to reduce the vibration levels. However, structural stiffness requirements may necessitate the use of a metal, or the environment, like hot engine oil might attack plastics.

A plastic damping layer can be applied to one or both sides of the metal (unconstrained damping), or the damping layer can be sandwiched between the panel and a thin metal skin (constrained damping; Fig. 7.14). Often the metal skin, typically 0.5 mm thick and the polymer layer 0.2 mm thick are supplied with a self-adhesive layer.

For unconstrained damping, the addition of the low-modulus polymer hardly changes the position of the neutral surface of the panel. Consequently, the strain in the polymer alternates from tensile to compressive as the panel bends. The energy absorbed by the polymer per cycle is proportional to its E'' value.

The constrained layer geometry is a form of asymmetric sandwich beam (Section 4.3.1). The damping treatment may cover the whole of the panel or there may be a series of patches which may be applied. This affects the geometry of deformation. Given the complex nature of vibration modes in structural panels (higher modes have closer spaced nodes and higher frequencies), it is not easy to give general rules for the optimisation of damping.

For patches of length less than the vibration wavelength, beam end effects influence the deformation. Deformation can be by extension of the metal skin, by shear of the core, or a combination of the two. For a beam of length L and width W, the tensile stiffness (force/extension) of a skin of thickness t and Young's modulus E_s is $E_s wt/L$. Similarly the core, of shear modulus G_c and thickness b, has a shear stiffness $G_c wL/b$. Consequently, the average shear strain in the polymeric core can be high if the skin tensile stiffness is much higher than the core shear stiffness.

For large patches, or a continuous layer of constrained damping, the optimisation must use computer methods. Figure 7.15 shows the effect of

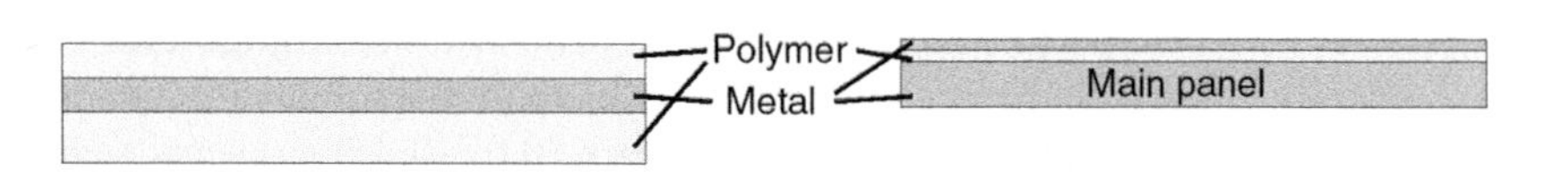

Figure 7.14 Sections of (a) Floor pan of a car, showing the external and internal damping layers. (b) Constrained layer damping of a flat metal panel.

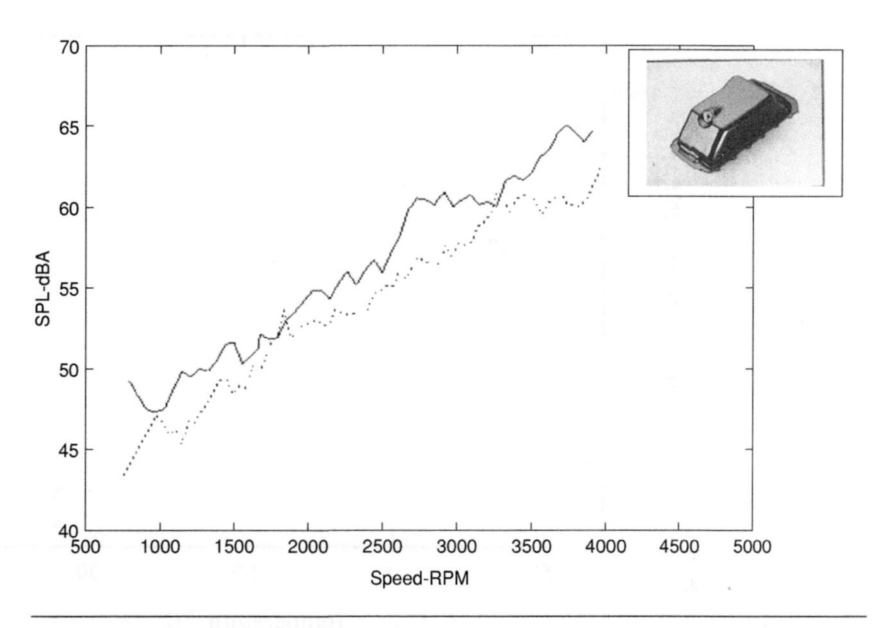

Figure 7.15 Sound level at the driver's ear for a car fitted with a regular and with a damped oil pan (Rao, M. D., *J. Sound Vibration*, **262**, 457, 2001) © Elsevier.

using a damped oil pan on the measured sound level at the driver's ear in a car. There is a significant reduction across the range of engine speeds. Such damping measures are part of the gradual improvement of noise and vibration in cars.

7.4.4 High damping polymers

Vibration damping applications require polymers with high $\tan\delta$ values. If damping is required at a fixed temperature, a polymer with its T_g at that temperature will have a high $\tan\delta$ over a reasonably wide frequency range. However, it is not easy to achieve high damping over a wide temperature range, as required for a car door panel. A blend of two incompatible polymers has separate $\tan\delta$ peaks at the two T_gs, but neither peak will be as high as in the original polymers. Improvements in vibration damping have concentrated on interpenetrating networks (IPN), which have two-phase structures on a 100 nm scale. A complex process of crosslinking produces a composition gradient on a smaller scale. The aim is to achieve a single broad $\tan\delta$ peak, as in Fig. 7.16. There is a relationship between the slope dG/dT and the magnitude of $\tan\delta$. There is a peak in the latter when the modulus slope is greatest.

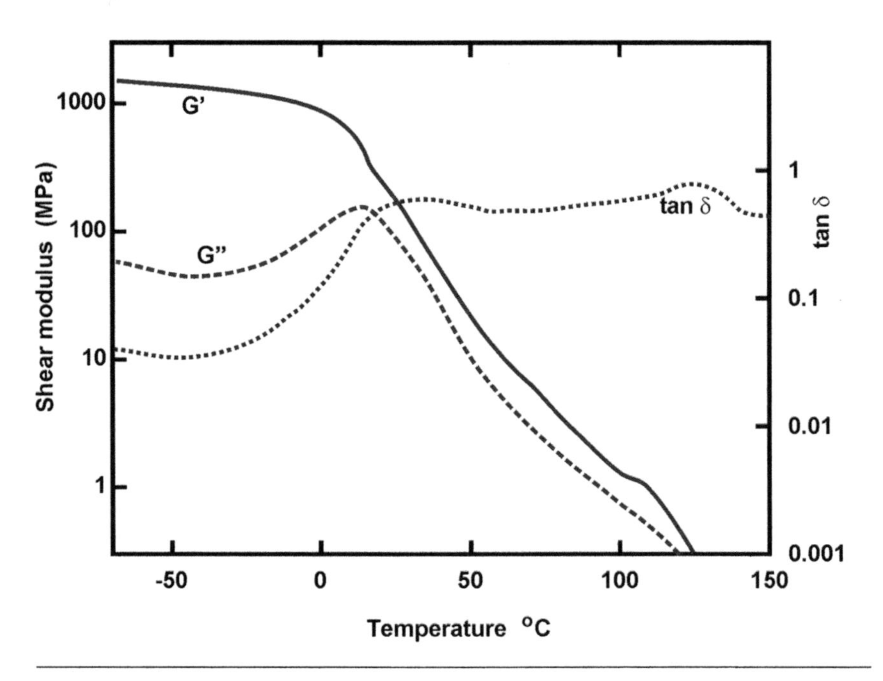

Figure 7.16 Complex shear modulus and tan δ vs. temperature for an IPN containing inverted core-shell particles of SAN/poly ethyl hexyl methacyrlate (El-Aasser, M. S. *et al.*, *Colloids Surf. A*, **153**, 241, 1999) © Elsevier.

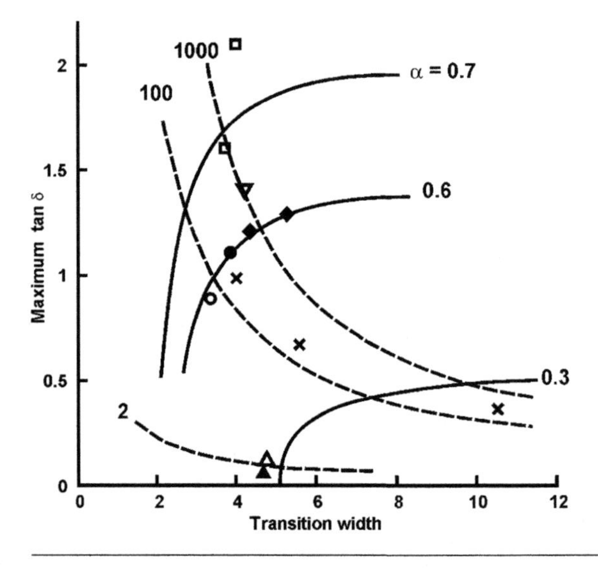

Figure 7.17 Maximum damping and transition half-width for commercial damping polymers x polyurethanes, (Pritz, T., *J. Sound Vibration*, **246**, 265, 2001) © Elsevier.

Figure 7.17 shows the peak $\tan\delta$ values for a range of commercial damping polymers, as a function of the transition half-width (the factor by which frequency must change for $\tan\delta$ to drop to half the peak value). The dashed curves are theoretical relationships for different values of the transition magnitude, while the solid curves are for constant values of the Cole-Cole transition width factor α. The most promising materials have IPN structures.

Chapter 8

Yielding

Chapter contents

In this chapter, a brief survey of the molecular mechanisms of yielding is followed by analyses of yield phenomena in tension, compression and bending. Some deformation mechanisms are specific to plastics products; for instance buckling in compression rather than uniaxial yielding. Localised yielding will be analysed, to explain scratching and film penetration. Rate dependence, introduced in the last chapter, also affects yield stresses. Consequently, products must be designed using the yield stress for the appropriate timescale. Orientation hardening, which enhances the strength of some products, will be explored. Finally, yielding mechanisms on a micron scale will be described; crazing in bulk polymers and yielding in the thin faces of polymer foams.

8.1 Molecular mechanisms of yielding

The molecular mechanisms of yielding in polymers are less well established than that of dislocation motion in metal crystals, because the microstructure is complex, and no techniques exist to directly observe deformation processes on a scale <10 nm. However, deformation processes are expected to minimise the breaking of covalent bonds, which are much stronger than the van der Waals forces between chains. In isotropic polymers, the yield stress can only be changed by a small factor by heat treatment. Consequently, there is less need to understand the yielding mechanisms. This contrasts with metals, where the yield stress can be significantly increased by minor additions of a second element, mechanical working (the rolling of plate or the drawing of wire) or heat treatment (to precipitate a second phase).

8.1.1 Glassy polymers

Dislocation motion, which allows yield at low average strains, can only occur in materials with regular crystal lattices. Dislocations have no meaning in a polymer glass. However, yield can proceed by the propagation of a small, highly sheared region through the material. For some glassy polymers, the initial stages of compressive yielding involve inhomogeneous deformation—the propagation of *shear bands* within an undeformed bulk. There is a shear strain of ~1 unit in the bands, which are approximately 10 μm thick (Fig. 8.1). Shear band patterns will be analysed in Chapter 9 to estimate the stresses for crack initiation. However, many glassy polymers yield homogeneously in compression, as the conditions for strain localisation (a reduction in the yield stress—see Section 8.2.1) are not met.

During the yielding process:

(a) The free volume of the glass increases. Section 6.3.1 showed how quenched PVC has a lower initial yield stress than a PVC with a near-equilibrium structure. The diffusion constant increases to that typical at T_g.

Figure 8.1 Pattern of shear bends in a block of polystyrene, under a strip indenter.

(b) Conformational changes occur in the polymer chains. Molecular modelling suggests that the relative number of *trans* rotational isomers increases slightly, as the chains stretch from the initial random coil shape (Fig. 8.2).

(c) Some chains break, creating free radicals that can be detected using electron spin resonance.

(d) The entanglement network, created by the inter-twining of neighbouring polymer chains, survives. This can be demonstrated by heating a necked tensile specimen of a glassy polymer to a temperature just exceeding T_g, whereupon it slowly regains its original shape. The tensile specimen should be cut from compression moulded or extruded sheet, because injection mouldings contain molecular orientation. The entanglement network reverts to its maximum entropy, unstrained state. However, details of how chains move relative to one another, have not been determined.

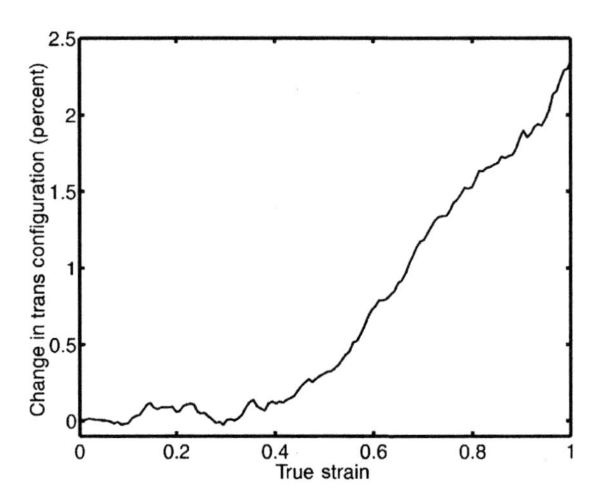

Figure 8.2 Per cent of *trans* isomers vs. strain from molecular modelling of glassy polyethylene (Capaldi, F. M. *et al.*, *Polymer*, **45**, 1391, 2004) © Elsevier.

8.1.2 Semi-crystalline polymers

In semi-crystalline polymers, the crystal lamellae play a part in the yielding process. They form too large a fraction of the microstructure, and are too well connected by inter-lamellar links, to act as rigid inclusions in a deformable matrix. However, the lamellae are too thin for their internal structure to be observed during yielding. Consequently, the crystal deformation mechanisms are inferred from electron microscopy and X-ray diffraction of deformed samples. Deformation mechanisms that occur in metal crystals also occur in polymer crystals: Slip, twinning and stress-induced phase changes. Figure 8.3 shows a side view of a lamellar crystal that is undergoing 'chain-slip'—the slip plane contains the covalently bonded polymer chains and the slip direction is parallel to the crystal **c** axis. If this process occurs on a number of parallel, but widely separated, slip planes, the lamellar crystals eventually break up into blocks, without polymer chains breaking. This happens at tensile strains of 200–600%, when the polymer is drawn into a fibre (Fig. 8.15).

For semi-crystalline polymers above their glass transition temperatures, the volume fraction crystallinity has the largest effect on the yield stress. Figure 4.20b showed a near linear variation of yield stress with crystallinity, for some commercial polyethylenes. Although, for metals, a decrease in grain size increases the yield stress, there is no parallel effect of spherulite size on the yield stress of a polymer. Grain boundaries in metals are major obstacles to dislocation movement, but polymer yield stresses are determined by the sub-spherulitic microstructure. Average spherulite sizes larger than 50 μm are usually avoided because the polymer tends to become

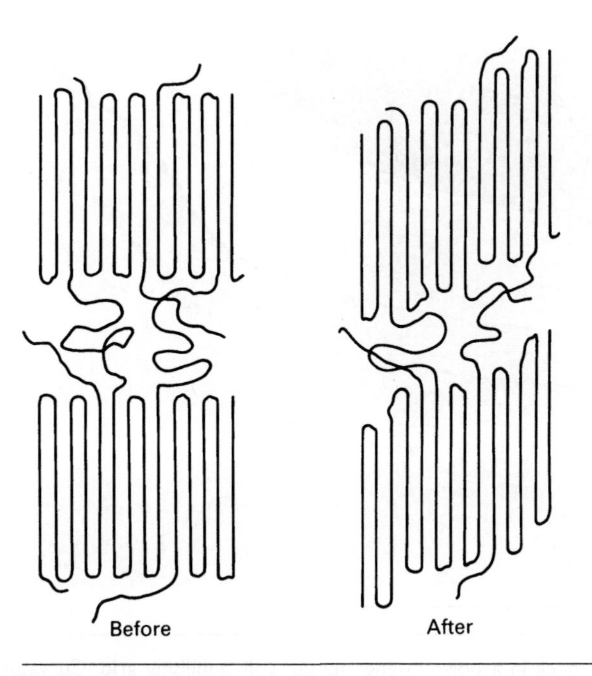

Figure 8.3 'Chain-slip' in lamellar crystals, seen edge-on. The slip direction is parallel to the **c** axis and occurs on many parallel slip planes.

brittle. Such large spherulites only form if crystallisation is slow or nucleating agents are absent. Fracture occurs at the spherulite boundaries either because weak low molecular mass fractions migrate away from the growing spherulites, or because the volume contraction on crystallisation causes voids to form at spherulite boundaries.

8.2 Yield under different stress states

8.2.1 Tensile instability and necking

Most mechanics-of-solids textbooks analyse the necking instability that occurs in a tensile test. The analysis will be extended here to deal with products made by the stable propagation of a neck (textile fibres and the 'Tensar' soil stabilising grids shown in Fig. 8.4). The analysis starts with two assumptions:

(a) The yield processes do not affect the density. However, if crazing occurs, the density decreases. Hence, the theory does not apply to polymers that craze (Section 8.5.1).
(b) The slope of the true stress σ versus true strain ε curve decreases as the true strain increases. This is no longer true at high strains for most polymers.

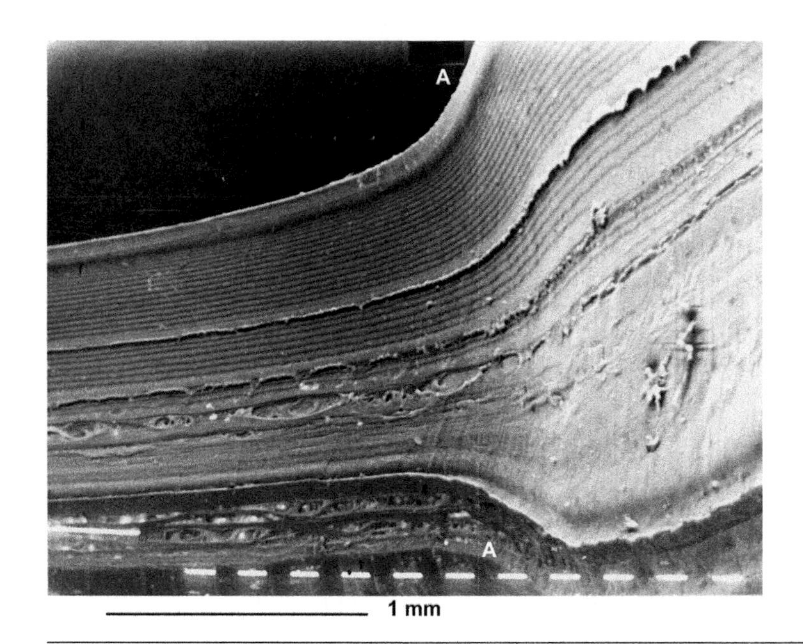

Figure 8.4 A tensile neck in a polyethylene 'Tensar' soil stabilising grid. Curvature of the principal stress directions increases the average yield stress on the section AA.

While the specimen is extending uniformly (Fig. 8.5), the true tensile strain can be defined by

$$\varepsilon \equiv \ln\left(\frac{L}{L_0}\right) \tag{8.1}$$

where L is the current and L_0 the initial length. A slice across the specimen, of initial cross-sectional area A_0, has a constant volume so the cross-sectional area A is given by

$$\frac{A_0}{A} = \frac{L}{L_0}$$

Taking natural logarithms and using Eq. (8.1) gives

$$\ln\left(\frac{A_0}{A}\right) = \varepsilon$$

Differentiation leads to

$$\frac{d\varepsilon}{dA} = -\frac{1}{A} \tag{8.2}$$

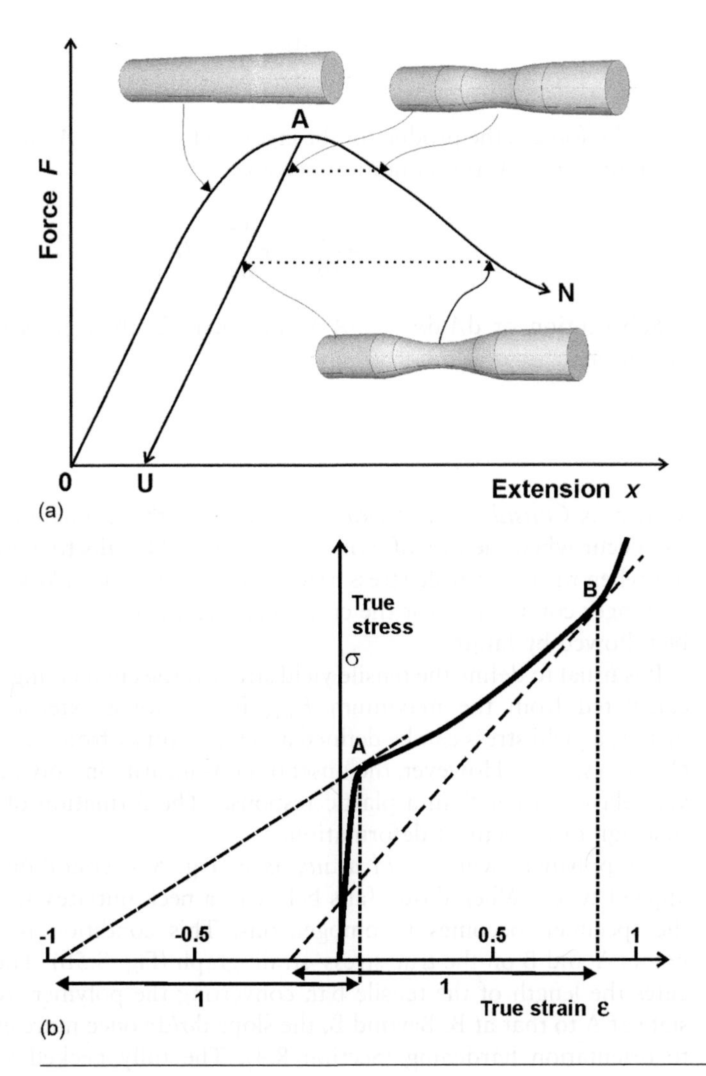

Figure 8.5 (a) Force vs. elongation in a tensile test. From 0 to A the specimen extends uniformly. Beyond A the end parts unload elastically along the path AU, while the necked portion proceeds along path AN. (b) True stress vs. true strain relationship for a polymer that cold draws; at A, a neck forms and at B, it stabilises.

At the peak force (position A in Fig. 8.5a), there are two possibilities for the next strain state: Elastic unloading along path AU, and further plastic straining along the path AN. A non-uniform strain state develops, as parts of the specimen elastically unload, and the plastic strain in one region increases to form a *neck*. The plastic deformation of the neck is partially driven by elastic energy release from the rest of the specimen. The condition that A is at the maximum in the force–extension or force–strain curve can be written

$$\frac{dF}{d\varepsilon} = 0 \tag{8.3}$$

As the force is the product of the cross-sectional area A and the true stress σ, defined as F/A, this condition becomes

$$A\frac{d\sigma}{d\varepsilon} + \sigma\frac{dA}{d\varepsilon} = 0$$

Substitution of $dA/d\varepsilon = -A$ from Eq. (8.2) then gives the instability condition

$$\frac{d\sigma}{d\varepsilon} = \sigma \tag{8.4}$$

known as *Considere's criterion*. It shows that the ultimate tensile strength can occur when the rate of work hardening $d\sigma/d\varepsilon$ falls to a critical value σ, not when the true tensile stress reaches a critical value. Work hardening can no longer compensate for the cross-sectional area decrease, so necking can be followed by failure.

It is usual to define the tensile yield stress as the engineering stress F_{max}/A_0 calculated from the maximum F_{max} in the force–extension curve. For metals, a yield stress can be defined as a 0.2% offset from the initial straight elastic response. However, the onset of non-linearity in polymers indicates a viscoelastic rather than a plastic response. The formation of a neck is the first sign of permanent deformation.

For polymers which *cold draw*, as in Fig. 8.5, condition (8.4) can be applied twice. When $d\sigma/d\varepsilon$ falls below σ, a neck initiates and the strain in the specimen becomes inhomogeneous. This condition is true between points A and B on the true stress–strain graph (Fig. 8.5b). The neck propagates the length of the tensile bar, converting the polymer from the strain state at A to that at B. Beyond B, the slope $d\sigma/d\varepsilon$ once more exceeds σ, due to orientation hardening (Section 8.4). The fully necked specimen then strains homogeneously again.

The total differential $d\sigma/d\varepsilon$ can be split into four partial differentials, representing factors that affect neck stability

$$\frac{d\sigma}{d\varepsilon} = \frac{\partial\sigma}{\partial\varepsilon} + \frac{\partial\sigma}{\partial\dot{\varepsilon}}\frac{d\dot{\varepsilon}}{d\varepsilon} + \frac{\partial\sigma}{\partial T}\frac{dT}{d\varepsilon} + \frac{\partial\sigma}{\partial G}\frac{dG}{d\varepsilon} \tag{8.5}$$

In an incipient neck, the strain rate ε increases. The yield stress increases with increasing strain rate (Eq. 8.17), so the second term on the right-hand side of Eq. (8.5) is a stabilising influence. The work input to the yielding process is converted to heat, which is only slowly conducted down the specimen or convected into the surrounding air. As the yield stress is a decreasing function of temperature T, the third term in Eq. (8.5) is a destabilising influence. The final geometrical factor G is associated with the shape of the neck. For a material obeying Tresca's yield criterion (Section 8.2.6), the curvature of the lines of principal stress in the neck (Fig. 8.4) cause the average stress across that cross section to be

$$\overline{\sigma} = 2k\left(1 + \frac{a}{3R}\right) \tag{8.6}$$

where $2k$ is the uniaxial tensile yield stress, and $2a$ the thickness of the flat strip specimen. The radius of curvature R is negative at the initial shoulder of the neck, but positive at the ends of the fully formed neck (A in Fig. 8.5), so the fourth term in Eq. (8.5) is a destabilising factor as the neck forms, but is stabilising later.

When a neck propagates at the order of $1\ \mathrm{mm\,s^{-1}}$ along a specimen, the large geometry changes and heating means that the true stress and temperature differ from the nominal values. Computer modelling predicts that the maximum temperature in the neck is an almost linear function of the specimen extension speed. Figure. 8.6a shows the temperature distribution along a 0.5 mm thick PET sample, extended at $1.6\ \mathrm{mm\,s^{-1}}$, measured with an infrared camera. The peak temperature rise of 40°, which occurs just after necking, is sufficient to reduce the yield stress. Careful control of heat transfer is necessary when the necking process is used to make highly oriented products. If the neck propagation is too fast, the heat generated can soften the polymer to the point where *thermal runaway* occurs. Figure 8.6b shows such a failure in a thin-walled polyethylene liquid container that was dropped from 5 m onto a hard surface.

8.2.2 Yield in bending

A beam of ductile material can be bent until it stays permanently deformed. The yielded region acts as a very stiff hinge, so is referred to as a *plastic hinge* (Fig. 8.7). In this region, one side of the beam yields in tension while the other side yields in compression. The material is assumed not to work harden after yielding (a good approximation for polymers for strains up to 50%), so the longitudinal stress is σ_0 on the tensile side and $-\sigma_0$ on the compressive side, where σ_0 is the initial yield stress. The longitudinal strain e varies linearly through the beam according to Eq. (C.1) of Appendix C. Hence, yielding spreads inwards from the beam surfaces as the radius of curvature R decreases. Figure 8.7a shows a partly yielded beam in three-point bending, where the bending moment M increases linearly from the ends to the central point. When the two yielded zones meet at the neutral surface, the stress distribution is shown in Fig. 8.7b, and the bending moment at the central load point reaches a limiting plastic moment M_{pl}. For a beam, w wide and d deep, M_{pl} can be calculated by summing the internal moments of the forces acting on the cross section, as

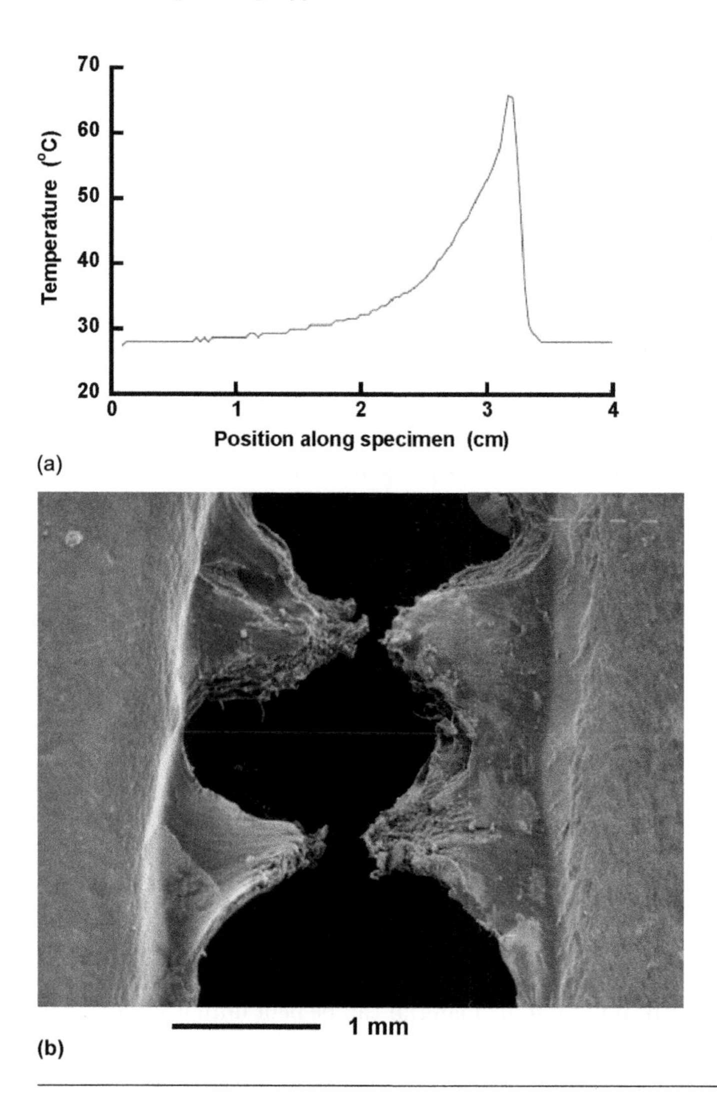

Figure 8.6 (a) Temperature distribution along a PET tensile specimen in which a neck is propagating (Toda, A. *et al.*, *Polymer*, **43**, 947, 2002) © Elsevier. (b) Unstable neck in the wall of an HDPE container, which has been impacted.

$$M_{\mathrm{pl}} = \frac{wd^2}{4}\sigma_0 \tag{8.7}$$

Necking will not occur on the tensile side of the beam, because of the support of the compressive side. For metals, the initial stages of yielding occur as in Fig. 8.7a, and the beam remains very slightly bent if the loads are removed. There is no evidence of permanent deformation in polymer beams before a plastic hinge forms. The non-linearity in the early part of the

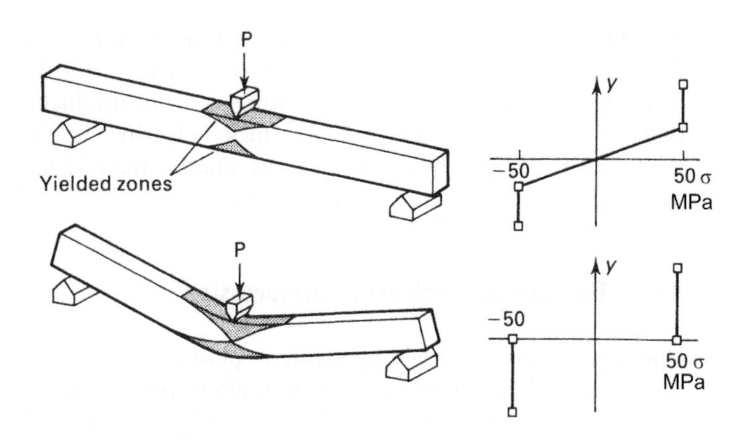

Figure 8.7 Yielded zones in a three-point bend test when: (a) The centre of the beam remains elastic and (b) a plastic hinge forms. The stress distributions, on the beam section under the load *P*, are shown for a material of infinite Young's modulus, and constant yield stress of 50 MPa.

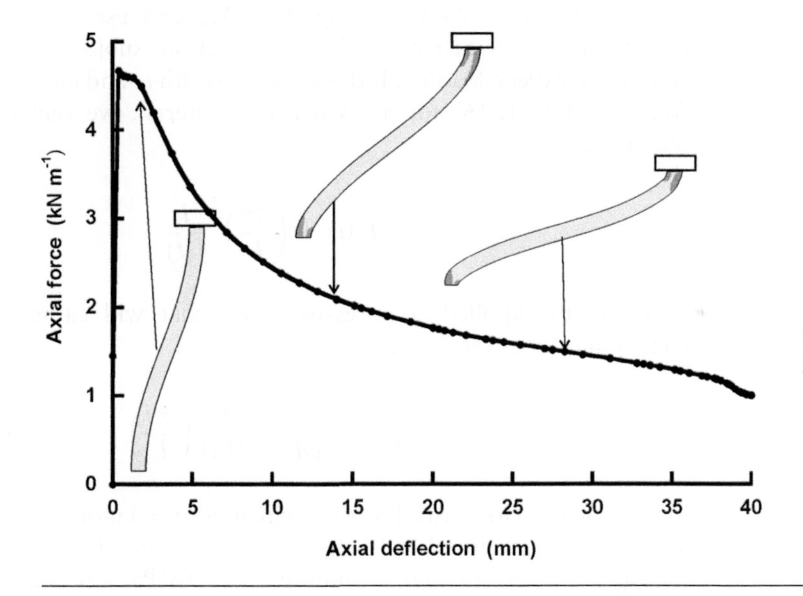

Figure 8.8 Predicted axial force vs. deflection for a 100 mm long strut. The yielded regions in the half strut are shown in dark grey, at three deflections.

force–deflection relationship is due to non-linear viscoelasticity. Since polymer yield strains are approximately 5%, there is always a small elastic region near the neutral surface in the plastic hinge region (Fig. 8.8).

The recommended short-term surface strains that can be used without causing yielding are higher for semi-crystalline plastics (ranging from 4% for polyamides to 8% for polyethylene) than for glassy plastics (1.8% for polystyrene to 4% for polycarbonate). To cause these large surface strains, the beam must be bent to a small radius of curvature R. Equation (C.3) of Appendix C gives the radius of curvature as the beam thickness d divided by twice the yield strain. For a 2 mm thick polyethylene, the deflections involved in bending a beam to a radius smaller than 12 mm would be unacceptable for the function of any product.

8.2.3 Buckling and yielding in compression

Uniform compressive yielding is rare in plastics products, which are usually thin-walled. It is more likely that regions under excessive compressive forces, such as the vertical sides of a bottle crate at the base of a stack, will fail by elastic *buckling* (Section C.1.5 of Appendix C). These regions are built into the rest of the moulding, so their ends cannot rotate or move laterally. The simple Euler buckling theory predicts that a built-in strut of length L collapses at a critical axial force F_c given by Eq. (C.16), and the buckling mode is shown in Fig. C.3. We can use this equation for a polyethylene strut of rectangular cross section, subjected to 3 months of compressive creep loading. If the strut has width w and depth d, $I = wd^3/12$. Adapting Eq. (C.16) for a viscoelastic material, we replace E by $1/J(t)$, obtaining

$$F_c(t) = \left(\frac{2\pi}{L}\right)^2 \frac{I}{J(t)} \tag{8.8}$$

Hence, the applied compressive stress that will cause buckling of a rectangular cross section strut at time t is

$$\sigma_b(t) = \frac{F_c(t)}{wd} = \frac{1}{3J(t)}\left(\frac{\pi d}{L}\right)^2 \tag{8.9}$$

This equation for elastic buckling with mode $n = 1$ is only valid for $L/d > 33$. Substituting the values $L = 100$ mm, $d = 3$ mm and $1/J$ (3 months) $= 0.3$ GPa for a polyethylene gives σ_b (3 months) $= 0.9$ MPa, low compared with the creep rupture stress (Section 8.3.1) of about 10 MPa.

Euler buckling theory predicts collapse at a constant force. However, finite element analysis (FEA) shows that the onset of buckling causes the load bearing capacity to decrease (Fig. 8.8). At high axial deflections, plastic hinges develop at mid-length and the ends of these slender struts.

For products under a constant applied load, such strut collapse causes failure. The struts should be redesigned with L- or U-shaped cross sections to increase their bending stiffness. The introduction of diagonal cross-ribs (Chapter 13) reduces the effective length L of the struts, so increases the buckling load. For struts with $L/d < 30$, yielding occurs before the strut buckles. FEA predicts that for $L/d = 25$, the buckling mode is $n = 2$. Tests on the product should be used to confirm such failure modes, since the deformation of the rest of the product may change the buckling mode.

The polypropylene arm of a folding garden chair is an example of a product which collapsed by a combination of elastic deformation and yielding. The L-section beam, probably copied from an aluminium product, was intended to provide adequate bending stiffness. When a load was applied to the centre of the arm, the arm both twisted and bend, and the lower part of the L deformed plastically. Figure. 8.9 shows an FEA prediction of the deformation and the yielded region.

8.2.4 Localised yield in compression—hardness

Hardness tests are widely used as a non-destructive method of estimating the yield stress of metal products, to check whether heat or surface treatments have been carried out correctly. The test is less common for plastics, partly because such treatments are not used, and partly because viscoelasticity makes the indentation size decrease with time. Recently, nano-indentation has been used to examine microstructural variation in polymers. This section considers the case where the indentation depth is much smaller than the product thickness, whereas Section 8.2.6 considers the case of the indenter penetrating the product.

The high elastic strains in polymers affect the pattern of plastic flow in hardness tests. The analysis for metals often assumes an infinite Young's modulus, so the plastically deformed material must flow to free surfaces at the sides of the indenter. For polymers, the yielding process largely occurs directly below the indenter, with elastic expansion of the surrounding region. Figure 8.1 showed the shear band patterns when a strip indenter

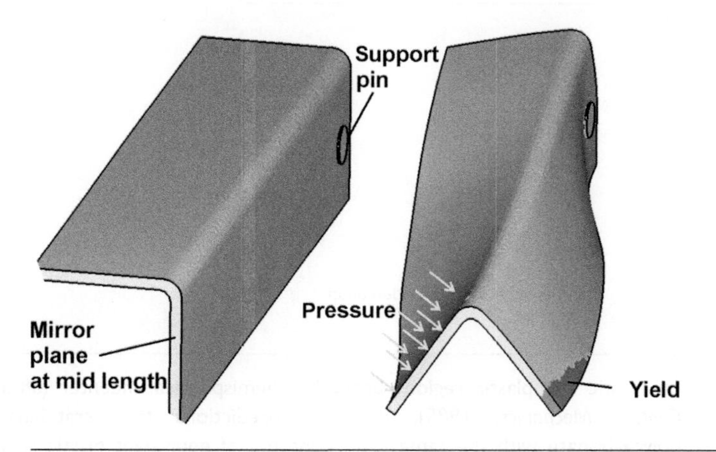

Figure 8.9 FEA of garden chair collapse: Before loading, and with a central load of 150 N.

was pressed into a block of polystyrene. The polymer flowed towards the interior of the block, causing large elastic strains in the surrounding region, with an indentation pressure approximately three times the uniaxial compressive yield stress.

Johnson (1985) assumed that the tip of an indenter (hemispherical or conical) is encased in a hemispherical core region (Fig. 8.10a) of radius a, in which the material has a hydrostatic pressure $\bar{p}$. The upper surface of this region is in contact with the indenter. Outside the core, he used the solution for the expansion of a spherical cavity, in an elastic–plastic solid with a tensile yield stress Y. The stresses and displacements have radial symmetry. In the core, the mean pressure is

$$\frac{\bar{p}}{Y} = \frac{2}{3} + 2\ln\left(\frac{c}{a}\right) \tag{8.10}$$

where c, the outer radius of the yielded zone, is given by

$$6(1-\nu)\left(\frac{c}{a}\right)^3 = \frac{Ea}{YR} + 4(1-2\nu) \tag{8.11}$$

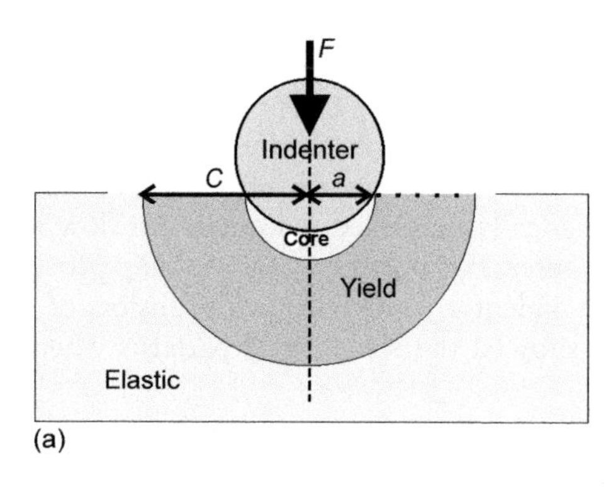

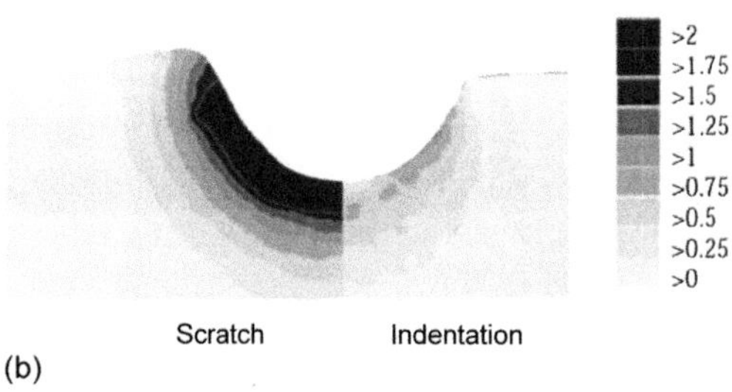

Figure 8.10 (a) Elastic and plastic regions beneath a hemispherical indenter (after Johnson, K. L., *Contact Mechanics*, 1985). (b) FEA predictions for scratching and indenting polycarbonate with the same load, contours of equivalent plastic strain (Bucaille, J. L. et al., *Trans. ASME J. Tribol.*, **126**, 372, 2004) © ASME. reprinted with permission.

where ν is the Poisson's ratio. The initial indentation hardness is $1.1\times$ the tensile yield stress Y, but, by the time the factor Ea/YR reaches 30, the hardness $\bar{p} = 3Y$. FEA of the indentation of polycarbonate (Fig. 8.10b) shows that Johnson's model is reasonable.

8.2.5 Localised yield—scratching of surfaces

Most surface damage to plastics is caused by moving rather than stationary objects. Scratches, produced when hard particles drag across a surface, have irregular shapes. Figure 8.11 shows the damage when a 1 mm diameter steel ball, loaded with a force of 7 N, was moved across a polypropylene surface at $100\,\mathrm{mm\,s^{-1}}$. The material, displaced from the central groove by the steel ball, flows to form raised regions about 1 μm high at the sides of the groove.

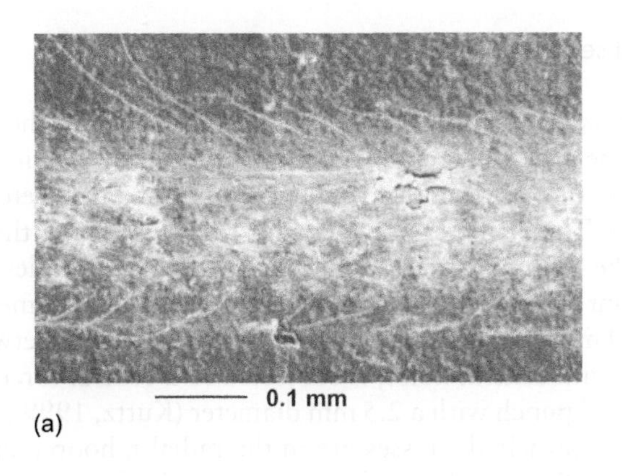

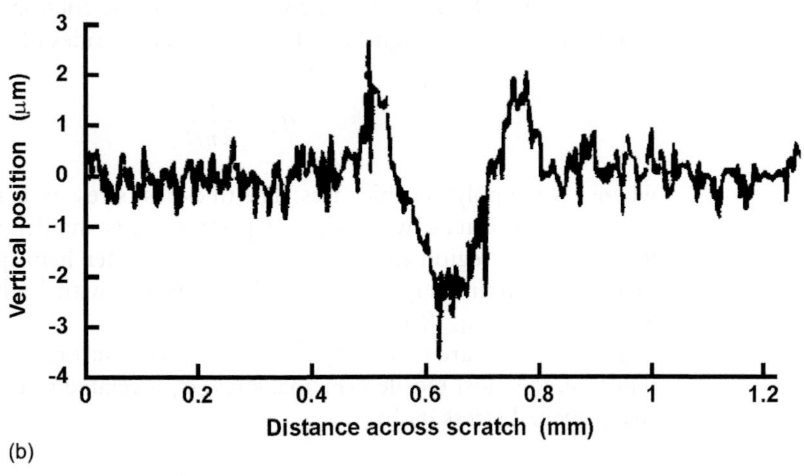

Figure 8.11 Scanning electron micrograph of a scratch on a PP with 25% talc, and profile across a similar specimen (from Xiang, C. *et al.*, *Polym. Eng. Sci.*, **41**, 23, 2001) © John Wiley and Sons Inc. reprinted with permission.

Such visible scratches are detrimental to the product appearance. Stress analysis shows there are tensile stresses on the polymer surface, downstream of the moving hemisphere. However, examination of the scratched surface (Fig. 8.11a) shows cracks (the white curves) only at the sides of the groove. Plastic deformation within the groove region probably removes the risk of cracking.

For the scratching of polycarbonate, FEA (Fig. 8.10b) shows that the plastic strains are higher than for indentation with the same load. The groove left behind the moving hemisphere cannot support the hemisphere, so the contact diameter must be larger on the upstream part of the indenter. Also the frictional forces on the contact surface increase the shear stresses in the surface layers of the polymer. The assumed coefficient of friction was 1.2.

8.2.6 Localised yield—film or sheet penetration

Convex objects can deform plastic products locally, with the yielded region penetrating the product. Examples are a small stone impacting the injection-moulded body of a hover-mover, and the corner of a tin stretching an LDPE shopping bag. The diameter of the yielded region is greater than the product thickness. The 'small punch test' apparatus, used on samples of UHMWPE from hip joint implants (Chapter 15), has a related geometry. A disc of diameter 6.4 mm and thickness 0.5 mm, constrained between two flat metal plates with central holes, is subjected to a central force from a hemispherical ended punch with a 2.5 mm diameter (Kurtz, 1999). At the sides of the punch, the principal stresses are in the radial r, hoop θ and thickness z directions (Fig. 8.12). When a force F acts, if the friction at the polymer–metal interface can be ignored, the stress σ_z at the sides of the punch is

$$\sigma_z = \frac{F}{2\pi rt} \tag{8.12}$$

where t is the polymer thickness. The thickness stress σ_r is small because of nearby free surfaces, while the hoop stress σ_θ is smaller than σ_r but still tensile. In the region in contact with the indenter hemispherical surface, there is approximately biaxial in-plane tension, with both tensile stresses being given by Eq. (8.12).

These stresses are related by a *yield criterion*. According to Tresca's yield criterion, the most tensile principal stress σ_1 is related to the most compressive principal stress σ_2 by

$$\sigma_1 - \sigma_2 \geq 2k \tag{8.13}$$

where the constant k is the shear yield stress of the material. At the cylindrical sides of the punch, yielding occurs in the polymer when

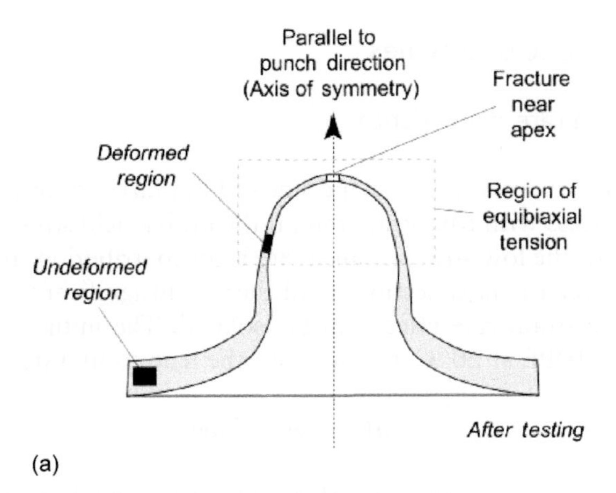

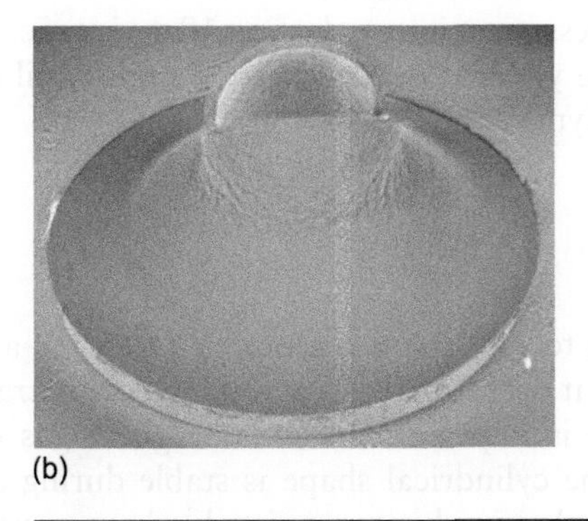

Figure 8.12 (a) Penetration of a UHMWPE sheet by a hemispherical punch, showing the polar coordinates used for stress analysis. (b) Specimen shape after the test (Kurtz, S. M. *et al.*, *Biomaterials*, **20**, 1449, 1999) © Elsevier.

$$\sigma_z - \sigma_r = 2k \tag{8.14}$$

causing it to thin. Figure 8.12b shows the UHMWPE enveloping the punch tip. The sheet initially yields by stretching radially while reducing in length in the hoop direction. When the sheet contacts the punch sides, it can no longer compress in the hoop direction. Ultimately fracture may occur at the base of the yielded cylindrical region or at the top of the punch.

Plastic products with biaxial orientation (such as the wall of a PET carbonated drink bottle) have a greater resistance to puncturing than initially isotropic or uniaxially oriented film, because there is no weak direction in the plane of the film. In such anisotropic films, the value of $\sigma_z - \sigma_r$ to cause thinning is larger than $\sigma_z - \sigma_\theta$ to cause constant-thickness yielding, so the latter occurs.

8.3 Yield on different timescales

8.3.1 Strain rate dependence

The second term on the right-hand side of Eq. (8.5) describes the increase in true yield stress with true strain rate. If the initial yield stress is measured in a tensile test, the low strain means there is no contribution from orientation hardening (see the next section), and there is insignificant heating. Consequently, the strain rate effect can be isolated. The initial yield stress was found, for HDPE at 20 °C, to vary with the true strain rate according to

$$\sigma(\dot{\varepsilon}) = A + B \log \dot{\varepsilon} \tag{8.15}$$

where the constants have the values $A = 44.7$ and $B = 4.67\,\text{MPa}$. This means that the yield stress in an impact lasting 10 ms is twice as large as in a slow tensile test, where yielding occurs after 10 min. Similar relationships are found for other polymers.

8.3.2 Creep rupture

If a plastic is subjected to a constant high creep stress for a long time, the *creep rupture* failure that occurs may be by yielding, or by crazing and crack growth. Creep rupture is important for the design of gas or water pipes (Chapter 14), where the cylindrical shape is stable during creep. It is less important for products that are bent or twisted, where excessive deflection is likely before any rupture process starts.

8.4 Orientation hardening

When metals *strain harden*, the cumulative plastic strain affects the yield stress. A copper rod, bent plastically then straightened, is harder to bend a second time. This is the result of the dislocation network that develops during the plastic deformation. However, if polymers are cyclically strained, they tend to soften, as a result of heat build-up. The increase in true yield stress with true strain in polymers (the first term on the right-hand side of Eq. 8.5) should be called *orientation hardening*, because it relates to orientation of either amorphous chains or of crystals. It is impossible to roll plastic sheet into thin film, in the same way as aluminium sheet can be rolled into 0.1 mm foil, by using intermediate annealing stages. Annealing of a rolled polymer sheet causes a partial recovery of the original shape, but does not allow complete molecular relaxation to the equilibrium state.

To quantify orientation hardening in a tensile test, the strain rate and temperature should be kept constant, since changes in these variables may

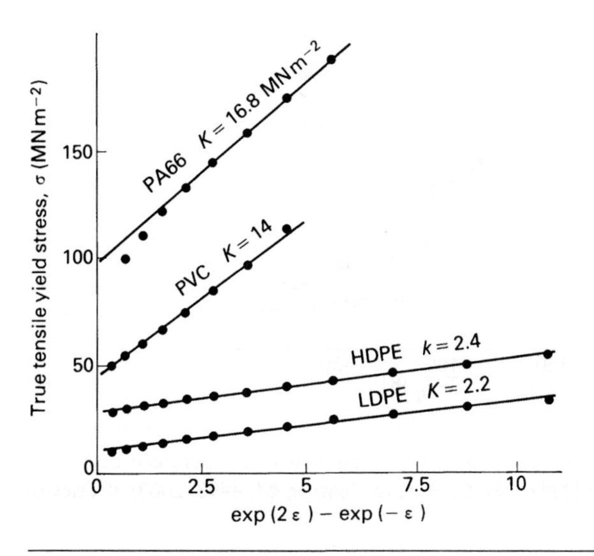

Figure 8.13 Variation of true tensile yield stress with true strain, fitted by Eq. (8.20), for various polymers. K is the slope of the lines in MPa (From G'Sell C and Jonas, J.J. *Mater. Sci.*, **16**, 1966, 1981, Chapman and Hall).

mask the effect. If the strain in the neck is monitored, and the cross-head speed controlled, the strain rate in the neck can be kept constant. Under these conditions, the true stress versus extension ratio λ relationship (Fig. 8.13) follows

$$\sigma = \sigma_0 + K(\lambda^2 - \lambda^{-1}) \tag{8.16}$$

From Eq. (8.1), $\lambda = \exp \varepsilon$, where ε is the true strain. Equation (8.16) is Eq. (3.21) for a crosslinked rubber network, plus a constant term σ_0, suggesting that an entanglement network in solid thermoplastic acts like a rubber network. The constant K is the effective modulus of the entanglement network in the glass. However, this modulus is hundredfold larger than that of a melt entanglement network, discussed in Section 2.3.3. Furthermore, K decreases as temperature increases (Fig. 8.14), so it appears to be inversely related to the chain segment mobility. K is much higher for polycarbonate than for polystyrene, which makes the latter more susceptible to crazing (Section 8.5.1).

For semi-crystalline polymers, the average orientation function P_2 for the crystal **c** axes can be calculated from X-ray diffraction measurements (Chapter 3). Figure 8.15 shows how P_2 increases linearly with the draw ratio, for polypropylene fibres and films, while the spherulitic microstructure survives. At $P_2 = 0.9$, where the spherulites are destroyed and replaced by a microfibrillar structure, there is an increase in the slope of the P_2 versus true strain relationship. It is impossible to achieve perfect **c** axis orientation

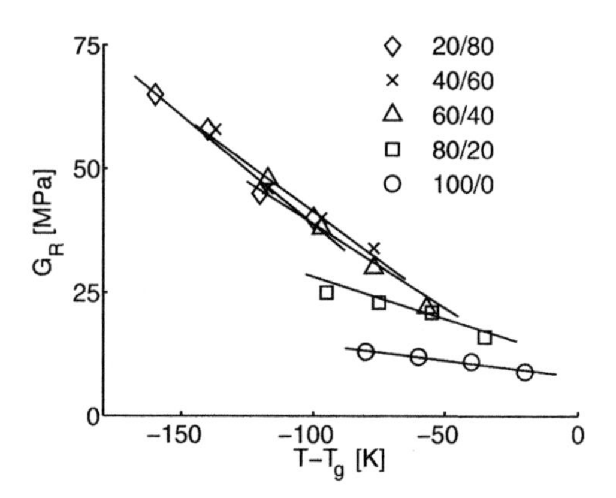

Figure 8.14 Variation of strain hardening modulus with temperature below T_g, for PS/PPO blends (from van Mellick, H. G. H. *et al.*, *Polymer*, **44**, 4493, 2003) © Elsevier.

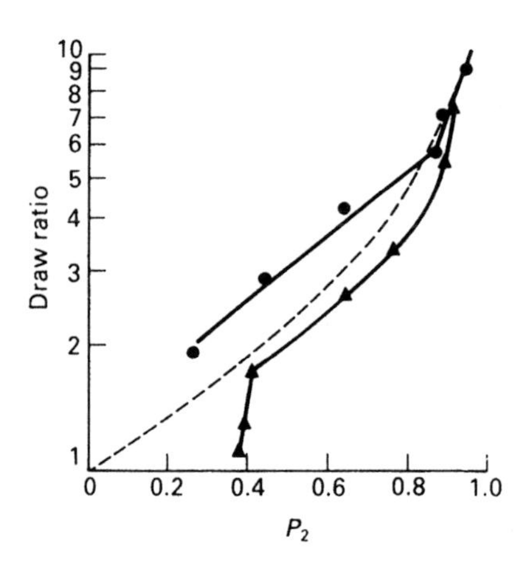

Figure 8.15 The crystal orientation of polypropylene fibres and films as a function of the true strain in the deformation process: draw temperatures (•) 135 °C, (▲) 110 °C. The prediction of the pseudo-affine model is shown as a dashed line (from *Samuels R.J. Structured Polymer Properties*, Wiley, 1974).

by mechanically stretching a fibre or film, since there is a limit on the achievable extension ratio. The pseudo-affine deformation model of Section 3.4.10 predicts that P_2 initially increases almost linearly with true strain, but becomes non-linear when $P_2 \cong 0.8$. Although the model ignores the spherulitic microstructure and deformation of the crystalline lamellae, it is a good approximation to the experimental data.

Molecular and/or crystal orientation increases the yield stress in the direction of drawing (Fig. 8.16). However, the yield stress at right angles to the draw direction hardly changes or may even decrease. The resulting

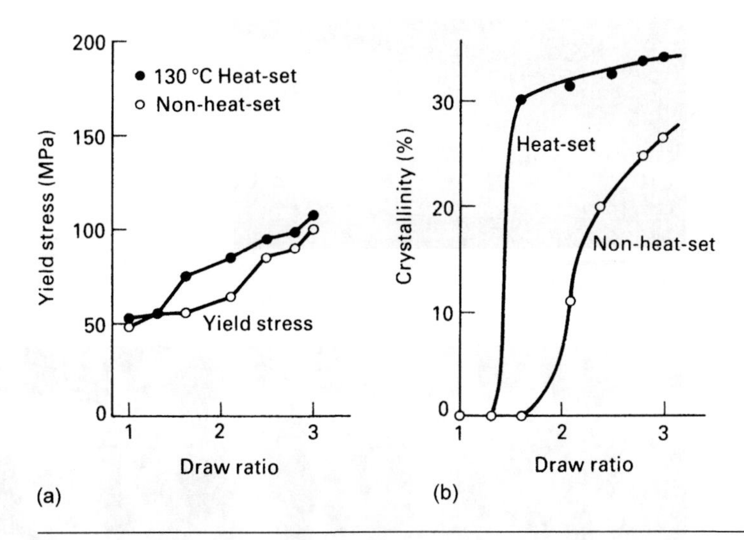

Figure 8.16 Variation of: (a) The yield stress and (b) the per cent crystallinity of biaxially stretched PET with the draw ratio (from Maruhashy, *Polym. Eng. Sci.*, **32**, 481, 1992).

strength anisotropy is acceptable in a fibre, but not in film products. Consequently, biaxial orientation is more common in film. Figure 8.16a shows how the yield stress of biaxially stretched amorphous PET increases with the draw ratio. The drawing process also causes the crystallinity to increase (Fig. 8.16b). However, the biaxial orientation of the glassy phase, rather than the crystallinity increase, is the main cause of the yield stress increase. Some of the polymer is *heat set* by contact with a heated mould at 130 °C for 10 s. This provides dimensional stability to PET blow mouldings that are subsequently filled with liquids at 90 °C.

8.5 Micro-yielding

8.5.1 Crazing

Crazing is most readily observed in transparent glassy plastics (the biro bending experiment in Chapter 1). It also occurs in semi-crystalline plastics such as polyethylene, but the opaque nature of the plastic makes the crazes more difficult to observe. There are usually a large number of crazes in close proximity, since crazes nucleate on surface defects. They are usually less than 1 mm in length. Crazes in glassy polymers, which reflect light, look like cracks that have grown from the surface. However, when a thin section of a craze is examined in a transmission electron microscope (Fig. 8.17a), fibrils of oriented polymer are seen bridging the craze. When a fracture surface is examined in a scanning electron microscope (Fig. 8.17b), the

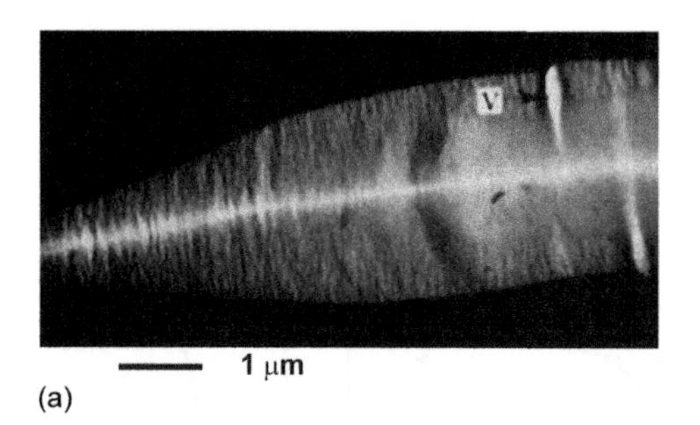

(a)

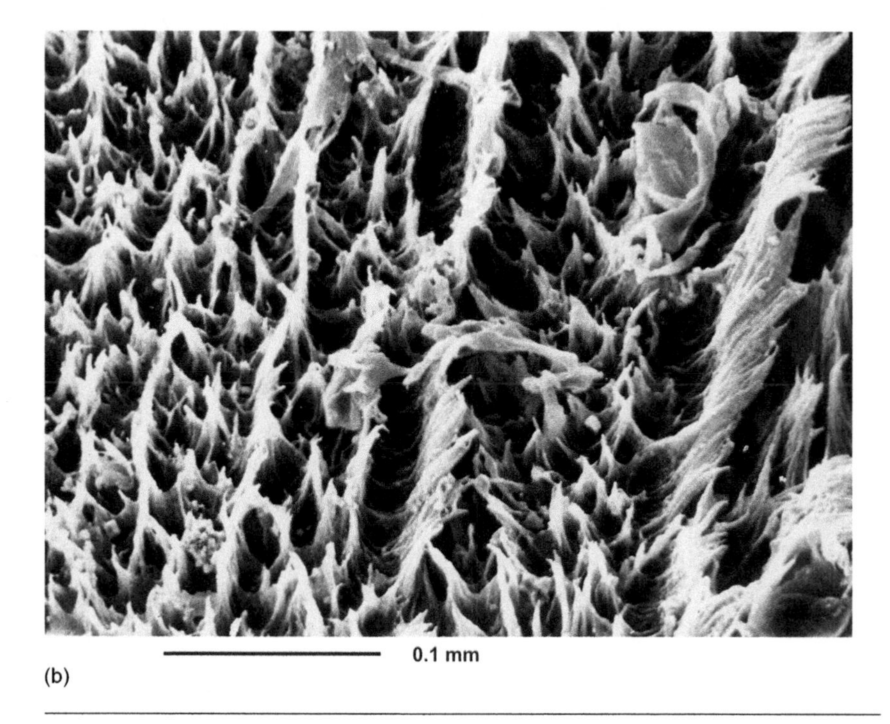

(b)

Figure 8.17 (a) Craze in polystyrene seen in cross section in the transmission electron microscope (from Beahan P. *et al.*, *Proc. R. Soc.*, **A 343**, 530, 1975). (b) Craze remnants seen on the fracture surface of a polyethylene.

broken fibrils halves are seen. Fibril formation from bulk polymer involves both yielding and the void formation, since a craze typically has a 50% void content. Crazes form on planes perpendicular to the largest tensile principal stress; in this they are similar to cracks. There is a nearly uniform tensile stress across a loaded craze, the value being characteristic of the polymer, the environment and the timescale. Chapter 10 will explore the environmental factors further. For polystyrene in air at 20 °C crazes appear in 30 s if a constant stress of 25 MPa is applied, and in 24 h for a stress of 10 MPa. Instrumented impact tests (Section 8.5.1) suggest that the craze stress rises to 100 MPa on a timescale of 5 ms. This is further evidence for the time-dependence of yielding processes.

Crazes thicken by drawing in new material, across the craze–bulk interface that stretches to a characteristic extension ratio. The plastic deformation occurs under 'plane strain' conditions (Appendix C), in the plane containing the tensile stress direction and the craze advance direction. The bulk polymer, above and below the craze, remains elastic, but it does not constrain the craze opening, because the void creation means that the craze Poisson's ratio is zero.

Crazes do not form unless the tensile strain exceeds a critical value, approximately 2% for polycarbonate and 0.4% for polystyrene, in air. We will see in Chapter 10 that these values are reduced when the polymer is exposed to certain liquids. If the applied strain barely exceeds the critical value, the crazes are widely spaced. The craze spacing decreases as the applied strain increases. This is further evidence that crazing is a yield process.

The mechanism for craze tip advance has been the subject of speculation, since there have been no direct observations at sufficient magnification. The stress across the craze is smaller than the tensile yield stress of the bulk polymer (40 compared with 90 MPa for polystyrene). A hydrostatic tensile stress, three times the uniaxial tensile yield stress, is required to expand an isolated spherical air bubble in a polymer. However, the mechanism of craze advance does not involve such high stresses. Figure 8.18a shows the fracture surface after liquid carbon tetrachloride has advanced through a craze in polycarbonate—the walls of the liquid channels fracture last and have a speckled appearance. A similar process could occur on a 1000 times smaller scale when air advances into a bulk polymer (Fig. 8.18b), except the channel walls fibrillate into strands of oriented polymer, rather than fracturing. The co-operative growth of parallel finger-like cracks is a phenomenon observed in the fracture of materials. In a high molecular mass polymer, the fingers advance too rapidly for polymer molecules to disentangle and flow apart. Consequently, there is considerable chain scission as the craze advances. Crazes therefore are a locally weakened part of the microstructure.

8.5.2 Energetics of craze growth

Once crazes reach a critical size, they play an important part in fracture (Chapter 9). The energetics of craze growth depends on whether the craze is isolated, or it interacts with its neighbours. Figure 8.19a shows single craze with a tensile stress σ_c across it, in a large sheet of material with a tensile stress σ applied to the ends. This stress analysis problem can be decomposed into two parts—a uniform tensile stress σ_c in an uncracked sheet, plus a cracked sheet with a stress $(\sigma - \sigma_c)$ applied to the ends. Therefore, craze growth energetics are a modified form of crack growth energetics, so long as the applied stress exceeds σ_c. We will investigate how the elastic energy W

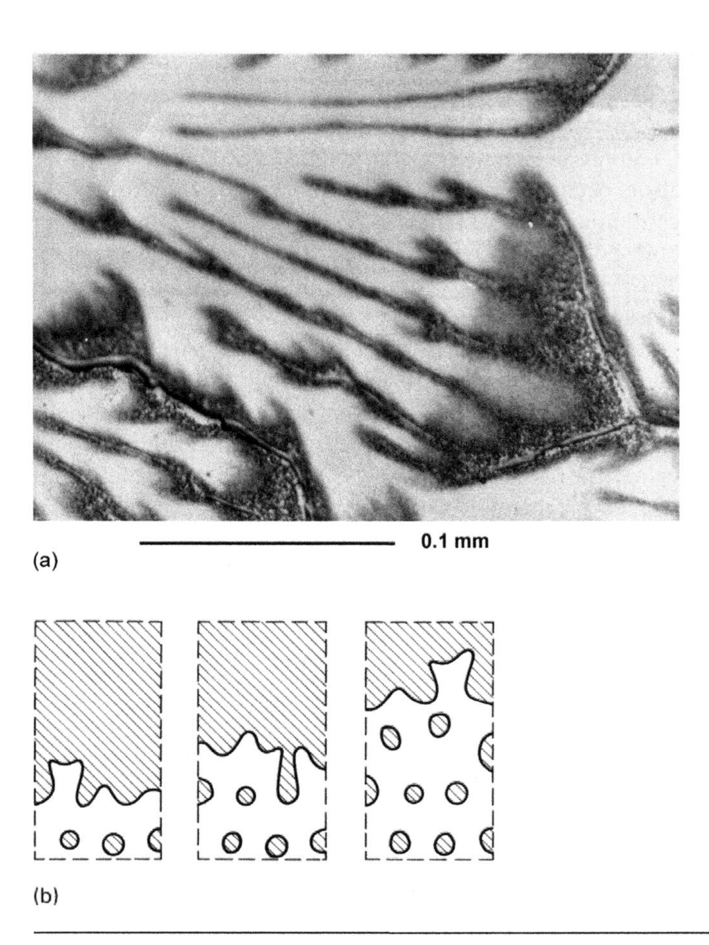

Figure 8.18 Finger-like advance of cracks lying in a plane: (a) Surface of polycarbonate fractured in a carbon tetrachloride environment; the liquid has advanced down parallel channels. (b) Sketch of a proposed method of craze advance.

stored in the sheet changes with the craze area A, when the sheet ends are held a fixed distance apart. The craze growth velocity V is assumed to be an increasing function of the *strain energy release rate* $\partial W/\partial A$.

The calculation of $\partial W/\partial A$ is complex; a reasonable approximation is obtained by assuming that a craze of length a totally relieves the stress in two triangular areas with total area βa^2 while leaving the stress distribution unchanged elsewhere. β is a constant. The elastic energy density in a uniform tensile stress field σ is $\sigma^2/2E$. If W_0 is the stored energy in the sheet prior to crazing, the stored elastic energy in the crazed sheet of thickness t is

$$W = W_0 - \frac{\beta a^2 t}{2E}(\sigma - \sigma_c)^2 \tag{8.17}$$

Differentiation gives

$$-\frac{\partial W}{\partial A} = \frac{\beta a}{E}(\sigma - \sigma_c)^2 \tag{8.18}$$

showing that the strain energy release rate is directly proportional to the craze length. Therefore, the growth rate of an isolated craze accelerates as its length increases, so long as the applied stress exceeds σ_c.

The model for multiple crazing (Fig. 8.19b) has parallel crazes, regularly spaced with separation s. The problem is split into two parts, as before. In

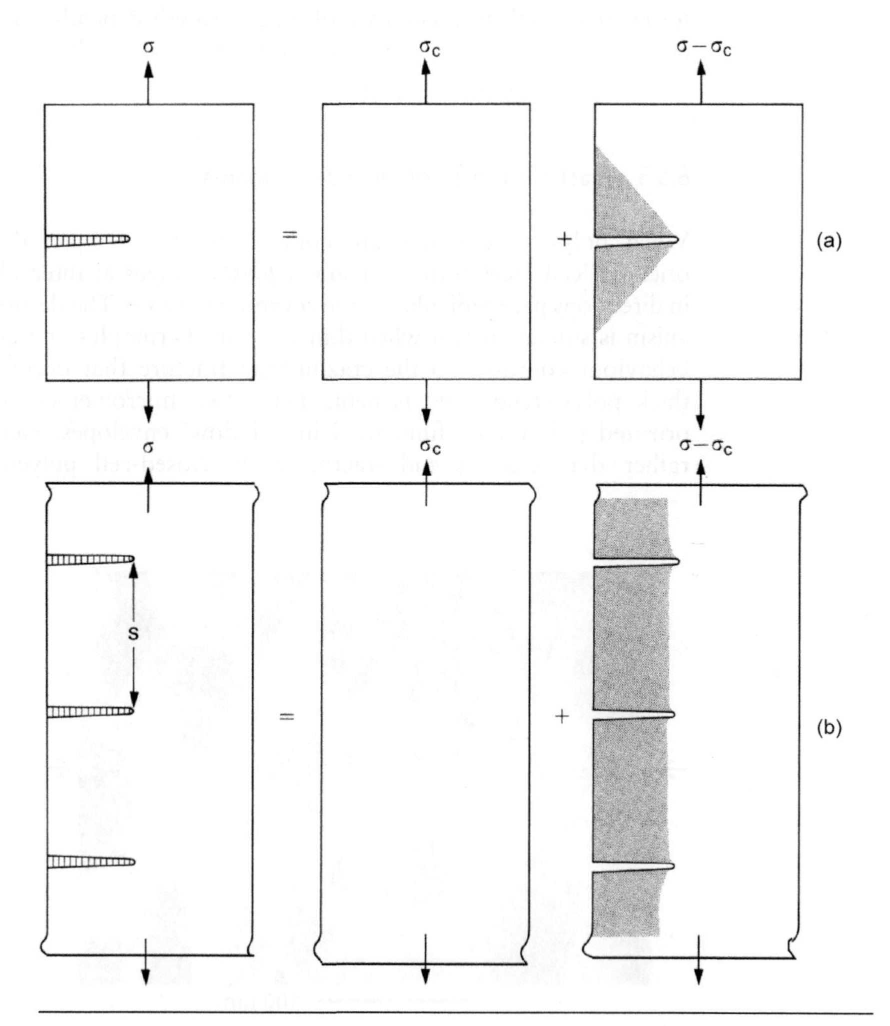

Figure 8.19 Stress analysis of craze growth in a tensile stress field, with stress-free areas shown as shaded: (a) Single craze of length a and (b) array of equal length crazes, separated by s in the direction of the tensile stress.

the multiple crack problem, the region ahead of the cracks is at the uniform stress $\sigma - \sigma_c$, whereas, to the left of the crack tips, the stress is zero. The advance of each craze by a length *da* transfers a volume *st da* from a stress of $(\sigma - \sigma_c)$ to a stress free state. Hence

$$-\frac{\partial W}{\partial A} = \frac{s}{E}(\sigma - \sigma_c)^2 \tag{8.19}$$

showing that the strain energy release rate remains constant. Consequently, the parallel crazes can grow in a stable manner.

In the next chapter we see that either a single craze or a bunch of crazes forms at a crack tip. The ease of crack growth depends on which of this occurs. Therefore, it is important to observe the number and geometry of the crazes involved in a fracture process.

8.5.3 Plastic collapse of closed-cell foams

When polystyrene foams are compressed, the 1–5 μm thick, biaxially oriented, cell faces form permanent *plastic hinges* at intervals (Fig. 8.20), in directions perpendicular to the compression axis. The deformation mechanism is similar to that when thin sheet steel crumples in a car crash. This behaviour contrasts to the crazing and fracture that occurs when 2 mm thick polystyrene sheet is bent. Thirty-two micrometers thick, biaxially oriented polystyrene film, used in 'window' envelopes, yields in tension rather than crazing and fracturing. In closed-cell polyethylene foams

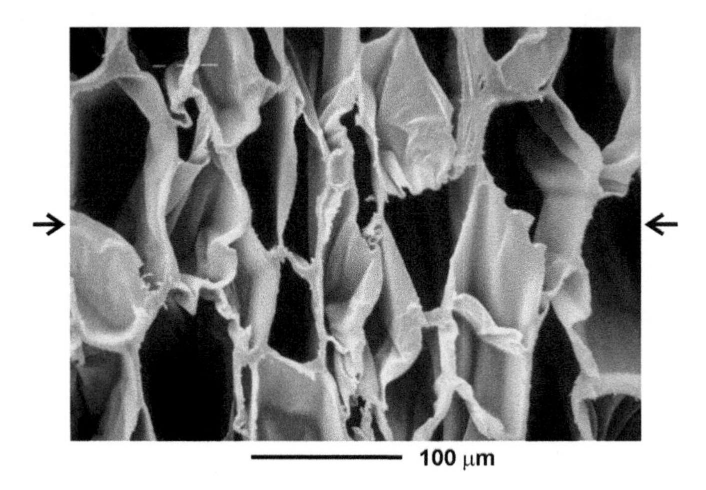

Figure 8.20 SEM micrograph of a polystyrene closed-cell foam after compression in the direction indicated to a strain of 80%.

permanent deformation does not occur, rather there is a viscoelastic recovery to nearly zero strain, even after severe compressive impacts. This contrasts with the plastic hinges that form in severely bent 2 mm thick HDPE sheet. The cell faces in HDPE foams are also biaxially oriented, and have different crystal orientations than in the spherulitic bulk mouldings. Consequently, thin cell faces are more ductile than the corresponding bulk polymer.

Any pressure differential between neighbouring cells would cause bowing of the intervening face. The face deformation mode should cause equal pressure rises in neighbouring cells, for instance by having an even number of plastic hinges across buckled faces. If cell faces concertina as in a bellows, it allows the foam Poisson's ratio to be zero.

There is no fully developed micro-mechanics model for the compressive yielding of low-density, closed-cell foams. Less than 10% of the polymer is in the foam edges, so the cell faces support the great majority of the load. The plastic hinge pattern in polystyrene foam faces is similar to a model for crushing aluminium closed-cell foam (Fig. 8.21); the repeating structure is a truncated cube, that contains a cruciform section and two pyramidal sections. The former is assumed to form plastic hinges, at a moment given by Eq. (8.7), when the model is compressed along a cube axis. The rate of plastic energy dissipation is minimised to obtain the plastic hinge spacing h. The predicted stress to crush the foam is

$$\sigma_f = 4.43\sigma_0\left(\frac{t}{b}\right)^{1.5}$$

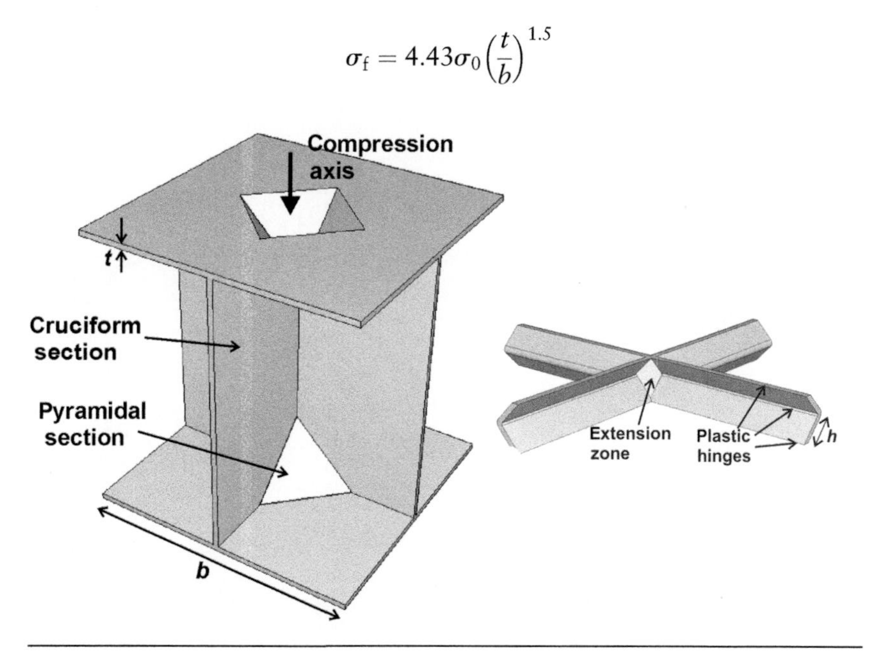

Figure 8.21 Truncated cube model of aluminium foam, and folds that develop in the cruciform section on compressive yield (adapted from Santosa S. and Wierzbicki T. *J. Mech. Phys. Solids*, **46**, 645, 1998).

where t is the face thickness, b its breadth and σ_0 the initial yield stress. If further terms for crushing the pyramidal sections are ignored, the result, expressed in terms of the foam relative density R, becomes

$$\frac{\sigma_f}{\sigma_0} = 0.63R^{1.5} \tag{8.20}$$

Plots of compressive yield stress versus density on logarithmic scales, for polystyrene and HDPE foams, have slopes $\cong 1.5$. If the lines are extrapolated to the density of the solid polymer, the yield stress is close to σ_0, that measured for the solid. Consequently, the form of Eq. (8.20) is confirmed. Cell faces in HDPE foams behave in a non-linear viscoelastic manner when bent. The stress distribution, resembles that in a plastic hinge (Fig. 8.7b), so it is not surprising that the exponent in the yield stress–relative density relationship is the same as for polystyrene. We will return to these materials in the cycle helmet case study in Chapter 14.

Chapter 9

Fracture

Chapter contents

9.1 Introduction

Polymer fracture is extremely varied in nature, covering the range from brittle to extreme toughness and extreme flexibility. Ceramics are nearly always brittle, while metals do not display time-dependent fracture phenomena in air at room temperature. The possibility of high anisotropy, in fibre and thin film products, makes a comprehensive treatment difficult. The design of plastic products to avoid fracture involves estimation of the loads in normal use. This may not always be possible. Extreme impacts can always initiate fracture, so some products are designed to avoid the risk of fast crack growth. For example, polyethylene pipes for natural gas distribution are designed to avoid slow crack growth from defects at welded joints. As it is always possible to fracture such pipes by the careless use of a mechanical excavator, the polymer is selected so that rapid crack growth down the length of the pipe is impossible (Chapter 14).

Failure investigation covers a number of disciplines, and the investigator's background may affect the range of failure causes considered. For example, a mechanical engineer would be aware of the effects of stress concentrations, whereas a polymer technologist might be inclined to blame poor processing. A polymer chemist might check the molecular weight distribution of the polymer, to see whether it had been reduced by environmental degradation. Several of these factors may be involved in a particular failure. Environmental changes will be analysed in Chapter 10, but their mechanical consequences will be examined here.

We start with fracture surface examination, which can provide clues to the causes of the fracture. The mechanical causes of crack initiation are then described. Enough fracture mechanics theory is included to explain crack growth criteria, and to link branches of the theory to various fracture phenomena. Finally, impact tests, widely used to characterise polymers and products, are analysed. The product environment, which may affect the failure mechanism, is considered further in Chapter 10.

9.2 Fracture surfaces and their interpretation

Failure investigations usually involve a detailed examination of fracture surfaces. The crack front geometry, at various stages of the fracture process, can usually be deduced, providing information about the type of loads that were acting. The basic principle is that the *crack plane is perpendicular to the most tensile principal stress*. Markings on the fracture surface can indicate the direction of crack growth and the approximate crack velocity, while crack arrest lines can show the sequence of crack front positions.

Figure. 9.1a shows the fracture surface, from a tensile test on a brittle plastic, is perpendicular to the length of the bar, the direction of the principal tensile stress. As there is initially a uniform stress in the bar, the crack could initiate

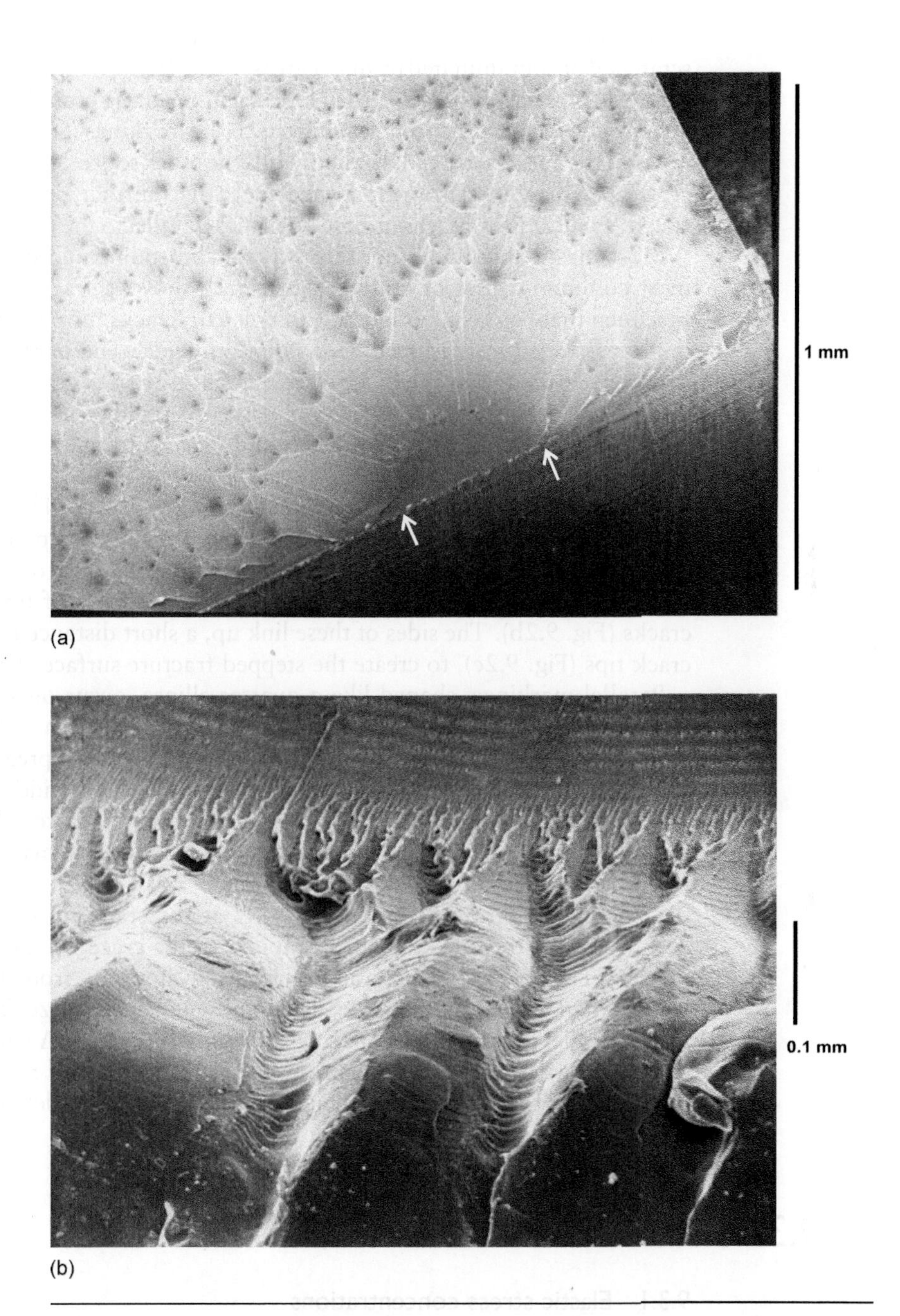

Figure 9.1 (a) Fracture surface of a PMMA tensile specimen. A craze has formed on one edge, then failed, leaving a flat region. (b) Bending fracture surface in PC showing the splitting of the crack, as it moves from top to bottom of the picture.

at any site. However, initiation is most likely at a surface, which can be scratched or contaminated by finger grease. The flat, mirror-like area indicates where a craze has formed and then failed. The resulting crack has propagated at a rapidly increasing speed, leaving parabolic markings (Section 9.4.5). The noses of the parabolae point back towards the crack source, so the direction of crack growth is radial from the craze. Eventually, the fracture surface becomes rough, as subsidiary cracks initiate on planes parallel to the main crack plane.

Most plastic mouldings are thin-walled, so bending and twisting are the most common causes of failure. Figure 9.1b shows a fracture surface of a specimen that has been both bent and twisted. Cracks formed on the surface that was placed in tension by a bending moment. The parallel horizontal markings at the top of the picture are remnants of crazes that formed at the crack front. A single crack has moved down the fracture surface towards the viewer. The crack has then twisted and divided into many parallel cracks, under the influence of a tensile stress T from the bending moment, and a shear stress S from the torsional loading (Fig. 9.2a). The crack plane attempts to rotate about the growth direction, to remain perpendicular to the tensile principal stress σ. However, rotation can only occur in a short growth distance if the crack fragments into an echelon of finger-like sub-cracks (Fig. 9.2b). The sides of these link up, a short distance back from the crack tips (Fig. 9.2c), to create the stepped fracture surface (Fig. 9.1b).

Parallel markings, shaped like a quarter ellipse, occur on some fracture surfaces (Fig. 9.3a). A surface crack has initiated when a blunt object pressed on the product surface (Fig. 9.3b). As this crack spreads sideways, the object penetrates the product and twists the two sides in opposite directions. This double torsion loading causes the crack to advance more rapidly on the lower surface in tension. The characteristic markings are due to momentary hesitations of the crack front.

Sometimes a craze grows fairly slowly at the crack tip while the crack hesitates, then the crack advances rapidly to the craze tip and the process repeats. The crack advances at one location along its front by the craze length, then this crack step spreads laterally along the craze. This tends to leave parallel markings at a spacing of less than 0.1 mm. A similar process can occur when an edge dislocation line spreads through a metal crystal. Hull's book *Fractography* (Further Reading) gives further details of fracture surface interpretation.

9.3 Crack initiation

9.3.1 Elastic stress concentrations

Holes or sharp corners are familiar causes of stress concentrations. Near such a feature, the most tensile principal stress reaches a maximum value,

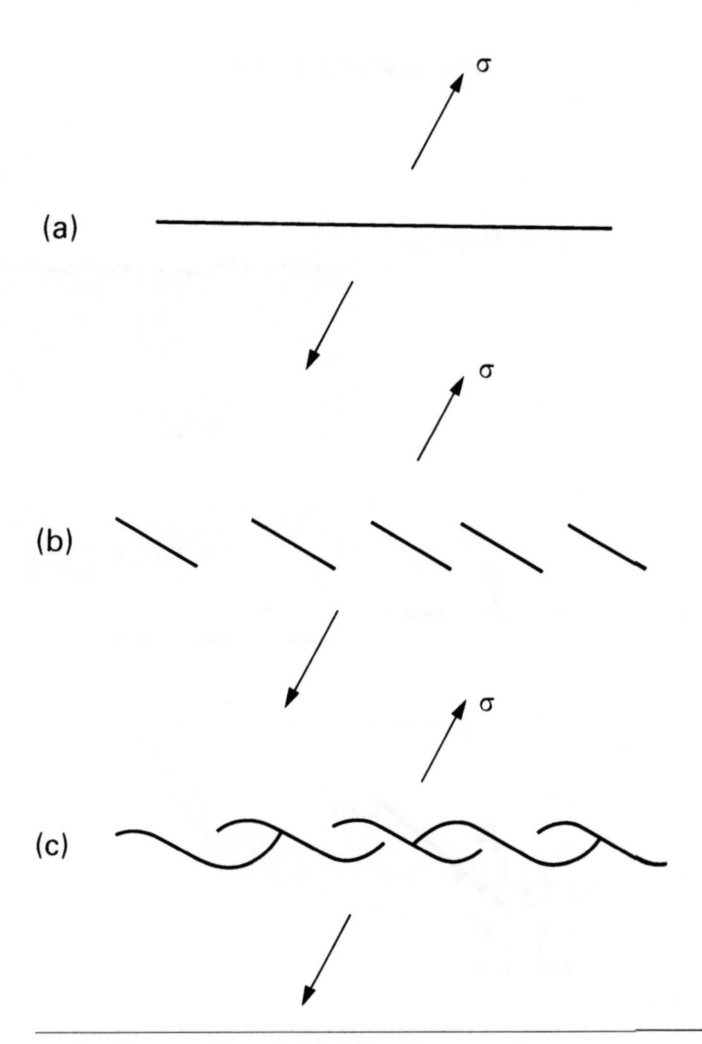

Figure 9.2 (a) A crack growing into the paper with the crack plane at an angle to the principal tensile stress σ; (b) it splits into finger-like cracks that advance in echelon; (c) these link up behind the crack tips.

which is higher than the tensile stress that occurs if the hole, etc. was absent. The *stress concentration factor* q is defined as

$$q = \frac{\text{Local maximum stress}}{\text{Stress in absence of feature}} \tag{9.1}$$

q is dimensionless. It should not be confused with the stress intensity factor K, with dimensions $\mathrm{N\,m^{-1.5}}$, used in Section 9.4.1 to characterise crack tip stress fields. For a cylindrical hole in a wide plate subjected to a tensile stress σ, the maximum stress is $3\,\sigma$ at the side of the hole, so $q = 3$. Finite element

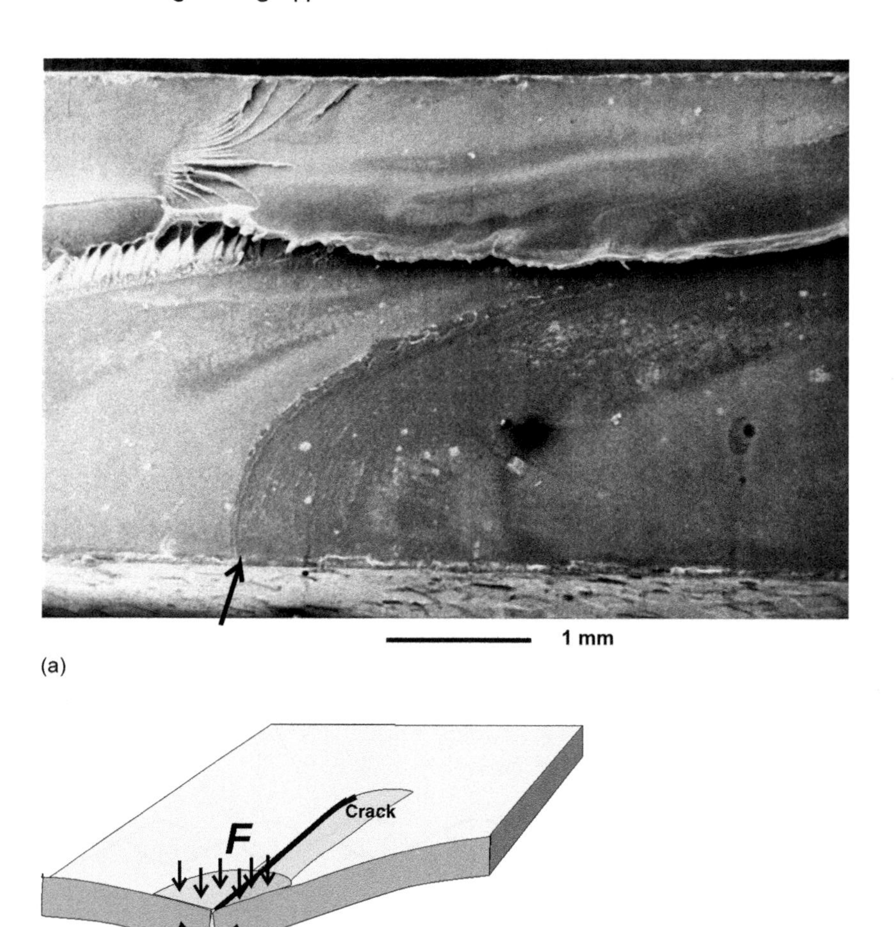

Figure 9.3 (a) Elliptical markings (arrowed) due to momentary crack arrests on a polycarbonate fracture surface. (b) Double torsion loading that occurs when a crack is driven by a surface force *F*.

computations for plastic products give slightly different values, since there are significant changes in product geometry under load. For a 2 mm radius circular hole in a 16 mm wide strip (Fig. 9.4a), subjected to a distant 10 MPa stress, $q = 2.8$. At moulded-in holes, the melt divides to pass on either side of a steel pin, and the weld line downstream (Section 5.5) leaves a surface groove. The surfaces of drilled holes tend to be rough, as a result of the localised heating and tearing of the plastic. Both these effects tend to increase the q value. Internal corners in products should have a generous fillet radius r (Fig. 9.4b), but this may be omitted to reduce mould machining costs. A machined 'sharp' corner would have a radius $r \approx 0.05$ mm. When such a part is bent, in a direction to cause tensile stresses at the fillet

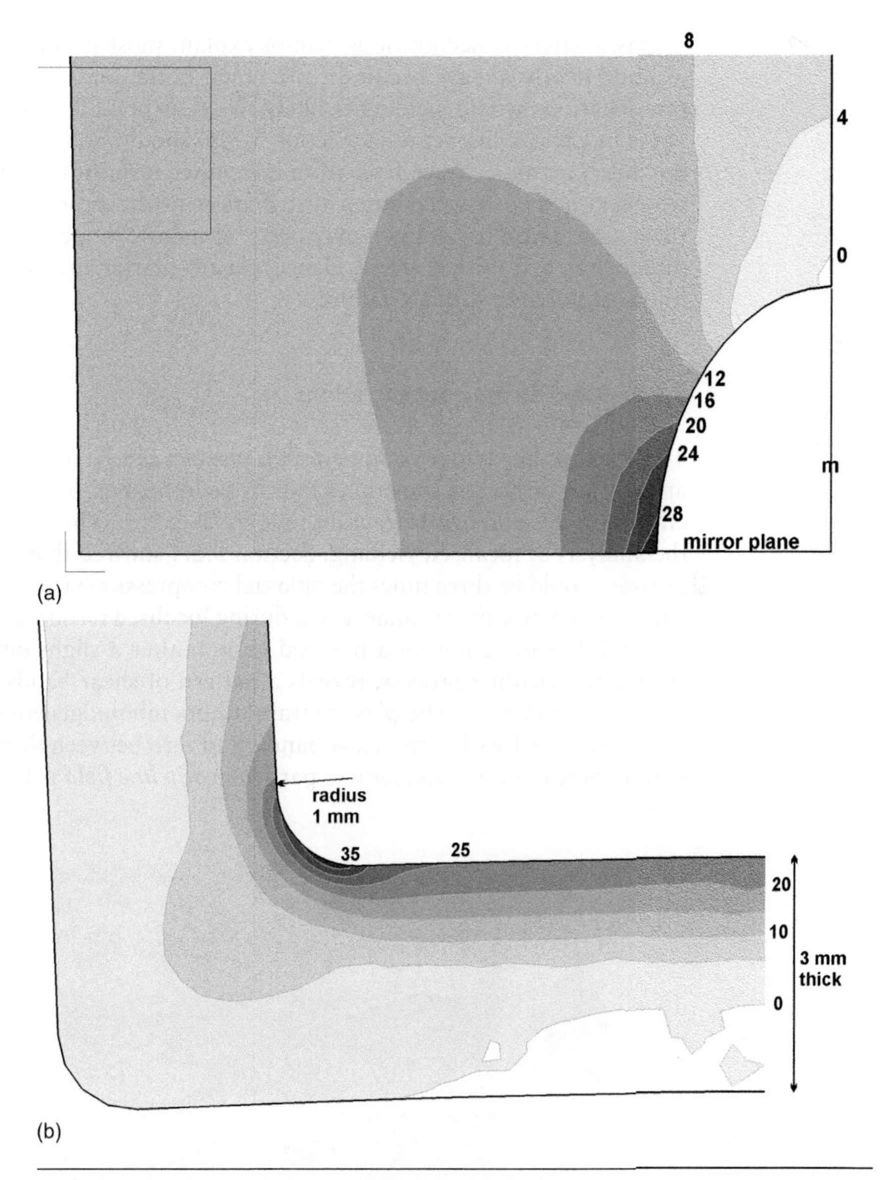

Figure 9.4 Stress concentration factors: contours of principal tensile stress (MPa) for (a) a central hole in a plate under tension—contours of relative stress; 10 MPa tensile stress at the ends (b) 1 mm radius internal corner in a product subjected to bending.

radius, we expect that $q \propto r^{-0.5}$. In Fig. 9.4b, an L-shaped part of a product is bent; the left-hand end, 10 mm above the corner, was moved 0.5 mm to the left while the right-hand end, 10 mm to the right of the corner, was clamped. A comparison, of the peak stress at the corner, with the inner surface stress due to bending, shows that $q \approx 1.4$.

Elastic stress concentrations cannot explain most product failures, since yielding nearly always occurs before crack initiation. However, they indicate locations where yielding is likely to occur first. Therefore, the failure stress in Charpy impact tests (Section 9.5.1) should not be calculated using the notch q value. Craze formation is another form of (localised) yielding, which also modifies the stress distribution in the product. Section 9.4.4 shows that craze breakdown may occur at a critical opening displacement, rather than at a critical stress. Hence, elastic–plastic analyses must be used for most polymer product failures.

9.3.2 Yield stress concentrations

The fracture behaviour of some tough plastics can be inconsistent. For example, polycarbonate sometimes fails by yielding, but, on other occasions, a crack initiates and brittle fracture follows (Fig. 9.5). The explanation lies in the analysis of localised yielding. Section 8.2.4 showed that the indentation pressure could be three times the uniaxial compression yield stress. The yield stress can change by a similar factor during localised tensile deformation.

Careful examination of a notched region, after a slight impact load has started the yielding process, reveals a pattern of shear bands (Fig. 9.6a) in some glassy plastics. The plastic strain occurs inhomogeneously. The shear strain is about 1 within the shear bands, and zero between them. The overall pattern is remarkably similar to a particular *slip line field* pattern (Fig. 9.6b).

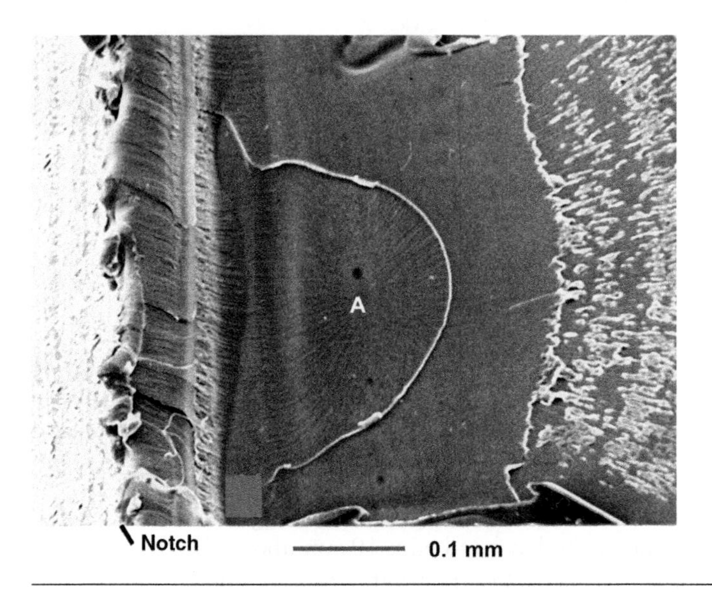

Figure 9.5 Fracture surface of polycarbonate in which a crack has initiated at A, 0.55 mm below the root of a notch of radius 0.25 mm, in a sheet 6 mm thick.

Slip line fields are used to analyse metal plasticity under plane strain conditions (see Appendix C). The slip line field consists of two families of logarithmic spirals, with equations in polar coordinates r, θ

$$\ln r = \ln a \pm \theta - \theta_0 \tag{9.2}$$

a is the notch radius, while the constant θ_0 differs for each spiral. The α slip lines are orthogonal to the β lines, and is α chosen so that the more tensile principal stresses lies in the first quadrant of the $\alpha\beta$ axes, that occur at every slip line intersection.

The Mohr circle representation (Fig. 9.6c) is a graphical method of relating stress components in different sets of axes. When the axes in the material rotate by an angle θ, the diameter of the circle rotates by an angle 2θ. If the material yields, the circle has radius k, the constant in the Tresca yield criterion. The axes of the Mohr diagram are the tensile and shear stress components. Thus, in the left-hand circle, representing the stresses at A in Fig. 9.6b, the ends of the horizontal diameter are the principal stresses. The principal axes are parallel and perpendicular to the notch-free surface. There is a tensile principal stress $2k$ parallel to the surface, and a zero stress perpendicular to the surface. The points at the ends of the vertical diameter represent the stress components in the $\alpha\beta$ axes, rotated by 45° from the principal axes. In the $\alpha\beta$ axes, the shear stresses have a maximum value k, and there are equal biaxial tensile stresses of magnitude $\bar{\sigma} = k$ (the coordinate of the centre of the circle).

To find out how $\bar{\sigma}$ changes along an α slip line, we consider a special case where the β slip lines are straight, and the α slip lines have a radius of curvature r. Figure 9.6d shows the stress components, on the surface of the prism marked out by neighbouring α and β slip lines, that contribute to the moment about the point O. The prism length perpendicular to the paper is unity. The prism is in static equilibrium, so the moment of the forces on it about O is zero, hence

$$kr^2 \mathrm{d}\phi - k(r + \mathrm{d}r)^2 \mathrm{d}\phi + (\bar{\sigma} + \mathrm{d}\bar{\sigma})r\mathrm{d}r - \bar{\sigma}r\mathrm{d}r = 0$$

where $\mathrm{d}\phi$ is the angle between the two α lines. Expanding this and ignoring the $\mathrm{d}r^2$ term gives

$$\mathrm{d}\bar{\sigma} - 2k\,\mathrm{d}\phi = 0$$

so on integration

$$\bar{\sigma} - 2k\phi = \text{constant} \tag{9.3}$$

along an α line. We use this equation to relate the stress state at A (on the notch surface) to that at B (the tip of the yielded zone) in Fig. 9.6b. The

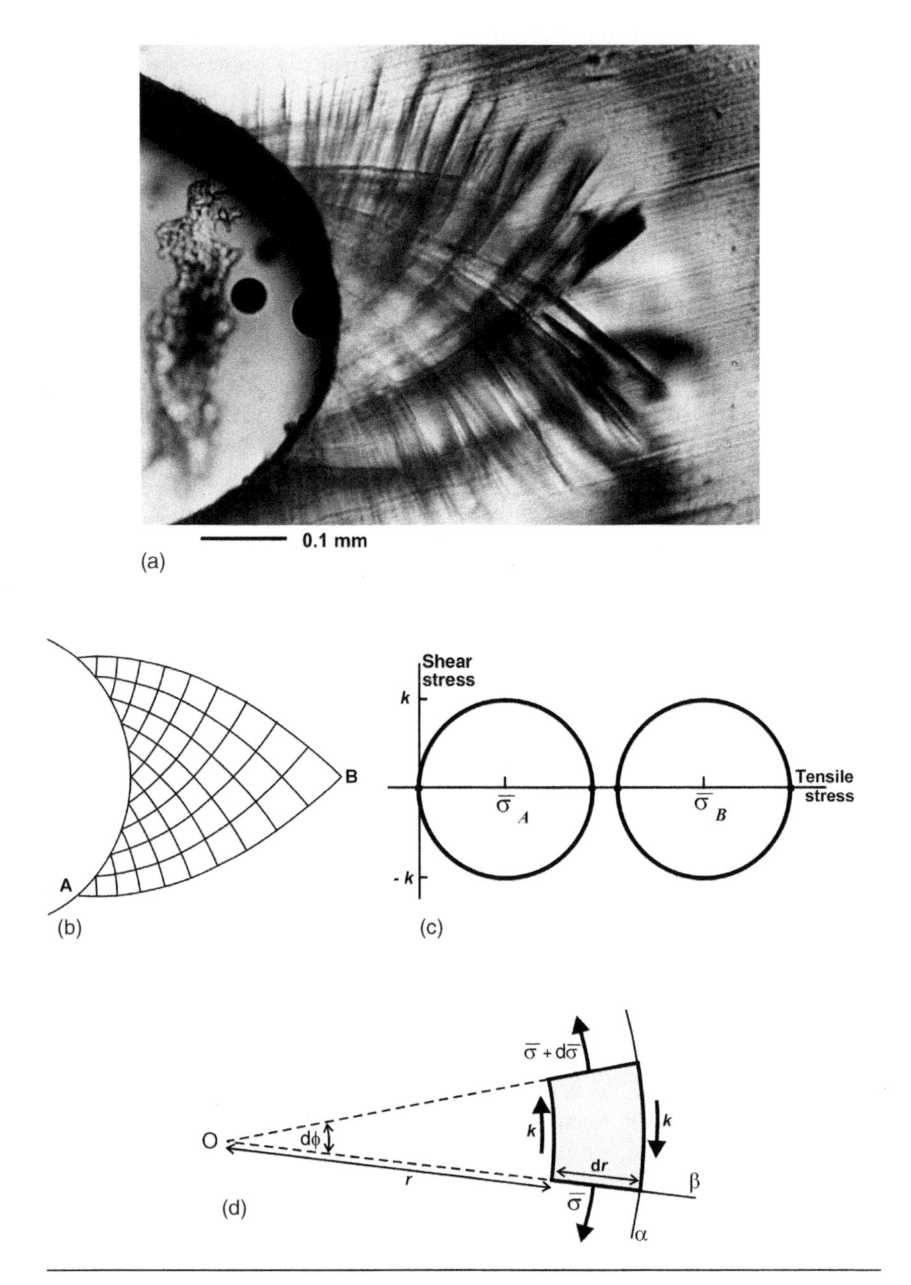

Figure 9.6 Analysis of yielding at a notch: (a) shear band patterns seen in a thin section cut from a polycarbonate specimen; (b) slip line field pattern for yielding; (c) Mohr circle diagram for the states of stress at points A and B in (b); (d) stress components on the surface of the prism marked out by neighbouring α and β slip lines.

angle of the slip line increases by θ between A and B (the slip line remains at 45° to the radius), so from Eq. (9.3)

$$\bar{\sigma}_A - 2k\phi_0 = \bar{\sigma}_B - 2k(\phi_0 + \theta)$$

so

$$\bar{\sigma}_B = \bar{\sigma}_A + 2k\theta \tag{9.4}$$

The stress state at B is represented by the right-hand circle in Fig. 9.6c, so the most tensile principal stress

$$\sigma_{max} = 2k(1 + \theta) \tag{9.5}$$

The usual notch angle for impact tests is 45°. This means that the largest possible θ value is 1.18 rad. By Eq. (9.5) the tensile stress at the tip of the yielded zone could rise to 2.28 times the unrestrained tensile yield stress. The yielded zone would be large at this stage with the point B at a distance 3.25 *a* from the centre of the notch. In practice, the yielded zone rarely becomes this large because either

(a) a crack initiates and yielding ceases, or
(b) the yielded zone length exceeds 50% of the specimen thickness *t*, and some plane stress (through thickness) yielding occurs.

We conclude that crack initiation can occur in notched plastics if the notch tip radius is smaller than *t*/10 (experimental data given in Section 9.5.1).

An analysis of the shear band pattern in Fig. 8.1 predicted high tensile stresses parallel to the surface, at the point where the shear bands penetrate deepest into the block. The figure shows the crack that has formed at this location. Therefore, a compressive force, applied on the surface of a material that is brittle in tension, can cause fracture.

9.3.3 Cracks in brittle surface layers

In order to improve the surface gloss and appearance, thin layers of polystyrene or chromium have been applied to rubber-toughened plastics like ABS. However, these surface layers are brittle. Outdoor exposure to ultraviolet radiation (Chapter 10) can also convert the surface layer of a plastic into a brittle state. The effects of such layers are the most marked when the product is bent, with the brittle layer being in tension. The tensile failure strain of the surface layer is smaller than that of the substrate, so it fails first. A series of sharp cracks forms perpendicular to the surface tensile stress; each relieves the surface stress over a limited length. The cracks start at the outer surface, and accelerate rapidly through the surface layer. If the substrate is tough, the cracks can be arrested, but if it has a low resistance to fast crack growth, the cracks will continue, causing a brittle failure.

9.3.4 Residual stresses

Chapter 6 dealt with residual stresses that occur when products are cooled rapidly from both the sides. There are biaxial compressive stresses in the surface layers and biaxial tensile stresses in the interior. If a hole is drilled through such a product, it cuts through the tensile stress region, and acts as a stress-concentrating feature with a q value of 2. If there is ingress of a stress cracking fluid, radial cracks may form from the bore of the hole, perpendicular to the residual circumferential stresses. These cracks will be at the mid-thickness of the product.

Residual stresses can also arise if a hole is drilled, with a blunt bit, through a product that is initially stress-free. The drilling operation generates enough heat to melt a thin annulus of plastic surrounding the hole. When this cools down, it contracts, so has a residual tensile circumferential stress (the effect is the converse of shrink-fitting a metal rim on a wooden wagon wheel). Consequently, cracks may start in a radial direction, but they will turn to follow the boundary of the overheated layer. These two examples show that the crack patterns can reveal the type of residual stress field in a product.

9.3.5 Summary

Cracks tend to initiate on the surface of products for a number of reasons—bending or torsion loading causing high surface stresses, surface scratches causing stress concentrations or surface degradation. However, in some circumstances (yield stress concentrations at a notch or weak interfaces) crack initiation is internal.

9.4 Crack growth

9.4.1 Fracture mechanics: The stress intensity factor of a crack tip stress field

Once a sharp crack has formed, it is possible to analyse its growth, using the concepts of fracture mechanics. The subject was developed for the failure of large metal structures. *Linear elastic fracture mechanics*, the simplest theory, considers the stress and strain fields around the crack tip in elastic materials. In the majority of cases, the crack faces move directly apart (mode I deformation in the jargon) rather than sliding over each other in one of two directions (mode II or III). In Fig. 9.7a, the crack front lies along the negative z axis and the crack faces move apart in the $\pm y$ direction. The origin of the xy axes, at the crack tip, remains there when the specimen is loaded. As all points in the body are displaced from their unloaded positions by a vector **u**, with components (u_x, u_y), $\mathbf{u} = 0$ at the crack tip.

There are two *boundary conditions* for crack opening (Fig. 9.7a). Firstly, due to the crack opening displacement δ, there is a sudden jump in u_y from $+\delta/2$ to $-\delta/2$ on crossing the crack. In mathematical terms, u_y becomes two-valued along the crack. Secondly, the shear stress on, and the tensile stress normal to, the crack plane are zero; a condition for all free surfaces. If position in the stress field is given by the complex number $z = x + iy$, we seek suitable functions of z to describe the stress and displacement fields. We can try

$$u_y = K^* \mathrm{Im} z^{0.5} \tag{9.6}$$

for the discontinuous part of the displacement field, where K^* is a constant. Along the crack, where $z = -x$, the imaginary part of $z^{0.5}$ takes the values $\pm x^{0.5}$, so the function is suitable, and the crack tip has a parabolic shape as required. Higher powers of z, such as $z^{1.5}$ or $z^{2.5}$, are also suitable, but the $z^{0.5}$ solution dominates close to the crack tip, so is used to describe the crack tip strain field.

The strains are obtained from the displacement by differentiation, using equations such as

$$e_y = \frac{\partial u_y}{\partial y} = \frac{K^*}{2} \mathrm{Im} z^{-0.5} \tag{9.7}$$

Therefore, the strains vary as $z^{-0.5}$. For an elastic material, the stress components are linearly related to the strain components, so they too vary with $z^{-0.5}$. For convenience, the mean tensile stress and maximum shear stress $\tau_{\max}$ (used to generate the pattern in Fig. 9.7a) are expressed in terms of the polar (r, θ) coordinates, giving

$$\bar{\sigma} \equiv \frac{\sigma_{xx} + \sigma_{yy}}{2} = \frac{K_I}{\sqrt{2\pi r}} \cos\frac{\theta}{2}$$

$$\tau_{max} \equiv \sqrt{\left(\frac{\sigma_{yy} - \sigma_{xx}}{2}\right)^2 + \sigma_{xy}^2} = \frac{K_I}{2\sqrt{2\pi r}} \sin\theta \tag{9.8}$$

The scaling constant K_I for mode I crack opening, which occurs in these equations, is known as the *stress intensity factor*. It is proportional to the K^* of Eq. (9.7). The inclusion of the $\sqrt{2\pi}$ term is a result of the 1939 definition of K_I—logically it could be omitted. Both stress components given by Eq. (9.8) are zero along the crack surface, as required by the boundary conditions. The stress intensity factor has the units of [stress × (length)$^{0.5}$], in contrast to the stress concentration factor for a notch (Section 9.2) which is dimensionless.

Equation (9.8) states that the stress field close to the crack tip has a particular form. This can be revealed using the *photo-elastic effect*. The refractive indices n_1 and n_2 in the directions of the principal stresses σ_1 and σ_2 are related by

$$n_1 - n_2 = C(\sigma_1 - \sigma_2) = 2C\tau_{max} \tag{9.9}$$

The magnitude of the stress-optical coefficient C (Table 11.6) varies slightly with wavelength, so monochromatic light is used for photography (Fig. 9.7b). Two-dimensional models are cut from cast epoxy resin, or carefully annealed PC sheet which is optically isotropic. The *isochromatic fringes* (of a uniform colour if white light is used for illumination) are contour levels of the maximum shear stress τ_{max}. Each dark fringe is assigned an integral fringe order f, with the first fringe to appear on loading having $f = 1$. Since the maximum shear stress is proportional to the fringe order, the radial distance r of any fringe from the crack tip is given by Eq. (9.8) as

$$r = A\left(\frac{\sin\theta}{f}\right)^2 \tag{9.10}$$

where A is a constant. Consequently, each fringe has a nearly elliptical shape, with its major axis perpendicular to the crack (Fig. 9.7a). The characteristic stress field can be recognised in the isochromatic pattern of any cracked body, loaded so as to open the crack, such as a *compact tension* specimen (Fig. 9.7b). The stress intensity factor K_I can be calculated by plotting the fringe number f versus ${r_{max}}^{-0.5}$, where r_{max} is the maximum excursion of fringes, near the crack tip, away from it. However, the more distant stress field is dominated by the point loading and the bending of the specimen as a C-shaped beam. The crack tip stress field has parallels with other characteristic fields, such as the magnetic field of a bar magnet. The latter is visualised by scattering iron filings on a sheet of white paper on top of the magnet. The shape of the pattern is always the same, but the magnetic field at any point is determined by the pole strength of the magnet, the analogue of K_I.

9.4.2 Stress intensity factors for certain specimen geometries

For certain simple specimen geometries and modes of loading, there are analytical solutions for the stress intensity factor. For a central crack of length $2a$ in an infinitely wide sheet, subjected to a distant tensile stress σ

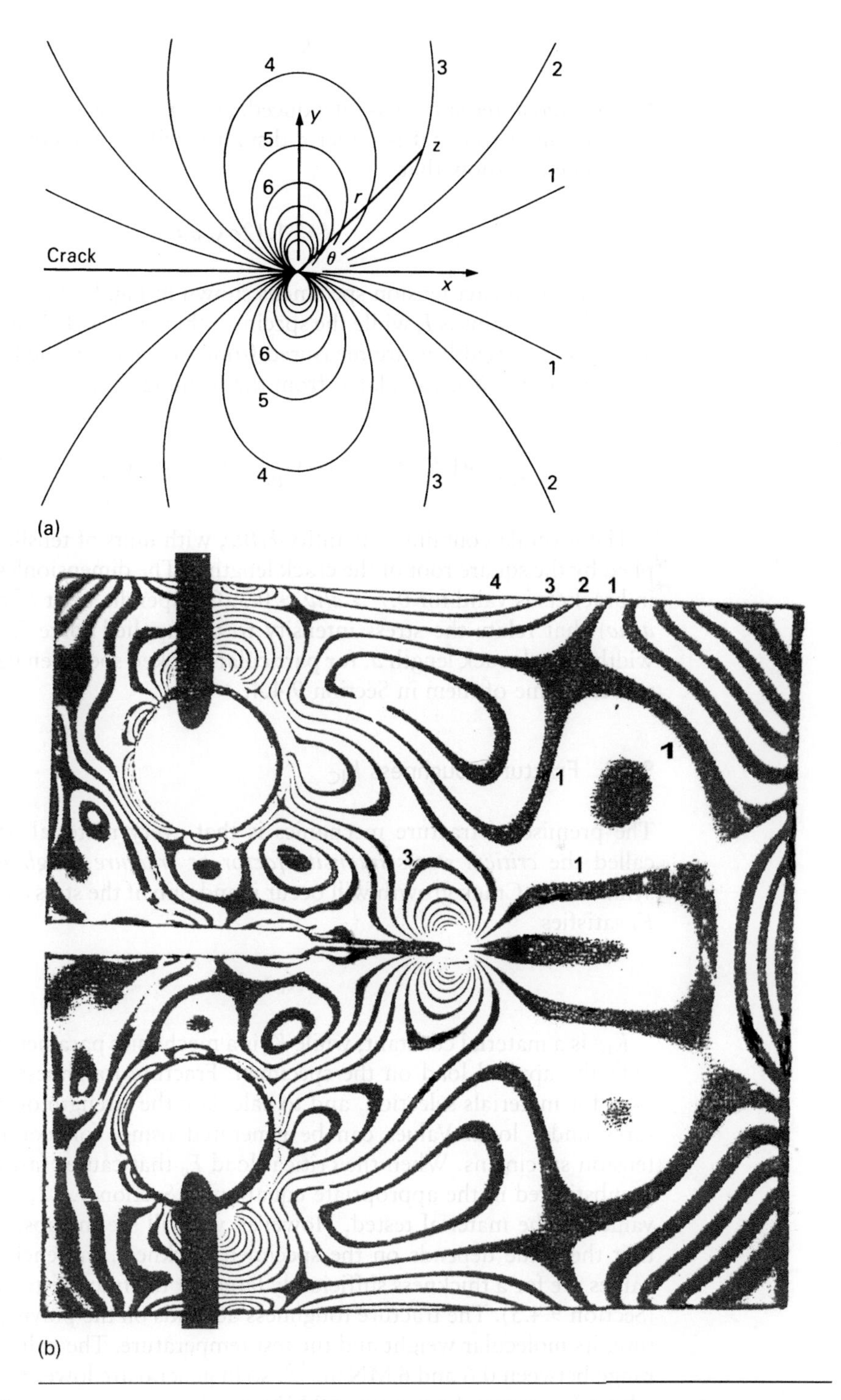

Figure 9.7 (a) Theoretical isochromatic photo-elastic pattern at a crack tip, with contours of maximum shear stress. (b) Experimental isochromatic pattern in a loaded polycarbonate compact tension specimen, with some fringe orders marked.

$$K = \sigma\sqrt{\pi a} \tag{9.11}$$

However, if a free surface is introduced, i.e. for an edge crack of length a in a semi-infinite sheet that is under a distant tensile stress, computer methods are needed to show that

$$K = 1.12\sigma\sqrt{\pi a} \tag{9.12}$$

For the compact tension specimens shown in Fig. 9.7b, the force applied to the loading pins is F, while the specimen thickness is B. The crack length a and specimen width w are measured from an origin on the line joining the loading points. K is calculated from the formula

$$K = \frac{F}{Bw}\sqrt{a}\left(29.6 - 185.5\left(\frac{a}{w}\right)^{0.5} + 655.7\left(\frac{a}{w}\right)^{1.5} - \ldots\right) \tag{9.13}$$

The formula contains a quantity F/Bw, with units of tensile stress, multiplied by the square root of the crack length a. The dimensionless polynomial in brackets is valid for $0.7 > a/w > 0.3$. Compendia exist of functions $K(F, a, w)$ that relate the stress intensity to the applied force F, the specimen width w and crack length a, for particular cracked specimen geometries. We will see some of them in Section 9.4.4.

9.4.3 Fracture toughness K_{IC}

The premise of fracture mechanics is that every material has a property called the *critical stress intensity factor* or *fracture toughness*, given the symbol K_{IC}. Crack growth will occur if and only if the stress intensity factor K_I satisfies

$$K_I > K_{IC} \tag{9.14}$$

K_{IC} is a material constant, while K_I is a mechanics parameter that changes with the applied load on the specimen. Fracture toughness values can be used for materials selection, and to calculate the strength of cracked structures under load. Values can be generated using, for example, compact tension specimens. When the critical load F_c that causes fast crack growth is substituted in the appropriate equation of Section 9.4.2, it gives the K_{IC} value for the material tested. However, we will see in subsequent sections that the value depends on the specimen thickness. In general, the quoted values are for a thickness sufficiently large for the fracture to be plane strain (Section 9.4.5). The fracture toughness depends on the polymer microstructure, its molecular weight and the test temperature. The values in Table 9.1 range between 0.6 and 6 $\text{MN m}^{-1.5}$, so in general are lower than for metals, where K_{IC} ranges from 5 to 100 $\text{MN m}^{-1.5}$.

Table 9.1 Typical K_{IC} values for polymers at 20 °C

Polymer	K_{IC} *(MN m$^{-1.5}$)*
Polyester thermoset	0.6
Polystyrene	0.7–1.1
Polymethyl methacrylate	0.7–1.6
Polyvinyl chloride	2–4
Polycarbonate	2.2
Polyamide (nylon 6,6)	2.5–3
Polyethylene	1–6
Polypropylene	3–4.5
Polyoxymethylene	4

9.4.4 Crack tip yielding and Dugdale's model

High stresses near the crack tip cause yielding in all polymers. This makes crack propagation more difficult, because the crack must grow through oriented polymer. In Fig. 9.7b there is yielding at the crack tip, yet the photo-elastic pattern follows Eq. (9.10) for radial distances of 0.1–10 mm. Therefore, K_I describes the form and magnitude of the elastic stress field surrounding the yielded region. Hence, if the yielded zone is small compared to the specimen width, and the polymer isotropic, the concept of K_{IC} is valid.

The elastic region can release strain energy to drive crack growth; an example of such a calculation is given in Chapter 14 in the gas pipe case study. However, if the yielded zone extends across the specimen width before crack growth occurs, or the polymer is anisotropic, the stress field implied by the use of K_I no longer exists. Other methods must be used to characterise fracture resistance.

Elastic–plastic fracture mechanics considers yielding near a crack tip. Dugdale in 1960 proposed a simple geometry for the yielded zone shape in steel, that is appropriate for some types of polymer behaviour. He assumed that the yielded zone is an extension of the crack plane, with height much smaller than its length. The tensile strain e_y in the yielded zone decreases with the distance x from the crack tip. The material is assumed not to work harden after tensile yielding, so the stress σ_{yy} across the yielded zone is constant at σ_0. Since the yielded zone shape and the boundary stresses were known, the stresses in the surrounding elastic region could be calculated. The elastic stress field surrounding at the crack tip before yielding is quantified by K_I, calculated as in Section 9.4.2. A yielded zone of length R forms directly ahead of the crack of length a (Fig. 9.8), relieving any stresses greater than the yield stress σ_0. A hypothetical crack, R longer than the real crack, is made by cutting around the boundary of the yielded zone. $R = a$, so the small increase in crack length does not significantly

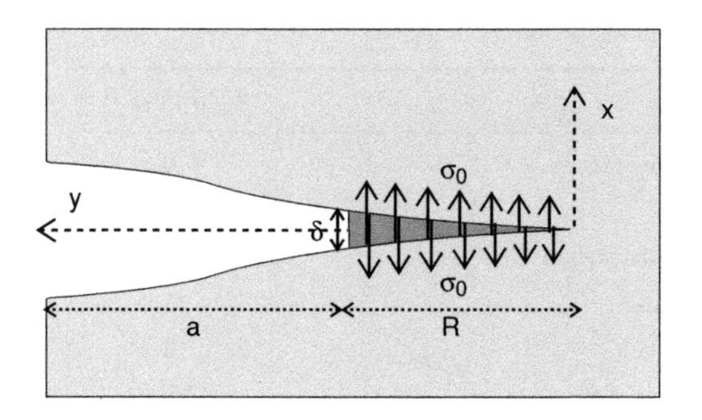

Figure 9.8 Dugdale's model for crack tip yielding under plane stress conditions. R, yielded zone length; δ, crack tip opening. The shaded region is the yielded zone, while the arrows show the closure force on the hypothetical crack.

increase K_I. Pairs of closure forces are applied to the cut boundary to prevent the shape of the surrounding elastic region changing.

The stress intensity factor due to a pair of closure forces F, acting a distance x from the tip of a crack is

$$K = -\frac{F}{t}\sqrt{\frac{2}{\pi x}} \tag{9.15}$$

The sum of such values, from pairs of closure forces acting on length elements dx between 0 and R, on the sheet of thickness t is

$$\begin{aligned} K_{close} &= \int_0^R \frac{\sigma_0 t}{t}\sqrt{\frac{2}{\pi x}}dx \\ &= -\sigma_0\sqrt{\frac{2}{\pi}}\left[2\sqrt{x}\right]_0^R \\ &= -\sigma_0\sqrt{\frac{8R}{\pi}} \end{aligned} \tag{9.16}$$

The total stress intensity K_{total} at the hypothetical crack tip must be zero, to avoid infinite stresses occurring there, according to Eq. (9.8), i.e.

$$K_{total} = K_I + K_{close} = 0 \tag{9.17}$$

The value of K_{close} increases as the yielded zone grows in length, until the condition of Eq. (9.17) is obeyed. Combining this with Eq. (9.16) gives

$$R = \frac{\pi}{8}\left(\frac{K_I}{\sigma_0}\right)^2 \tag{9.18}$$

Analysis also gives crack tip opening displacement δ_0 (Fig. 9.8), the height of the thick end of the yielded zone, as

$$\delta_0 = \frac{K_I^2}{\sigma_0 E^*} \tag{9.19}$$

where E^* is equal to the Young's modulus E under plane stress conditions, or $E/(1 - \nu^2)$ under plane strain conditions. Equation (9.19) is the basis of an alternative failure criterion that crack growth occurs when

$$\delta_0 > \delta_{0C} \tag{9.20}$$

where δ_{0C} is the critical crack tip opening displacement. If this failure criterion applies, Eq. (9.16) shows that the K_{IC} criterion of linear elastic fracture mechanics is still valid, if the length of the yielded zone is small compared to the specimen width.

9.4.5 Plain strain fracture in thick sheet

Some fracture mechanics jargon can be confusing, because similar expressions have different meanings elsewhere in mechanics. In Appendix C, *plane strain elastic deformation* means that the non-zero strains (in a pipe wall) occur in one plane. In *plane strain fracture*, the non-zero *plastic* strains in the yielded zone occur in the xy plane (Fig. 9.9a), that is perpendicular to the crack tip line. The strain $e_{zz} = 0$, so the sides of the specimen do not move inwards, and the fracture surface appears macroscopically flat. If a crack grows through a craze, a plane strain fracture will result. Voiding in the craze allows it to open, while the strain e_{zz} in the craze remains zero, and the surrounding material remains elastic.

The Dugdale model can be applied to a craze at a crack tip, because the craze plane is an extension of the crack plane, and the tensile stress across the craze (Chapter 8) is approximately constant. If the craze length is measured, in a specimen in which K_I is known, Eq. (9.13) can be used to calculate the craze stress σ_0. Craze thickness profiles, calculated from the interference fringe patterns seen when crazes are viewed normally using reflected monochromatic light, are in good agreement with the Dugdale model predictions (Fig. 9.10).

The crack growth condition of Eq. (9.15) can be used: A craze fails when its opening displacement reaches a critical value. However, this does not explain the failure mechanism. It could be by failure of the entanglement network in the craze fibrils. Crazes in some polymers fail at their mid-planes, and in other polymers at the bulk–craze interface. For viscoelastic materials, in which both the craze stress and the Young's modulus vary with the strain rate, Eq. (9.19) predicts that the crack tip opening displacement is no longer proportional to the stress intensity factor. Figure 9.11 shows that

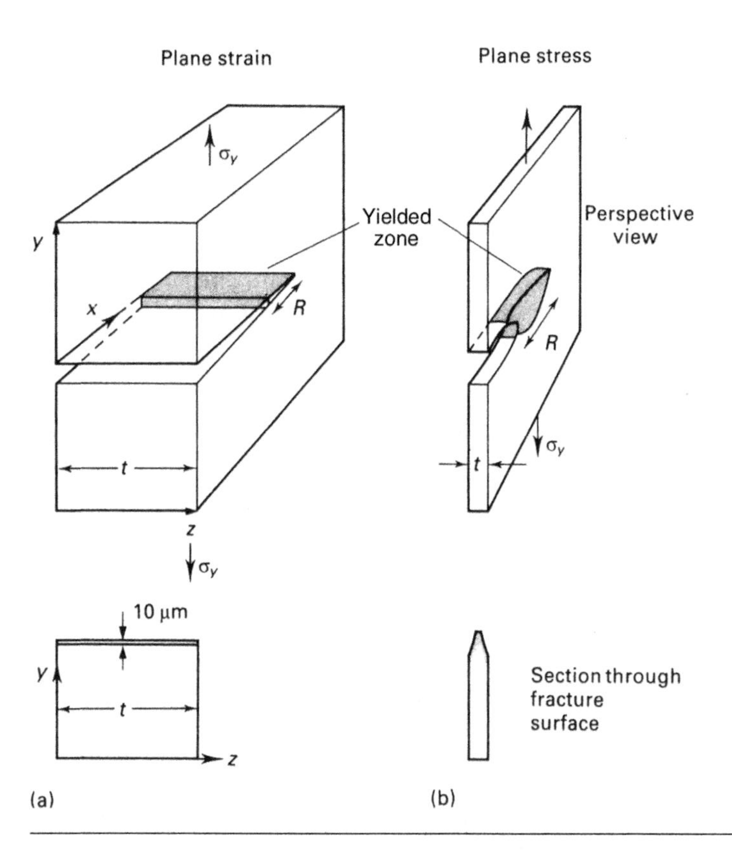

Figure 9.9 Yielding at a crack tip and its effect on the fracture surface appearance. The two limiting cases are: (a) Plane strain fracture in thick specimens and (b) plane stress fracture in thin specimens.

the crack opening displacement criterion of Eq. (9.20) is better obeyed for PMMA than the K_{IC} criterion.

The δ_{0C} criterion applies when the fracture mechanism is unchanged. For PMMA, slowly growing cracks progress down the mid-plane of a growing craze. When the crack velocity exceeds $0.05\,\mathrm{m\,s^{-1}}$, it suddenly accelerates to over $100\,\mathrm{m\,s^{-1}}$ without an increase in K_I. The fracture surface appearance changes from being flat to having a set of parabolic markings (Fig. 9.12). These are ridges where the surface changes level by about $10\,\mu\mathrm{m}$. The ridges result from the nucleation of penny-shaped cracks, ahead of the main crack front, nearly in its plane. Before the penny-shaped cracks grow large, they are overrun by the main crack front. Continued crack nucleation causes a 'leap frog' effect, and allows a much higher crack speed than possible with a single crack. The critical δ_{0C} value will be different for this situation than it is for slow crack growth.

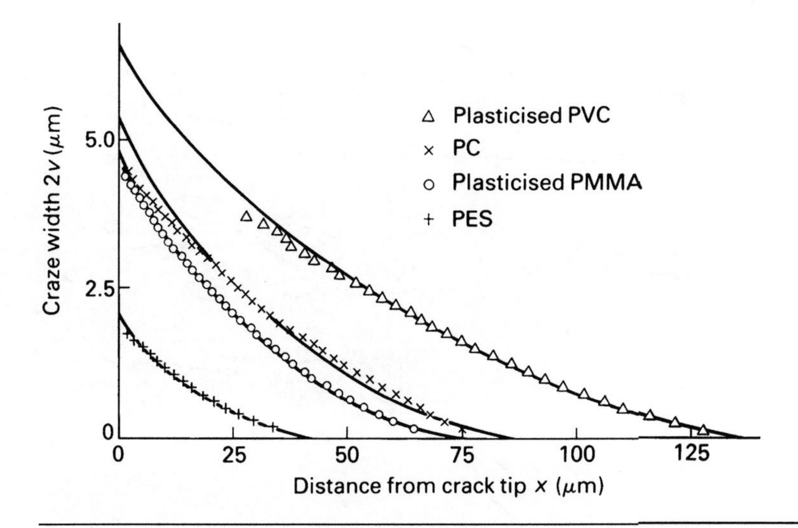

Figure 9.10 Thickness profile of crazes in PES, PMMA, PC and plasticised PVC, fitted by the prediction of Dugdale's model (from Doll, W., *Adv. Polym. Sci.*, **52/3**, 119, 1983, Springer Verlag). The vertical scale is exaggerated.

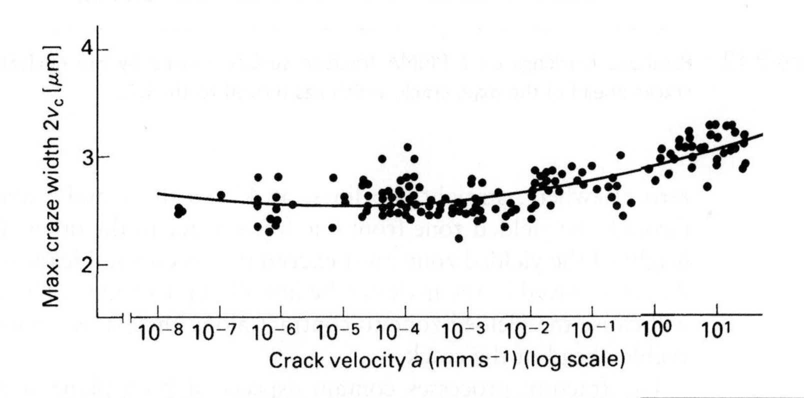

Figure 9.11 Variation of crack tip opening with crack velocity in PMMA (from Doll, W., *Adv. Polym. Sci.*, **52/3**, 120, 1983, Springer Verlag).

9.4.6 Plane stress fracture in thin sheet

The mechanics concept of a *state of plane stress* means that the only non-zero stress components act in a particular plane. If this is the *xy* plane, there are non-zero tensile stresses σ_{xx} and σ_{yy} and shear stress σ_{xy}. In a *plane stress fracture* there is a state of plane stress *in the yielded zone* ahead of the crack. The tensile stress σ_{zz}, parallel to the crack front is zero, while the non-zero stress components are σ_{xx}, σ_{yy} and σ_{xy}. The stress component σ_{zz} will always be zero at the free surfaces of a specimen. For σ_{zz} to be equal to

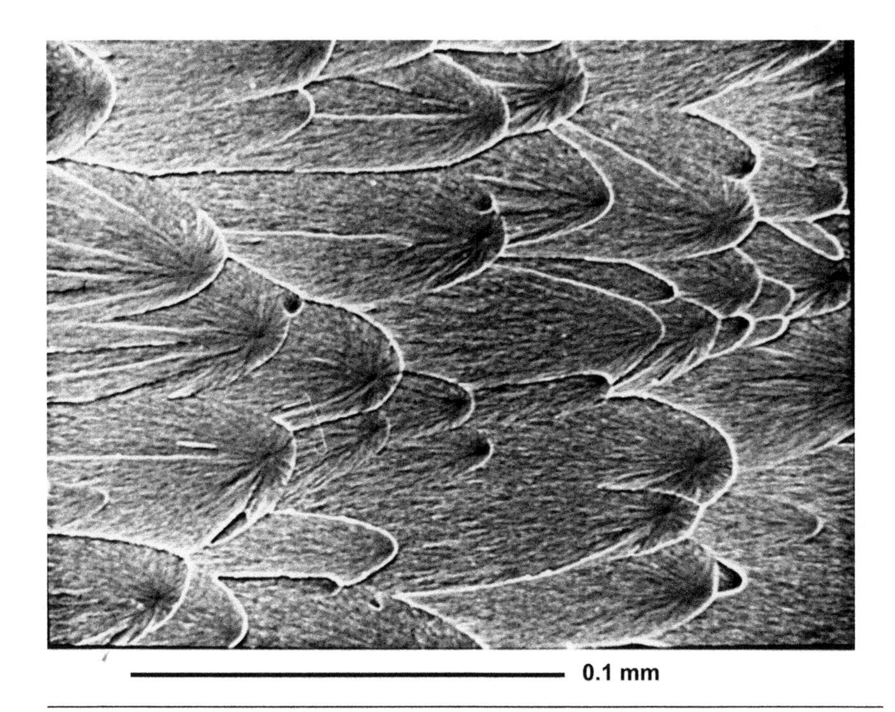

Figure 9.12 Parabolic markings on a PMMA fracture surface caused by the nucleation of disc like cracks ahead of the main crack, which has moved to the left.

zero elsewhere, straight slip lines, at 45° to the y and z axes, must pass through the yielded zone from one free surface to the other. Therefore, the height of the yielded zone must exceed the specimen thickness t (Fig. 9.9b). Points involved in the neck can be initially up to t apart. Slip on these lines will cause the yielded zone to contract along the z axis, so necking will be visible ahead of the crack.

The fracture processes contain aspects of both plane stress and plane strain behaviour. The initial stages of crack growth will always be under plane strain conditions, because the yielded zone must grow to a certain size before plane stress conditions can develop. However, plane stress conditions always apply near the free surfaces. Consequently, plane stress 'shear lips' are often observed at the edges of flat fracture surfaces.

The opening displacement δ of a yielded zone (or neck) is the distance that a point just above the neck has moved apart from a point just below the neck. It is less than the height of the neck h. Figure 9.13 shows the necked region ahead of a crack in 0.2 mm thick polycarbonate, after some crack growth. The polarised monochromatic light distinguishes the yielded region from the elastic material. Although the yielded zone shape is less elongated than what Dugdale assumed, Eq. (9.18) predicts its length to within 20%. The post-yield tensile stress–strain curve of

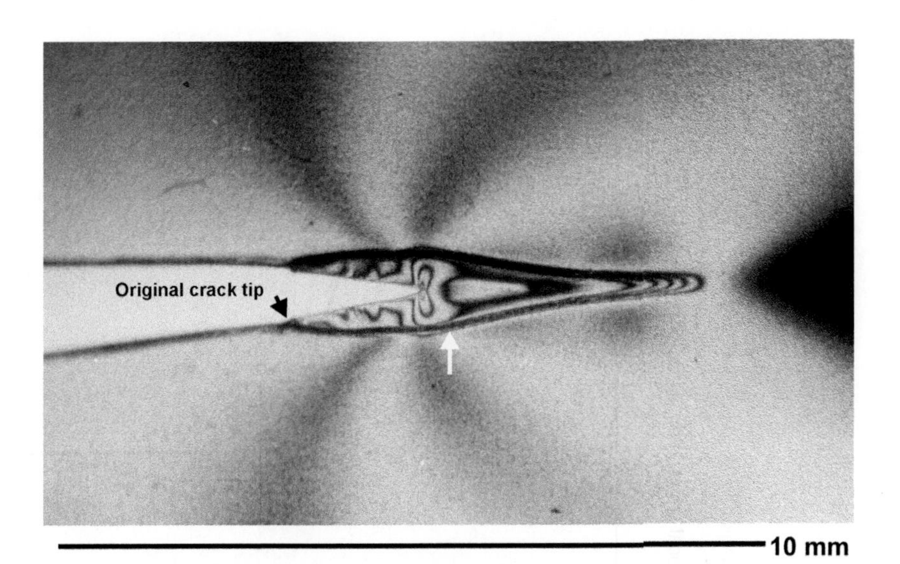

Figure 9.13 Plane stress yielding at a crack tip in 0.2 mm thick polycarbonate, viewed with circularly polarised light. The white arrow indicates the boundary of the yielded zone.

polycarbonate has insignificant orientation hardening until the strain exceeds 1 unit, so the constant stress condition of the Dugdale model is met. If the experiment is repeated with isotropic PVC sheet, or with biaxially stretched PET from a carbonated drink bottle, the yielded zone has a quite different shape. Biaxially stretched PET orientation hardens very rapidly after yield, and the crack plane tries to turn to be parallel to the sheet surface. The material prefers to 'delaminate' rather than to allow through-thickness crack growth. Therefore, the model does not apply to these materials.

Necking produces an orientation-hardened material with anisotropic yield properties. In the neck, the shear strain in the through-thickness yz plane is higher than that in the xy plane. It is easier for further shear to occur at the crack tip in the xy plane than the yz plane, causing the crack faces to open to an acute angle. The advancing crack must tear through stretched material, in which the covalent carbon–carbon bonds are preferentially oriented across the crack path. Therefore, the fracture toughness under plane stress conditions is several times higher than under plane strain conditions. Figure 9.14 shows that K_{IC} approximately doubles when the polycarbonate specimen thickness is less than 4 mm.

It is possible to estimate the specimen thickness at which the level of K_{IC} falls rapidly. Equation (9.18) is used to calculate the yielded zone length when $K = K_{IC}$. It is observed that to achieve plane strain conditions, the specimen thickness should exceed five times the yielded zone length. Hence, the critical thickness is

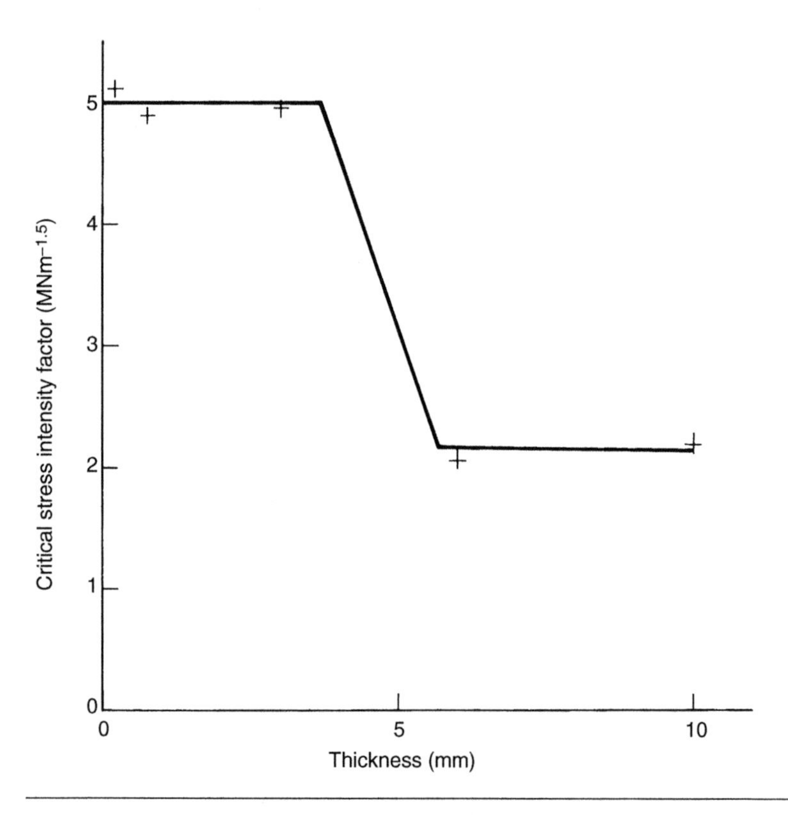

Figure 9.14 Variation in the critical stress intensity factor with specimen thickness for slow fracture tests in polycarbonate.

$$t_c = 2\left(\frac{K_{IC}}{\sigma_0}\right)^2 \tag{9.21}$$

If we substitute $K_{IC} = 2.2\ \mathrm{MN\,m^{-1.5}}$, and $\sigma_0 = 57\ \mathrm{MPa}$ for polycarbonate to yield on a 1 min timescale at room temperature, the predicted critical thickness is 3 mm. On an impact loading timescale, the yield stress is higher, so the critical thickness is only 0.7 mm. The critical thickness t_c is a useful polymer selection criterion. It varies from 0.6 mm for a PMMA with $K_{IC} = 1.6\ \mathrm{MN\,m^{-1.5}}$, $\sigma_0 =$ 90 MPa, to 195 mm for an MDPE with $K_{IC} = 5\ \mathrm{MN\,m^{-1.5}}$, $\sigma_0 = 16\ \mathrm{MPa}$. Consequently, the polyethylene is more suitable than PMMA for a thick-walled gas pipe, which must not fail by plane strain fracture.

9.4.7 Strain rate and crack velocity effects

Chapters 7 and 8 showed that the moduli and yield stresses of plastics are time-dependent. Consequently, we expect that the fracture properties will

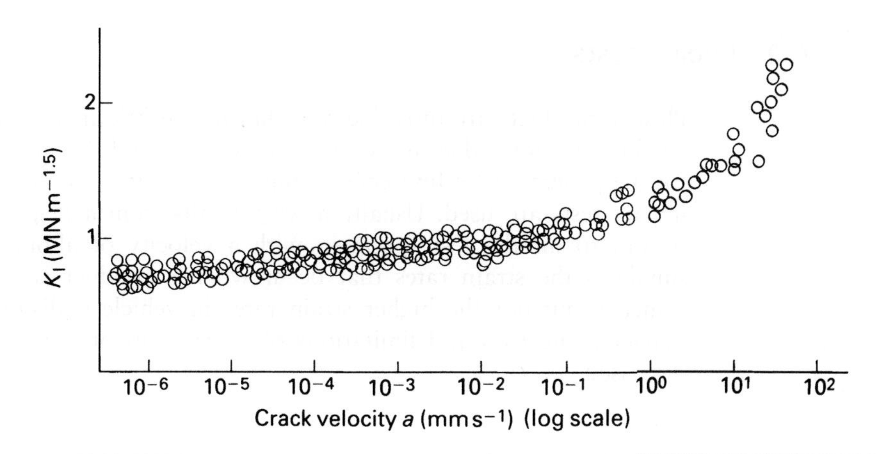

Figure 9.15 Variation of stress intensity factor with crack velocity in PMMA (from Doll, W., *Adv. Polym. Sci.*, **52/3**, 119, 1983, Springer Verlag).

also be time- or crack velocity-dependent. Figure 9.15 shows crack velocity data versus K_I for a glassy polymer, determined using the double torsion loading mentioned in Section 9.1. Slow crack growth is possible at much lower K_I values than is rapid crack growth. Above a velocity V of about 50 m s^{-1}, crack growth becomes unstable, and the velocity jumps to about half the speed of sound without any further increase in the applied K_I. Rapid crack growth is therefore taken as being at a speed exceeding 100 m s^{-1}.

To allow for the slow crack propagation at stress intensity factors $K_I < K_{IC}$, the criterion of Eq. (9.14) is re-expressed as

$$\text{Rapid crack growth occurs if and only if } K_I > K_{IC} \qquad (9.22)$$

Hence, the fracture toughness K_{IC} relates to rapid crack growth, while the function $K_I(V)$ characterises slow crack growth. Both $K_I(V)$, and the function relating the stress intensity factor to the load and geometry (Section 9.4.2), must be known if the life of a cracked product is to be calculated.

The failure load of a fracture toughness specimen depends on the rate of load application. If a cracked compact tension specimen of 5 mm thick polycarbonate is loaded in a tensile testing machine, there is time for a neck to develop from the crack tip, so plane stress fracture occurs at a crack velocity of 5 m s^{-1}. However, if the load is applied in 1 ms by impact loading, plane strain fracture occurs at a low K_{IC} value, and the crack velocity exceeds 200 m s^{-1}. Special instrumentation, having a quartz crystal force gauge that responds in 0.1 ms, and a grid of resistance lines on the surface to monitor the crack velocity, are needed to measure the K_{IC} value.

9.5 Impact tests

Plastics products are most likely to fail in a brittle manner under impact conditions, both due to strain rate effects and because large forces can be generated by low energy impacts on stiff structures. A variety of impact tests are used. Usually a weight falls from a height of the order of 1 m to hit the test specimen with a velocity of about $5\,m\,s^{-1}$. This simulates the strain rates that occur when a product is dropped about a metre, but not the higher strain rates in vehicle collisions or ballistic impacts. The uses and limitations of three types of impact tests will be discussed.

9.5.1 Izod and Charpy impact tests on bars

The Izod test is a variant of the notched bending test. A swinging pendulum hits a clamped bar (Fig. 9.16) and loses kinetic energy. The bar has a width of 12.5 mm and a thickness representative of the plastic product considered. The 2.5 mm deep, 45° notch has a tip radius of 0.25 mm. Section 9.3.2 showed that the tensile stress in the yielded zone at the notch tip can exceed the uniaxial tension yield stress by up to 118%. It is common to use 3.2 mm thick specimens, which restricts the value of the data to products of similar thickness. The lower half of the bar is clamped in vice, and the upper part is struck 22 mm above the notch by the pendulum. The results can be quoted as the absorbed energy divided by the specimen thickness, in $J\,m^{-1}$, or as the absorbed energy divided by the area of one fracture surface, in $J\,m^{-2}$.

Charpy specimens are also notched in the centre, but the bar is freely supported at the ends, and struck in the centre of the face opposite the

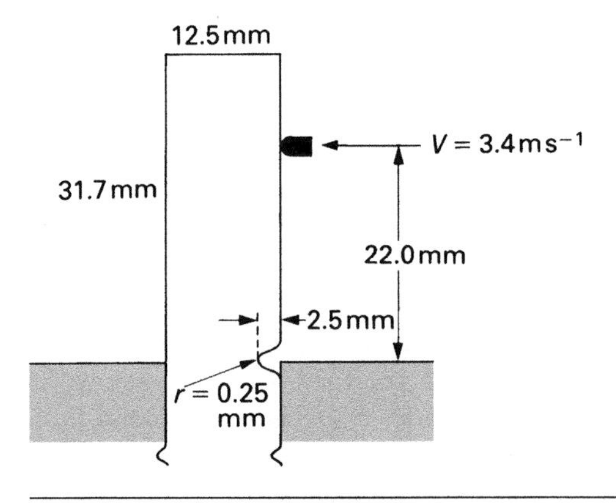

Figure 9.16 Izod impact notched bar clamped for testing.

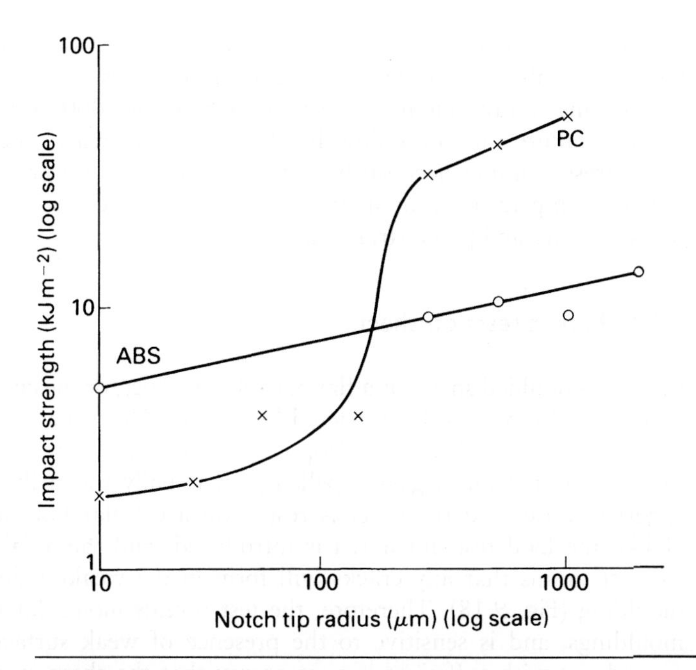

Figure 9.17 Variation of the energy absorbed in a Charpy impact test with the radius of the notch tip in 3 mm thick bars (from Ogorkiewicz, R. M., Ed., *Thermoplastics—Properties and Design*, Wiley, 1974).

notch. Consequently, the bar undergoes a three-point bending. Figure 9.17 shows the effect of notch tip radius on the impact energy for Charpy impact tests. The ABS, which contains dispersed rubber particles, is much less sensitive to the notch radius than the glassy polycarbonate. The notch tip radius needs to be less than 100 μm to reveal the low impact strength of 3 mm thick polycarbonate, whereas the standard test uses a notch tip radius of 250 μm.

Izod or Charpy impact strengths are quoted by materials manufacturers, as these tests are quick and easy to perform. They are also used for quality control purposes. However, the values do not correlate well with fracture toughness. The analysis in Section 9.3.2 shows that a crack may initiate beneath the notch after some yielding has occurred. The Izod or Charpy energy is a combination of crack initiation and propagation energies, in unknown proportions. If the data is quoted in $\mathrm{J\,m^{-1}}$, it suggests that the result is independent of the specimen thickness. However, there is a plane stress to plane strain fracture transition as the thickness increases, so the impact strength in $\mathrm{J\,m^{-1}}$ decreases when the transition thickness is exceeded. Hence, the toughness should be measured for the same thickness as the proposed product. If the data is quoted in units of $\mathrm{J\,m^{-2}}$, as in Fig. 9.17, it suggests that the crack propagation requires a constant amount of energy per unit area of crack surface, whereas the energy consumption rate may be a strong function of the crack velocity. Consequently, the test results are only suitable for quality control purposes.

The 5 ms failure time in a Charpy impact test is 2000 times smaller than the 10 s failure time in a slow bend test. This increases the value of the yield stress, without necessarily increasing the stress for crazing. For some polymer/temperature combinations there may be a changeover from yielding to crazing and plane strain fracture.

9.5.2 Impact tests on sheet

Injection-moulded sheet, or a flat part of a product, or the curved surface of an extruded pipe, can be impacted by a mass. The impact velocity can be varied from 1 to 100 $\mathrm{m\,s^{-1}}$ by dropping the projectile from different heights, or by firing it from a gas-propelled gun. Usually the striker has a hemispherical nose, and the sheet is rested on a circular hole in a platform. Unlike the Izod test, no notch is introduced, and the axial symmetry of the test means that any cracks will form in the weakest direction of the moulding (Fig. 9.18). Therefore, the test reveals molecular orientation in mouldings, and is sensitive to the presence of weak surface layers. It is carried out with sufficient impact energy that the sheet must either yield locally and stop the mass, or crack and allow the mass to pass through, or yield then fail by ductile tearing.

Figure 9.18 Crack along the flow direction in a weathered, 100 $\mathrm{mm^2}$, ABS moulding, due to a central impact.

The stress analysis of the test depends on the phenomena that occur

(a) if the disc remains elastic
The lower surface of the sheet is subjected to a balanced biaxial tensile stress, which is maximum at the centre of the sheet. If a central force F is applied as a uniform pressure to a disc of radius a, on a sheet of thickness t supported at a radius R, the central biaxial tensile stress is

$$\sigma_\theta = \sigma_r = \frac{3F}{2\pi t^2}(1+\nu)\left(\ln\frac{R}{a} + \frac{a^2}{4R^2}\right) \tag{9.23}$$

Outside the contact radius the stresses are proportional $\ln(a/r)$, so high tensile stresses are limited to a small central region of the lower surface of the plate. If the plate fails due to crazing or crack propagation from some defect in the high stress region, the failure load could be small.
If the outside rim of the sheet is clamped, or the plate is a flat part of a larger product, large deflections cause in-plane stretching of the sheet and membrane stresses to be set up. These are constant in value through the thickness of the sheet in contrast with the bending stresses.

(b) if the central region of the plate yields
The yield geometry is similar to that described in Section 8.2.6, so the stress analysis of that section can be used. The force on the striker goes through a maximum value given by

$$F_{max} = \pi D t \sigma_y \tag{9.24}$$

where D is the diameter of the cylindrical portion of the punch, and σ_y is the tensile yield stress.
Cracks may initiate if a critical level of plastic strain is exceeded.
However, a plate that deforms to this degree is not in danger of brittle fracture at low impact energy. Hence, the value of the test is to screen injection mouldings for brittle failure, at forces well below F_{max}, due to orientation, weld lines or a weak surface structure.

9.5.3 Impact tests on products

The failure mechanisms that occur in small laboratory specimens often bear no relationship to service failures. Hence, product tests have been developed, by the British Standards Institution and others, to simulate product service conditions, using simple, reliable apparatus. For example, BS 6658:1985 'Protective helmets for vehicle users' requires impact tests to be carried out on motorcycle helmets after *crack initiation* is encouraged by wiping the outer surface of the helmet shell with a toluene/iso-octane mixture. This can rapidly produce small crazes and cracks at highly stressed places, such as where webbing chin strap were riveted to the shell of polycarbonate helmets manufactured prior to 1981. In real life, such cracks developed over a period of years (Fig. 9.19) under the influence of less severe environmental agents. The cracks that radiated from the rivet hole, were often hidden from sight. The webbing acted as a wick to the high stress region, where liquid could not easily evaporate. The problem was cured by using rubber-toughened polycarbonate that was less susceptible to stress cracking (Section 10.5), and by sewing the webbing strap to a steel hanger plate, which is riveted to the shell.

Rapid crack propagation is encouraged by the high energy impact. The helmet is cooled to $-20\,°C$ before falling at $7\,m\,s^{-1}$ on to a hemispherical steel anvil with a 50 mm radius. The kinetic energy of the head-form and the helmet is of the order of 150 J. The test temperature is at the lower limit of tolerable weather conditions, compensating for the sometimes higher impact speeds in crashes and the blunt nature of the anvil. The test is much simpler than ones using vehicles and dummies. Thermoplastic helmet shells should buckle inwards without yielding, and the polystyrene foam liner underneath should crush. If any small cracks propagate rapidly, the shell could split into two or more sections, and the load-spreading function of the shell would be lost. The shell normally absorbs about 30% of the impact energy, so if it is fractured the peak acceleration of the head-form would exceed the 300 'g' test limit. The large size of the product gives space for cracks to accelerate, and the high impact energy provides the driving force for such high-speed crack growth.

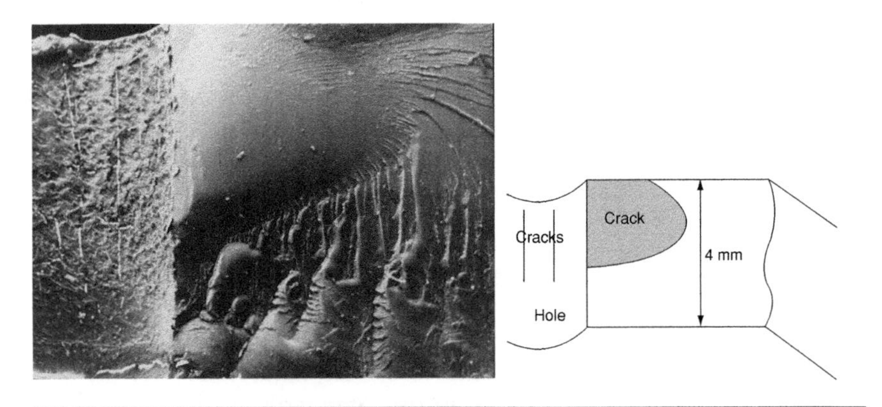

Figure 9.19 Crack at the side of a drilled hole in a polycarbonate helmet shell-fracture surface and schematic of features. The hole was loaded by impact forces from the chin strap rivet.

Many product standards contain *performance tests* rather than specifying the materials to be used, their fracture toughness or product design details. It is up to the manufacturers to reach the required performance level by whatever means possible. It is possible to design a helmet shell without rivet holes, or to use a plastic with higher fracture toughness. It is difficult to design a laboratory fracture test with the crack geometry of Fig. 9.19. Molecular orientation, residual stresses and local shape details differ between laboratory specimens and helmet shell mouldings, so the correlation between laboratory fracture tests and helmet performance is poor. Consequently, product testing, under conditions that simulate in-service failures, is essential.

9.5.4 Instrumented impact tests

The instrumentation of impact testing machines aims to record the force–time, and possibly the deflection–time, responses. These give clues to the nature of the fracture process. Whether there is yielding before crack initiation or during crack growth. However, several phenomena may make the traces difficult to interpret. Striker forces can be measured using a stiff, quartz piezoelectric force cell beneath the striker, and digitizing the data at time intervals of the order of 10 μs. The contact stiffness k between the striker and the specimen is defined as the slope of the force–deflection graph, for direct compressive loading in the absence of dynamic effects. k is usually orders of magnitude greater than the specimen bending stiffness. In the impact test there is an initial force peak (Fig. 9.18) of magnitude

$$F_1 = V\sqrt{mk} \tag{9.25}$$

where m is the effective specimen mass and V the striker velocity. The effective mass is the mass, moving with the velocity of the contact point, which has the same momentum as the bending beam. For a cantilever beam struck at the end, m is 39% of the specimen mass.

This initial force peak, from the elastic 'collision' between the striker and the specimen, excites flexural vibrations in the beam (Fig. 9.20). In an Izod test, where the specimen breaks near the clamped end, these oscillations are not transmitted to the failure region, and are an unwanted addition to the recorded force signal. They can be largely removed by reducing the contact stiffness, using a thin layer of high hysteresis polyurethane rubber on the striker face. It is not acceptable to remove the force oscillations by electronic filtering frequencies >200 Hz, as this distorts the shape of the force–time data. If un-notched bars are impacted it is possible to generate stress–strain curves for the maximum stress region on a 5 ms timescale. Figure 9.20 shows that the fracture stress of polystyrene has risen from 40 MPa in a conventional tensile test to 100 Mpa as a result of the time dependence of the craze stress.

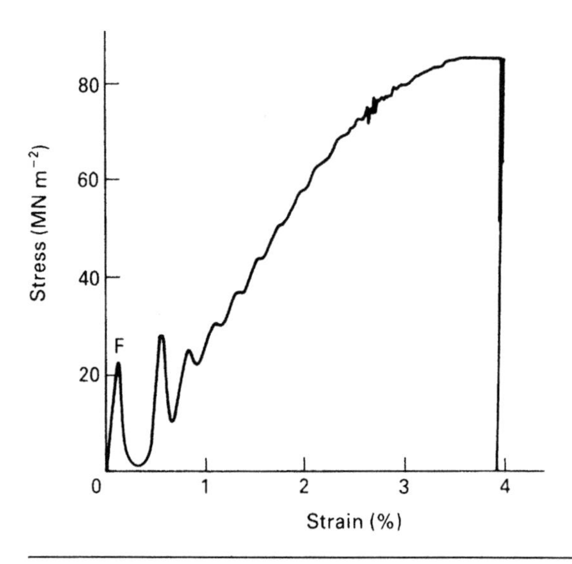

Figure 9.20 Stress–strain graph from an instrumented bend test on an un-notched polystyrene bar. The force corresponding to the initial peak F is predicted by Eq. (9.25) (from Mills N. J. and Zhang P. *J. Mater. Sci.*, **24**, 2099, 1989).

Figure 9.21a shows rapid force oscillations during crack growth in a 5 mm thick Charpy specimen impacted at $1\,\mathrm{m\,s^{-1}}$, which contained a sharp pre-crack. The HDPE had a density of $955\,\mathrm{kg\,m^{-3}}$. A number of crack arrest locations are visible on the fracture surface (Fig. 9.21b), so the crack appears to advance in an unstable manner. However, as the crack advances fastest at the specimen mid-thickness, it is not possible to monitor its position by photography, which would only record the surface crack growth through the shear lips (S).

It is preferable to identify and remove the cause of the oscillation, if this does not interfere with the impact test. Aggag *et al.* (1996) found that a 0.5 mm thick rubber sheet, on the surface of a polycarbonate Charpy bar impacted at $2.8\,\mathrm{m\,s^{-1}}$, was more effective at removing unwanted oscillations than was electrical filtering at 3 kHz (Fig. 9.22). In the striker (dart) signal, proportional to the striker force, the oscillations has almost disappeared before the fracture load. If 1 or 1.5 mm thick rubber layers were used, oscillations were suppressed, but there is a gentle initial rise in the signal as the rubber is compressed. The signal from a strain gauge close to the notch contained only a small oscillatory component, even when no 'mechanical' filtering was applied. Thus, the large striker force oscillations, observed in the absence of mechanical damping, are unlikely to influence the failure process.

However, some products contain heavy components that accelerate in impact tests. The shell of a motorcycle helmet can have a 700 g mass. When it is accelerated to $7\,\mathrm{m\,s^{-1}}$ in 5 ms in a typical impact test, significant forces

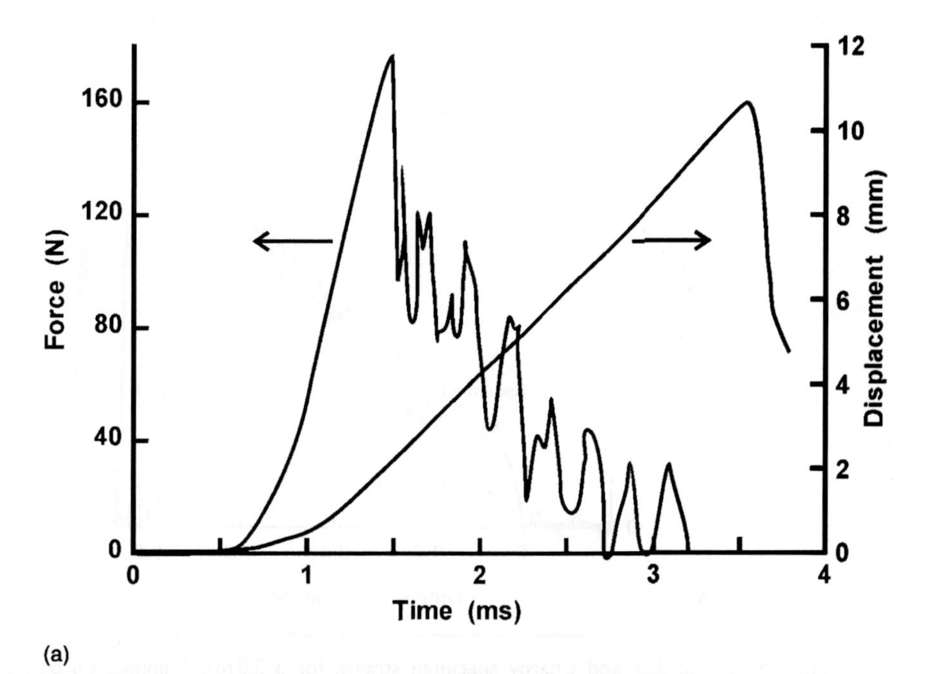

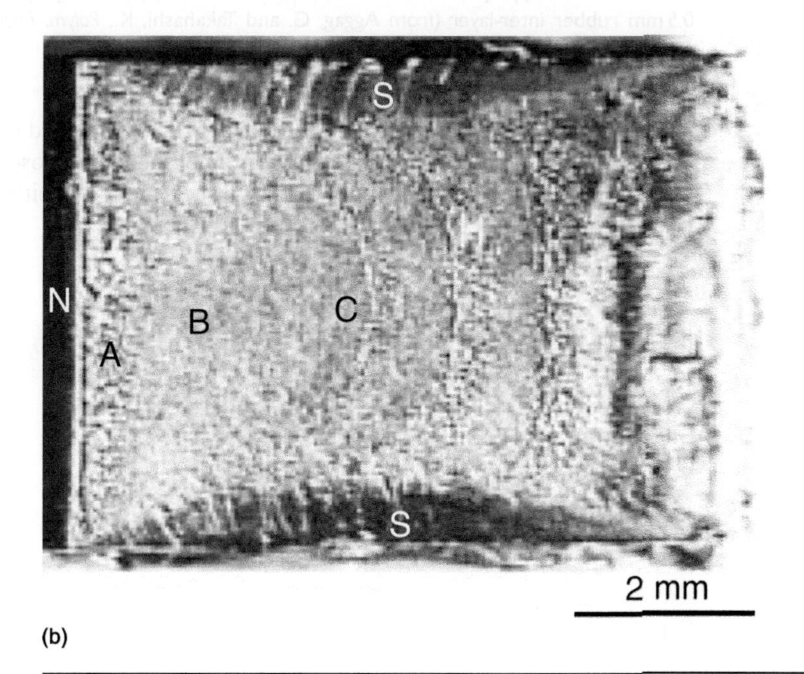

Figure 9.21 Charpy impact on an HDPE specimen: (a) Force oscillations when crack growth starts; (b) fracture surface markings (from Ravi, S. and Takahashi, K., *Polym. Eng. Sci.*, **42**, 2146, 2002) © John Wiley and Sons Inc. reprinted with permission. N, notch; A, craze; B, hackle; C, interior; S, shear lip zones.

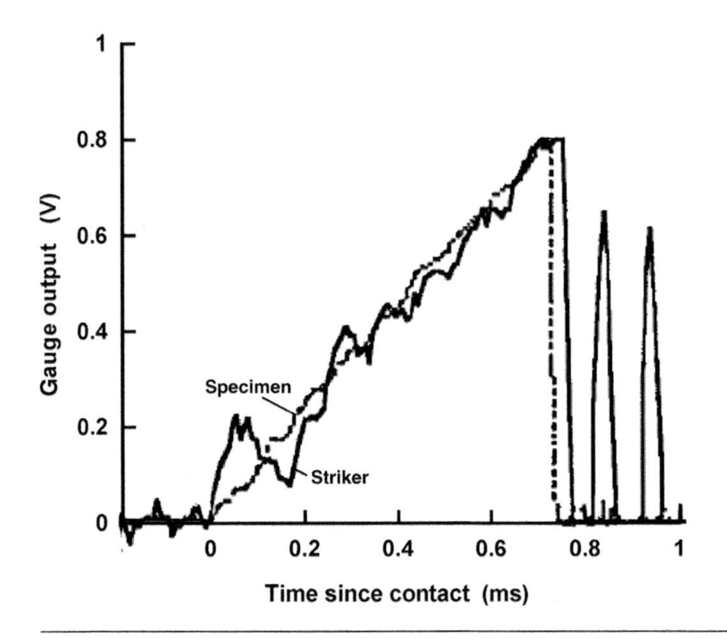

Figure 9.22 Striker and Charpy specimen strains, for a 2.8 m s^{-1} impact on polycarbonate with a 0.5 mm rubber inter-layer (from Aggag, G. and Takahashi, K., *Polym. Eng. Sci.*, **36**, 2260, 1996) © John Wiley and Sons Inc. reprinted with permission.

are required. Therefore, the impact force acting on the head (or head-form) differs from that acting on the striker. The shell mass oscillates on the polystyrene foam liner; these dynamic effects should neither be filtered out, nor attenuated by the use of compliant inter-layers.

Chapter 10

Degradation and environmental effects

Chapter contents

10.1 Introduction

The degradation of a polymer, when exposed to environmental factors such as oxygen, heat, UV light and moisture depends on its chemical structure. Examples will be given of the range of phenomena and the methods used to combat deterioration. Details for particular polymers are given in specialised texts. We start with the effects of melt processing, since the chemical changes can affect subsequent degradation. The effects of heat and oxidation are described before the more complex phenomena of outdoor weathering and environmental stress cracking (ESC).

An approximate ranking of the stability of polymers in a vacuum can be found by heating them at a constant rate until half of the initial mass has been lost. Figure 10.1 shows that this temperature T_h correlates reasonably well with the estimated dissociation energy of the weakest bonds in the polymer. The figure indicates the relatively low decomposition temperatures of polymers, compared with most other materials, and the relative stability of the fully fluorinated structure of PTFE.

10.2 Degradation during processing

The conditions during melt processing are extreme compared to those during the subsequent product life. Melt temperatures are high, as is the diffusion coefficient for oxygen, and mechanical stresses can be applied. Processing conditions and additives should be chosen so that the chemical

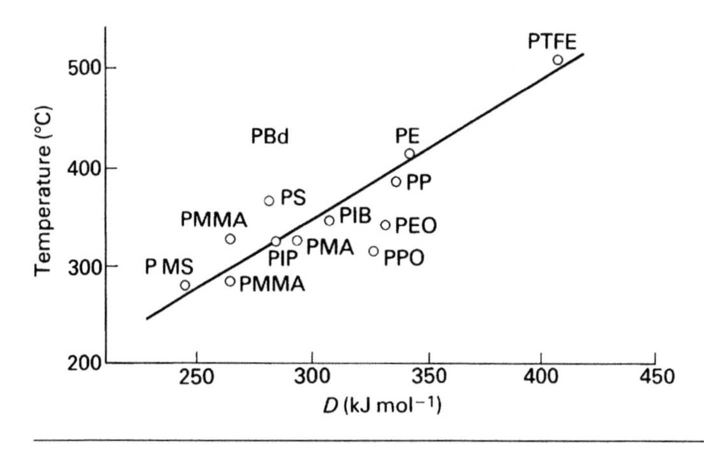

Figure 10.1 Temperature at which half the mass of a polymer is lost on heating in a vacuum vs. the bond dissociation energy (from Kelen, T., *Polymer Degradation*, Van Nostrand Rheinhold, 1983).

structure and the molecular weight of the polymer are not significantly changed.

10.2.1 Polyolefins

Polypropylene is more susceptible to melt degradation than polyethylene, because of the presence of more reactive *tertiary* hydrogen atoms (attached to the carbon atom that is bonded to three other C atoms). At a temperature of 270 °C in injection moulding, tertiary alkyl free radicals $R^{\bullet}$ are generated thermally. If oxygen is present, a rapid reaction ($R^{\bullet} + O_2 \rightarrow ROO^{\bullet}$) produces a peroxide radical, which reacts further to form hydroperoxides ($ROO^{\bullet} + RH \rightarrow ROOH + R^{\bullet}$). When the dissolved oxygen is used up, there is a greater chance of the chain scission reaction

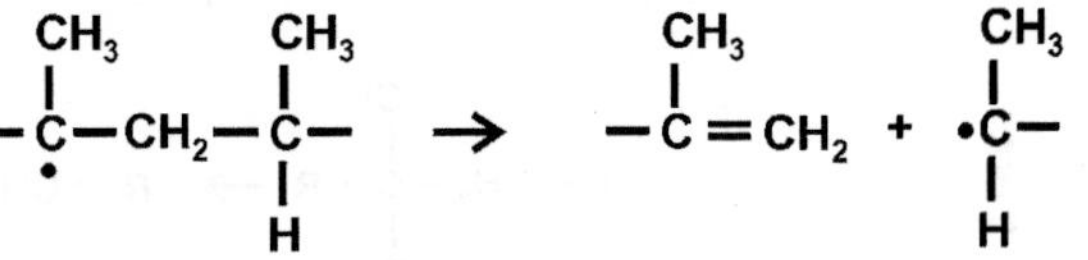

$$-\overset{\displaystyle CH_3}{\underset{\bullet}{C}}-CH_2-\overset{\displaystyle CH_3}{\underset{\displaystyle H}{C}}- \;\rightarrow\; -\overset{\displaystyle CH_3}{C}=CH_2 \;+\; \bullet\overset{\displaystyle CH_3}{\underset{\displaystyle H}{C}}-$$

Hindered phenol stabilisers *scavenge* alkyl radicals and prevent chain scission. An example is butylated hydroxytoluene (BHT)

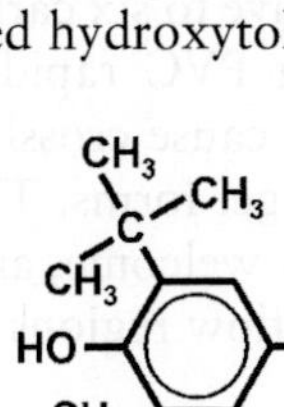

$$HO-C_6H_2(CH_3)\left[C(CH_3)_3\right]_2$$

Used in concentrations of less than 1%, it can increase the period of melt stability by an order of magnitude. Oxidation during melt processing produces carbonyl (−COO) and hydroperoxide (−COOH) groups, which accelerate photo-oxidation in outdoor weathering. As a fraction of the melt stabiliser is consumed on processing, only a low percentage of reground polymer, from faulty mouldings, can be mixed with virgin polymer if product quality is to be considered.

Melt degradation is used commercially to narrow the molecular weight distribution, from a starting value of $M_W/M_N \cong 5$, to make it more suitable for fibre spinning and blow moulding. Random chain scission at tertiary C−H bonds eventually produces a 'most probable' molecular weight distribution with $M_W/M_N = 2$. In practice, the distribution, produced by deliberately degrading the polypropylene with added peroxides, is broader, but higher molecular weight molecules are preferentially degraded. As M_W/M_N decreases, the melt viscosity becomes less non-Newtonian.

10.2.2 PVC

PVC is not stable at temperatures of 220–230 °C at which the crystalline phase melts. As the crystallinity of PVC is only of the order of 10%, processing in the semisolid state is not an insuperable problem, but the apparent viscosity is much higher than for most other polymer melts. Tertiary chlorine atoms, which occur at long-chain branches in PVC, are weak points, where the elimination of a hydrogen chloride molecule can occur.

R represents part of a PVC chain. The double bond formed is another weak site, so further HCl loss occurs from the neighbouring units. The resulting conjugated double bond polyene structure

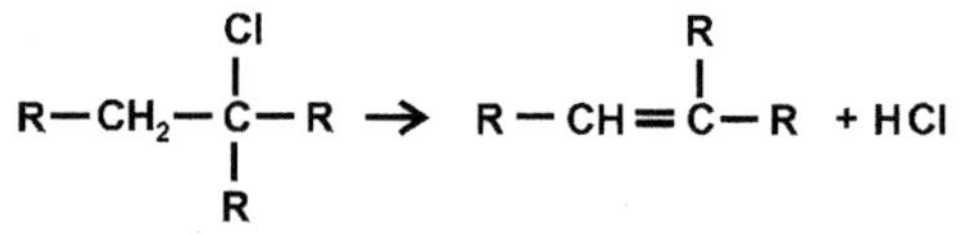

$$R{-}CH_2{-}\underset{R}{\overset{Cl}{C}}{-}R \rightarrow R{-}CH{=}\overset{R}{C}{-}R + HCl$$

is on an average five to six carbon atoms long. Polyenes are highly coloured, so the degrading PVC rapidly goes brown and then black. The highly reactive polyenes cause crosslinking, so the average molecular weight rises and eventually a gel forms. The resulting increase in the already high melt viscosity is hardly welcome, and, without careful design, degraded PVC can build up in slow flow regions of the processing equipment.

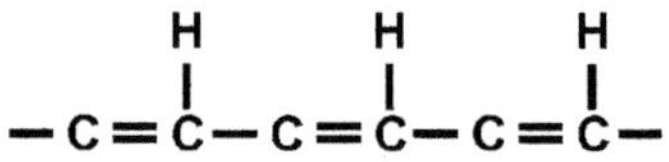

$$-C{=}\overset{H}{C}{-}C{=}\overset{H}{C}{-}C{=}\overset{H}{C}-$$

PVC is often compounded in large internal mixers. Laboratory scale stability tests use a small internal mixer, in which the torque is measured as a function of time. Figure 10.2 shows how the molecular weight average M_W increases, with mixing time at 190 °C, until it is no longer measurable. The torque falls from its initial peak after fusion, but rises again later. G indicates the gel point where an insoluble fraction occurs first. If oxygen was allowed free access to the mixer, degradation would be faster, as the polyenes readily oxidise into peroxides, which in turn decompose and liberate further HCl.

Many types of additives are used to stabilise PVC. For example, mixed metal salts of fatty acids, such as barium and cadmium stearates, are added at a 2 or 3% level. Metal stearates (MSt_2) react with and remove free hydrogen chloride, slowing down the degradation.

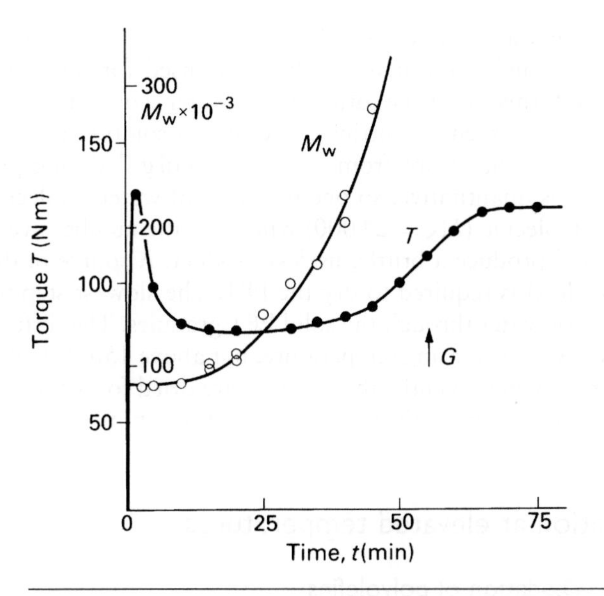

Figure 10.2 Degradation with time in the mixing chamber of a torque rheometer at 190 °C. The torque T and molecular weight M_W are shown, with the gel point G being shown by an arrow (from Kelen, T., *Polymer Degradation*, Van Nostrand Rheinhold, 1983).

$$MSt_2 + HCl \rightarrow MStCl + HSt$$

$$HCl + MSCl \rightarrow MCl_2 + HSt$$

Cadmium stearate alone is less effective, since $CdCl_2$ acts as a catalyst for dehydrochlorination. The additional barium stearate reacts with any $CdCl_2$ formed, and renders it harmless

$$CdCl_2 + BaSt_2 \rightarrow CdClSt + BaClSt$$

The more expensive organotin compound stabilisers replace tertiary chlorine atoms in PVC by more stable groups, as well as reacting with HCl. Therefore, they can prevent dehydrochlorination and the formation of polyenes.

10.2.3 Water and step-growth polymers

The melt stability of step-growth polymers is affected by water absorption prior to processing, which can reverse the polymerisation equilibrium reaction in the melt. For example, polyethylene terephthalate (PET) has an equilibrium water content of about 0.3% at 25 °C and 50% relative humidity. The high molecular weight grade used for injection moulding of bottle preforms must be dried to a water content of about 0.003% (30 ppm). Even then the intrinsic viscosity (a solution viscosity measure of molecular weight) drops from 0.73 to 0.71 $\mathrm{dl\,g^{-1}}$ during processing. PET hydrolysis is quantitative, so one molecule of water (molecular weight 18) per PET molecule ($M_N = 24\,000$) will halve the number average molecular weight and produce a brittle, useless product. A source of dry air (10 ppm water or less) is required to dry the PET. The slowest step in drying is the diffusion of water through the solid PET granules. The diffusion coefficient increases with increasing temperature, but above 150 °C hydrolysis starts at a slow rate. Consequently, the granules are dried for 4 h at 170 °C and kept blanketed in dry air until they enter the injection-moulding machine.

10.3 Degradation at elevated temperatures

10.3.1 Oxidation of polyolefins

In most polymer applications oxygen is present. Consequently, the stability is less than in a vacuum (Fig. 10.1) or that in an inert gas. Many oxidation studies use films, a few μm thick, in which there is a uniform oxygen concentration. The oxidation of polyolefins is auto-catalytic, since the main product (hydroperoxides) initiates the reaction (Section 10.2.1). An induction period is followed by a constant rate of oxygen consumption (Fig. 10.3). In the latter, the rates of hydroperoxide destruction and formation are equal. Antioxidants increase the induction period, but they are eventually consumed. The process is exactly the same as in the melt, except that the rate is lower. The activation energy for the maximum oxidation rate in polyethylene is 146 $\mathrm{kJ\,mol^{-1}}$. It appears that all hydrogen atoms on the chain are equally vulnerable to oxidation.

After the induction period, the oxygen concentration C is determined by competition between oxygen use and diffusion from the polymer surface. Gas diffusion into polymers, dealt with in Section 11.1.3, is a slow process. According to Eq. (11.9), it takes 6 h for the oxygen concentration at a 1 mm depth in LDPE to reach 50% of the surface concentration (assuming the polymer was initially oxygen free). Diffusion effects are still significant for 100 μm thick polypropylene film at 110 °C. Oxygen concentration profiles are predicted to develop once the induction time (28 h) is exceeded (Fig. 10.4).

The rate of change of oxygen concentration C is related to the diffusion constant D by

$$\frac{\mathrm{d}C}{\mathrm{d}t} = D\frac{\mathrm{d}^2C}{\mathrm{d}x^2} - kC \qquad (10.1)$$

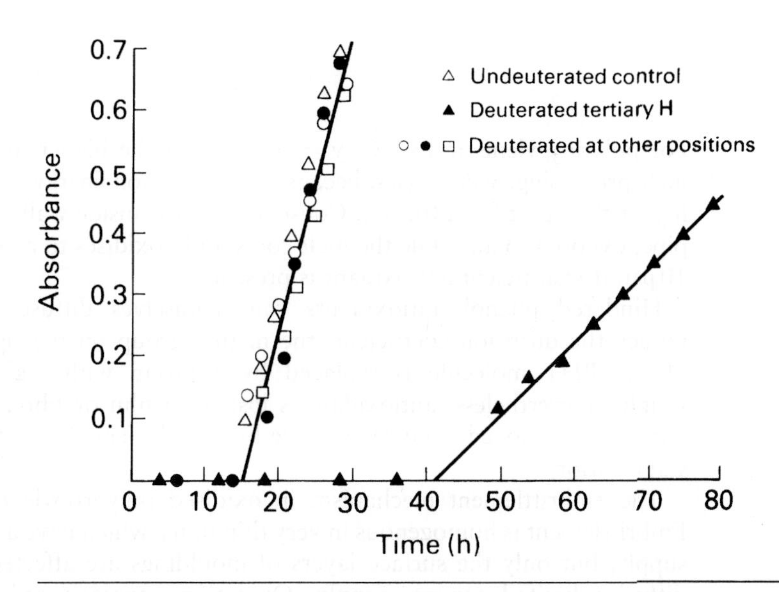

Figure 10.3 Carbonyl absorption of polypropylenes vs. the thermal oxidation time at 100 °C (from Kelen, T., *Polymer Degradation*, Van Nostrand Rheinhold, 1983).

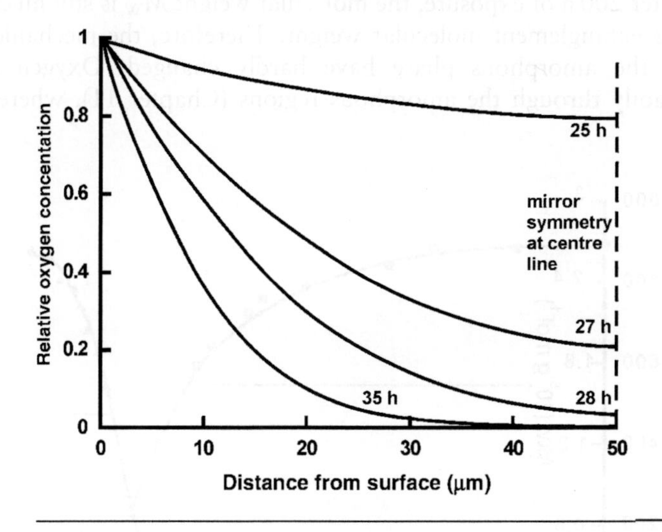

Figure 10.4 Predicted oxygen concentration profiles in a PP at 110 °C (Rincon-Rubio L. M. *et al.*, *Polym. Degrad. Stabil.*, **74**, 177, 2001) © Elsevier.

where k is a constant, and the x axis is normal to the sheet surface. This equation has a steady-state solution for thick sheet, with the oxygen concentration falling exponentially with distance from the surface. The concentration falls by a factor e in a distance y given by

$$y = \sqrt{\frac{D}{k}} \tag{10.2}$$

For polypropylene at 130 °C, $y = 0.1$ mm. At the higher temperatures in melt processing, y decreases, because the activation energy for oxidation is higher than that for diffusion. Consequently, the inside wall of a polyolefin pipe, exposed to air while the melt cools, only oxidises to a depth of about 10 μm if insufficient antioxidant is present.

Hindered phenol antioxidants can themselves diffuse. In order to reduce the diffusion coefficient, the methyl group on the right-hand side of the BHT molecule is replaced by a group with higher molecular weight. Nevertheless, antioxidant is lost from film or fibre, as it diffuses out and is removed from the surface by liquids. This limits the long-term stabilisation.

The embrittlement mechanism in oxidised polypropylene is complex. Embrittlement is homogenous in very thin films, which have a ready oxygen supply, but only the surface layers of mouldings are affected, due to the diffusion-limited oxygen supply. Oxidation causes a reduction in the elongation at break, when a thin film is tensile tested (Fig. 10.5). However, this only means that the neck fails to propagate the length of the specimen. After 200 h of exposure, the molecular weight M_W is still much higher than the entanglement molecular weight. Therefore, the mechanical properties of the amorphous phase have hardly changed. Oxygen only diffuses readily through the amorphous regions (Chapter 11), where it can reach

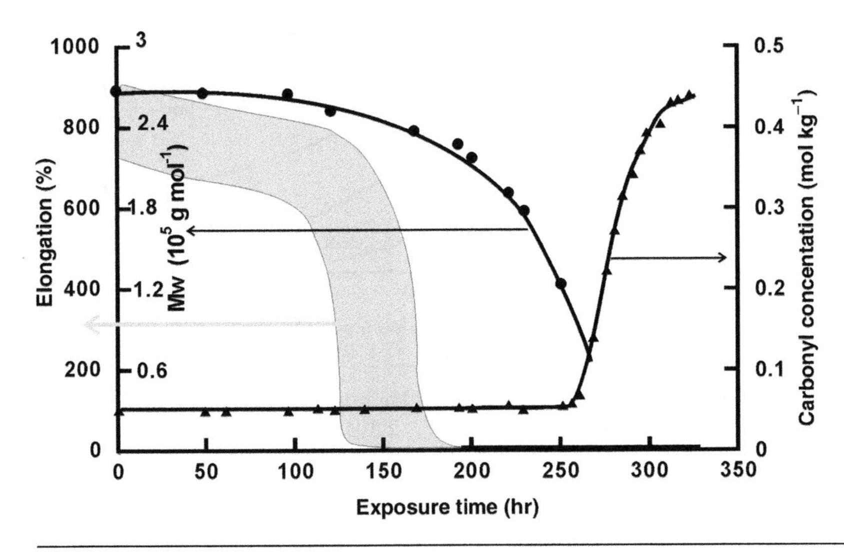

Figure 10.5 Tensile elongation at break, weight average molecular weight and carbonyl concentration of polypropylene vs. the exposure time to air at 110 °C (from Fayolle, B., *et al.*, *Polym. Degrad. Stabil.*, **70**, 333, 2000) © Elsevier.

inter-lamellae links. It is likely that a damage to these causes embrittlement. The oxidation induction time, measured from the concentration of carbonyl groups, is 250 h. Consequently, embrittlement occurs when the oxidation process is only 0.05% complete, and there has been very little chain scission. The surface layer of the necked region is covered with cracks at 90° to the tensile stress, and the ductile growth of one of these cracks terminates the necking process.

10.3.2 Hydrolysis

The effect of molecular weight on the tensile strength of two glassy polymers is shown in Fig. 10.6. The strength, of narrow molecular weight distribution polysulphone and polystyrene, reaches a plateau value when M_N exceeds the entanglement molecular weight (Table 3.1) by a factor of about 4. Hence, the effect of degradation is only marked when the molecular weight falls to a level where stress transfer between molecules no longer occurs via an entanglement network.

Hydrolysis can be a problem with polycarbonate, or polyurethane foam. Prolonged storage under hot damp conditions causes random chain scission, and a reduction in the average molecular weight. Polycarbonate has a high melt viscosity, and to facilitate melt processing, the average molecular

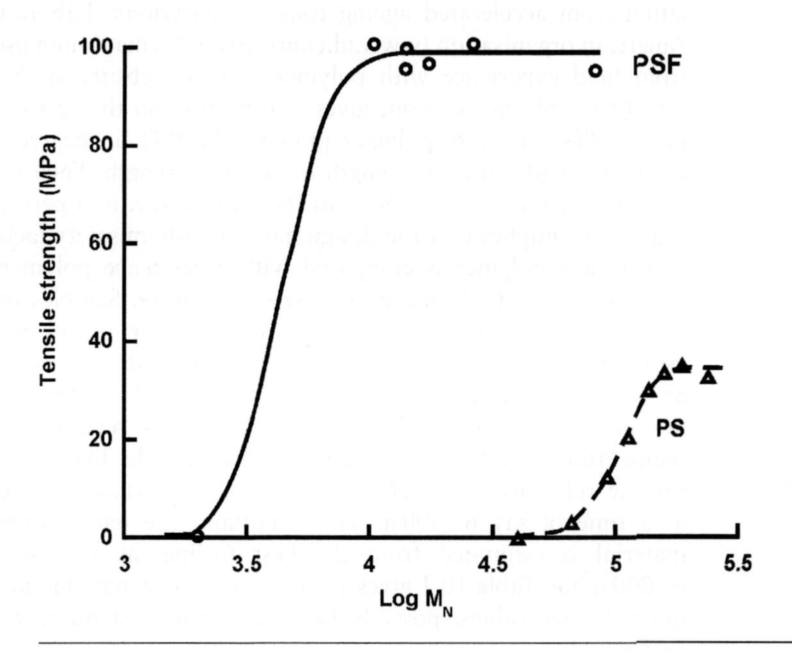

Figure 10.6 The tensile strength of polystyrene and the compressive strength of polysulfone vs. M_N, for narrow MWD samples (from Mills, N. J., *Rheolog. Acta*, **13**, 185, 1974).

weight only slightly exceeds that necessary to establish an entanglement network (Table 3.1 gives the entanglement molecular weight as 1870). Consequently, a modest drop in molecular weight reduces M_W to the level (14 000) at which brittle failure occurs in a tensile test.

A common method of predicting product lifetimes assumes a single, thermally activated degradation process, with rate given by Eq. (11.8). If failure occurs when oxidation or hydrolysis reaches a critical level, the failure time t_r is given by

$$t_r = A \exp\left(\frac{E}{RT}\right) \tag{10.3}$$

Consequently, a graph of log (failure time) versus $1/T$ should be a straight line. Figure 10.7 shows this as the case for the hydrolysis of polycarbonate. Extrapolating the line, with activation energy 8.5 kJ mol^{-1}, predicts a 5 year life at 38 °C. As extrapolation is by a time factor >10, if the degradation process changes in the intervening temperature range, the prediction may be inaccurate.

10.3.3 Maximum use temperature

It is often necessary to estimate the maximum use temperature, by extrapolating from accelerated ageing tests. Underwriters' Laboratories Inc., an American organisation (www.ul.com), estimates maximum use temperature from field experience with polymers. Their website, in the online tools *UL iQ for plastics* section, gives information on the relative temperature index (RTI) of many polymer grades. The RTI, in degree celsius, is for tensile strength, impact strength or electric strength. Each is based on the property deteriorating to 50% of its initial value, in a period of about 10 years. This implies that the design has a minimum safety factor of 2.

Any new polymer is compared with a reference polymer of a similar type, with an established upper use temperature. Samples of both, of the same thickness, are exposed in circulating air ovens at temperatures estimated to give lives of roughly 1, 3, 6 and 12 months. Specimens are removed at intervals for the relevant test to be performed. The results are plotted as log (half-strength time) versus reciprocal absolute test temperature, similar to Fig. 10.7, and a straight line is fitted. The line for the reference material passes through its maximum use temperature at a time of say 60 000 h. The maximum use temperature of the new material is estimated from the best fit-line, as that which allows a 60 000 h life. Table 10.1 gives some values. Some manufacturers' brochures quote higher values, possibly because a restricted number of tests were performed.

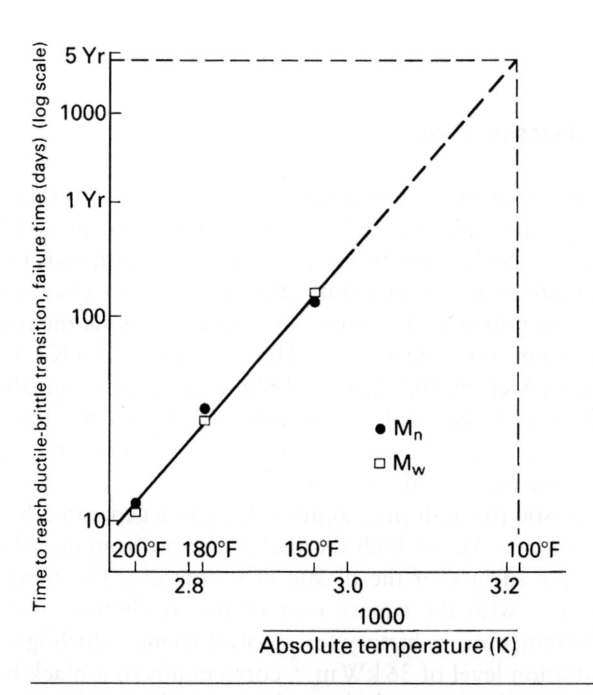

Figure 10.7 The exposure time at 100% RH necessary to make polycarbonate brittle in a tensile test, vs. the reciprocal absolute test temperature (from Gardner R. J. and Martin J. R. , *J. Appl. Polym. Sci.*, **24**, 1275, 1979).

A single maximum use temperature cannot apply to all products; in particular, film and fibre products may oxidise faster. It does not allow for outdoor use, or the effects of chemicals in the environment, topics that will be dealt with later in this chapter. However, it shows that degradation occurs well below the glass transition or melting temperature of the polymer.

Table 10.1 Maximum use temperatures in dry air

Plastic	*Maximum use temperature (°C)*
ABS	67
POM	87
PPO/PS	96
Nylon 6,6	96
PBTP	116
PC	120
PSO	145
PPS	165

10.4 Fire

10.4.1 Stages in a fire

As a plastic heats in a fire, it goes through the stages of thermal decomposition. Polyolefins decompose by random chain scission to liberate alkanes and alkenes as fuel for the flame. If the polymer chain ends fail, they 'unzip', liberating monomer. Consequently, they do not leave char or generate much smoke. Some polymers decompose by *chain stripping*, the release of a small molecule from the side chains. The liberation of HCl from PVC was explained in Section 10.2.2. Crosslinking tends to occur after chain stripping. The crosslinked polymer is more stable, so the likelihood of char formation increases. When char forms it can act as a barrier between the flame and the unaffected polymer.

For a plastic to catch fire, combustible gases generated by its decomposition must *ignite*. Under high thermal radiation from neighbouring red-hot material, the surface of the plastic heats rapidly. The temperature is predicted to rise with the square root of the irradiation time. Figure 10.8a shows the temperature history for a polyethylene, which ignites after 126 s. The irradiation level of 36 kW m^{-2} corresponds to a black body at 613 °C, by Eq. (5.7). The material *thermal inertia*, the product $k\rho C_p$ of thermal conductivity, density and specific heat, is a scaling constant for the time dependence of temperature rise. Consequently, foams, with low densities, heat more rapidly than solid polymers.

If the irradiation level is reduced, eventually the maximum polymer temperature will be insufficient to cause ignition. The hot layer is relatively thin, due to the low thermal diffusivity of the plastic. PMMA ignites at temperatures in the range 250–350 °C, while polyethylene ignites between 330 and 370 °C. The irradiation level should be representative of fires, since surface events, such as the formation of blisters and char, change with the irradiation level. If solid char seals the polymer surface, it prevents oxygen diffusing from the flame into the polymer, and gaseous polymer decomposition products passing as fuel to the flame.

When polymers burn, they act as fuel and the *heat release rate* affects the spread of the fire. *Flame spread* in a fire affects the amount of material involved. Flame spread often occurs along the ceiling of corridors. If a thermoplastic melts, and burning drops fall from a ceiling onto other combustible material, the fire spreads more rapidly.

Smoke and toxic gases are responsible for more fire deaths than the effects of burns. The major toxic hazard is carbon monoxide, produced by incomplete combustion of hydrocarbons. Nitrogen-containing polymers (polyamides, polyurethanes, polyacrylonitrile) can produce hydrogen cyanide.

Finally, the *ease of extinction* depends on the proportion of oxygen in the atmosphere of the fire. The limiting oxygen index (LOI) is the minimum percentage by volume of oxygen, in an oxygen/nitrogen mixture, that

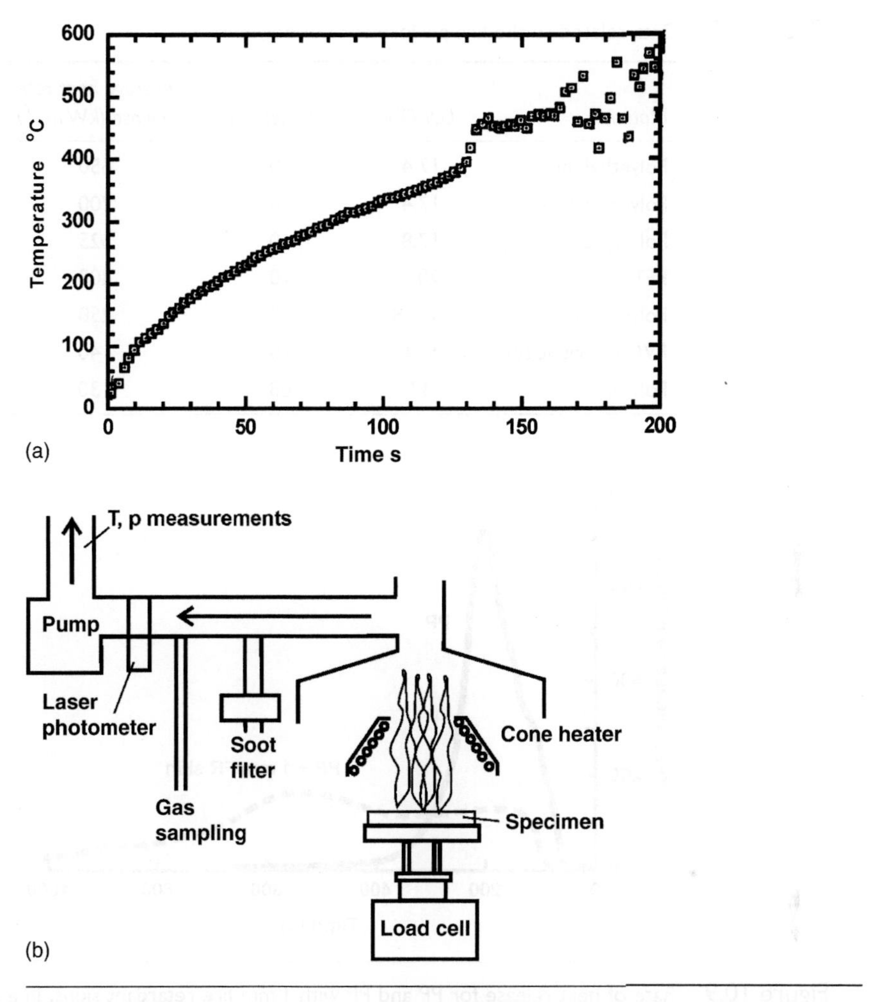

Figure 10.8 (a) Variation of surface temperature of a polyethylene with time: (from Hopkins D. and Quintiere J. G., *Fire Safety J.*, **26**, 241, 1996) © Elsevier; (b) cone calorimeter.

supports combustion of a polymer. A plastic strip, approximately 10 mm wide and 100 mm high, is ignited at the top, and it is found that the gas mixture will allow the flame to burn for just 3 min, or downward by 50 mm. The LOI (Table 10.2) does not correlate well with fire performance, since irradiation levels are extremely low.

10.4.2 Fire tests

Laboratory scale tests, such as the cone calorimeter (Fig. 10.8b), provide basic information about the fire performance of plastics. The flames from

Table 10.2 Data related to the fire performance of polymers

Material	*LOI (%)*	*Char yield (%)*	*Average heat rate release (kW m^{-2})*	*Smoke density rating*
Polyethylene	17.4	0	650	15
Polypropylene	17.4	0	700	32
Polystyrene	17.8	0	625	94
PET	20	10	400	84
Polycarbonate	22–28	25	250	
PVC unplasticised	45–49	10	40	97
Polyimide	37	68	30	

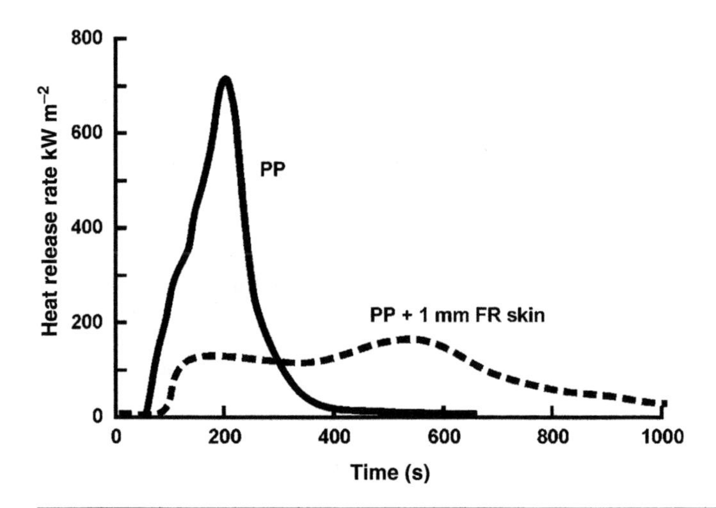

Figure 10.9 Rate of heat release for PP and PP with 1 mm fire retardant skins, in a cone calorimeter (redrawn from Ahmadnia A. *et al.*, *ANTEC*, 2755, 2003) © Society of plastics Engineers.

the burning plastic are above the bank of radiant heaters, so do not affect the irradiation level on the plastic. The mass of the plastic is monitored during the test, and the heat release rate measured (Fig. 10.9). The peak heat release rate is high for polypropylene.

The performance of furniture, mattresses and the seating in aircraft or trains is evaluated in realistic settings. In the ISO standard for furniture, the heat release rate is measured in a simulated room, after burning has been initiated. However, with increasing use of fire retardant cloth and foams, some fires do not propagate—they remain in the vicinity of the ignition source, then go out. Therefore, the correlation between furniture tests in a room and cone calorimeter tests on small samples, is only good for high irradiance levels that cause fire propagation.

British Standard BS 476 part 7 uses an intermediate scale test panel 900 mm wide and 225 mm tall. This is exposed edgeways on to a radiant heat source, so that the radiation intensity decreases from 40 to 8 $kW\,m^{-2}$ across the width of the panel. Materials are classified by the rate of flame spreads; less than 165 mm in 10 min for class 1, and greater than 710 mm in 10 min for class 4. PMMA is rated as class 3, whereas PVC and polycarbonate are class 1. Stringent fire regulations for glazing in public buildings have meant that PMMA, which has a good weathering resistance, can no longer be used.

Fire tests on HDPE fuel tanks for cars simulate a fire following a crash. When a fire is lit under a stationary car, the tank must remain intact for 2 min—sufficient time to escape from the vehicle. Although the HDPE surface burns, the tank remains intact, as the thermal inertia of the thick tank wall prevents the petrol from boiling.

10.4.3 Fire and flame retardants

A typical fire retardant for polypropylene is magnesium hydroxide, surface-modified to improve adhesion. When heated above 320 °C, it begins to decompose, absorbing heat and liberating water. The decomposition temperature is above that used in polymer processing. When 60% by weight is added to 1 mm surface layers of a 3 mm moulding, not only is the initial heat release rate reduced (Fig. 10.9), but the beneficial effect persists after the fire retardant layer has combusted. As large amounts of filler reduce toughness of PP, the pure PP core is necessary to optimise the impact toughness of the moulding.

A typical flame retardant is a mixture of chlorinated alkane (a compatible substance that provides HCl to the flame) and antimony oxide. These have a synergistic effect, forming antimony trichloride in the flame. Hydrogen chloride, formed in the flame, reacts with the high energy hydroxide free radicals formed by polymer decomposition

$$HCl + OH^{\bullet} \rightarrow H_2O + Cl^{\bullet}$$

The chloride radical then reacts with more fuel to regenerate hydrogen chloride

$$Cl^{\bullet} + RH \rightarrow R^{\bullet} + HCl$$

10.4.4 Fires involving cable and foam

The copper conductor in electric cable initially acts as a heat sink in a fire. Later on, it transfers heat to other parts of the insulation. Plasticisers, in flexible PVC electrical insulation, do not contain chlorine, so reduce the

LOI. A 60 pph addition of dioctylphthalate reduces it to 22. Flame retardants, such as antimony oxide or tricresylphosphate, generate large amounts of smoke. Since the smoke density rating is already high (Table 10.2), this is a problem. PVC degradation liberates hydrogen chloride which can corrode neighbouring electrical equipment. Some countries have banned the use of halogen-containing polymers for certain wire insulation applications. There are alternative flexible polymers. A mixture of calcium carbonate, silicone elastomer, ethylene butylene acrylate/ethylene copolymer, with the acronym CASICO, effervesces in a fire, and the ceramic layer formed provides some protection. Polymers made from fluorinated hydrocarbons (FEP copolymer) have a good fire resistance but are very expensive. Consequently, there is no ideal solution to the problem.

Polyurethane foams, once widely used as flexible foam in furniture and as rigid foam in insulated building panels, have a low LOI and burn rapidly because of the high surface-to-volume ratio. The use of halogen and/or phosphorus additives to reduce flammability is expensive, and adds to the release of toxic gases on combustion. A change in one of the constituents to form polyisocyanurates (PIR) increases the char yields to about 50% and improves the fire retardation. However, rigid PIR foams are friable and do not bond well to the surfaces of laminated building panels. Chemical changes of this kind, together with the addition of glass fibres to prevent the char from cracking, have greatly improved the fire rating of wall and roof linings. The fire performance of flexible polyurethane foams in furniture has similarly been improved by using a fire-retarding cotton inter-layer between the foam and the fabric covering.

10.5 Weathering

10.5.1 Ultraviolet wavelengths and absorption coefficients

Plastics, used outdoors, are exposed to solar radiation. Figure 10.10 shows the short wavelength end of the solar spectrum. However, absorption alone is insufficient to damage the polymer. Unless a specific wavelength raises a covalent bond to an excited state, the absorbed radiation merely heats the polymer. The energy E of a photon of light of wavelength λ is given by

$$E = \frac{hc}{\lambda} = h\nu \tag{10.4}$$

where h is Planck's constant, ν the frequency and c the velocity of light. Consequently, the ultraviolet (UV) end of the solar spectrum contains the most energetic photons. Any damage to the ozone layer in the stratosphere increases the amount of UV light at the ground level. Table 10.3 lists the

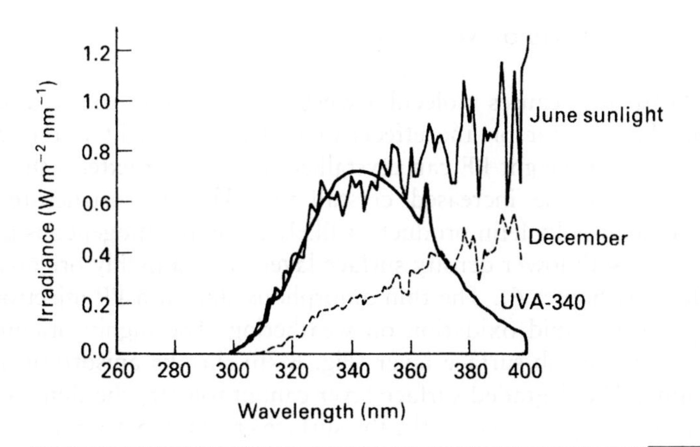

Figure 10.10 UV spectrum from sunlight at noon at Cleveland, Ohio, in June and December, compared with the output from a fluorescent UVA tube (from the Q Panel Company, Cleveland).

Table 10.3 Ultraviolet wavelengths which cause photodegradation

Polymer	*Wavelength (nm)*	*Energy (kJ mol^{-1})*
Polyethylene	300	400
Polypropylene	310	384
Polystyrene	318	376
PVC	320	372
Polycarbonate	293, 345	405, 347
SAN copolymer	290, 325	414, 368

UV wavelengths which damage certain polymers. The absorbed photons raise electrons to an excited state and cause bond dissociation reactions. An example is the photodecomposition of hydroperoxide groups introduced during melt processing.

$$\text{ROOH} + h\nu \rightarrow RO^{\bullet} + {}^{\bullet}OH$$

A high absorption coefficient for a particular wavelength means that the light penetrates the polymer to a limited distance. The intensity falls exponentially with penetration distance x, according to

$$I = I_0 \exp\left(-\frac{x}{L}\right) \tag{10.5}$$

For unpigmented HDPE irradiated at a wavelength of 310 nm, the constant L is equal to 1.25 mm, which explains why the photodegradation is confined to the surface layers of a 4 mm thick injection moulding.

10.5.2 Effects of weathering

Weathering causes molecular weight degradation, so the discussion earlier in this chapter on the effects of molecular weight is relevant. As lower molecular weight PE can crystallize to a greater extent, the weathering of PE can cause increased crystallinity. The microstructure of injection-moulded polyolefin products is likely to be inhomogeneous prior to weathering, with lower density surface layers, and a highly oriented near-surface layer (Chapter 6). The thin amorphous skin of a PP injection moulding is prone to rapid oxidation on weathering. The highly oriented and highly crystalline sub-surface layer (Fig. 6.7b) shrinks anisotropically on weathering. The degraded surface layer cannot tolerate the dimensional changes, so it cracks. Consequently, the surface cracking pattern indicates the orientation of the sub-surface crystalline orientation.

The weathering-induced formation of brittle surface layers is similar to that formed by oxidation, explained in Section 10.3.1. However, surface roughening is unique to weathering. It leaves filler particles exposed on the surface. There is more scattering of reflected light, causing the surface gloss to decrease. Eventually, weathering can cause polymers to loose mass. Figure 10.11 shows data for PVC: The initial low loss rate occurs while stabilisers absorb HCl (Section 10.2), then the loss rate increases when the stabiliser is exhausted.

Weathering is critical with film products, such as the 250 μm thick LDPE film coverings of agricultural poly-tunnels. It is accepted that the film will only last a couple of seasons, but the lightweight aluminium tube support structure has a low capital cost, in comparison to the complex and strong structure needed to support glass panes in greenhouses.

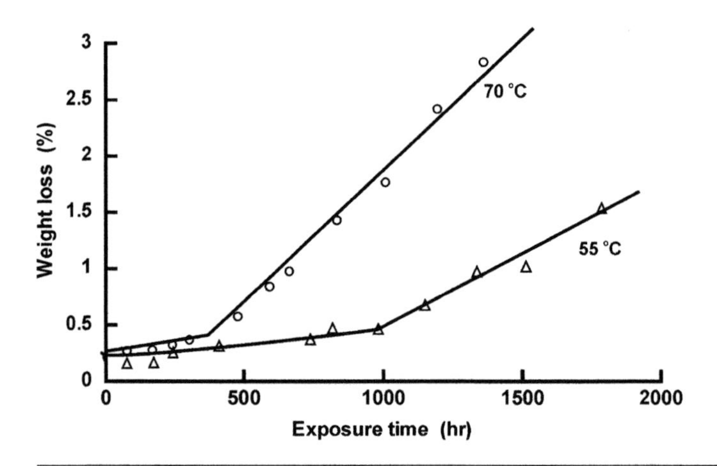

Figure 10.11 Mass loss of 80 μm thick PVC film vs. exposure time to UV at 40 and 70 °C (redrawn from Anton-Prinet, C. *et al.*, *Polym. Degrad. Stabil.*, **68**, 265, 1998).

Consequently, the running costs plus interest charges are lower than that for glass greenhouses. Some black polyethylene film is used directly in contact with the soil, with holes cut for plants. Attempts have been made to make such films biodegradable, so that they can be ploughed in at the end of each growing season.

10.5.3 Protection against photo-oxidation

Carbonyl groups are efficient absorbers of UV light. If the free radicals produced subsequently react with oxygen, the process is referred to as *photo-oxidation*. This is much faster than the thermal oxidation discussed in Section 10.3.1, even though the latter occurs at a higher temperature. Figure 10.12 compares the oxygen uptake by the two mechanisms for polyethylene. An oxygen uptake of $50\,cm^3\,g^{-1}$ greatly exceeds the solubility of oxygen in polyethylene ($\sim 0.04\,cm^3\,g^{-1}$). Once the dissolved oxygen is depleted, it must diffuse in from the surface to allow further photo-oxidation. This slow process, and the strong surface absorption of UV radiation, means that photo-oxidation is confined to a thin surface layer. For unpigmented ABS exposed outdoors for 3 years all the butadiene is oxidised in the surface 125 μm layer, then the butadiene deficit decreases exponentially

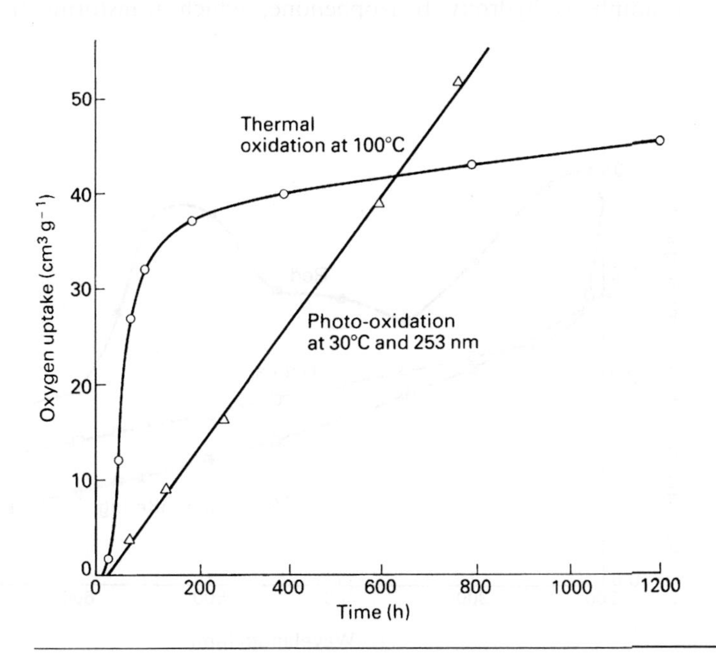

Figure 10.12 Uptake of oxygen by a polyethylene film sample during thermal oxidation at 100 °C and photo-oxidation at 30 °C (from Kelen, T., *Polymer Degradation*, Van Nostrand Rheinhold, 1983).

further into the moulding. Since a certain UV dose is required per unit volume of polymer to oxidise the butadiene, the depleted surface layer thickness increases with the logarithm of the exposure time. When the butadiene rubber is converted to a glassy material, the toughening mechanism for the glassy SAN matrix is lost, with the consequences shown in Fig. 9.19.

Methods for protecting polymers against photo-oxidation interfere with the UV photon at various stages in the sequence of photo-oxidation events. *Protective layers* prevent photon transmission. However, the paint film may become brittle or be damaged. The cheapest and most effective way of *absorbing* UV radiation is to use well-dispersed carbon black piment in the polymer. White pigments such as TiO_2 and ZnO are effective at scattering UV radiation, hence in causing diffuse reflection. Figure 10.13 shows the absorption increase as a result of adding a red organic pigment to HDPE. The attenuation in the natural HDPE is mainly due to Rayleigh scattering (Section 13.4.3) from small crystals. The red colour is achieved by absorbing the violet end of the visible spectrum. The broad absorption peak also absorbs in the UV region. This pigment itself photodegrades, so the protection has a limited life. The surface layers of HDPE traffic cones and industrial helmets loose their red colour after several years of exposure. This indicates that they need replacement!

Some organic *additives* absorb UV radiation and convert it into heat. An example is hydroxy benzophenone, which transforms from the usual

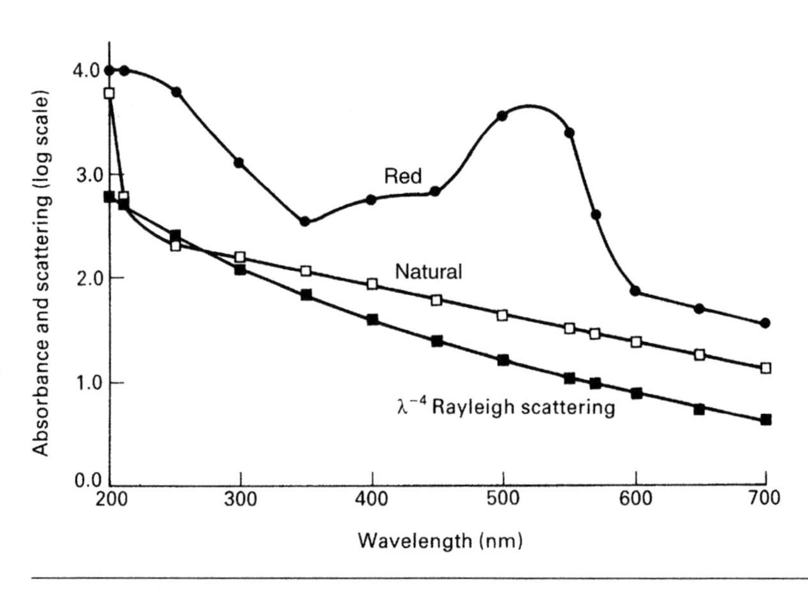

Figure 10.13 Absorption coefficient of natural and red pigmented HDPE in the UV region (from Hulme A. Birmingham University, unpublished).

keto form to the *enol* form when a photon is absorbed; then decays back to the keto form releasing heat.

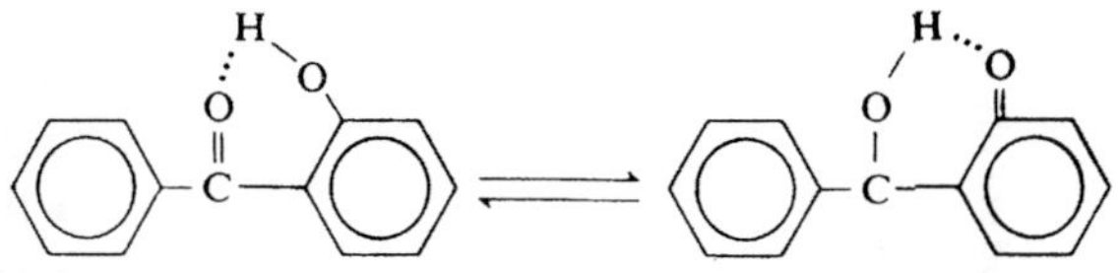

When UV is absorbed by a polymer, free radicals may be formed. These can be scavenged by hindered amine light stabilisers (HALS) before they cause chain scission or oxidation. An example is

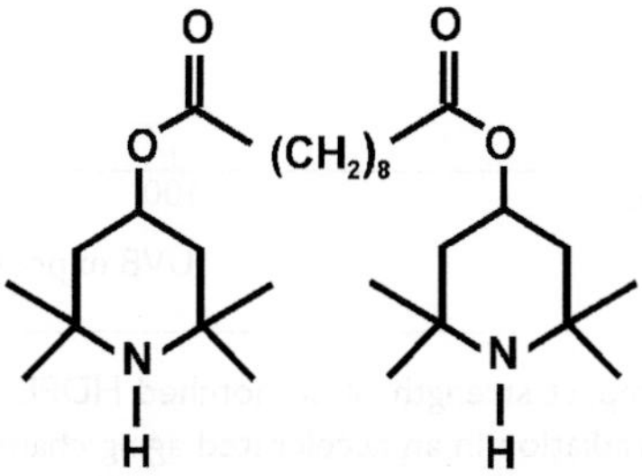

where methyl groups occur on the ends of 'free' bonds. The −NH group is oxidised into a nitrosyl radical ($NO^\bullet$), which can scavenge polymer radicals and then regenerate itself. Although HALS is highly effective at a 1% level in stabilising ABS, there is a synergistic effect if it is used in combination with a UV absorber. Accelerated exposure tests with a xenon light source showed that the impact strength of an unstabilised ABS fell to a specified level after 150 h exposure. With 1% HALS added, the lifetime was increased to 1000 h, but with 0.5% HALS and 0.5% of a UV absorber it was increased to 1650 h. One per cent of the UV absorber on its own, only gave a 350 h life.

10.5.4 Accelerated exposure tests

Accelerated UV exposure tests are used to predict the performance of outdoor exposed polymers, as natural weathering takes several years. It is not possible to increase the intensity of sunlight, as the polymer would get too hot and thermal degradation would dominate. If high pressure xenon sunlamps are used, the infrared part of the spectrum must be filtered out to prevent the plastic overheating. Fluorescent tubes with suitable phosphorus can produce a spectrum in the UVB region, defined as 280–320 nm (Fig. 10.10), similar to sunlight at the Earth's surface, without producing a significant infrared component. Environmental test chambers, using such tubes, expose the plastic to levels of UVB of about 2 mW cm^{-2}. The

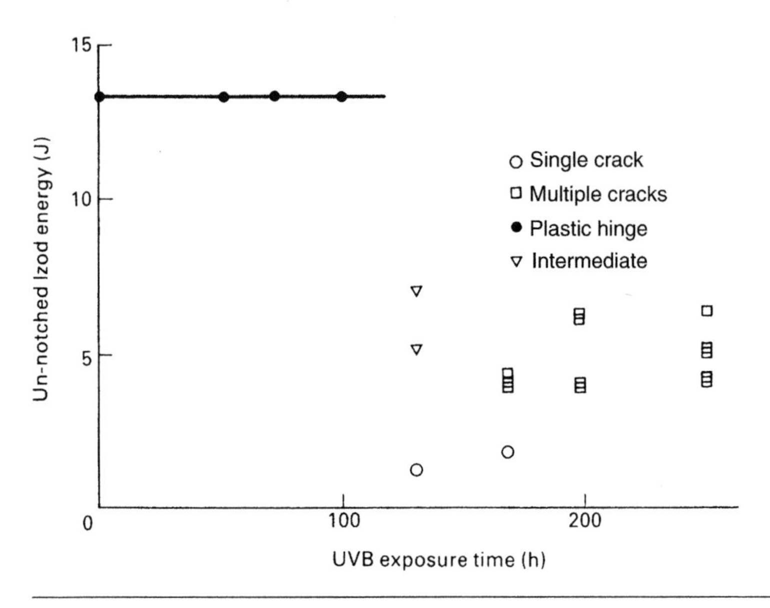

Figure 10.14 The impact strength of un-notched HDPE bars as a function of the time of exposure to UVB radiation in an accelerated aging chamber (Mills, unpublished).

maximum level of outdoor UV in England in summer is 2–3 mW cm^{-2}, so the accelerated test achieves its ends, not by having a high radiation level, but by having continuous output. If the outdoor UV levels are averaged over a year (Davis and Sims), the acceleration factor of the UVB chamber is found to be about 80. Figure 10.14 shows the results of exposing natural HDPE of MFI = 5 in a UVB chamber. The bars were subjected to impact bend tests and the energy absorbed in the tests plotted against the exposure time. The type of failure changed from plastic hinge formation to crazing at an exposure time of 120 h. At higher exposures the toughness increases again, as a result of the formation of more crazes in the high stress region. The estimated outdoor life of this unpigmented and non-UV stabilised HDPE is 120×80 h = 1.1 years.

The effects of accelerated exposure (no visible cracks before impact testing) differ from outdoor exposure. Figure 10.15 shows the pattern of surface cracks on a polyethylene garden chair after several years of use. Some cracks may be due to the low-cycle fatigue loads when the chair was used, but others may be due to the presence of a degraded amorphous layer, and the anisotropic shrinkage of the underlying structure (Section 10.4.2). Once surface cracks have formed, dirt particles can enter. These wedge cracks open when the surface heats up and expands, and liquid pollutants can diffuse more readily into the polymer. Hence, accelerated UVB exposure can only be a guide to one aspect of weathering. If an accelerated exposure chamber is used with a cycle of 8 h UVB followed by 4 h of 'rain', it is possible to cause surface

Figure 10.15 SEM of a polyethylene garden chair surface showing cracks due to weathering and fatigue loads.

cracking in polyethylene. However, it is impossible to accelerate the diffusion of oxygen into the polyethylene, so degradation tends to be closer to the surface than in outdoor exposed products.

10.6 Environmental stress cracking

10.6.1 ESC phenomena

Environmental stress cracking is similar, but not identical to, stress corrosion cracking of metals. *Corrosion* involves chemical reactions that produce corrosion products, whereas, in ESC, a liquid is absorbed by the polymer, promoting crazing and crack formation. Corrosion reactions are rare in polymers. ESC can typically cause a factor-of-ten reduction in strength. The two conditions for it to occur are that

(1) an *active liquid* is in contact with the surface of the plastic,
(2) the plastic *surface* is under *tensile stress* (either from externally applied loads, or residual tensile stresses from processing).

The sequence of events for ESC can be observed in transparent glassy plastics, loaded in three-point bending, with the surface in tension in contact with the active liquid. They are that

(a) crazes (Section 8.5.1) appear relatively rapidly, in about 10 s. The small gaps between the fibrils mean that capillary forces at the liquid/air interface suck liquid into the craze;
(b) liquid diffuses into the craze fibrils, weakening them. Crazes therefore grow in both penetration, surface length and in number;
(c) one or more crazes fail, forming a crack, still with a craze at its tip;
(d) the crack grows across the bar, at a rate of the order of 0.1 mm s^{-1};
(e) the stress intensity factor of the crack tip stress field reaches the fracture toughness, so the crack propagates at high speed (Chapter 9).

As there is approximately 50% air in a dry craze, their refractive index is intermediate between that of air (1.0) and solid polymer (1.5). Consequently, crazes can be distinguished from cracks by obliquely incident transmitted light. Figure 10.16 shows the schematic arrangement of multiple crazes on a surface affected by ESC.

10.6.2 Craze swelling in liquids

Liquids affect the mechanics of crazing by plasticising the craze fibrils, and hence reducing the tensile stress across the craze. Table 10.4 shows the effects of various alcohols on crazes in high molecular weight PMMA. The values in the second and third columns show that methanol markedly reduces the glass transition temperature of PMMA swollen to equilibrium. The diffusion coefficient in the swollen surface layer is orders of magnitude larger than that of the glassy core. The position of the sharp boundary, between the swollen and unswollen material, advances linearly with time, taking about 5 days to penetrate 1 mm, a non-Fickian diffusion process. In a short laboratory test, methanol acts as an ESC agent for PMMA by reducing the crazing stress, without affecting the

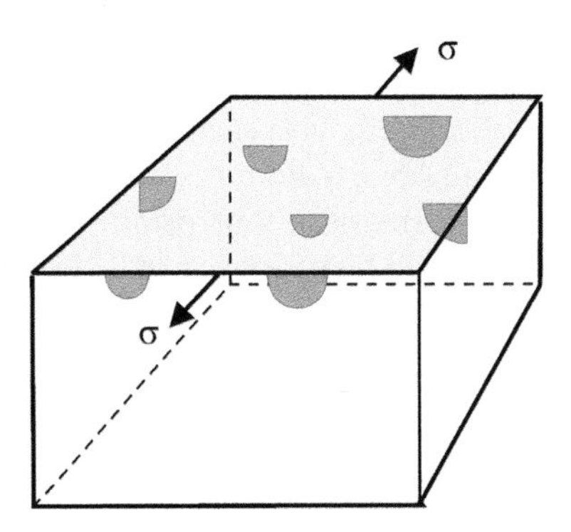

Figure 10.16 Craze planes, perpendicular to the axis of the maximum tensile stress, that initiate at the product surfaces in contact with the active liquid.

Table 10.4 Craze parameters for PMMA for equilibrium swelling

Environment	*Volume fraction liquid*	T_g (°C)	*Craze stress (MPa)*
Air	0	115	100
Methanol	0.23	36	70
Ethanol	0.24	30	52
N-propanol	0.24	32	53

bulk of the material. However, continuous immersion in methanol eventually changes the mechanical properties of the polymer, a quite different result.

The craze tip growth velocity V can be limited by the liquid flow velocity within the craze. Figure 10.17 shows a craze containing a length L of liquid. The liquid pressure p_1 at the crack tip is atmospheric, but p_2 at the liquid/air interface due to the capillary attraction. If the liquid moves inside the craze with the same velocity V as the advancing craze tip, and the pore area A of the craze cross section is constant, D'Arcy's law for the flow of a liquid of viscosity η through a porous medium

$$V = -\frac{A}{12\eta}\frac{\mathrm{d}p}{\mathrm{d}x} \tag{10.6}$$

can be applied. Substituting the pressure gradient $\mathrm{d}p/\mathrm{d}x = (p_2 - p_1)/L$ and $V = \mathrm{d}L/\mathrm{d}t$, integration gives

$$L^2 = \frac{A}{6\eta}(p_1 - p_2)t \tag{10.7}$$

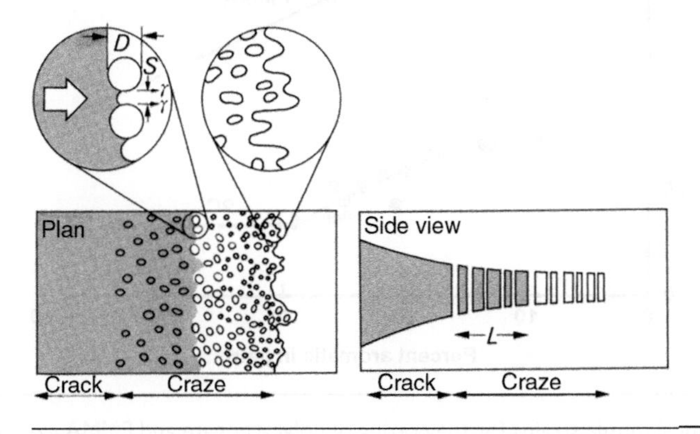

Figure 10.17 Plan and side views of liquid flow into a craze, with air near the craze tip. Liquid is shown in grey.

Experimental data on the growth of methanol-filled crazes in PMMA confirms Eq. (10.7). The length increases with $\sqrt{t}$ until the equilibrium length according to Dugdale's model (Eq. 9.15) is reached.

10.6.3 Crack and craze initiation

ESC is a problem for plastics that are normally tough, such as polycarbonate, ABS and polyethylene. For example, unexpected failures occurred when hospital equipment was disinfected with sterilisation solutions. There is not a priori method of knowing whether a liquid is an ESC agent. Consequently, simple laboratory tests are used to screen plastics for susceptibility to a range of chemicals. Strips of polymer are bent on to an elliptical former and the surface in tension exposed to various liquids. The elliptical shape means that the surface strain varies by a factor of about 10 typically from 0.3 to 3%. After exposure for a fixed time, the surface can be examined with a microscope to find the minimum strain at which crazes or cracks form. Figure 10.18 shows results for two glassy polymers exposed to petrol (gasoline) with different aromatic contents. Polycarbonate is very susceptible to petrol with a high aromatic content, hence motorcycle helmets are wiped with a 50% iso-octane–50% toluene mixture prior to impact testing (Section 9.5.3).

Semi-crystalline polymers such as polyethylene are less affected by organic liquids, but nevertheless, the amorphous phase is susceptible to attack. Both alcohols and surface active agents can eventually lead to crack formation. Severe conditions are used for laboratory quality control tests of the

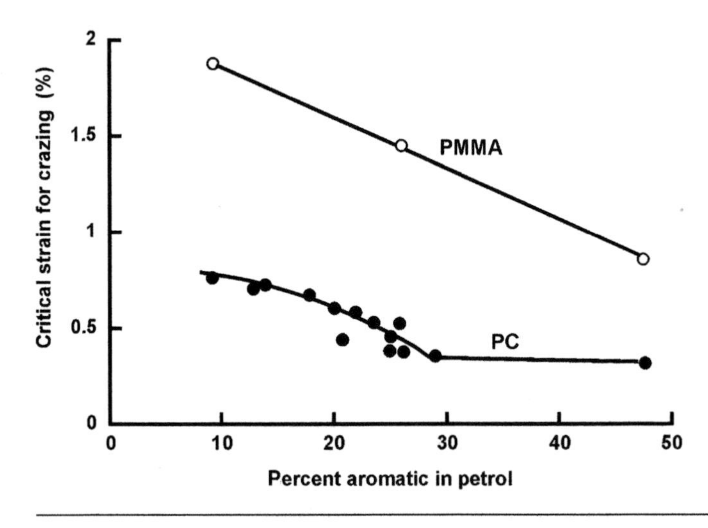

Figure 10.18 Critical tensile strains for the crazing of polycarbonate and PMMA exposed to petrol with different aromatic content (from Wysgoski M. G. and Jacques C. H. M., *Polym. Eng. Sci.*, **17**, 858, 1977).

ESC resistance. In the Bell Telephone test, a razor cut is made parallel to the length of rectangular $38 \times 13 \times 3$ mm specimen. It is bent into a U shape, with maximum surface tensile strain of about 12% and placed in concentrated surface active agent at 50 °C. The time until cracks appear, at the corner between the cut and the tensile surface, is measured. The cut does not create a stress concentration, because it is parallel to the bending stresses, but the two free surfaces may assist the crack opening process.

Attempts have been made to rationalise the susceptibility of glassy polymers to ESC. The critical strain for crazing is a minimum when the solubility parameter δ of the liquid is the same as that of the polymer (Fig. 10.19). The former is calculated from the energy of vaporisation and molar volume v of the solvent

$$\delta = \sqrt{\frac{\Delta E_{\text{vap}}}{v}} \tag{10.8}$$

The solubility parameter of the polymer is calculated from bond energies. The sensitivity to ESC is affected by the polymer molecular weight and by molecular orientation at the surface of injection mouldings.

10.6.4 Crack growth in a liquid environment

Even if a liquid promotes craze growth in a stressed polymer, growth may not be fast. For rapid crack growth, the crack should be preceded by a single craze. However, when a cracked PMMA product is exposed to methanol,

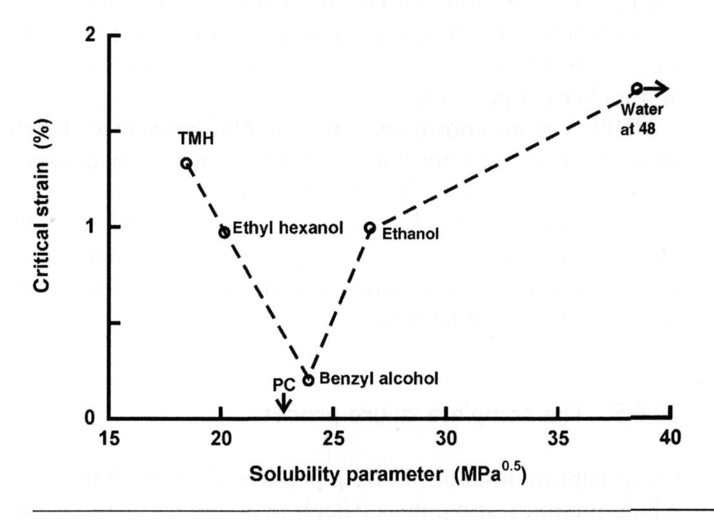

Figure 10.19 Critical strain for crazing polycarbonate vs. solubility parameter of liquids (data from Arnold, J. C. and Taylor, J. E., *J. Appl. Polym. Sci.*, **71**, 2155, 1999).

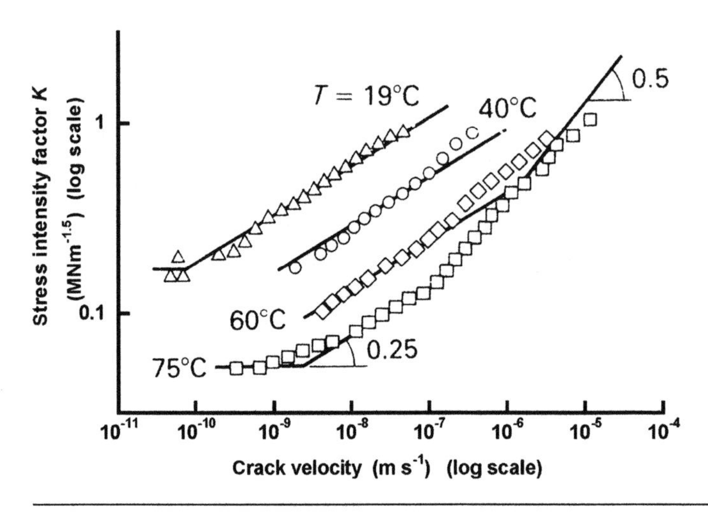

Figure 10.20 Variation of the crack speed with the stress intensity factor for slow crack growth in an HDPE in a detergent environment (from Williams, J. C., *Fracture Mechanics of Polymers*, Ellis Horwood, 1984).

multiple crazes form at the crack tip. These are more effective than a single craze at blunting the crack tip, and the K_{IC} value increases above the value in air.

The presence of a liquid modifies the relationship between the stress intensity factor K_I and the crack velocity V. Figure 10.20 shows that a liquid can allow slow crack growth at very much lower K values than in air. The threshold Kth value, at which the crack velocity is 10^{-9} m s^{-1}, is relevant for the design of structures that are intended to last for years. A crack, that grows 1 mm in 12 days, rapidly acquires a size that can cause fracture. The weakening effect is greater when the polyethylene is exposed to a detergent at a higher temperature.

HDPE has an enormous range of ESC resistance. Figure 10.21 shows creep rupture data for blow moulding and pipe grades under extreme conditions, of 80 °C in 2% Arkopal 110 surfactant solution, which allow the reasonably rapid selection of promising candidate materials. The samples have a 10 × 10 mm cross section, with a circumferential, 1.6 mm deep, razor blade notch. The samples with the longest lives have more tie links between the crystal lamellae.

10.6.5 The complete failure process

Creep rupture tests of plastic pipe, described in Chapter 14, are typical of whole-lifetime testing. It is difficult to manufacture plastic products without incorporating foreign particles, of size about 0.1 mm, such as undispersed pigment or stabiliser, or metal wear fragments from extruder screws. The

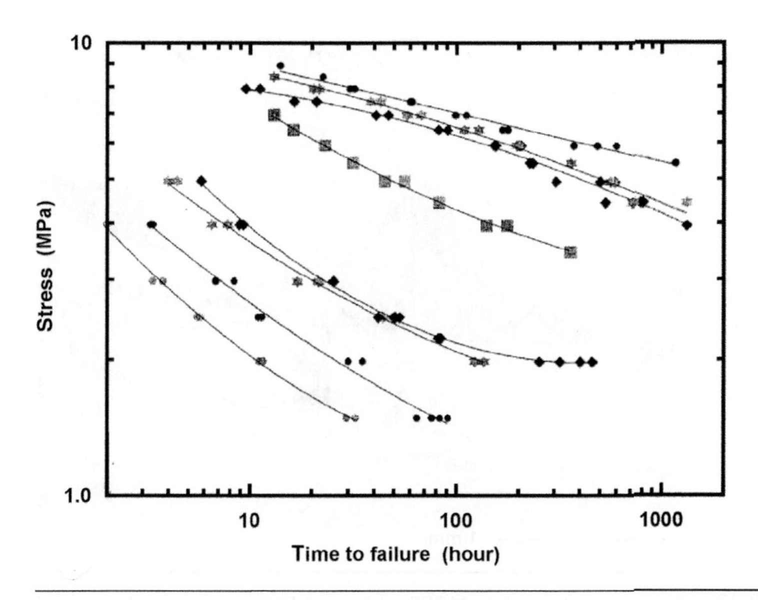

Figure 10.21 Creep rupture data, for a range of circumferentially notched HDPEs. The top three graphs are for 5%, and the rest for 2% Arcopal 110 surfactant (Fleissner, M., *Polym. Eng. Sci.*, **38**, 330, 1998) © John Wiley and Sons Inc. reprinted with permission.

adhesion between such particles and the polymer is so poor that a crack could be considered at the interface. Consequently, small cracks are assumed to be present in the pipe from the start of its service life.

The lifetime of a structure, under a constant load or stress, can be calculated if we know: (i) The initial crack size and location, (ii) the K_I versus crack length relationship and (iii) the variation in the crack velocity V with K_I. This will be illustrated for an elliptical-shaped surface crack at the bore of an HDPE pipe under internal pressure p. The crack plane is normal to the hoop stress σ_H, while its ellipse minor axis a is perpendicular to, and major axis b parallel to, the bore. The stress intensity factor of such a crack is given by Ewalds and Wanhill (1984) as

$$K_I = C_1 \sigma_H \sqrt{\pi a} + C_2 p \sqrt{\pi b} \tag{10.9}$$

where the constants C_1 and C_2 depend on the ellipticity of the crack and the size of the crack relative to the wall thickness. The crack velocity V versus stress intensity factor graphs for certain HDPEs (Fig. 10.20) have the form

$$V = AK_I^n \tag{10.10}$$

where A and n *are* constant for velocities between 10^{-6} and 10^{-2} mm s^{-1}. The pipe lifetime is the sum of time increments $\Delta t = V/\Delta a$ for the crack

Figure 10.22 Fracture surface of an MDPE with good ESC resistance, after crack growth from left to right at 80 °C.

length to grow by an increment Δa. After each increment, Eq. (10.9) gives the new K_I value and Eq. (10.10) the new velocity.

Tougher medium density polyethylenes used for gas pipes no longer have K–V relationships that can be expressed by Eq. (10.10), but failure by slow crack growth is unlikely. The fracture surfaces show crack arrest markings running vertically (Fig. 10.22). The crack stops while a craze grows ahead of it. When the craze fails, a rapid increment of crack growth occurs and the process repeats. The coarse fibrils on the fracture surface are the final products of craze breakdown.

Chapter 11

Transport properties

Chapter contents

This chapter considers the transport of gases, liquids and solids through polymer structures, as well as the transmission of light and heat. It explains how these processes can be controlled. The aim may be to minimise water transport through an LDPE damp-proof membrane, or to maximise the transport of light through a fibre optic. Intermediate cases, such as the separation of salt from water in a desalination plant, require selective transport. Solid transfer through polymer sheet or film is impossible, but woven polymer tapes or fibres, or polymer grids, allow selective transfer through holes in the product. The optical transmission of films is important, while the transmission of light via fibre optics is explored because of its use in telecommunication links. Heat transmission through solid polymers uses many of the same concepts as gaseous diffusion, so is conveniently dealt with here. Electromagnetic screening and dielectric properties are dealt with in Chapter 12.

11.1 Gases

11.1.1 Solubility

The transport rate of gas through a polymer film depends both on its solubility and the diffusion coefficient. Gas solubility is affected by the strength of the inter-molecular forces between gas molecules. In Chapter 2, the strength of van der Waals forces was characterised by the depth E_0 of the potential energy (Fig. 2.3). Table 11.1 gives values of E_0 for some common gases. The higher the value of E_0, the greater is the propensity of the gas to condense into a liquid, hence the higher is the boiling point of the liquid.

For the gases down to oxygen in the table, the gas concentration C in a polymer is related to the gas pressure p by Henry's law

$$C = Sp \tag{11.1}$$

Table 11.1 Energies constants of van der Waals forces between gas molecules, and solubility in the amorphous phase of polyethylene

Gas	$E_o(10^{-23}J)$	$S^*(10^{-6} mol\ m^{-3}Pa^{-1})$
He	14	5.4
H_2	52	–
N_2	131	18.4
O_2	163	34.3
CH_4	204	90.6
CO_2	261	201

where S is the solubility constant for the gas. Gas concentration can be expressed in various units; $\mathrm{mol\,m^{-3}}$ in SI units, but often as m^3 of gas at standard temperature and pressure (STP) per m^3 of polymer. The molar volume (at 0 °C, 1 bar) is $22.4 \times 10^{-3}\,m^3$. Solubility is expressed in SI units as $\mathrm{mol\,m^{-3}\,Pa^{-1}}$ (where $1\,Pa = 1\,N\,m^{-2} = 10^{-5}$ bar).

In semi-crystalline polymers above their glass transition temperatures, the solubility constant is proportional to the volume fraction V_{am} of the rubber-like amorphous phase

$$S = V_{am}S^* \tag{11.2}$$

The solubility constant for 100% amorphous material S^* increases exponentially with the energy constant E_0 of the van der Waals forces.

For more soluble gases like methane and carbon dioxide, the solubility, in glassy polymers with a large difference between the volume expansion coefficients of the liquid and glassy states, is non-linear. The values of $\alpha_L - \alpha_G$ for PET, PC and PMMA are 8.0, 4.3 and $1.3 \times 10^{-4}\,C^{-1}$, respectively, and only the first two polymers show this anomalous effect. Some gas is physically adsorbed on the surface of sub-microscopic holes in the polymer. Figure 11.1 shows how the concentration of CO_2 increases with pressure p in PET, a polymer used in carbonated drinks containers. The solubility is described by

$$C = Sp + \frac{abp}{1 + bp} \tag{11.3}$$

The second term on the right is the *Langmuir adsorption isotherm*, which describes the equilibrium concentration of gas molecules on a surface. Some of the gas molecules which bombard the surface stick, while other molecules, absorbed on the surface, escape. a is the hole saturation constant

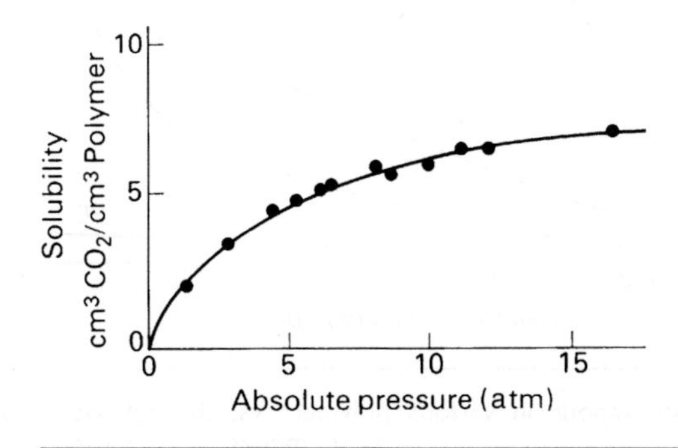

Figure 11.1 Solubility of CO_2 at 25 °C in polyethylene terephthalate vs. the gas pressure (from Hopfenberg H. B., Ed., *Permeability of Plastic Films and Coatings*, Plenum Press, 1974).

(the internal hole area per unit volume of polymer) and b is the ratio of the adsorption to desorption rates. $a = 7.9\,\text{cm}^3$ (STP) m^{-3} and b is $0.35\,\text{bar}^{-1}$ for CO_2 in PET. The Langmuir term dominates the solubility for pressures below 5 bar. The gas adsorbed on internal surfaces plays little part in gas transport through the polymer, so can be ignored in calculations of gas permeation.

Water vapour has anomalous solubility characteristic, because of the strong hydrogen bonding between the molecules. Figure 11.2 shows that the sorption isotherms can curve steeply upwards as the relative pressure approaches 1. However, hydrophobic polymers such as polyolefins still obey Henry's law.

11.1.2 Steady-state gas diffusion

The mathematical description of gas diffusion through a polymer is the same as that for heat diffusion considered in Section A.2 of Appendix A. Two material constants, *diffusivity* D and *permeability* P, are defined in terms of steady-state flow from a gas at a pressure p_1, on one side of a polymer film of thickness L, to a pressure p_2 on the other side (Fig. 11.3). The gas concentration in the polymer is constant at C_1 and C_2, respectively at the two surfaces. The flow rate Q through an area A of film is then given either by

$$Q = DA\frac{C_1 - C_2}{L} \tag{11.4}$$

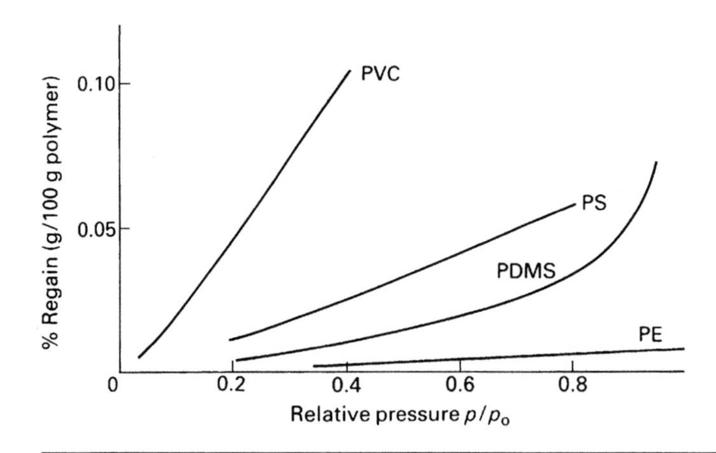

Figure 11.2 Solubility of water vapour in various polymers vs. the relative pressure (= vapour pressure/saturation vapour pressure) at 25 °C (PDMS at 35 °C) (from Crank J. and Park G. S., Ed., *Diffusion of Polymers*, Academic Press, 1968).

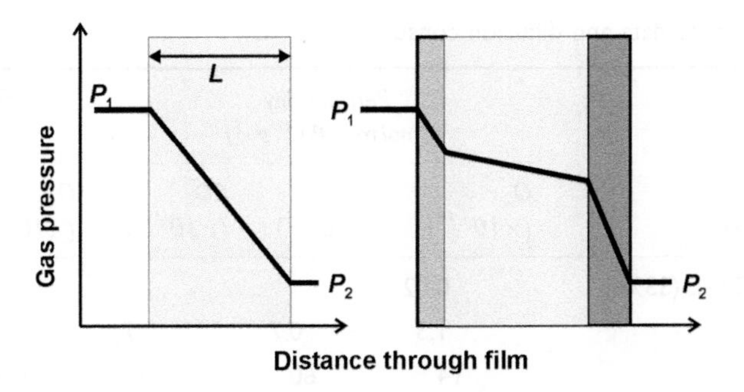

Figure 11.3 Variation of gas pressure through a single polymer film, and a multi-layer film, in which the three polymers have different permeabilities, for steady-state gas flow.

or

$$Q = PA\frac{p_1 - p_2}{L} \tag{11.5}$$

Permeability is quoted in many different units, because the equipment may measure gas volumes or mass changes, while pressure and time units vary. Diffusivity has units $m^2\,s^{-1}$ if the same units for amount of gas are used in Q and C. Table 11.2 gives some diffusivities for pure polymers in the unoriented state.

The gas diffusivity is related to the permeability by

$$P = DS \tag{11.6}$$

so the SI units of permeability are $mol\,m^{-1}\,Pa^{-1}\,s^{-1}$. As 1 mol of gas at STP occupies 22.4 l, it is also possible to use the units m^3 (STP) $m^{-3}\,bar^{-1}\,s^{-1}$. American permeability data uses 'mil' (0.001 in.) for thickness, and a standard test area of 100 in.^{2}. The conversion factor is

$$\frac{\text{mol}}{\text{m Pa s}} = 4.91 \times 10^{17} \frac{\text{cc mil}}{100\,\text{in}^2\,\text{atm day}}$$

For semi-crystalline polymers above T_g, the permeability is proportional to the nth power of the amorphous volume fraction; n lies between 1.2 and 2. The gas must diffuse between the lamellar crystals, and the detailed morphology depends on the polymerisation route, thermal history and whether orientation is present. The permeability of gases of molecular weight M is approximately inversely proportional to $\sqrt{M}$. However, Table 11.2 shows that the ratio of CO_2 to O_2 permeability in glassy polymers is higher than in semi-crystalline polymers.

Table 11.2 Permeability data and diffusion constants at 25 °C

	Permeability *(mol m^{-1} Pa^{-1} s^{-1})*			*Diffusion constant* *(m^2 s^{-1})*	
Polymer	O_2 ($\times 10^{-18}$)	H_2O ($\times 10^{-15}$)	CO_2 ($\times 10^{-18}$)	O_2 ($\times 10^{-12}$)	CO_2 ($\times 10^{-12}$)
Dry EVAL (33%E)	0.02				
PVDC	1.3	0.7	7		0.001
PET	14	60	30	0.36	0.054
PVC rigid	23	40	98		
Nylon 6	30	135	200		
Polyethersulphone	340		1900		2.0
HDPE	400	4	1000	17	12
LDPE	1100	30	5700	46	37
PP	400	17	1000		
PC	500	470	2900		5.3
PS	580	330	4000	12	1.3
Polyphenyleneoxide	780		3000		
Butyl rubber	370		1500	80	5.8
Natural rubber	7000	770	37 000	158	110
Silicone rubber	205 000	14 500	1 095 000	1700	

11.1.3 Transient effects in gaseous diffusion

Transient effects often occur when a plastic container is filled with gas. Fick's second law, derived in Section A.2 of Appendix A, applies if the diffusion coefficient is independent of the gas concentration C. The differential equation for one-dimensional diffusion along the x axis is

$$\frac{dC}{dt} = D\frac{d^2C}{dx^2} \tag{11.7}$$

where t is the time and D is the diffusion coefficient. The variation of D with temperature is described by

$$D = D_0 \exp\left(-\frac{E_D}{RT}\right) \tag{11.8}$$

where E_D is the activation energy for diffusion, R is the gas constant and T the absolute temperature. The diffusion coefficients are highest in semi-crystalline polymers above their glass transition temperature T_g. The crystalline phase has a negligible diffusion coefficient, so LDPE, with a higher amorphous content, has a higher diffusion coefficient than HDPE (Table 11.2). Equation (A.19) shows that, if a sheet was initially oxygen free, the thickness x of the layer, in which the oxygen concentration exceeds 50% of its surface level, is related to time t by

$$x = 0.94\sqrt{Dt} \tag{11.9}$$

When Fick's second law is obeyed, the analytical methods in Appendix A can be used to predict the total gas flow through the film. If a constant gas concentration C is applied at time $t = 0$ to one surface of an initially gas-free film of thickness L, the total volume V that passes through unit area is given by

$$\frac{V}{LC} = \frac{Dt}{L^2} - \frac{1}{6} - \frac{2}{\pi^2}\sum_{n=1}^{\infty}\frac{(-1)^n}{n^2}\exp\left(-\frac{Dn^2\pi^2 t}{L^2}\right) \tag{11.10}$$

A steady permeation rate develops (Fig. 11.4). If this straight line is extrapolated back, it cuts the dimensionless time axis at 1/6, which means that the *time lag* t_L is given by

$$t_L = \frac{L^2}{6D} \tag{11.11}$$

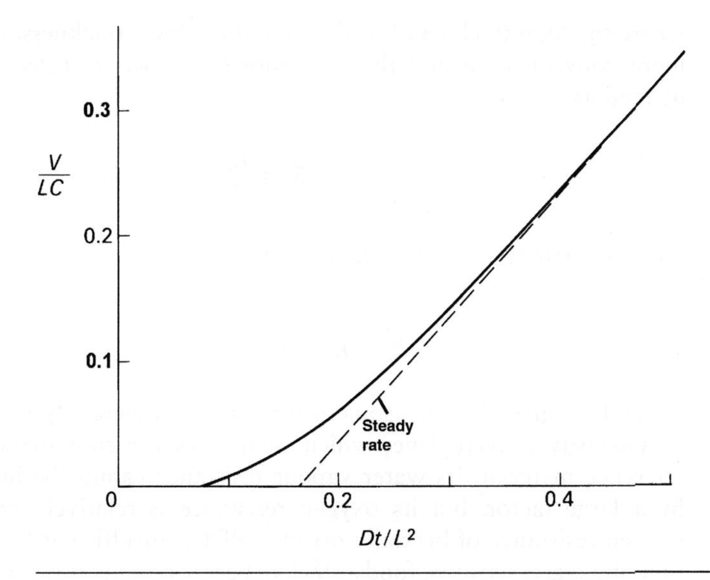

Figure 11.4 Volume of gas passing through 1 m^2 of sheet of thickness L vs. time, according to Eq. (11.10).

For natural gas (methane) passing through a 3 mm thick MDPE pipe wall, the time lag is 10.4 days, as $D = 1.0 \times 10^{-11}\,\text{m}^2\,\text{s}^{-1}$. Consequently, a steady-state permeation test with this pipe is a long-term experiment.

When the diffusion coefficient D increases with the gas concentration, Eq. (11.7) is replaced by

$$\frac{\partial C}{\partial t} = \frac{\partial}{\partial x}\left(D\frac{\partial C}{\partial x}\right) \tag{11.12}$$

This equation is best solved by finite difference methods on a computer. It applies when organic vapours, diffusing through a rubber, cause it to swell. It also applies when a high concentration of gases swells a glassy polymer (Fig. 11.1) and alters the T_g value.

11.1.4 Packaging applications

Biaxial orientation is often used to improve the in-plane tensile strength and toughness of polymer films. The crystallinity of PET increases on stretching (Fig. 8.16), so the permeability decreases. For multi-layer films made by co-extrusion or coating, the total permeability P_{TOT} is related to the layer value P_i by

$$\frac{P_{TOT}}{L} = \frac{P_1}{L_1} + \frac{P_2}{L_2} + \ldots \tag{11.13}$$

where the total thickness L is the sum of the layer thicknesses L_i. It is often more convenient to add the *resistance to the gas transfer* of each layer, defined as

$$R_i \equiv \frac{L_i}{P_i} \tag{11.14}$$

then calculate the steady-state gas flow using

$$Q = \frac{p_1 - p_2}{R_1 + R_2 + R_3 + \ldots.} \tag{11.15}$$

Table 11.3 gives film resistances for some commercially important films. Low-density polyethylene, which dominates the film market, acts as a reference material. Its water vapour resistance cannot be improved upon by a large factor, but its oxygen resistance is relatively poor. Even the oxygen resistance of biaxially oriented PET is insufficient for the preservation of oxygen-sensitive foodstuffs like beer. Consequently, PET bottles may have an outer coating of a PVDC copolymer. The oxygen resistance of each 4 μm thick PVDC coating can be calculated from the data in Table 11.3 as

Table 11.3 Film strength and gas resistance

Polymer	*Film thickness (mm)*	*Tensile strength ($kN\,m^{-1}$)*	*Oxygen resistance ($GN\,s\,mol^{-1}$)*	*Water vapour resistance ($GN\,s\,mol^{-1}$)*
LDPE	25	0.35	23	0.8
Chill cast PP	32	0.9	45	1.4
Biaxially oriented PP	14	3.5	45	1.2
Biaxially oriented PET	12	2.0	1700	0.32
Biaxially oriented PET + 4 μm coat of PVDC	20	2.5	8000	2.5

3150 $GN\,s\,mol^{-1}$. The PVDC coating is sprayed on to the outside of completed bottles, which consequently are difficult to recycle. The initial pressure loss, in a carbonated drink bottle after filling, results from transient diffusion and the Langmuir adsorption of CO_2 in the PET.

Active packaging systems exist, in which a chemical, present inside the container, absorbs oxygen or removes the products of food oxidation.

Extruded ethylene vinyl alcohol copolymer (EVAL) is used as an oxygen-barrier packaging material. Dry EVAL has extremely low oxygen permeability, but the vinyl alcohol part of the copolymer is hydrophilic, and in the swollen wet state its permeability becomes higher than that of PVDC (Fig. 11.5).

The packaging technology for squeezy bottles, for foods like ketchup that are sensitive to oxygen yet need to be sterilised in the container, has changed with time. Biaxially stretched PET, with a T_g of 80 °C, is insufficiently form-stable at 100 °C. Polypropylene has adequate form stability but too high an oxygen permeability. In the 1990s multi-layer bottles were used, with a layer of EVAL sandwiched between inner and outer layers of polypropylene. The oxygen resistance was achieved by keeping the relative humidity of the EVAL below 75%. However, the permeability of all polymers rises rapidly with temperature (Fig. 11.6). Consequently, after sterilisation, the EVAL layer is at nearly 100% RH. By placing the EVAL layer near the outside of the polypropylene sandwich, water diffuses through the outer polypropylene layer, and the EVAL humidity drops to an acceptable level within a week or so of filling the container. The five layer container has a 0.6 μm thick adhesive layer between the polypropylene and EVAL.

For a film of a single material, the water vapour permeation rate increases more rapidly with temperature than the permeability, because the vapour pressure of water increases rapidly with temperature (2.3 kPa at 20 °C, 7.4 kPa at 40 °C, 19.9 kPa at 60 °C, 47.2 kPa at 80 °C). From the definition

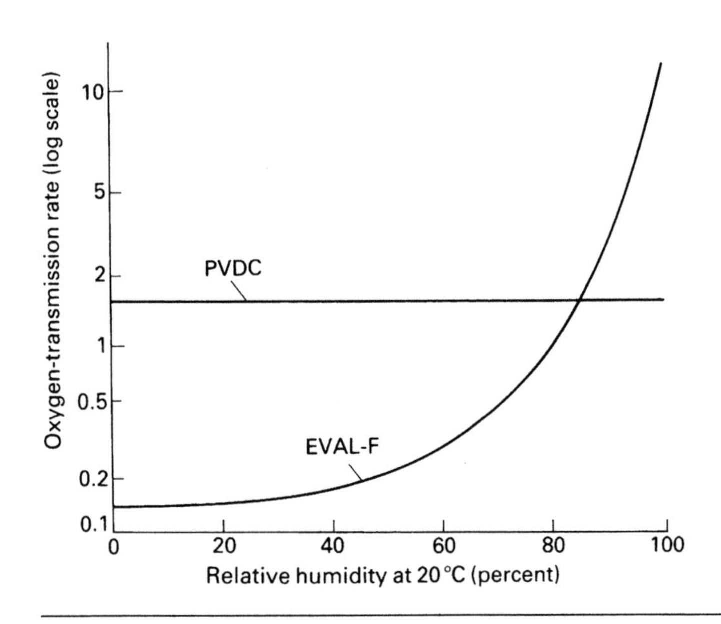

Figure 11.5 Variation of the oxygen transmission rate ($cm^3 m^2 day^{-1} bar^{-1}$ for 20 mm film) with relative humidity for the high barrier PVDC and EVAL films (from *Plastics Engineering*, May, 43, Soc. Plastics Eng. Inc., 1984).

$$\text{Relative humidity} = \frac{\text{Partial pressure of water vapour}}{\text{Vapour pressure of water at that temperature}} \tag{11.16}$$

it follows that the water transport rate through the film is proportional to the permeability, to the water vapour pressure and to the relative humidity difference across the film.

11.1.5 Metal and ceramic coatings

To increase the barrier to oxygen, polymer packaging is coated with impermeable layers of metal or silicate glass. Aluminium, if vacuum deposited, is less than a micron thick. Table 11.4 shows that it reduces the permeability of a PET/PE composite film by a factor of 10. Such collapsible bags are used inside wine boxes, as a method of increasing the shelf life of the partly consumed wine.

11.1.6 Gas separation

Gas separation membranes can separate nitrogen from air. Liquefaction at high pressure and low temperature, followed by fractional distillation, is

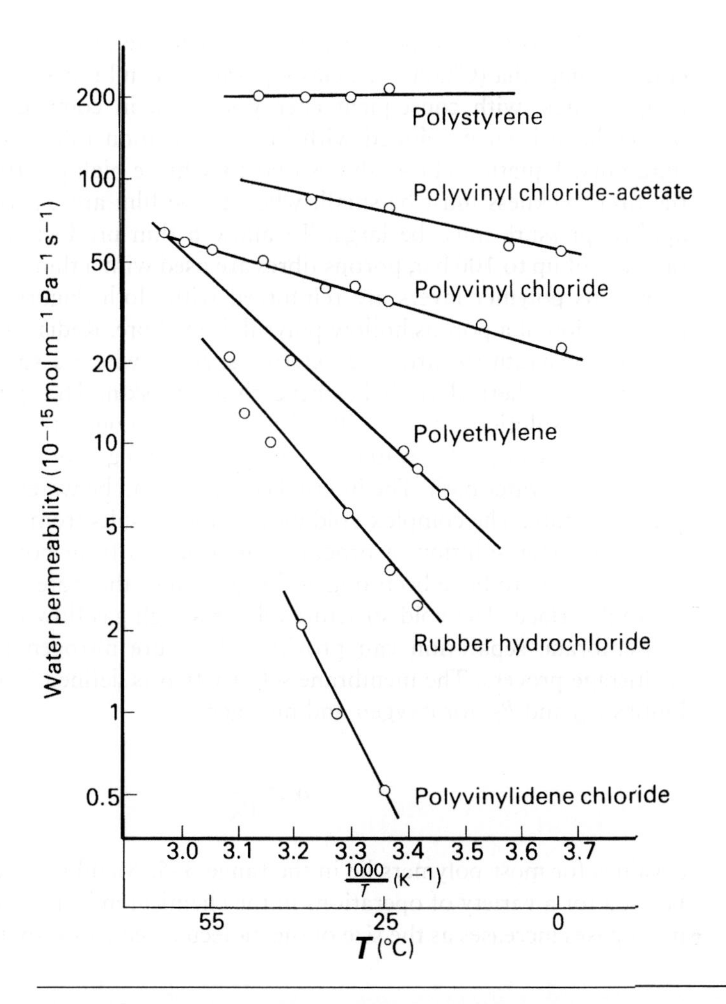

Figure 11.6 Variation of water permeability with reciprocal absolute temperature (from Hennessy B. J. *et al.*, *The Permeability of Plastics Films*, Plastics and Rubber Inst., 1966).

Table 11.4 Effect of an inter-layer between 12 μm PET film and 50 μm PE film on transmission rates at 23 °C

Inter-layer	*O_2 at 50% RH (cm^3/m^3 day atm)*	*Water vapour at 85% RH (g/m^3 day)*
None	15–20	4–6
PVDC 4 μm	5	2
Vacuum metallised	1–2	0.1–0.5
Al 9 μm	0	0

also used. This set of processes is used to separate the gaseous products of cracking naphtha (Chapter 2). Gas separation membranes work at ambient temperatures, with consequent energy savings, in compact plant. Membranes have been developed with high permeation rates and mechanical durability. Equation (11.4) shows that to achieve high gas transport rates, the film thickness must be small, whereas the film area, permeability and applied pressure must be large. To allow a thin product to cope with a pressure of up to 100 bar, porous fibres are used with a thin membrane skin, or porous polymer layers are reinforced with cloth. Figure 11.7 shows a cross section of a porous hollow polysulphone fibre. Radial columnar voids increase in width towards the external surface, where a sub-surface layer with nearly spherical voids lies beneath a thin skin. The spinning dope is a viscous solution of typically 30% polysulphone, 60% of *n*-methyl 2-pyrrolidone and 10% water. This is spun, with a gap of a few centimetres in air, into a water bath. The internal coagulant can be water or an ethanol–water mixture. The complex void morphology results from phase inversion of the polymer solution, a process that proceeds from both the surfaces. There appears to have been finger-like growth of the water phase from the internal surface. The void structure allows a high gas flow rate.

Membrane separation can provide 99% pure nitrogen from air, in a multistage process. The membrane selectivity α is defined from its permeabilities P_{O} and P_{N} for oxygen and nitrogen

$$\alpha \equiv \frac{P_{\mathrm{O}}}{P_{\mathrm{N}}} \tag{11.17}$$

α values for most polymers lie in the range 3–5. Membrane separation can be used for a variety of operations in the chemical industry. The permeability of gases increases as the size of the molecule decreases, so it is possible to

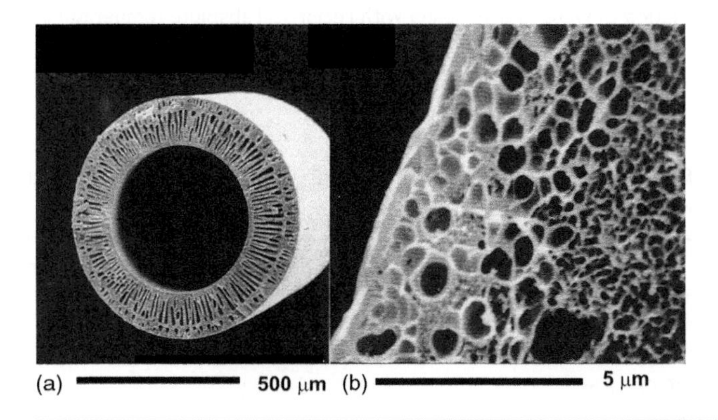

Figure 11.7 Polysulphone hollow fibre membrane. (a) Cross section. (b) Outer edge (from Wang, D. *et al.*, J. *Membr. Sci.*, **204**, 247, 2002) © Elsevier.

separate small amounts of hydrogen from a mixture of gases. Applications are reviewed by Spillman (1989).

11.2 Liquids

Liquid diffusion in polymers is generally slower than gas diffusion, with diffusivities of the order of $10^{-13}\,m^2\,s^{-1}$. The equilibrium solubility of liquids can be much larger than that of gases, and the liquid content can change the diffusion constant or even the physical state of the polymer. Semi-crystalline polymers are in general more resistant to organic liquids than glassy polymers, so the former are preferred.

11.2.1 HDPE fuel tanks

Hollow containers are often made by blow moulding polyolefins, especially HDPE (Fig. 5.17). They can contain a great variety of liquids, because there are few solvents for HDPE. Although high molecular mass homopolymer grades have good ESC resistance, MDPE copolymers have a better resistance (Chapter 10). However, aromatic or chlorinated hydrocarbons, which swell the amorphous phase of polyethylene, have relatively high permeabilities (Table 11.5). The corresponding figures for LDPE are a factor of 10 larger, since diffusion is through the amorphous rubbery phase. Ethyl acetate is a representative constituent of foodstuffs. The high permeability of alkanes is not surprising considering the chemical similarity with polyethylene.

Blow-moulded polyethylene petrol tanks provide considerable weight savings compared to steel tanks. Corrosion is eliminated, and the complex moulded shapes can fit into spaces above the rear axle, protected from a rear impact. This allows more of the boot space to be used for luggage.

The permeation loss per litre of liquid stored, is smaller in large containers than small ones, because they have a higher ratio of surface area to volume. This also applies to CO_2 loss from carbonated drinks bottles,

Table 11.5 Permeability of liquids through HDPE of density $950\,kg\,m^{-3}$ at 23 °C

Liquid	*Permeability* ($g\,mm\,m^{-2}\,day^{-1}\,bar^{-1}$)
Toluene	37.5
n-Heptane	17.1
97 Octane petrol	16
Ethyl acetate	1.6
Diesel oil	0.5–3
Methanol	0.15

which explains why 2 l PET bottles were introduced first. The container is designed to withstand an internal pressure of 3 bar, in spite of crash tests showing that internal pressures did not exceed 1 bar. Blow moulding does not produce a uniform wall thickness product; values ranging from 4 to 7 mm are found in a fuel tank. In 2003, stricter regulations in California limited the emission to 0.5 g per 24 h shed (sealed house emission determination) test, because hydrocarbon gas emissions can lead to photochemical smog. Since the 1990s, gasoline has been reformulated to improve air quality, but some changes (adding ethanol) have increased the permeability through polyethylene. Some constituents of gasoline swell polyolefins. The polyethylene is sulphonated (treated with concentrated SO_3) to decrease the permeability of a surface layer by a factor of 10, or fluorinated to decrease the permeability by 97%. Figure 11.8 shows the construction of fuel tank wall, which contains a barrier layer of an ethylene vinyl alcohol copolymer (32% ethylene). The solubility of gasoline in the barrier layer is much

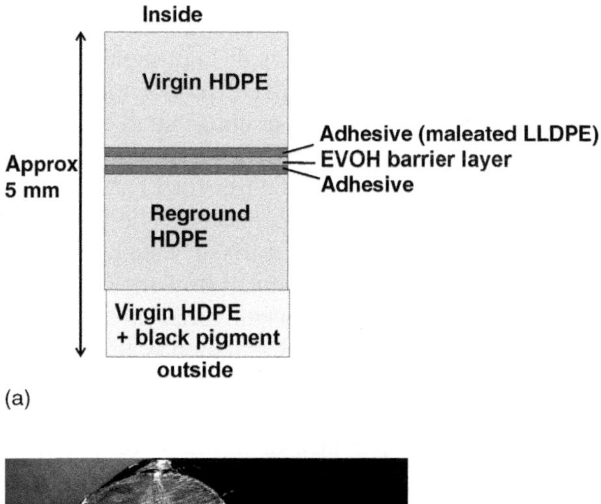

Figure 11.8 (a) Cross section of the wall of a polyethylene fuel tank. (b) Geometry of external part of weld, where the barrier layer meets the outside (from Ellis, T. S., *Soc. Auto. Eng.*, 2003-01-1121) © SAE International.

smaller than in HDPE, whereas the diffusion coefficient is about 20% of that in HDPE. Consequently, the permeability is reduced to a level where the main losses are near the seam of the tank, where the barrier layer comes to the surface of the tank.

11.2.2 Extraction of additives by food liquids

When polymeric containers are used to store food, some additives may be extracted from the polymer. Certain food constituents, such as fats or oils, can diffuse quite readily into polyethylene. If the fat has a strong affinity for a polymer stabiliser or antioxidant, the equilibrium concentration in the fat will be much higher than that in the polymer. Two-way diffusion occurs, with the food component entering the polymer and the polymer additive entering the foodstuff. Figure 11.9 shows experimental results for the fat tricapyrlin in contact with HDPE containing 0.25% of the hindered phenol antioxidant BHT. The results, for diffusion at 40 °C, are normalised by using $x/\sqrt{t}$ as the horizontal axis.

The concentrations of both diffusants remain low and the results fit the theory for a constant diffusion coefficient of $5 \times 10^{-13}\,\mathrm{m^2\,s^{-1}}$ for the fat, and $1 \times 10^{-14}\,\mathrm{m^2\,s^{-1}}$ for the antioxidant. Consequently, only certain non-toxic additives are permitted in polymers used as food containers. If a plastic container is reused with another foodstuff, constituents of the first foodstuff may diffuse back out of the polymer into the second foodstuff. This is noticeable if polyethylene beakers or bottles are filled with orange squash, then reused with water; the water develops an orange flavour.

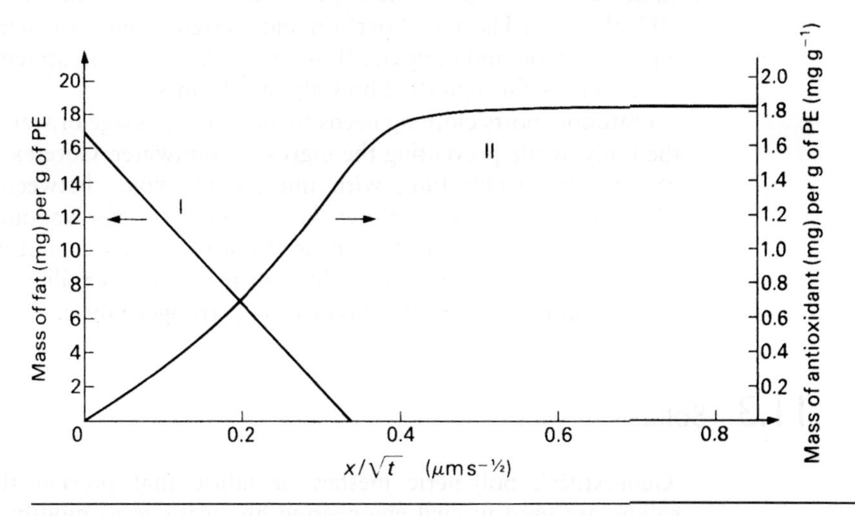

Figure 11.9 Diffusion of a fat into HDPE and of an antioxidant additive from the HDPE, plotted against *x/t*, where *x* is the distance from the surface and *t* the time (from Figge, K, and Rudolf, F, Angewandte Macromol Chemie, ***78***, *179, 1979, Heuthig and Wepf, Verlag).*

11.2.3 Reverse osmosis and dialysis membranes

Seawater or brackish water can be purified by *reverse osmosis*. To maximise the flow of water through a polymer membrane, the polymer must have a high water permeability, yet a low permeability for the salts. To maximise efficiency, the membrane area must be large and its thickness as small as possible consistent with a lack of pinholes. A high pressure is applied to the salt water side of the membrane. Because it is thin, a cellulose triacetate membrane is supported on a porous cellulose nitrate–cellulose acetate support structure to resist the pressure. To make the unit compact the composite membrane is spirally wound on to an inner cylinder, and the edges glued together. When a pressure of 70 bar is applied to the seawater side, NaCl rejection levels in excess of 99.7% can be achieved.

Blood *dialysis* is used for patients with kidney malfunction. In a disposable dialysis unit, about 10 000 hollow polysulphone fibres are mounted in a 30 cm long cylindrical unit. The fibre microstructure (Fig. 11.10) is related to that shown in Fig. 11.7. However, the internal diameter is smaller at 185 or 200 μm, and the solid skin occurs on the interior of the fibre. Blood passes through the fibres at a velocity of 1.3 cm s^{-1}, while the saline solution outside the fibres removes metabolic waste products from the blood and neutralises excess acids. The fibres can have a waviness along their length, in an attempt to achieve more uniform packing, hence to have a nearly uniform dialysate flow rate. A slotted plastic part distributes the dialysate uniformly to the perimeter of the fibre bundle (Fig. 11.10c). The dialyser performance is, in principal, determined by its mass-transfer area coefficient, the product of its surface area A and urea permeability K_0. Typical values are $A = 1.2\,\text{m}^2$ and $K_0A = 960\,\text{ml min}^{-1}$, while the blood flow is $400\,\text{ml min}^{-1}$. The actual performance is significantly smaller, but increases with the blood and dialysate flow rates. During the treatment, each litre of blood passes through the fibres about 10 times.

Outdoor sports clothing needs to allow the passage of water vapour from the body, while preventing the ingress of rainwater. Goretex fabrics consist of stretched PTFE film, with microscopic voids between PTFE fibrils (Fig. 11.11). The voids allow the passage of water vapour but the very low surface energy of PTFE means that liquid water cannot wet its surface. Hence, rain cannot be drawn through the film by capillary action. As the film is relatively weak, it is bonded to a stronger fabric.

11.3 Solids

'Geotextiles', polymeric meshes or fabric that prevent the passage of solids, are used in civil engineering applications to modify the properties of soil. The two main product types are based on highly oriented polyolefins, the polymer being chosen for its low cost and high strength. One

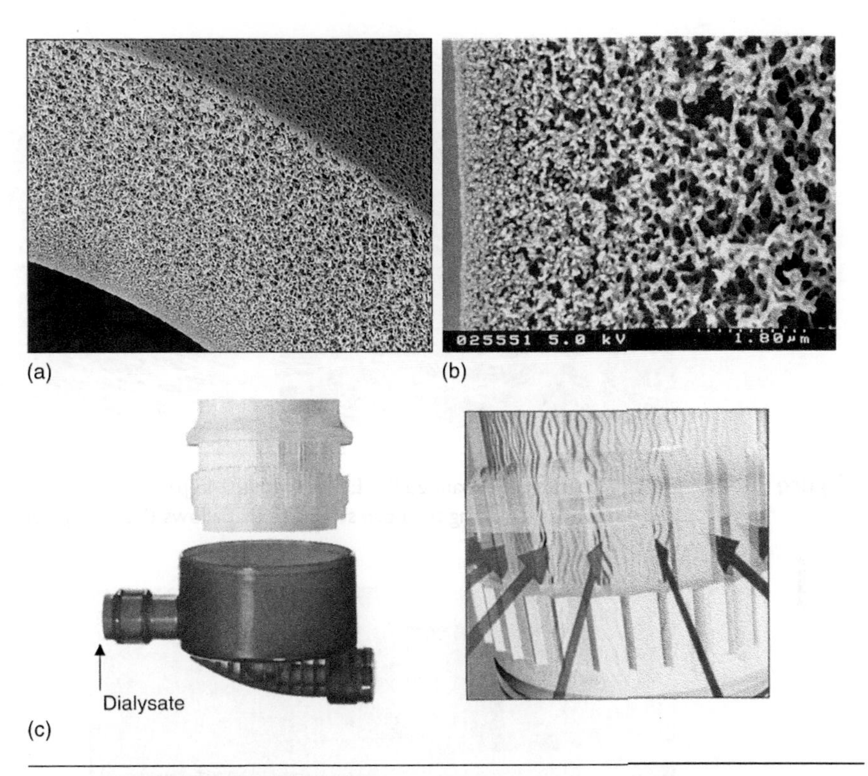

Figure 11.10 Hemodialyser: (a) Cross section of 35 μm thick fibre wall; (b) detail of the inner region of the fibre; (c) dialysate distributor to the wavy polysulphone fibres in the unit (Ronco, C. *et al.*, *Kidney Int.*, **60** (Suppl. 80), S126, 2002) © Blackwell.

type utilises uniaxially drawn polypropylene film, fibrillated to produce a low cost substitute for fibres. This is woven into a coarse textile with a mass of between 100 and 300 $g\,m^{-2}$ and a thickness of 0.3–0.7 mm (Fig. 11.12). The second type is 'Netlon' products, based on the uniaxial or biaxial drawing of perforated HDPE sheets. The holes, orders of magnitude larger than those in the woven film, prevent the passage of stones.

Geotextiles have a high in-plane tensile strength, so are ideal for soil reinforcement. Soils have zero tensile strength, and steep sided soil embankments can fail by shear on surfaces at 45° to the vertical, especially if they contain clay and become waterlogged. Horizontal layers of geotextile, with a tensile strength of 50–100 kN per metre width, can be incorporated into embankments at 1 m vertical separation, while the embankment is constructed. The plane of the geotextile is chosen to coincide with the tensile principal stress direction in the soil, to provide reinforcement in the optimum direction. Because the geotextile is buried it cannot be degraded by UV radiation.

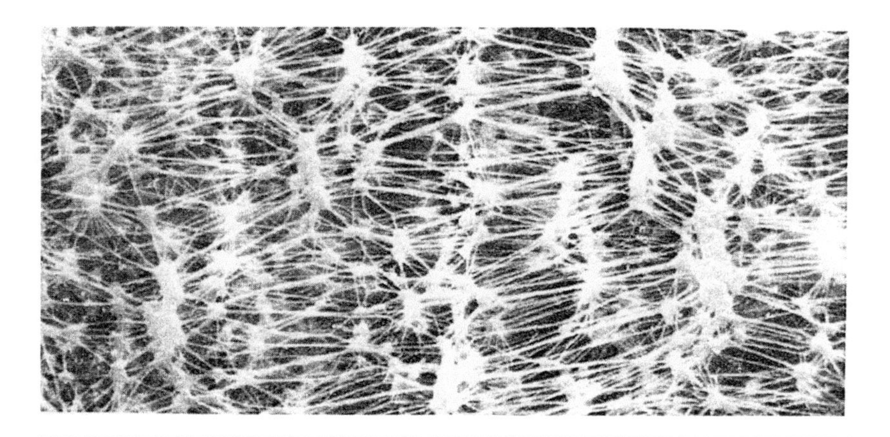

Figure 11.11 SEM micrograph of expanded PTFE fitter with a 0.45 μm pore size (courtesy of W. L. Gore and Associates) showing the pore structure that allows the passage of water vapour.

Figure 11.12 Woven mesh of fibrillated polypropylene film that allows water permeation but prevents the ingress of fine soil particles.

Figure 11.13 shows an application of unidirectionally oriented Netlon. Vertical rods couple the faces of the cellular structure, which can then be filled with granular material to form a 1 m thick stable 'mattress' at the foot of an embankment, on top of soft soil.

Woven geotextiles prevent the loss of fine soil particles across the fabric plane. If unidirectional water flow occurs through the textile, a filter cake of fine particles builds up on the textile, aiding the filtration process. By preventing road stone from being punched into soft underlying soil, the total amount of road stone used can be reduced. The shear strength

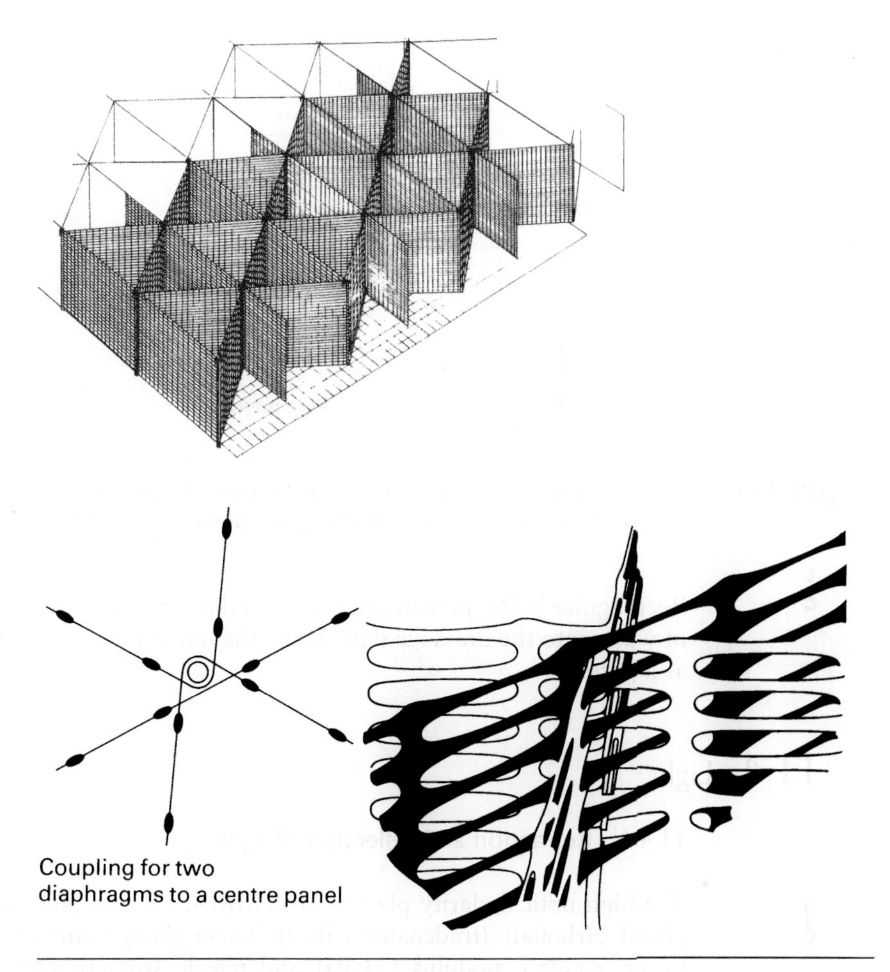

Figure 11.13 Use of unidirectionally drawn perforated HDPE sheet as the vertical walls of triangular cells, that form a reinforcing mattress at the base of an embankment (from 'Tensor Geocell Mattress' pamphlet, Courtesy of Netlon).

of the road stone layer is preserved, by preventing the ingress of fine soil particles. Geotextiles also allow drainage in the plane of the fabric, so water can drain to the sides of a newly constructed embankment, aiding its consolidation.

Perforated corrugated pipe (Fig. 11.14) aids soil drainage of agricultural land, but lacks the soil-reinforcing properties of geotextiles. The hoop direction corrugations are produced by specially shaped cooling sections that move down the cooling section of an extrusion line on a caterpillar track. The corrugations increase the resistance to diametral crushing by soil loads, and yet allow flexibility for coiling small diameter pipes.

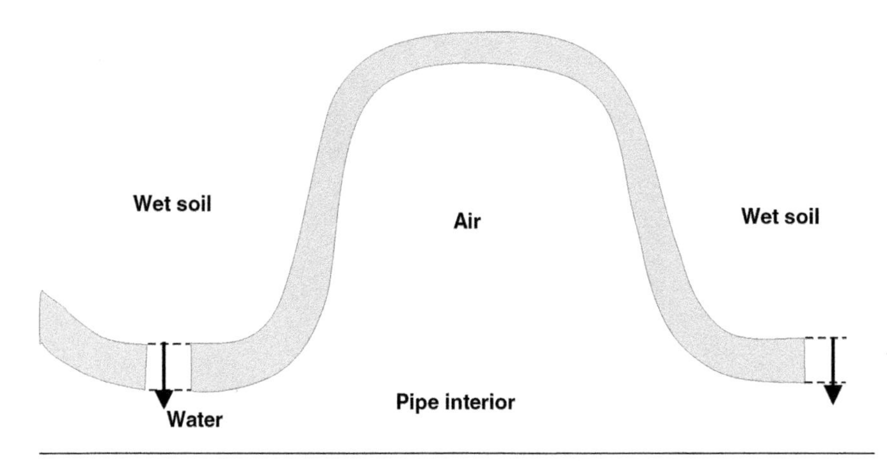

Figure 11.14 Section through the wall of a corrugated PVC pipe, of diameter 150 mm, for soil drainage. Slots of size 8 × 1 mm, at intervals in the base of the corrugation, allow water ingress.

Rectangular holes, punched in the small diameter regions, allow the ingress of water into the drainage pipe, while the intact lower part of the pipe acts as a gutter.

11.4 Light

11.4.1 Refraction and reflection of light

The high optical clarity plastics are PMMA, PC and thermosetting diallyl glycol carbonate (tradename CR39). These glassy materials have a much lower Young's modulus (3 GPa) and tensile strength (50–70 MPa) than conventional soda-lime glass(70 GPa and >200 MPa, respectively). Usually only infrared and ultraviolet light is absorbed by the polymer, unless there is a pigment present or the polymer contains conjugated double bonds. The advantage of these polymers lies in their lower density and higher toughness, and the fact that they can be moulded to high precision, obviating the polishing stages needed with silicate glasses.

The refractive index n and dispersive power D are important in lens applications. The refractive indices measured are n_B for blue (486 nm), n_Y for yellow (587 nm) and n_R for red (656 nm) wavelengths, and D calculated from

$$D = \frac{n_B - n_R}{n_Y - 1} \tag{11.18}$$

Table 11.6 shows that polycarbonate has a high refractive index, but its high dispersive power increases chromatic aberrations.

Table 11.6 Optical properties of glassy polymers

Material	Refractive index (n_Y)	Dispersive Power (!D)	Density (ρ)(kg m^{-3})	$\frac{\rho}{n_Y - 1}$	Stress optical coefficient ($\times 10^{-12}$ m^2 N^{-1})
PMMA	1.495	0.0189	1190	2400	4
CR-39	1.498	0.0172	1320	2650	34
PC	1.596	0.0333	1200	2010	78
PS	1.590	0.0323	1060	1800	9
Soda lime silica glass	1.520		2530	4870	2.7

The mass of a lens of a given diameter and focal length is proportional to its axial thickness and the material density. If the radius of curvature of both surfaces of a biconvex lens is R then the focal length f is given by

$$f = \frac{R}{2(n-1)} \tag{11.19}$$

As the axial thickness of the lens is proportional to $1/R$, Eq. (11.19) shows that it will also be proportional to $1/(n - 1)$. Hence, the mass of the lens

$$m \propto \frac{\rho}{n-1}$$

This parameter, given in Table 11.6, shows that plastic lenses provide considerable mass savings compared to a silicate glass lens.

When a beam of light meets a sheet of plastic at normal incidence, about 4% of the light intensity will be reflected back at the air/polymer and at the polymer/air interfaces (Fig. 11.15). The reflected intensity R_0 for normal incidence is related to the incident intensity I by

$$R_0 = I\left(\frac{n_1 - n_2}{n_1 + n_2}\right)^2 \tag{11.20}$$

where the refractive index of the polymer $n_1 \cong 1.5$ and that of air $n_2 = 1$. If a high reflectivity is required, (see CD manufacture in Chapter 14), a conducting coating is required. Metals have a complex refractive index, with real part n_R and imaginary part n_I. This means that the light wave, that penetrates the metal, has exponentially decaying amplitude. The magnitude E of the electric vector varies with the distance y from the surface of the metal as

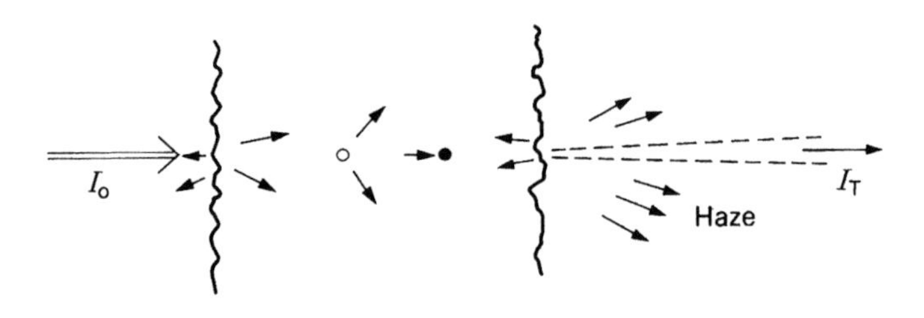

Figure 11.15 Processes in a plastic film that reduce the transmitted light intensity.

$$E = E_0 \exp(-\omega n_I y/c) \cos \omega(t - n_R y/c) \tag{11.21}$$

where ω is the frequency of the wave, and c the speed of light. For solid sodium at a wavelength of 589 nm, the components of the refractive index are $n_R = 0.04$ and $n_I = 2.4$. This leads to a reflectivity for thick films of $R = 0.9$. For thin films the amplitude of the transmitted light decreases, according to Eq. (11.21), as $\exp(-2\pi n_I y/\lambda)$. This means that the metal only needs to be a few wavelengths thick for the reflectivity to be high. Such a thickness can easily be applied by vacuum evaporation, and the main concern is the protection of the layer from abrasion, with a transparent lacquer.

11.4.2 Light scattering

Light scattering can occur at polymer/air interfaces, and internally in polymers (Fig. 11.15). The acceptable level depends on the application. If the transmitted light provides illumination, as in a ceiling light, or allows a liquid level to be inspected, as in a brake fluid reservoir, then a high level of light scattering can be tolerated. However, if a clear image is required to perform an eye–limb coordination task like driving a vehicle, a high level of optical clarity is required. Small angular deviations of the light path, caused by the lens effects of non-planar surfaces, will cause image distortion. High angle light scattering will cause *glare* from a bright light source in the field of view, such as oncoming car headlights at night.

Light scattering can be marked in semi-crystalline polymers with a spherulitic microstructure; for example, unpigmented polyethylene appears milky and opaque. However, it is negligible when the diameter, of inclusions in a matrix, are smaller than 10% of the wavelength of the light (Fig. 11.16).

The scattering coefficient is a maximum when the inclusion diameter is slightly smaller than the wavelength of light. The scattering also depends on the difference $(n_1 - n_2)^2$ according to Eq. (11.20).

The mean refractive index $\bar{n}$ of a phase is related to its density ρ and to the mean polarisability of the monomer unit $\bar{\alpha}$ by

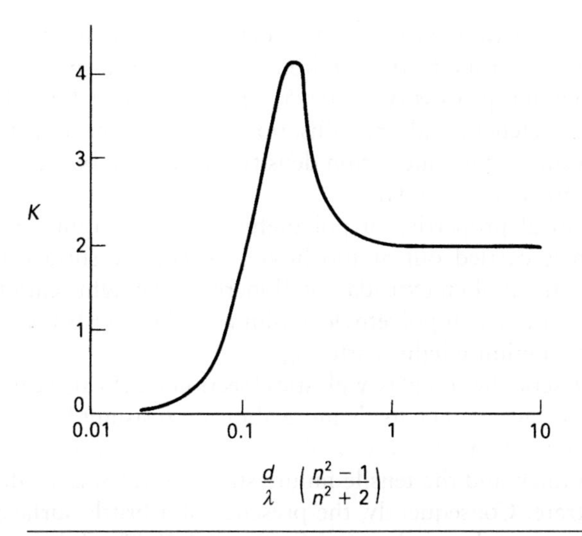

Figure 11.16 Scattering coefficient for a single sphere of relative refractive index n vs. the ratio of the sphere diameter D to the wavelength of the light λ.

Table 11.7 Densities of polyolefin phases at 20 °C

Polymer	*Crystal density* ($kg\,m^{-3}$)	*Amorphous density* ($kg\,m^{-3}$)
Polyethylene	1000	854
Polypropylene	940	850
Polymethylpentene	820	840

$$\frac{\bar{n}^2-1}{\bar{n}^2+2} = C\rho\bar{\alpha} \tag{11.22}$$

where C is a constant. For most semi-crystalline polymers, the crystalline phase density exceeds that of the amorphous phase (Table 11.7) and the width of the crystal lamellae is of the same order as the wavelength of visible light. The exceptional case is polymethylpentene which is transparent; the crystal has open helical chain conformations, so the crystal and amorphous densities are almost equal. Rubber-toughened polymers like ABS are opaque because the phases differ both in density and polarisability.

Even if a polymer could be made 100% crystalline, there would be light scattering from neighbouring crystals with different orientations. The anisotropy of bonding means that polymer crystals have a different refractive index n_c, for light polarised along the covalently bonded **c** direction, than for light polarised along the **a** or **b** directions. Stretching the product aligns the **c** axes of crystals, hence, reduces the range of crystal orientations.

However, reduction of the crystal size, to much smaller than the wavelength of light, is the most effective method of inducing transparency. In the stretch blow moulding process (Chapter 5), the walls of PET bottle preforms are biaxially stretched, and crystallisation occurs on heating from the glassy state, causing a high nucleation density. Hence, the bottles are transparent, yet have 50% crystallinity.

The optical properties of polyolefin packaging films are important. If extrusion is carried out at too high a speed, the surface roughness that occurs on the molten extrudate will increase the light scattering. The average spherulite size in polyethylene film must be kept below the wavelength of light to minimise light scattering.

Surface scratches on glassy plastics (Section 8.2.5) also cause light scattering. The surfaces can be made more abrasion resistant by coating them with a hard layer of a highly crosslinked silicone thermoset. The layers are 5–10 μm thick and the tensile failure strain at 1.2% is smaller than that of the substrate. Consequently, the presence of a brittle surface layer reduces the toughness of the product (see Section 8.2.3). This is less of a problem for spectacle lenses than it is for motorcycle visors, which are designed to cope with 145 km h^{-1} impacts of a 7 mm ball bearing.

11.4.3 Fibre optics

A fibre optic uses total internal reflection at the core/coating interface to restrict the light beam to the interior of a fibre (Fig. 11.17). The refractive index of the fibre core n_f must be less than that of the coat n_c so that the angle of incidence γ at the interface is greater than the critical angle θ_c, given by applying Snell's law

$$n_f \sin \theta_c = n_c \tag{11.23}$$

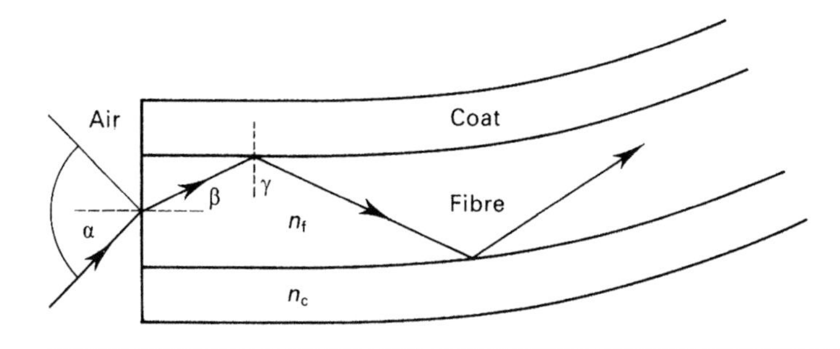

Figure 11.17 Ray diagram for light passing down a fibre optic cable, showing the limiting ray that just undergoes total internal reflection. $n_f = 1.65$, $n_c = 1.50$, numerical aperture = 0.69, limiting $\alpha = 43°$.

The brightness of fibre optic illumination depends on the amount of light that enters the flat end of the fibre, and is transmitted along it by total internal reflection. This is determined by the semi-acceptance angle α, given by applying Snell's law at the end face of the fibre

$$\sin\alpha = n_f \sin\beta \tag{11.24}$$

Assuming that there is a uniformly bright light source near end of the fibre, its light gathering power P is proportional to $(\sin\alpha_c)^2$ and to the fibre cross-sectional area. By Eq. (11.24) this gives

$$P \propto n_f^2 \sin^2\beta_c$$

Since $\beta + \gamma = 90°$, using Eq. (11.23) gives

$$P \propto n_f^2 - n_c^2 \tag{11.25}$$

The quantity on the right-hand side of Eq. (11.25) is the square of the numerical aperture of the fibre. Uncoated fibres, surrounded by air, would work well in the regions where the fibres do not touch, but light could pass between touching fibres, and the contact damage would cause losses.

There are several types of fibre optics: The requirement on losses (Table 11.8) determines the type of material used. The stringent requirements on transmission losses for long distance telecommunication applications can only be met if the light travels axially down the fibre as a single mode wave (similar to a waveguide for cm wavelength radar waves). Figure 11.18 compares the loss spectrum of high purity silica fibres with those of ordinary and deuterated PMMA. The loss peaks are due to various molecular vibrations. In the PMMA, these are harmonics of the C−H bond vibration, and the absorption becomes very strong in the infrared (IR). In the silica fibre, there is a loss peak at 1.4 μm due to absorbed hydroxyl groups. There is a background effect of Rayleigh scattering, due to variations in the density of the material, which decreases with λ^{-4}. Consequently, telecommunications fibres operate with laser light at 1.30 or 0.85 μm, where the Rayleigh scattering is a minimum. The silicate glass core is 8 μm diameter while the doped silicate glass coating is 125 μm

Table 11.8 Types of fibre optic system

Application	*Type of transmission*	*Losses (dB/km)*	*Core/coat diameter (μm)*
Telecommunications	Single mode wave	<0.4	8/125
Local area networks	Multi mode waves	~0.6	50/125
Endoscopes	Ray	>500	250

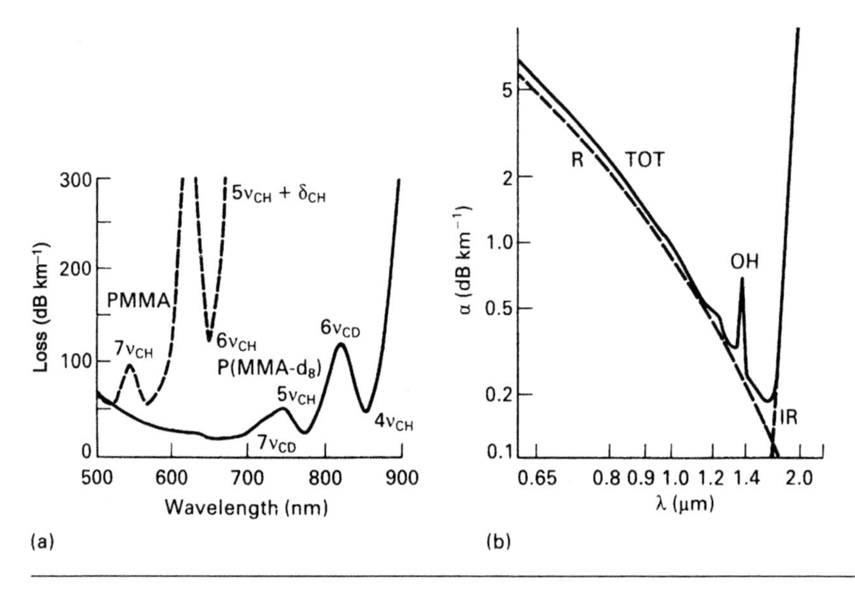

Figure 11.18 Transmission losses vs. wavelength for: (a) PMMA and deuterated PMMA (from Kaino T. and Katayana Y., *Polym. Eng. Sci.*, **29**, 2109, 1989; (b) pure silica single-mode fibre. *R* is the Rayleigh scattering contribution, IR the infra red absorption and OH the hydroxyl absorption (from Geittner P and Lydtin, H. *Philips Tech. J.*, **44**, 45, 1989).

diameter. A rubber inter-layer is used between this and the 900 μm diameter UV-cured glassy polyacrylate secondary coating. The secondary coating prevents damage to the silicate glass and provides some bending stiffness.

The transmission losses of polymer-cored fibres have been reduced by using fluorinated structures and avoiding the C−H bond. However, the lower limit to the transmission loss is thought to be 5 dB km^{-1} at a wavelength of 0.65 μm.

Polymer fibres are preferred for flexible short length fibre optics, where the total light output needs to be large. In medical applications, endoscopes are sufficiently flexible to pass into the stomach. They are used either for diagnosis, or, with cutting tools attached to the endoscope, for surgery. Flexibility of the several mm diameter cable is achieved by having many fibres of small diameter—the theory is covered in Appendix C, Section C.1.3. The PMMA fibres tolerate higher strains than does a silica fibre; if of equal diameter, PMMA fibres can be bent to a smaller radius than silica fibres.

11.5 Thermal barriers

The thermal insulation of buildings and refrigerators has become more important as energy costs have risen. A number of materials have low thermal conductivity, so the material selection depends on factors such as cost, water barrier properties and mechanical strength. For refrigerator insulation, it helps if the material bonds to the inner moulding and outer layer. We will look at the development of polyurethane and polystyrene foams for thermal insulation.

There are several contributions to the thermal conductivity of low-density, closed-cell foams; thermal conductivity of the polymeric cell walls and the cell gas, plus convection and radiation in the cells. The thermal conductivity of most solid polymers is within a factor of 2 of 0.3 $W\,m^{-1}\,K^{-1}$. For foams of density 30 $kg\,m^{-3}$, the cell wall contribution, which is proportional to the foam relative density, is small. The contribution from convection inside the cells is negligible for cell diameters smaller than 10 mm. The radiation contribution is linearly proportional to the cell size, because infrared radiation is absorbed at each cell face then re-radiated. Figure 11.19 shows the effect of reducing the cell size of polyurethane foams on the total thermal conductivity. In polystyrene foams, the cells are rarely larger than 0.5 mm, so the radiation contribution to the foam conductivity is minimal.

As gas conduction is the major contribution to the foam thermal conductivity, the gas should have a low cost, low thermal conductivity and not

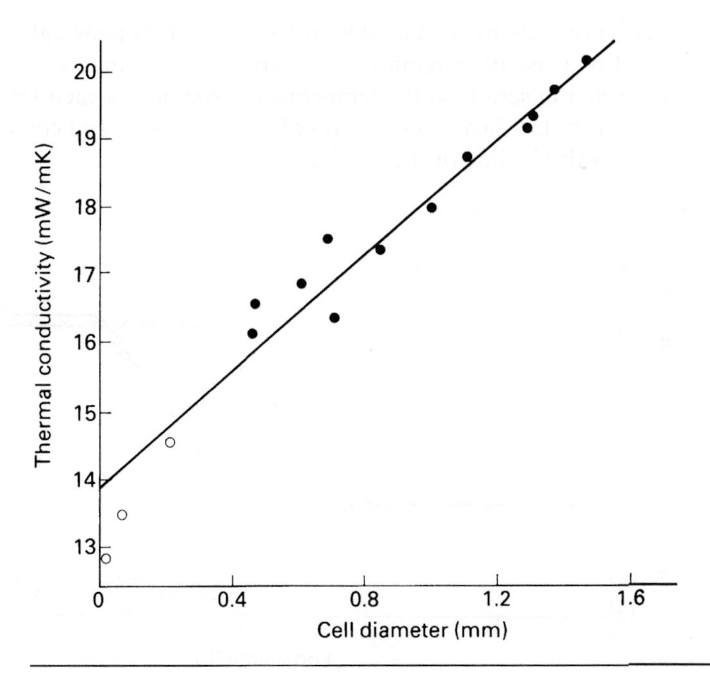

Figure 11.19 Thermal conductivity of a polyurethane foam of a fixed density, containing CFC-11, vs. the mean cell size (from Buist, J. M., Ed., *Development in Polyurethanes*, Elsevier Applied Science, 1978).

escape from the cells with time. Chlorofluorocarbon (CFC-11 is CCl_3F) gases were used, due to their low thermal conductivity and effectiveness as blowing agents in the foaming process; their rate of escape from the foam by diffusion was very low. However, the 1987 Montreal protocol phased out their use in developed countries by 1996, to stop damage to the ozone layer. The hydrochlorofluorocarbon HCFC-141b (CH_3CCl_3F) has a small diffusion coefficient through polyurethane foams, but its use was phased out in EU countries by 2003. Pentane has a higher thermal conductivity ($13\,mW\,m^{-1}\,K^{-1}$ at 25 °C) than CFC-11 ($8\,mW\,m^{-1}\,K^{-1}$), but lower than air ($26\,mW\,m^{-1}\,K^{-1}$). However, mixtures of between 1.5 and 8% of pentane in air are explosive. The increase in foam thermal conductivity with time (Fig. 11.20) is due to the ingress of air, with an increase in the total pressure of gas in the cells.

Heat loss calculations for buildings must be based on long-term values. UK Building Regulations specify the U-value of insulation, whereas in the USA, the thermal resistance R is specified. These quantities, the reciprocal of one another, are defined by Eq. (11.26), for the steady-state heat flow q across an area A, for inside T_i and the outside T_o temperatures

$$R = \frac{1}{U} = \frac{A(T_i - T_o)}{q} \tag{11.26}$$

Polyurethane foam is used with facings such as paper, glass fibre, plasterboard or glass fibre reinforced concrete. The facings and foam are layers connected in series, so the temperature drop across each layer is added for the steady-state heat flow. If layer i has thickness L_i and conductivity k_i then the overall U-value of the product is given by

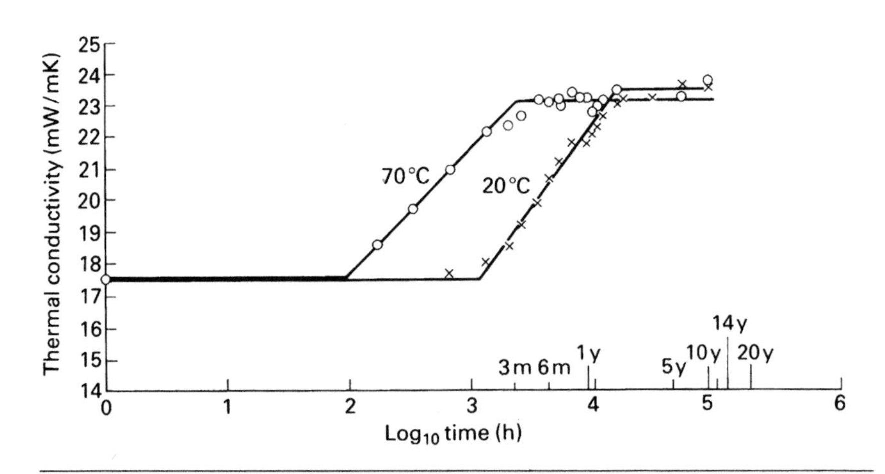

Figure 11.20 Variation of the thermal conductivity of polyurethane foam with storage time (from Buist, J. M., Ed., *Developments in Polyurethanes*, Elsevier Applied Science, 1978).

$$\frac{1}{U} = \frac{1}{h_i} + \frac{L_1}{k_1} + \frac{L_2}{k_2} + \ldots + \frac{1}{h_o} \qquad (11.27)$$

For a single material layer, ignoring the surface convection terms, U is the thermal conductivity divided by the thickness. The heat transfer coefficient h_o for forced convection at the outer surface is sufficiently high, so that its contribution to the U-value can be ignored. Current UK Building Regulations call require $U < 0.35\,\mathrm{W\,m^{-2}\,K^{-1}}$ for external walls, $0.35\,\mathrm{W\,m^{-2}\,K^{-1}}$ for ground floors and $0.16\,\mathrm{W\,m^{-2}\,K^{-1}}$ for pitched roofs. A value of $0.35\,\mathrm{W\,m^{-2}\,K^{-1}}$ can be achieved using a 15 mm layer of polyurethane foam of density $32\,\mathrm{kg\,m^{-3}}$ and thermal conductivity $0.024\,\mathrm{W\,m^{-1}\,K^{-1}}$. Losses from windows are significant—a single-glazed wooden frame window has $U \cong 5\,\mathrm{W\,m^{-2}\,K^{-1}}$, whereas a double-glazed PVC framed window must have $U < 2\,\mathrm{W\,m^{-2}\,K^{-1}}$.

Rigid polyurethane foam, injected as a liquid, forms a good adhesive bond to most surface layers, and it has a relatively high Young's modulus. Consequently, the sandwich structure performs efficiently in terms of bending stiffness per unit panel mass (Chapter 4). Lightweight sandwich panels are used in roofs and walls. If thermal insulation is of paramount importance, as in cold stores, the foam thickness can be increased to 125 mm.

Expanded polystyrene foam (EPS) has different applications, because of its physical form (beads) and properties (higher permeability to water and less effective adhesion to facing materials than polyurethane). The expansion gases, pentane and steam, escape fairly rapidly from the foam, so the thermal conductivity of the foam filled with air is about twice that of the best polyurethane foam—a 50 mm thick slab of foam has a U-value of 0.5–$0.6\,\mathrm{W\,m^{-2}\,K^{-1}}$. EPS mouldings can be used as shutters (formwork) for pouring concrete in a composite wall. The two EPS layers are connected at intervals to fix the thickness of the concrete. Extruded polystyrene foam (XPS) is used in plank form for insulation under the concrete floor of houses, and in roofing panels.

Chapter 12

Electrical properties

Chapter contents

The chapter begins with the topic of electrical insulation and proceeds to explore the dielectric response of polymers exposed to high frequency electrical signals. Finally, a few applications are described. However, specialist applications in the course of development, such as polymeric light-emitting diodes, are omitted for lack of space.

12.1 Volume and surface resistivity

Plastics have the great advantage over ceramics in being flexible, easily moulded, electrical insulators. They can be processed into thin insulating films, or polymeric layers can be used to isolate electrical circuits. With suitable additives they can conduct electricity, but they cannot compete with metals as low cost conductors for long distances. In many cases, the mechanical or thermal product requirements are more important than electrical strength in determining the required thickness of the insulating layer.

Figure 12.1 shows the areas occupied by various materials on a conductivity–temperature map; polymers are mainly low temperature insulators. The figure does not show the form of the materials: Semiconducting polymers are thin films, while polymeric insulators can have a variety of forms.

The concept of *volume resistivity* ρ implies that the resistance R of a bar is proportional to its length L and inversely proportional to its cross section A

$$R = \rho \frac{L}{A} \tag{12.1}$$

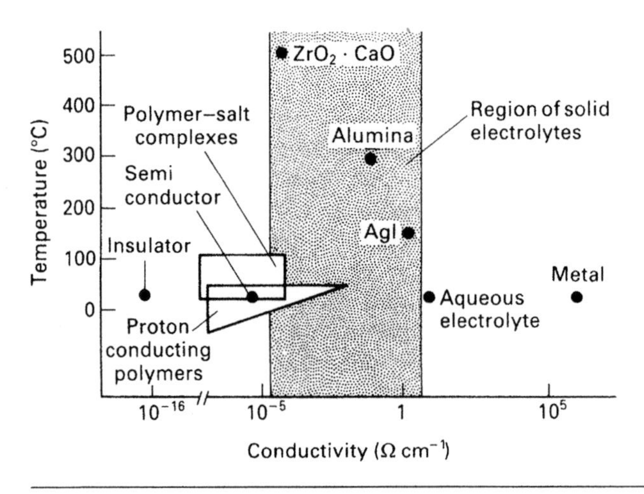

Figure 12.1 The conductivity of solids and their temperature range of use (from Margolis, J. M., *Conducting Polymers and Plastics*, Chapman & Hall, 1989).

It also implies that a constant applied voltage produces a constant current. However, for polymers with resistivities in the range 10^7 to $10^{16}\,\Omega$m, the extremely small currents decay with time after a voltage step is applied. Consequently, a time, often 1 min, is specified for the current measurement.

The composition of polymer surface layers often differs from the bulk, due to the migration of organic antistatic additives. The *surface resistivity* of the polymer can dominate the insulation resistance. The concept of surface resistivity implies the existence of a surface layer on top of an insulating substrate. The surface resistivity ρ_s is related to the layer thickness t and volume resistivity ρ by

$$\rho_s = \frac{\rho}{t} \tag{12.2}$$

It is also equal to the resistance in ohms between the opposite sides of a square of any size on the surface of the product.

The volume resistivity of polymers decreases with increasing temperature, in a way typical of semiconductors. If the logarithm of the resistivity is plotted against the reciprocal of the absolute temperature, a straight line results (Fig. 12.2), except when the polymer undergoes a phase change within the temperature range. Since the graph has the form used by Arrhenius for thermally activated chemical reactions, its slope is often interpreted in terms of the *activation energy* of the conduction process. Typical activation energies for amorphous polymers above T_g are between 0.2 and 0.5 eV. However, the charge carrier could be an electron, or one of many possible ions. The relationship between the volume resistivity and the concentration n of carriers of mobility μ is

$$\frac{1}{\rho} = qn\mu \tag{12.3}$$

The mobility, the ratio of the carrier velocity to the electric field, is of the order of $10^{-9}\,\mathrm{m^2\,s^{-1}\,V^{-1}}$ in polymers. Hence, one carrier, with an electronic charge q of 1.60×10^{-19} C, for every 10^9 monomer units causes the polymer resistivity to be $10^9\,\Omega$m. Ionic impurity levels of this magnitude have no effect on other physical properties.

A few polymers have lower resistivities. Polyamides, with hydrogen bonds that lie in parallel planes in the crystal structure, have resistivities a factor of 100 smaller than non-H-bonded polymers. At temperatures above 120 °C, at least half of the conduction in nylon 6,6 is due to protonic carriers, with hydrogen being liberated at one electrode. Nevertheless, the resistivity is still high.

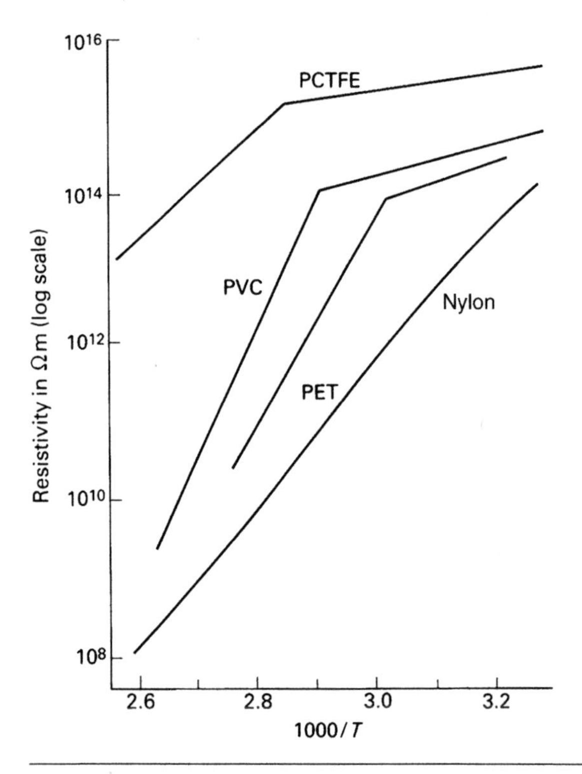

Figure 12.2 Log of direct current resistivity vs. reciprocal absolute temperature for various polymers used as insulators.

12.2 Insulation and semiconducting polymers

12.2.1 Low voltage electrical insulation

It is convenient to separate low and high voltage applications. For domestic and electronic applications, voltages do not exceed 500 V, and the electric strength of the insulation is not critical. The insulation separates two or more conductors, and provides mechanical support. Its thickness (Fig. 12.3a) is determined more by the safety aspects of abrasion and wear-and-tear than by insulation values. The cable bending stiffness can be reduced by replacing a single conducting wire of diameter D with n fine wires of diameter d (typically 0.2 mm) while keeping the same total cross-sectional area, to keep the resistance constant. Appendix C shows that the bending stiffness of a multi-core cable with n wires is a factor of n smaller than that of the equivalent single-core one. The insulating polymer sheath often has a diameter three times that of the core, therefore, although the polymer modulus may only be 0.1% of that of copper ($E = 100$ GPa), the polymer can add significantly to the overall bending stiffness.

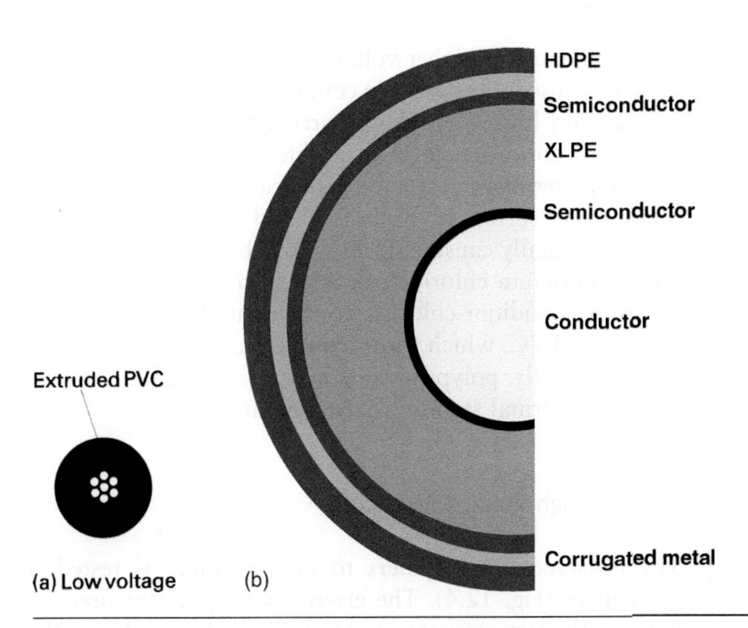

Figure 12.3 Construction of a) low voltage and b) extra high voltage cable. In the latter, the semiconducting layers equalise the electric field and prevent corona discharges.

The cross-sectional area of the copper conductors determines the current carrying capacity of the cable, because the ohmic heating of the wire must be limited. The temperature rating of the insulation, 60 °C for plasticised PVC, must not be exceeded. Excessive temperature could cause thermal degradation or oxidation (Chapter 10), so the insulation could crack when the cable is bent. Alternatively, the insulation may soften so much that it is penetrated by sharp objects. An empirical test of insulation failure uses a chisel with a 0.125 mm tip radius, pressed across the cable with a 3.5 N force. If electrical contact between the chisel and the conductor occurs within 10 min, the polymer fails the test. The test is repeated at higher temperatures until the insulation fails. The temperature rating can be increased by crosslinking the polymer, so that it no longer melts on heating above T_m or T_g. One to two per cent of a peroxide is incorporated into a polyethylene melt, before it is extrusion coated onto the cable. Crosslinking occurs when the cable is heated with steam or nitrogen in a long tube. This increases the temperature rating from 75 to about 120 °C.

Multi-pin edge connectors are used extensively in electronic circuits. The thermoplastic housing must position the pins accurately for correct mating with the connector, and withstand momentary high temperature peaks when soldered connections (at 300 °C) are made to the pins. Glass-fibre filled polycarbonate or polybutylene terephthalate mouldings survive ageing tests at 125 °C.

When mains or higher voltages are involved, tracking can occur. Surface moisture absorption or ionic contamination causes the surface resistivity to be relatively low. The leakage current heats and dries the surface. If narrow dry bands form on the surface, due to the surface layer contracting, they have a higher resistance and sparks can occur across them. The sparks heat the polymer surface above 500 °C and carbonaceous degradation products form, eventually causing flash-over. Laboratory tracking tests involve dropping ammonium chloride solution onto the surface between electrodes, or exposing a sodium chloride contaminated surface to an artificial fog. In such tests, PVC, which forms conducting conjugated structures on heating, performs badly, polypropylene performs quite well, whereas PTFE with its excellent thermal stability, is outstanding.

12.2.2 High voltage insulation

The resistance of polymers to high voltages is tested using a range of geometries (Fig. 12.4). The electric strength, measured in $V\,m^{-1}$, depends both on the test geometry and sample thickness, due to the influence of the metal/polymer interface, and arcing in air gaps. The results of repeated determinations follow Weibull statistics (Fig. 12.4). This skewed distribution applies to *extreme value* properties, where the weakest part of the product causes failure. The cumulative probability of failure $F(x)$, when an electric field strength x is applied, is given by

$$F(x) = 1 - \exp\left(-\left(\frac{x}{x_0}\right)^b\right) \tag{12.4}$$

where x_0 is the value at which 63.2% of the failures have occurred, and b is a distribution breadth parameter. Figure 12.4d shows that, for DC fields on low-density polyethylene, the cylindrical electrode geometry gives the highest x_0 values and the lowest b values.

Figure 12.3b shows the construction of high voltage cables. The conductors are first extrusion coated with a layer of carbon-black filled polymer. The polymer is ethylene vinyl acetate, or ethylene butyl acrylate which has a greater thermal stability at the 260–300 °C crosslinking temperature. This semiconducting layer equalises the electric field across the insulator and prevents corona damage. The main insulating layer, consisting of crosslinked low-density polyethylene (XLPE), is coated with a further semiconducting layer. The surfaces of the semiconducting layers must be smooth, as any protrusions act to concentrate the electric field.

The electric strength is high at 20 °C, but falls off at higher temperatures. Short-term strengths cannot be used to design high voltage cables, buried in the ground for many years. The electric strength falls with the time of voltage application (Fig. 12.5). This has a parallel in the creep rupture

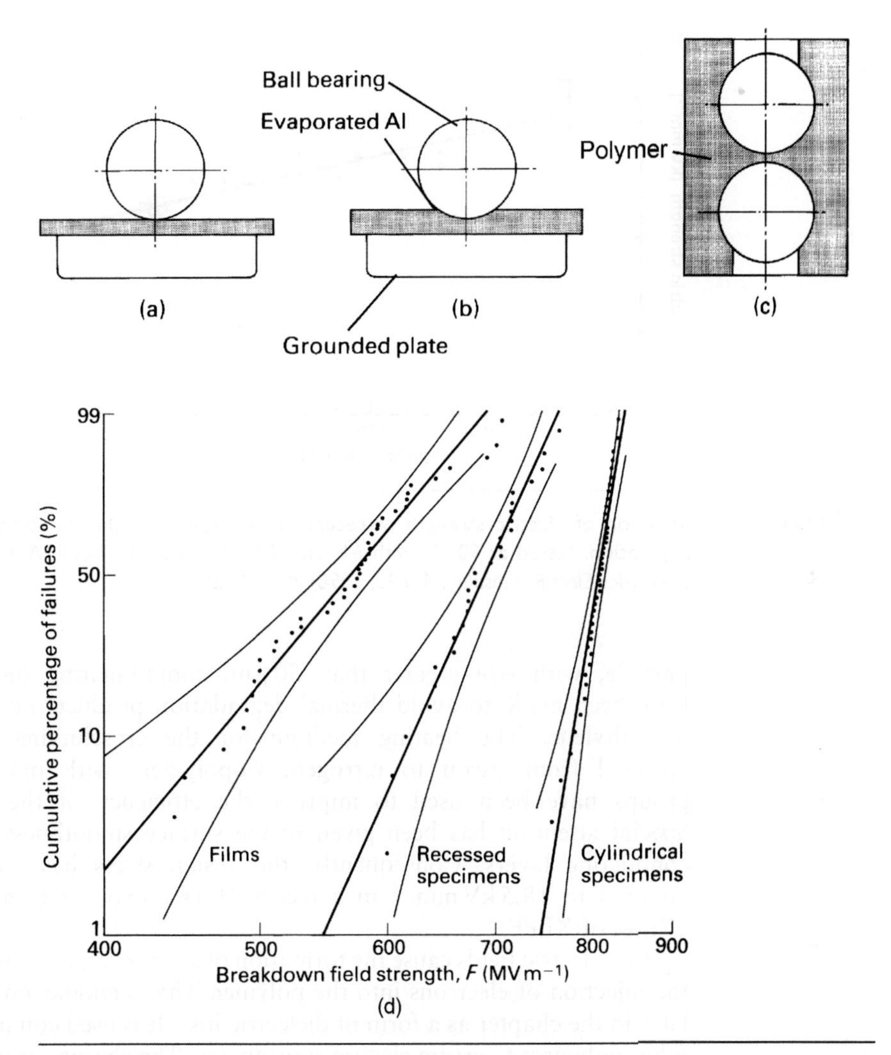

Figure 12.4 Geometries for the determination of electric strength: (a) Sphere on film; (b) sphere recessed into sheet; (c) cylinders embedded in plastic and (d) the corresponding Weibull plot of the results of many tests on LDPE at 20°C (from Seanor D. A., Ed., *Electrical Properties of Polymers*, Academic Press, 1982).

phenomenon (section 8.3.2). The overall level is affected by defects in the polymer, such as metallic wear particles from the extruder screw, fibrous contamination or voids, all of which act as electrical stress concentrators. The steam curing process can introduce dissolved water, which becomes supersaturated on cooling and nucleates as water-filled voids. Over the last 20 years, improvements to the quality of the XLPE have allowed the construction of cables for voltages up to 500 kV, for use on land. Clean-room type conditions have been used around the extruder to avoid any

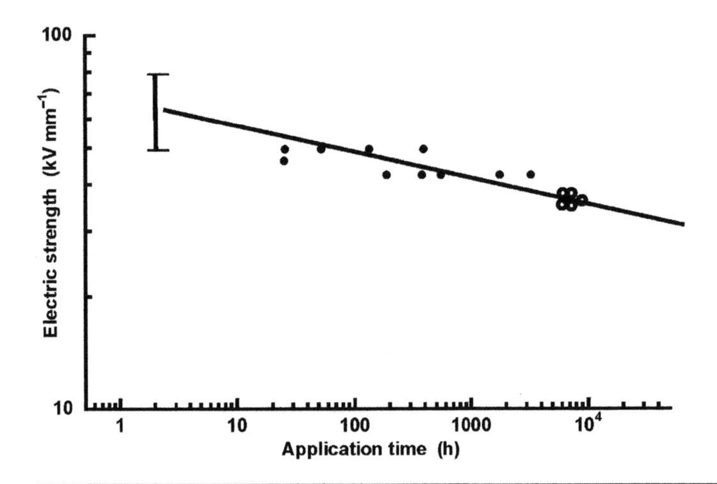

Figure 12.5 Variation of electric strength of recent XLPE cable with the duration of the voltage application, tested at 90 °C, with 6 h on, 18 h off. (From Ishibashi, A. *et al.*, *IEEE Trans. Dielectrics Electrical Insulat.*, **5**, 695, 1998) ©1998 IEEE

particle, with size greater than 50 μm, contaminating the PE. Efforts have been made to avoid thermal degradation products occurring in the polyethylene. The heating medium for the crosslinking process was changed from steam to nitrogen. Copolymers with unsaturated side groups have been used to improve the efficiency of the crosslinking. Special attention has been given to the surface smoothness of the semiconducting layers. Consequently, the design stress has increased from about 8 to 18.5 kV mm^{-1} in a recent 500 kV cable with a thickness of 27 mm of XLPE.

High electric fields cause the formation of *space charges* in polyethylene—the injection of electrons into the polymer. The phenomenon is seen again later in the chapter as a form of dielectric loss. It is used commercially with other polymers to create electret transducers. The charges probably form at the interfaces between the crystalline and amorphous phase. If the electric field is reversed, after a long period under a DC field, the space charges add to the new field and can lead to breakdown. Figure 12.6 shows how the charge increases with the electric field applied to 150 μm thick samples. The threshold field for XLPE is higher than that for LDPE or HDPE. The values appear to correlate with the design stresses for high voltage cable. The polymers differ in density; 922 kg m^{-3} for the LDPE, 945 kg m^{-3} for the HDPE and 922 kg m^{-3} for the XLPE.

Electrical breakdown is associated with the growth of *trees*, named after the structures that grow from charged metal needles in laboratory tests. Bow tie shaped trees grow in both directions from voids in the XLPE of high voltage DC cables (Fig. 12.7). The void acts as an electrical stress concentration, which initiates the electrical or electrochemical breakdown process.

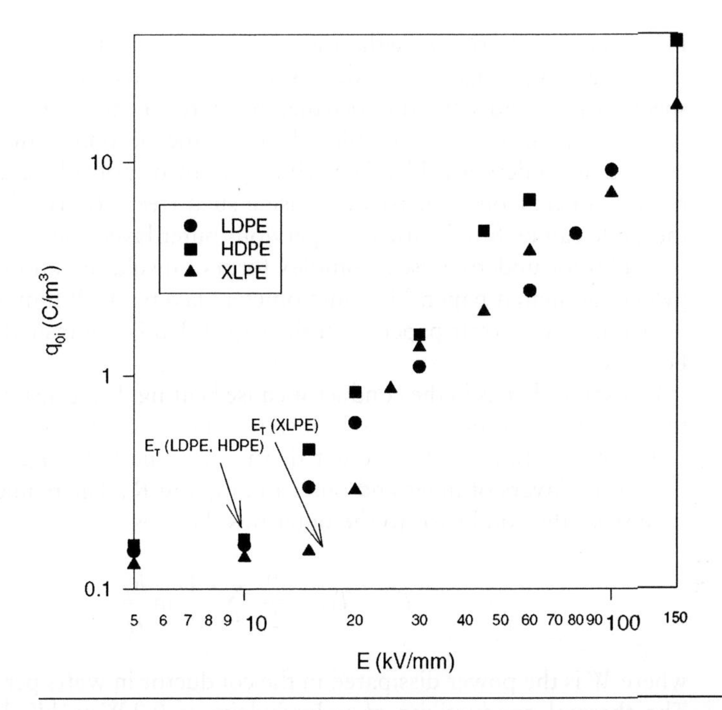

Figure 12.6 Threshold (arrowed) for the onset of space charge for different polyethylenes (Montanari, G. C. *et al.*, *J. Phys. D Appl. Phys.*, **34**, 2902, 2001) © IOP Publishing Ltd..

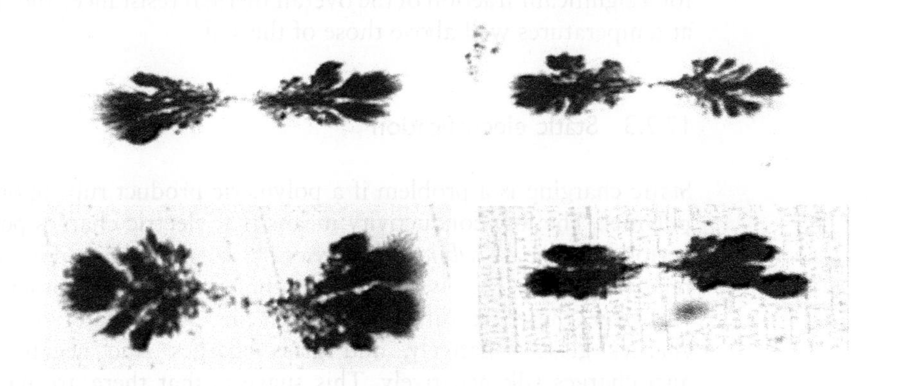

Figure 12.7 Four examples of bow tie water trees found in buried 35 kV electrical cable (from Ahmed, M. *et al.*, *Microscopy Anal.*, **Nov.**, 25, 2004) © John Wiley and Sons.

In the former, corona discharges occur in voids, causing hollow channels, lined with decomposed polymer, grow in the polymer. The field strength to cause corona discharge should be an inverse function of the void diameter, so breakdown should be avoidable if voids are smaller than 25 μm.

Electrochemical tree growth, the electrical equivalent of environmental stress cracking (Chapter 10), occurs at lower electric stresses. The chemical species vary. Lead salts from petroleum entered a cable buried near a petrol station, whereas hydrogen sulphide from the decomposition of seaweed entered an undersea cable. The whole gamut of polyethylene degradation reactions occur on a micro-scale inside such trees. To avoid such failures, the cable can be fitted with an impervious outer layer, such as a lead sheath.

Cables for undersea use a complex form of insulation known as polypropylene laminated paper. Fifty micrometers layers of PP film alternate with 36 μm layers of kraft paper, with the gaps filled by an insulating oil (alkylbenzene).

Resistance losses in the conductor cause heating. If the insulation becomes too hot, it promotes the growth of trees. The conductor is surrounded by a series of materials, with different thermal conductivities k_i, in the form of concentric layers of inner and outer radii r_i and R_i. The temperature difference from the conductor to the outermost layer is

$$T_C - T_O = \frac{W}{2\pi} \sum \frac{1}{k_i} \ln \frac{R_i}{r_i} \qquad (12.5)$$

where W is the power dissipated in the conductor in watts per meter length. The thermal conductivity of polyethylene at $0.2\,\mathrm{W\,m^{-1}\,K^{-1}}$ is low compared with sand bedding for the cable ($0.5\,\mathrm{W\,m^{-1}\,K^{-1}}$ when dry), and the polyethylene layer has $R_i \cong 2r_i$. Consequently, polyethylene is responsible for a significant fraction of the overall thermal resistance, and has to operate at temperatures well above those of the soil.

12.2.3 Static electrification

Static charging is a problem if a polymeric product rubs upon a dissimilar polymer. The low conductivity means that electric charges persist for long periods of time. A *triboelectric series* has been established by measuring the relative polarity of polymers when rubbed against each other (Table 12.1). However, there are difficulties in placing some materials in order: Silk charges glass negatively, and glass charges zinc negatively, but zinc also charges silk negatively. This suggests that there are mechanisms for charging other than electron transfer.

Large charge densities can build up by repeated contacts. The nylon end mouldings of conveyor belt rollers, driven by a polyurethane rubber belt, charged to such a voltage that spark discharges occurred to nearby workers.

Table 12.1 Polymer triboelectric series (positive first)

PU > POM > PC > PA > ABS > PS > PE > PP > PET > PVC > PVDF > PTFE

In an environment of less than 40% relative humidity, walking on a polypropylene carpet, or removing a sweater, can charge the body to 4 kV. High resistivity polyurethane soles on shoes prevent this charge leaking to the ground. As the capacitance of the body is around 200 pF, the energy available for a spark discharge is 0.8 mJ, four times greater than that necessary to ignite a petrol–air mixture. If an LDPE bag is picked up from a surface, when the relative humidity is 15%, the 20 kV generated could destroy a microchip. Consequently, antistatic additives are used in polyethylene bags for packaging microelectronics components. Electric charges on plastic products attract fine dust particles from the air, detracting from their appearance. Such dust patterns are often visible on the underside of polypropylene stacking chairs.

Most antistatic additives are polar waxes; the alkane chain part of the molecule is attracted to the polymer, while the hydrophilic end attracts water. This moisture forms a thin conductive film on the surface of the plastic. A charge decay half time of 0.1 s or less provides adequate protection against static electrification. To achieve this, the surface resistivity must be less than 3×10^{11} Ω/square. Although surface films are worn away by abrasion, they are replenished by the additive slowly diffusing to the polymer surface. They will not function adequately when the relative humidity is less than 15% (not a problem in the UK!), and cannot be used for specialised polymers with melt processing temperatures exceeding 300 °C. The use of conducting fillers (see the next section) is a more permanent solution to static electrification.

Electrostatic separation, of a granulated mixture of polymers from electronic products, can separate polymers of similar density. The particles are charged by passing through an inclined rotating drum, then separate during free fall between two charged plates.

12.2.4 Electromagnetic screening of plastic mouldings

Plastic mouldings have replaced sheet metal housings for most electronic equipments and computers. However, as non-conductors, they do not prevent electromagnetic interference (EMI). Radiation, such as from the thyristor switching of industrial heaters, can interfere with the operation of microprocessors, since the induced voltages can be large enough (>5 V) to be treated as signals by the computer. Conversely, radiation, escaping from the housing, may interfere with other equipments. Standards exist for screening; European Emissions Requirements for IT equipment calls for a 40 dB reduction of radiation in the 30–230 MHz frequency band, and a 47 dB reduction for 230–1000 MHz, at a distance of 10 m from the equipment. The dB is a logarithmic measure of attenuation; a 10 dB reduction represents a 10-fold reduction of field strength. US regulations for emissions from computers require the electric field strength, 3 m away from the computer, to be less than 100 $\mu V\,m^{-1}$, for the 30–88 MHz band.

The electric signal is mainly attenuated by reflection from the casing surfaces, and absorption inside the material; radiation scattering is insignificant. The absorption A depends on the conductor thickness t (m) and the signal frequency f (Hz) according to

$$A\,(\mathrm{dB}) = 131.4t\sqrt{f\sigma_r\mu_r} \quad (12.6)$$

where σ_r is the conductivity and μ_r is the permeability, relative to copper. Table 12.2 shows that the non-magnetic metals have similar behaviour. Magnetic metals have higher losses but corrosion problems prevent their use. The reflection loss R, given by

$$R\,(\mathrm{dB}) \propto 10\log_{10}(\sigma_r/\mu_r) \quad (12.7)$$

is much greater for non-magnetic than for magnetic metals.

Various strategies for achieving the requisite attenuation are shown in Table 12.3. The most common is to spray the inside of the moulding with a nickel containing paint. An alternative is to increase the polymer conductivity by incorporating conducting fillers, the most common being carbon black or conducting fibres. Figure 12.8 shows how the EMI shielding contributions vary with the resistivity of the filled polymer. For a 3 mm thick moulding to provide a shielding of 40 dB, the resistivity must be less than $10^{-2}\,\Omega$m; most of the screening is by reflection.

Table 12.2 EMI shielding properties of metals relative to copper

Metal	*Relative conductivity (σ_r)*	*Relative permeability (μ_r)*	$\sqrt{\sigma_r\mu_r}$	*Reflection loss (Eq. 12.7)*
Copper	1	1	1	0
Aluminium foil	0.53	1	0.73	−2.8
Zinc	0.31	1	0.57	−4.9
Pure iron	0.17	5000	29.2	−44

Table 12.3 EMI shielding methods

Method	*Thickness (μm)*	*Shielding efficiency (dB)*
Zn spraying	~70	60–90
Ni containing paint	~50	40–60
Al vacuum plating	2–5	40–70
Conducting filler	Dispersed in polymer	40–60

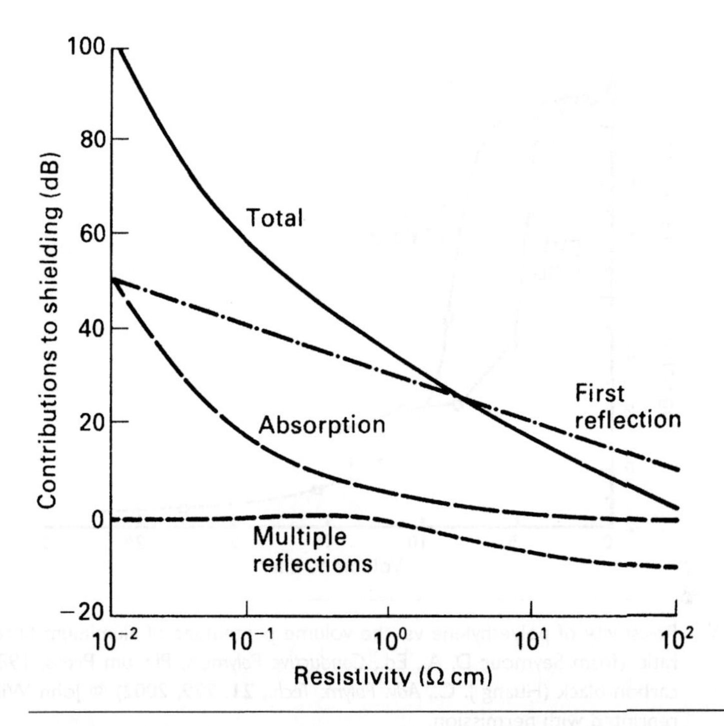

Figure 12.8 Contributions of reflection and absorption to the EM shielding efficiency of a 3 mm thick filled moulding, at a frequency of 100 MHz (from Möbius K. H., *Kunststoffe*, **78**, 31, 1978).

The resistivity changes with the filler content in a non-linear manner. When a conducting pathway is established across the polymer, the resistivity drops dramatically (Fig. 12.9). This limiting concentration is known as the percolation threshold. For carbon-black filled polymers, the percolation threshold is lower when the interaction between the polymer and the carbon black is low, i.e. in polypropylene (2%) rather than in polycarbonate (4%), and it is high in polyamides. Highly crystalline PP has a lower percolation threshold than an ethylene–octene copolymer with 10–15% crystallinity, because carbon black is rejected from the crystals. The filler content should be low as is consistent with the required resistivity, because most fillers bond weakly to polymeric matrices, and the toughness and tensile strength fall with increasing filler content. Fibres of a high aspect ratio are the most effective at creating conducting networks. Both brass and stainless-steel fibres have been used, but they are expensive, and cause wear in the moulds.

12.2.5 Semiconducting polymers for batteries and fuel cells

A solid polymer electrolyte (SPE), which contains mobile ionic salts, can be a semiconductor. The original systems used a high molecular weight

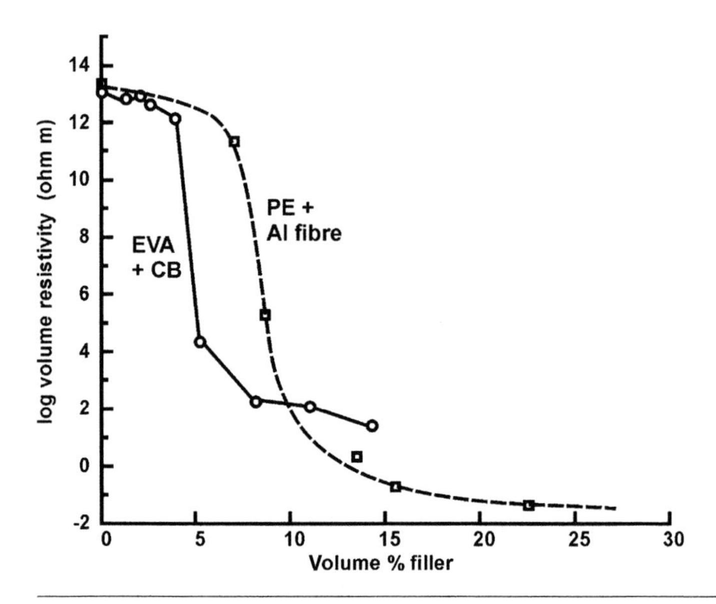

Figure 12.9 Resistivity of polyethylene vs. the volume percentage of aluminium fibres of 12:1 aspect ratio (from Seymour D. A., Ed., *Conductive Polymers*, Plenum Press, 1981) and EVA plus carbon black (Huang J. C., *Adv. Polym. Tech.*, **21**, 299, 2002) © John Wiley and Sons Inc. reprinted with permission.

polyethylene oxide plus the alkali metal salt $LiClO_4$. This has good mechanical properties but a high resistivity of $10^6\,\Omega$m at 20 °C. When heated to 110 °C, the resistivity reduced to $10^2\,\Omega$m. The conductivity of polymer electrolytes increases with temperature (Fig. 12.10), showing the advantage of greater-than-ambient temperature operation. The systems shown in the figure are crosslinked, and are based on polyethylene oxide of degree of polymerisation between 22 and 45. Crosslinking reduces the segmental mobility of the PEO, so it reduces the ionic conductivity to some extent. In an effort to increase the PEO mobility, liquid plasticisers such as propylene carbonate or ethylene carbonate have been used. Such gel SPEs combine the mechanical properties of swollen polymer networks with the high ion conductivities of liquid electrolytes.

Polymer electrolyte batteries have been used in implanted cardiac pacemakers since 1972. The system used is lithium/iodine–polyvinylpyridine. Although the conductivity of the Li ions in Li_2I is poor, the current requirements are very small, and the major consideration is the storage of a high energy density of nearly 1 W h cm^{-3}.

Electrodes and cell components must be thin to minimise the internal resistance of the batteries; the total cell can be less than 0.2 mm thick. Figure 12.11 shows the construction of a multi-layer film, rechargeable *lithium polymer* battery, using a solid polymer electrolyte. A thin lithium metal foil acts as an anode. The electrolyte is polyethylene oxide containing a lithium salt, and the cathode is a composite of the electrolyte and a

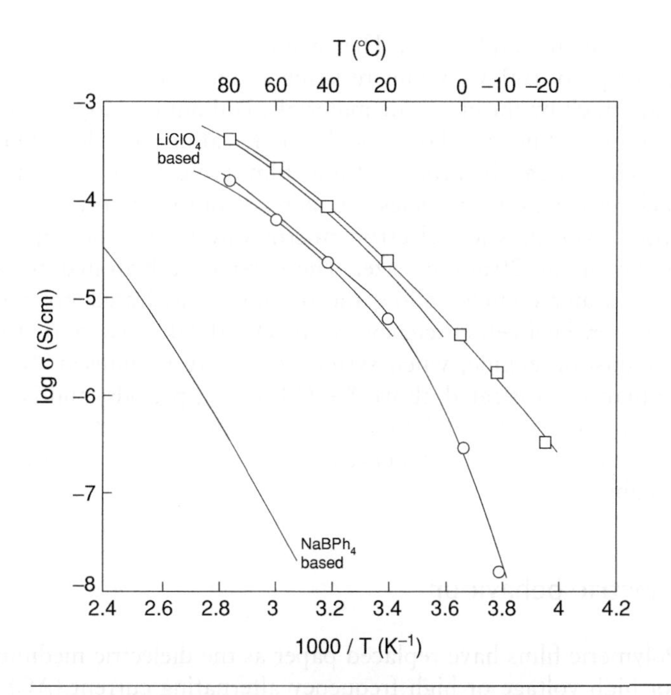

Figure 12.10 Temperature dependence of conductivity of typical solid polymer electrolytes (from Meyer, W. H., *Adv. Mater.*, **10**, 449, 1998) © Wiley-VCH.

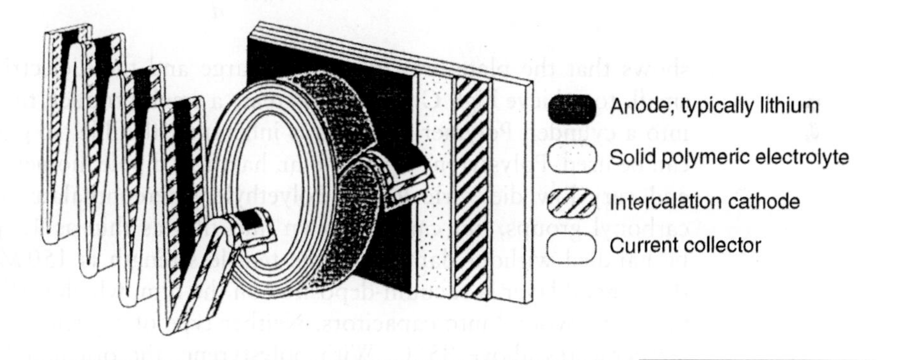

Figure 12.11 Construction of a polymer electrolyte battery (from Song, J. Y., *J. Power Sources*, **77**, 183, 1999) © Elsevier.

transition metal oxide or chalcogenide. The flexible, multi-layer films can be rolled up into a cylinder, or thin tablet-shaped batteries can fit into digital cameras and cell phones.

Fuel cells have been used since the 1970s for satellites, but lower cost fuel cells are being developed for use in road transport. They require a membrane that is an ionic conductor, but a barrier to the fuel (hydrogen), contaminants and water. The membrane must remain rigid at temperatures up to 100 °C, while resisting oxidation, reduction and hydrolysis, and attack by the electrode materials. DuPont developed *Nafion*, a copolymer of perfluorosuphonic acid and tetrafluoroethylene (TFE), a type of ionomer. It can be processed into film, typically 0.127 mm thick. In some fuel cells, Nafion particles are sintered onto a porous carbon-paper substrate. Transmission electron microscopy shows the suphonated regions to be about 20 nm in size. These regions, hydrated to approximately 50% water content, allow the motion of hydrated protons through the polymer. Fuel cell efficiency can reach 50–60%, twice that of the internal combustion engine, when system losses are minimised. Nafion loses performance if operated above 80 °C for long periods, and is expensive. It is also permeable to methanol, a component of some fuel cells. Hence, alternative polymers with higher temperature resistance and lower cost are being sought.

12.3 Dielectric behaviour

Polymeric films have replaced paper as the dielectric medium in capacitors for high voltage or high frequency alternating current (AC) use. The relationship for the capacitance of a parallel plate capacitor

$$C = 4\pi A \frac{\varepsilon\varepsilon_0}{d} \tag{12.8}$$

shows that the plate area A must be large and the dielectric thickness d small, to achieve high C values. To create a small product, the film is rolled into a cylinder. Polymer thin film is inherently flexible, so glassy polymers can be used. Polystyrene is non-polar, having a dipole moment of 0.3 Debye and very low dielectric losses. Polyethylene terephthalate contains polar carbonyl groups, but can be drawn into film as thin as 1.5 μm, that can be handled without damage as the tensile strength is 150 MPa. A 15 nm thick metal layer is vacuum-deposited on the film, which is slit into narrow tapes and wound into capacitors. Neither type of capacitor can be used at temperatures above 85 °C. With polystyrene, the oriented film begins to recover as the T_g of 100 °C is approached, and PET begins to crystallise above 100 °C (Fig. 12.12). Polyimide (Kapton) films can be used up to 250 °C, but they have much higher cost than PS or PET.

12.3.1 Dielectric constant and losses

When an AC voltage is applied across a polymer, its dielectric properties determine the current. The dielectric constant (or relative permittivity) ε^* is defined in terms of the electric field **E** and electric displacement **D** vectors by

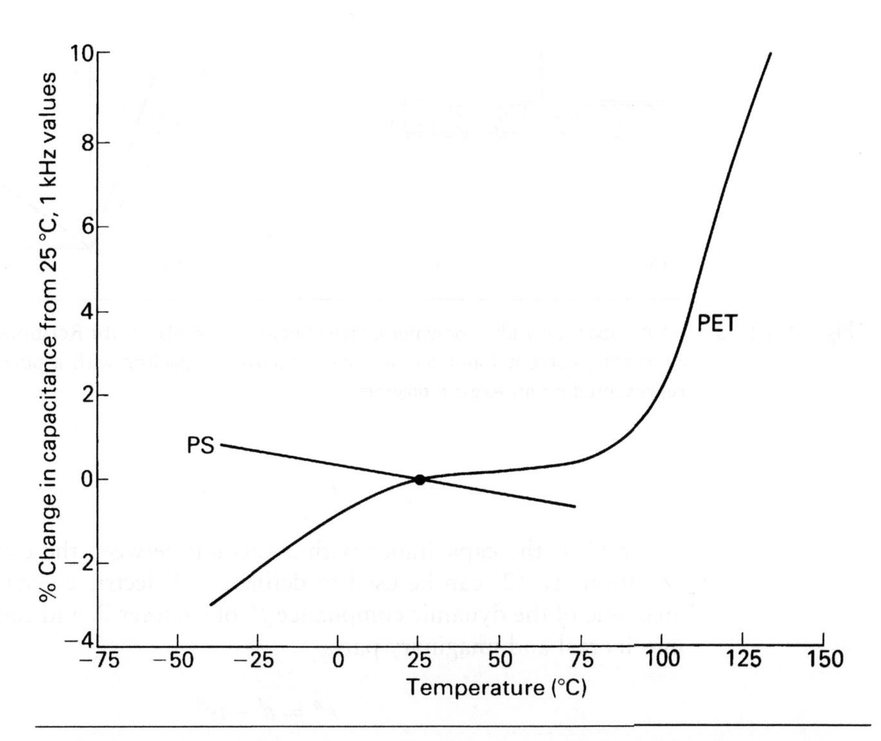

Figure 12.12 Percentage change in capacitance with temperature for capacitors with polystyrene and polyethylene terephthalate dielectric (from Bruins, P., Ed., *Plastics for Electrical Insulation*, Wiley Interscience, 1968).

$$\mathbf{D} = \varepsilon_0 \varepsilon^* \mathbf{E} \tag{12.9}$$

where the permittivity of free space $\varepsilon_0 = 8.85 \times 10^{-12}\,\mathrm{F\,m^{-1}}$. A more practical definition considers a capacitor with a polymeric layer between the plates (Fig. 12.13a). The applied voltage V varies sinusoidally with time

$$V = V_0 \exp{(i\omega t)} \tag{12.10}$$

V can be represented on an Argand diagram (Fig. 12.13b) by a vector of length V_0 rotating at an angular frequency ω. For an ideal capacitor, of capacitance C, the current is given by

$$I = C\frac{\mathrm{d}V}{\mathrm{d}t} = i\omega CV \tag{12.11}$$

In the Argand diagram, the current vector rotates 90° ahead of the voltage vector. For real dielectrics, the current leads the voltage by an angle $90° - \delta$, and is given by

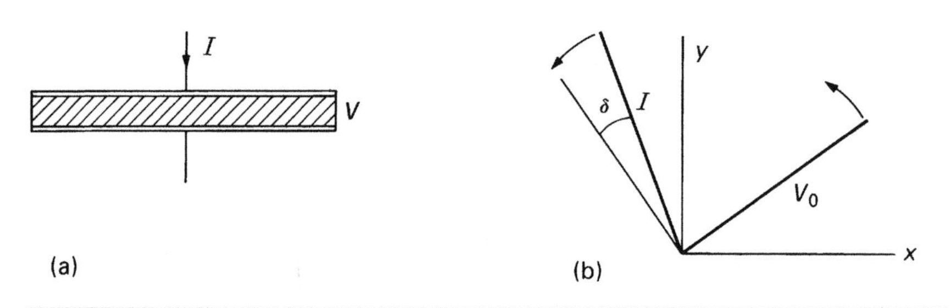

Figure 12.13 (a) A capacitor with a polymeric layer between the plates. (b) Relationship between an alternating current i and the voltage V, across a capacitor with a polymeric dielectric, represented on an Argand diagram.

$$I = i\omega\varepsilon^* C_0 V \tag{12.12}$$

where C_0 is the capacitance with a vacuum between the capacitor plates. Equation (12.12) can be used to define the dielectric constant ε^*. It is an analogue of the dynamic compliance J^* of Chapter 7, and can be expanded into its real and imaginary parts

$$\varepsilon^* = \varepsilon' - i\varepsilon'' \tag{12.13}$$

Dielectric data is usually presented as the real part of the dielectric constant ε' as a function of frequency, and the ratio

$$\tan\delta = \varepsilon''/\varepsilon' \tag{12.14}$$

Equation (7.19) shows that the energy dissipated per cycle proportional to $\tan\delta$.

12.3.2 Polarisation loss processes

If chemical groups with permanent dipoles are free to move in the polymer, these align with the electric field every half cycle, so long as the frequency is not too high. Such *orientation polarisation* is expected in PVC where the C–Cl bond has a permanent dipole moment of 1.0 Debye (1 Debye $= 3.33 \times 10^{-30}$ C m). *Space charge polarisation* occurs both in rubber-toughened polymers, and in polymers containing voids or inclusions. Charges accumulate at the interface between the two phases, if the product $\varepsilon'\rho$ of the dielectric constant and the resistivity differs for the phases.

In Debye's model of dielectric relaxation, the polarisation process has a single relaxation time. The model has both electrical circuit and viscoelastic model analogues (Fig. 12.14). The electrical circuit is the 'dual' of the mechanical model, because the voltages across the capacitor and resistor *in series* are added, whereas the forces on the *parallel* spring and dashpot are added. The differential equation corresponding to the mechanical model is

$$e + \tau\frac{de}{dt} = \left(\frac{1}{E_1} + \frac{1}{E_2}\right)\sigma + \frac{\tau}{E_2}\frac{d\sigma}{dt} \tag{12.15}$$

where e is the strain, σ the stress and the retardation time $\tau = \eta/E_1$. The compliances of the Voigt element and the spring E_2 in series are added to give

$$J^* = \frac{1}{E_2} + \frac{1}{E_1(1 + i\omega\tau)} \tag{12.16}$$

The dielectric equivalent of Eq. (12.16) relates ε^* to the relaxed low frequency dielectric constant ε_R and the unrelaxed high frequency value ε_U

$$\varepsilon^* = \varepsilon_U + \frac{\varepsilon_R - \varepsilon_U}{1 + i\omega\tau} \tag{12.17}$$

Figure 12.14b shows the variation of ε' and $\tan\delta$ with frequency.

Polymers have broader $\tan\delta$ peaks, with lower maxima, than the Debye model, indicating that they have a spectrum of retardation times. Space charge relaxations occur at lower frequencies than polar group relaxations. If either damping peak coincides with the frequency of the electrical signal, the signal will be strongly attenuated. If the signal is powerful, there will be significant heating effects. The electrical circuit in Fig. 12.14a, with a high resistance in parallel to model the true DC resistivity, can model the signal attenuation. The model also explains why the DC resistivity of polymers changes with time. When a constant voltage is applied, the polarisation current decays when the retardation time is exceeded.

12.3.3 High frequency insulation and capacitors

The dielectric for telecommunication cables must be non-polar to avoid orientation polarisation losses. Semi-crystalline polymers are used in preference to glassy polymers. They have lower Young's moduli and higher yield strains, so cables can be easily bent without damaging the dielectric. Polyethylene is the most commonly used material for cost reasons. Figure 12.15 shows the variation of $\tan\delta$ with temperature at a 10 kHz frequency. The loss peaks are labelled as the $\alpha, \beta, \gamma, \delta, \ldots$ peaks, starting at the highest temperature peak. The α loss peak at about 90 °C in HDPE is due to polar carbonyl groups (−C=O) with a dipole moment of 2.3 Debye units, introduced by degradation in melt processing (Chapter 10). The carbonyl groups can re-orient in the crystalline phase, some 40 °C below the melting point of the crystals. LDPE has a β peak at about 0 °C, due to

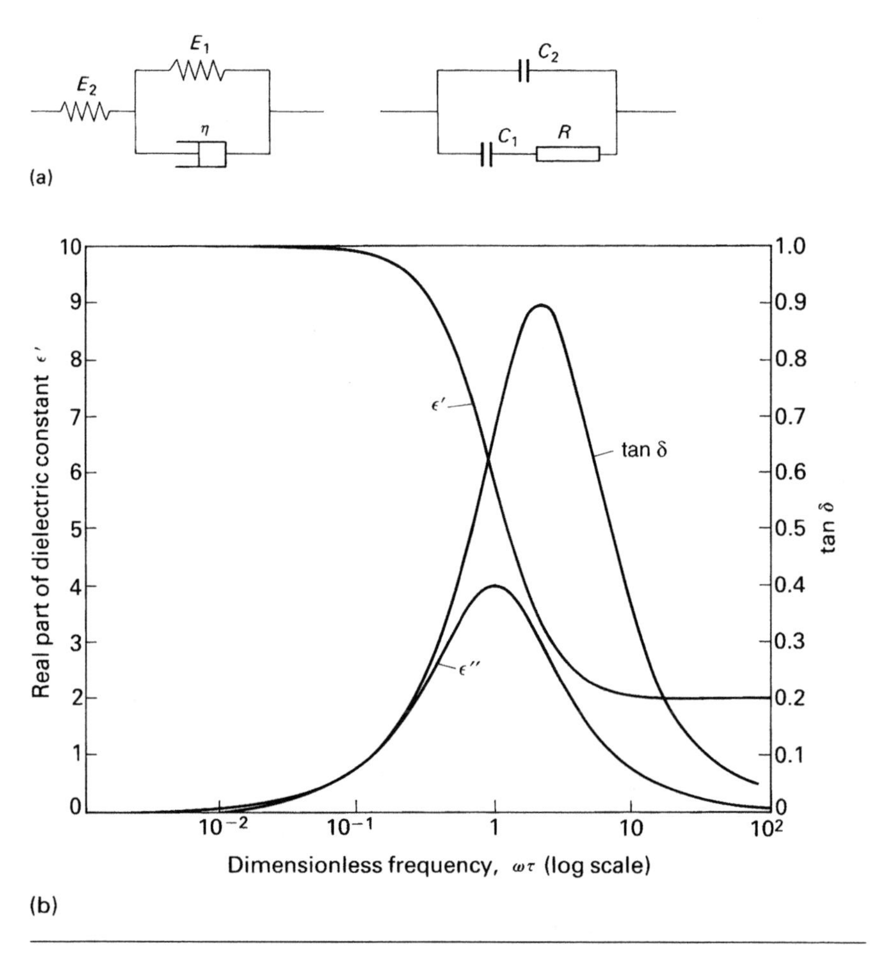

Figure 12.14 (a) Spring and dashpot mechanical model, and the equivalent electrical circuit that models a dielectric with a single relaxation time. (b) Predicted variation of the components of the dielectric constant with dimensionless frequency $\omega\tau$ for $\varepsilon_U = 10$ and $\varepsilon_R = 2$.

the re-orientation of carbonyl groups in the amorphous phase. Non-polar antioxidants can be used to minimise the oxidative degradation of polyethylene. This, and a reduction in the level of catalyst residues, has reduced the $\tan\delta$ values below those in Fig. 12.15. As the electrical frequency is increased, the loss peaks move to higher temperatures, so at 30 MHz used in submarine cables, the γ loss peak has moved to −28 °C, dominating the dielectric loss.

Polar polymers can be used for insulation at the 50 Hz mains frequency. Figure 12.16 shows how the dielectric properties of PVC change with temperature, for different amounts of diphenyl plasticiser. The imaginary part of the dielectric constant has a maximum value at the glass transition of the PVC. If the amount of plasticiser is excessive, there will be unacceptable

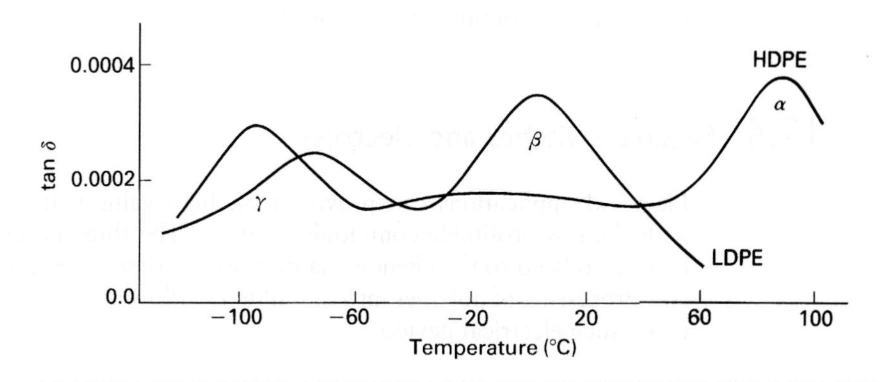

Figure 12.15 Variation with temperature of the tan δ at 10 kHz of oxidised polyethylene (from Baird, M. E., *Electrical Properties of Polymeric Materials*, Plastics Institute, 1973).

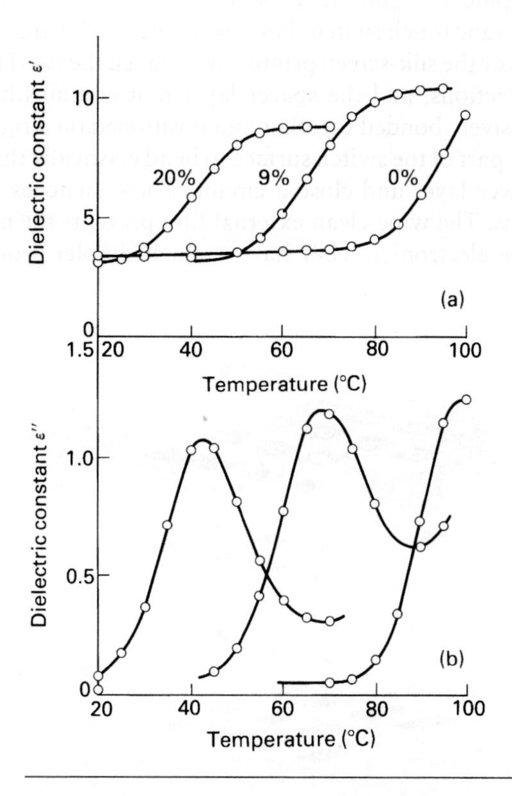

Figure 12.16 Components of the dielectric constant of PVC at 60 Hz vs. temperature, for: (a) Rigid PVC and (b) PVC plasticised with 9 and 20% diphenyl.

losses at room temperature. On the other hand, unplasticised PVC is too rigid to allow continual flexing of the cable.

12.4 Flexible switches and electrets

Electrical applications can provide new, high value markets, in contrast with the less profitable commodity plastics. The three products described here are related to the phenomena discussed earlier in the chapter. The low cost processing of polymer into thin film has allowed the radical redesign of many such electrical devices.

12.4.1 Film switches

Flexible polymer films can replace mechanical toggle switches in low voltage electronic equipment. Figure 12.17 shows one design of a touch-sensitive panel, or membrane touch switch. Polycarbonate or PET films, 125–250 μm thick, are used for the silk-screen printed outer layer, the two layers that carry the silver connections, and the spacer layer that contains holes. The four layers are adhesively bonded together and positioned on a rigid base. Finger pressure causes part of the switch surface to bend elastically through a hole to contact the lower layer and close a circuit. These switches survive 10^7 or more operations. The wipe-clean external film prevents the ingress of fluids or dirt onto the electronics. They have a much simpler geometry than the

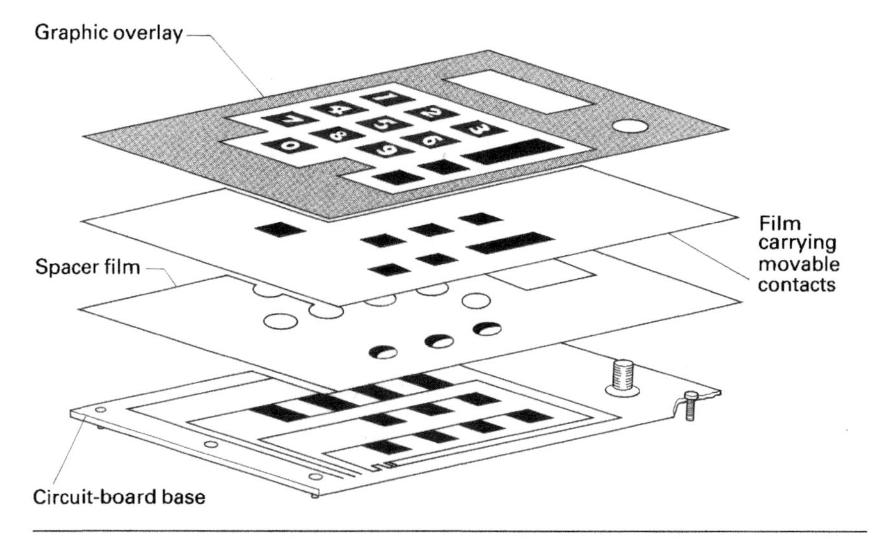

Figure 12.17 Four layer polycarbonate film switch for a keyboard (from *Modern Plastics International*, McGraw-Hill, 1983).

buckling rubber switches for telephone keypads (Chapter 1). However, they do not have the positive feel feedback of the rubber switches; the only feedback is a change in an optical display or a sound signal.

12.4.2 Electrets

Several electrical devices are based on *electrets*, the charge equivalents of magnets. The charges can be permanently separated by distances of the order of 10 μm, or there can be permanently oriented dipoles in polar polymers. Fluorinated ethylene propylene copolymer (FEP) is a semi-crystalline polymer with many of the properties of PTFE, yet it can be melt processed. Twenty-five micrometers thick FEP film is coated with 100 nm of aluminium on one surface, then other surface irradiated with a 20 kV electron beam. Twenty per cent of the electrons striking the surface cause secondary electron emission, leaving the surface positively charged. The primary electrons penetrate about 5 μm before they are sufficiently slowed down to be trapped (in scanning electron microscopy, secondary electrons are used to form an image of the surface, and surface charging is avoided by coating the specimen surface with a thin layer of gold). The charge stability can be studied by measuring the current as the electret temperature is gradually raised. Measurable currents are observed for FEP above 125 °C; at 20 °C, the charges have a half-life of 20 years.

If sound vibrations move an FEP diaphragm relative to another electrode, it acts as a microphone. In a telephone handset (Fig. 12.18), the metallised surface of an electret is exposed to the sound, while the other electrode is a metallised ABS moulding containing holes. Such microphones are insensitive to mechanical shocks and electromagnetic radiation, and are cheaper than condenser microphones.

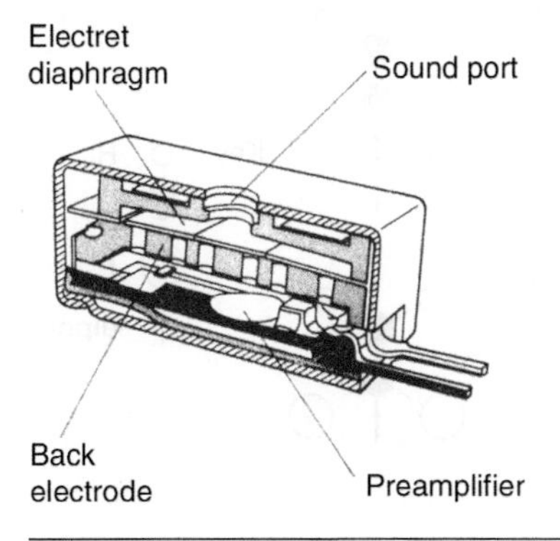

Figure 12.18 Cross section of an FEP electret microphone for telephone applications (from *Bell System Technical Journal*, Copyright 1979, American Telephone and Telegraph Company).

Polypropylene foam is another type of electret. The foamed film is biaxially oriented, so it contains disc-shaped voids of approximately 50 mm diameter. A corona discharge at 20 kV causes a breakdown inside the voids, leading to permanent charging. A typical piezoelectric constant is 220 pC N^{-1}. The foam electrets are being considered for biomedical applications; for these, having a flexible, large area detector is often more important than having absolute accuracy in the pressure measurement.

12.4.3 Piezoelectric film

Polyvinylidene fluoride (PVDF) is a polar polymer, of 50% crystallinity, in which the CF_2 group has a dipole moment of 2.1 Debye. There are at least two crystalline forms. Type II crystals, that form on the spherulitic crystallisation of unoriented PVDF, have no net dipole moment, because neighbouring polymer chains have opposite orientations of the polar CF_2 groups. If this material is stretched at 120 °C to a draw ratio of 4 or 5, type I crystals are formed in which the polymer chains have an all-*trans* conformation, and all dipoles are oriented parallel to the **b** axis (Fig. 12.19). In the film, there is almost complete alignment of crystal **c** axes along the draw direction (1 axis), but as many **b** axes are in the positive 2 direction (the transverse direction) as in the negative 2 direction. An electret is produced by *poling* the film; heating it to 100 °C, applying an electric field of 60 MV m^{-1} in the 3 direction (film normal) for 30 min, and then cooling the film in the field. This preferentially aligns the **b** axes towards

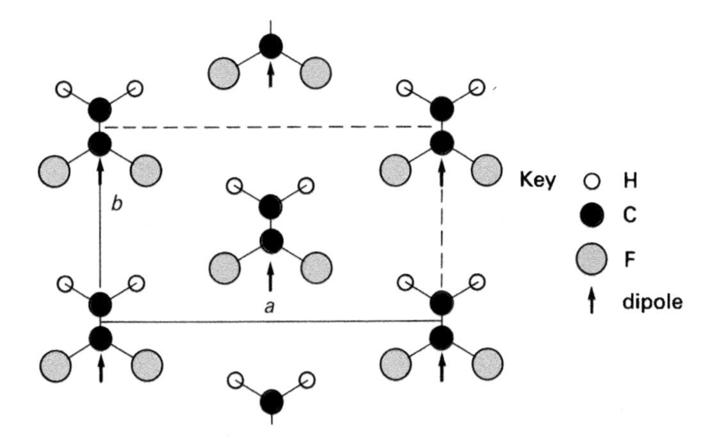

Figure 12.19 Type I crystal unit cell of PVDF, seen in the **a b** projection. The arrows represent dipoles.

the 3 axis, the average value of the cosine of the angle between **b** and the 3 axis being 0.84. The resulting film is *piezoelectric*; a charge density $Q_3/A\ \mathrm{C\,m^{-2}}$ appears on the upper and lower surfaces in the 3 direction as a result of stresses applied to the film. The crystals act as rigid dipoles embedded in a deformable matrix. When the film thickness contracts, charges appear on the surfaces. The largest piezoelectric stress coefficients are

$$d_{31} \equiv \frac{Q_3}{A\sigma_{11}} \quad \text{and} \quad d_{33} \equiv \frac{Q_3}{A\sigma_{33}} \tag{12.18}$$

for tensile stresses σ_{11} in the orientation direction, and for compressive stresses σ_{33} normal to the film. These coefficients depend on the degree of orientation of the crystal **b** axes, and temperature (Fig. 12.20). The Poisson's ratio ν_{31}, the contraction in the 3 direction divided by the tensile strain in the 1 direction, is high, causing d_{31} to be high.

Quartz and piezoelectric ceramic crystals have more temperature independent constants than PVDF, so they are used for force and acceleration transducers. However, PVDF films can be used for large area flexible transducers. Their sensitivity to stress or strain allows the construction of pressure sensors (using the d_{33} coefficient), and accelerometers by mounting a seismic mass on the film. PVDF electrets are particularly suited for large area hydrophones (Fig. 12.21) that detect underwater signals. Their

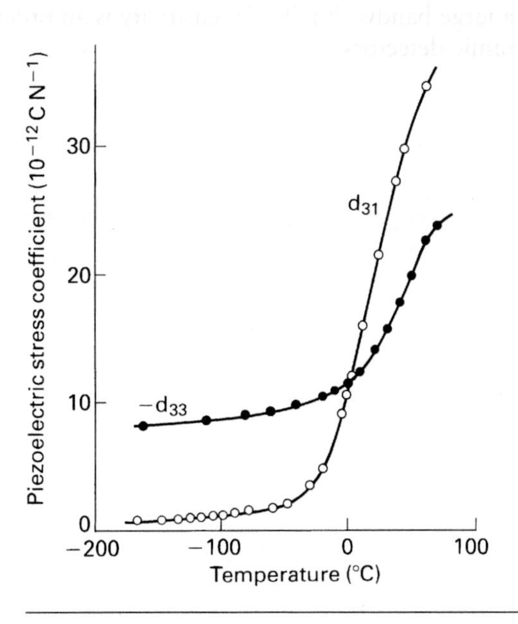

Figure 12.20 Typical piezoelectric constants of a PVDF electret vs. temperature (from Mort, J., Ed., *Electronic Properties of Polymers*, Wiley, 1982).

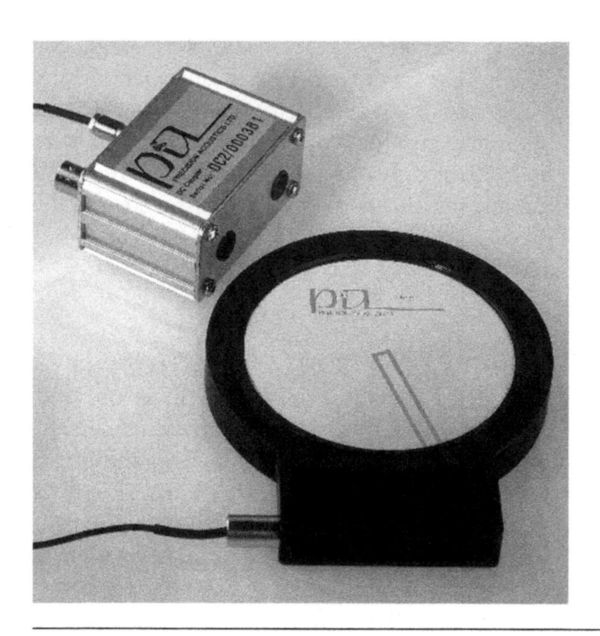

Figure 12.21 PVDF membrane hydrophone (© Precision Acoustics Ltd., Dorchester.)

compliance matches that of water, whereas ferroelectric ceramics need an intermediate matching layer. PVDF film hydrophones operate at high frequencies with a large bandwidth. Their sensitivity is an order of magnitude higher than ceramic detectors.

Chapter 13

Design: Material and shape selection

Chapter contents

13.1 Introduction

This chapter, covering the materials and shape selection aspects of design, builds on the simple exercises in Chapter 1. Ulrich and Eppinger (Further Reading) take a more general approach, showing how materials engineers interact with other disciplines in the design process. Polymer selection packages will be explored, before considering component shapes that maximise bending and torsion stiffness. These component shapes are influenced by the process used for manufacture, and the need to optimise its output. The reader is encouraged to observe the details of a variety of successful products, and to recognise why certain designs can be manufactured efficiently.

13.2 Polymer selection

13.2.1 Polymer selection packages

Materials selection packages have partly replaced tables of data for particular grades of plastics. They provide basic information. More help is available from the manufacturer's website or technical service department. CAMPUS, freely available from www.campusplastics.com, allows comparison of grades from a single manufacturer. An augmented version, available from www.mbase.de for an annual fee, allows data comparison between different companies. The properties are classified into types: Rheological, mechanical, thermal, electrical, other, processing and additives. Each menu contains many items; for instance, *mechanical* contains tensile modulus, charpy impact strength, . . . with a single value for each property. There are also multi-point data, in graphical form, such as

(a) shear modulus versus temperature;
(b) tensile stress–strain curves at a range of temperatures;
(c) shear viscosity versus strain rate at a range of temperatures;
(d) creep strain versus time at a range of creep stresses.

Not all manufacturers provide the same range of data, so the information for some grades of plastics may be incomplete. The selection process *ranks* polymer grades on the basis of one or more properties. Minimum and maximum values should be specified for the properties felt to be appropriate to the application. The programme then finds how many grades meet these requirements. It is unhelpful if the answer is hundreds of grades, or none. If the latter is the answer, one or more selection conditions must be relaxed. The number of grades manufactured is limited to reduce inventories. Thus, the level of glass fibre reinforcement or rubber toughening is limited to two or three levels. Consequently, there is unlikely to be a perfect match to a set of target specifications.

Single-value mechanical parameters lead to an unambiguous ranking of the polymer grades. However, this can mislead the unwary user into ignoring the time and temperature dependence of properties, issues emphasised in earlier chapters. A single modulus value relates to a specific time scale, strain rate and test temperature, and to the processing of the test specimen. Injection-moulded tensile bars are usually end-gated, therefore, polymer orientation along the length of the bar gives optimum values of the strength and modulus. Chapter 9 should have alerted the reader to the need for impact tests on a thickness, which is the same as that of the intended product. However, the data is for standard 3.2 mm (1/8 in.) thick bars. The lack of fracture mechanics data in the databases is another drawback. Such data does not appear to be part of the routine industrial characterisation of plastics.

It would be useful to sort grades according to price, so the cheapest grade that meets the requirements heads the list. However, the prices of plastics fluctuate with the price of oil (Chapter 2) and CAMPUS does not give prices. It indicates whether a grade is suitable for a particular process, but the database neither indicates the cheapest process route, nor gives the cost of the manufactured product.

Polar charts allow the comparison of polymers for multiple selection criteria. The properties for a particular grade are plotted on n scales that radiate from a central origin, and the points joined to form a closed polygon. Figure 13.1 compares five properties of a grade of polyamide 66, polyamide 612 and polyoxymethylene. There are different rankings for water uptake compared with notched impact strength. If it is possible to specify the minimum property values required for the product, a target polygon can also be drawn; only grades whose polygons enclose this are acceptable.

Plastic products should be easy to identify for recycling—hence, the marking of car components with polymer abbreviations such as 'PP' (Chapter 1). Products may be designed for ease of dismantling and separation into their component polymers at the end of their lives. This tends to reduce the number of plastics used, while discouraging the use of composite structures. A PBTP bumper skin, filled with polyurethane foam, attached to a steel sub-frame is more difficult to recycle than a bumper with a PP skin, PP foam core and a glass-reinforced PP mounting beam at the back. Recycling constraints are not at present part of selection packages.

13.2.2 Property combinations for materials selection

Some products or parts of products, such as rectangular cross section beams, are frequently loaded in bending or torsion. If the product must meet a mechanical property target, and its mass must be minimised, property combinations can be used to explain why some materials are feasible and others impractical. It is unlikely that the plastic was initially chosen on

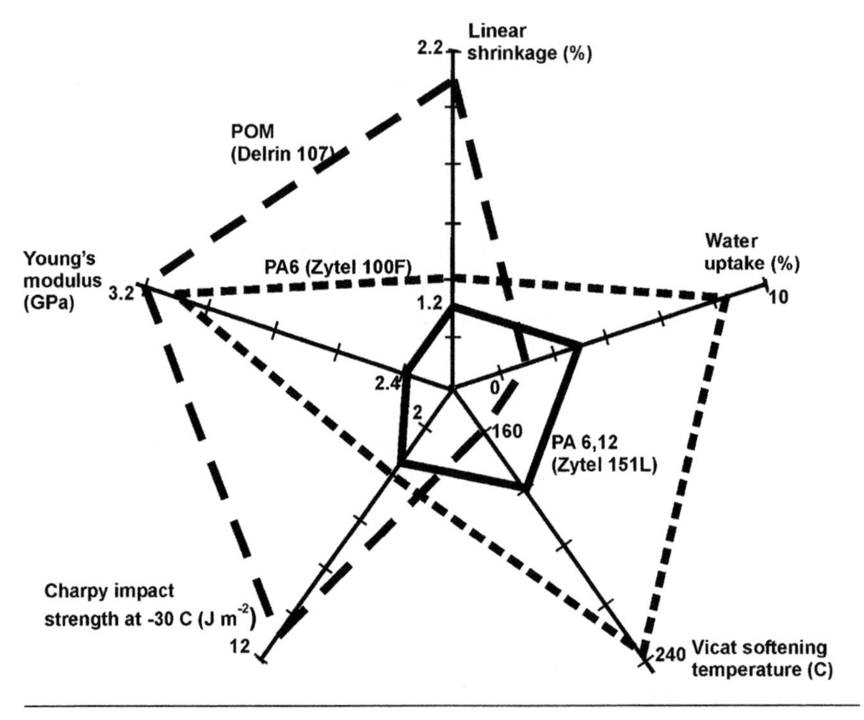

Figure 13.1 Comparison of five properties of un-reinforced grades of polyamide 66, polyamide 612 and polyoxymethylene, on a polar plot (redrawn from Du Pont data from CAMPUS).

the basis of property combinations. It is more likely that a plastic from a similar product was used, with low cost being the most important selection parameter.

Two examples are used to explain where the use of plastics is appropriate in car bodies, and where metals are more appropriate. Comparisons between plastics and wood are difficult, since different processing methods apply, and certain designs are optimum for each material. A design optimised for injection moulding in plastic (Sections 13.5 and 13.6), is neither suitable for anisostropic materials like wood, nor easy to construct.

13.2.3 The nearly flat skin of a car door

A car door skin is loaded in bending when pushed shut with the hand. If it deflects too much, it appears flimsy. To simplify the analysis, the panel deflection is calculated as for a constant cross section beam, with the strain varying linearly through the thickness (Appendix C). In reality, the curvature and styling ridges on metal door panels play an important stiffening role. For a flat panel of width w (treated as a constant) and thickness t, the bending stiffness is given by Eqs (C.7) and (C.9) as

$$MR = EI = \frac{Ewt^3}{12} \tag{13.1}$$

The skin panel must have a certain bending stiffness; the external loads fix M, while the deflection limit sets a minimum limit for R. The minimum MR value means that beam bending stiffness EI must exceed a minimum value. If the material, hence E, is varied while EI is kept constant, the necessary skin thickness t depends on $E^{-1/3}$. The panel must cover the door, so has a constant width. Hence, its mass m is directly proportional to the material density ρ and to t, so

$$m \propto \rho t \propto \frac{\rho}{E^{1/3}} \tag{13.2}$$

To minimise the panel mass, the combination of properties $E^{1/3}/\rho$ must be maximised. Hence, $E^{1/3}/\rho$ is the *selection parameter* for the panel bending stiffness requirement. The data in Table 13.1, for two types of fibre-reinforced plastics, polycarbonate and two metals, shows that sheet moulding compound (SMC) and aluminium are the best materials. However, the range of $E^{1/3}/\rho$ values is not large, so the material selection may be based on other factors, such as cost of processing into a slightly curved shape or corrosion resistance. A skin panel's resistance to denting (when another door edge hit it in a car park) is proportional to the material's elongation at yield. RRIM has a marked advantage in this respect.

13.2.4 The tubular frame of a car body

The bending and torsional stiffness of a car body is mainly provided by the hollow tubes that surround the passenger compartment. These have complex cross sections, but, to simplify the analysis, can be considered as

Table 13.1 Selection parameters for car body materials

Material *Density ρ(kg m^{-3})* *Young's modulus E (GPa)* *Property*	*Select on*	*Steel* *7800* *207*	*Aluminium* *2800* *69*	*PC* *1200* *2.2*	*RRIM* *1200* *1*	*SMC* *1800* *12*
Flat panel bending stiffness	$\sqrt[3]{E}/\rho$	0.76	**1.46**	1.08	0.83	1.27
Box. beam bending stiffness	E/ρ	**2.65**	2.46	0.18	0.08	0.67
Resistance to denting	e_{yield} (%)	0.15	0.2	6	**10**	1
Crash energy absorption	σ^*/ρ	**225***	223	95	49	101

*Value for high strength steel (Best results given in bold).

square-sectioned, thin-walled tubes of width w and wall thickness t. Their bending stiffness is given by the product of Young's modulus with

$$I = \frac{1}{12}\left[(w+2t)^4 - w^4\right] \cong \frac{2}{3}w^3 t \tag{13.3}$$

The maximum value of w is fixed by the design requirements that the door pillar does not obstruct vision and the sill does not obstruct entry. In a beam of constant mass, t is inversely proportional to the material density. Hence, the selection parameter for bending stiffness is E/ρ; steel and aluminium have high values (Table 13.1), so are used in preference to SMC or thermoplastics.

To provide protection for the occupants in a frontal crash, there are two fore-and-aft tubular 'rails', which support the engine, then make an S bend and connect to the door sills. These rails have a crushing resistance of approximately 200 kN, so that the passenger cage of a 1 tonne car decelerates at approximately 20 g. To minimise the mass of these rails, a material with a high ratio of yield stress σ_y to density is required. Table 13.1 shows that high strength steel is optimal for this component. For these two reasons, plastics are not used for the main structure of a car body.

The production engineering complexity of using one process technology for car panels and another for the tubular steel frame have been overcome for some American sports cars, but the majority of motor manufacturers prefer to continue to use steel throughout. The total production costs for sheet metal panels (tooling and materials) are smaller than for plastics for large volume production, but lower for RRIM and SMC plastics for lower production volumes.

13.3 Shape selection to optimise stiffness

13.3.1 Corrugations

A common way to increase the bending stiffness of products from melt-inflation processes is to use corrugations. It is not possible to create ribbed shapes, described in the next section, because there is no independent control of the shape of the inner surface of the melt. Corrugations are seen on thermoformed products, blow-moulded bottles and extruded pipes.

The diametral deflection of a buried pipe, due to soil loading, can be reduced by increasing the second moment of area I of the pipe wall. For plane-walled pipe, I is given by Eq. (C.9) with t being the wall thickness and w the length of pipe considered. The corrugated wall of the soil drainage pipe shown in Fig. 11.14 can be approximated as a trapezoidal wave, of wavelength L, wave amplitude H, crest length C and constant vertical thickness t. Its second moment of area I_C for a longitudinal section is related to that of the plane-walled pipe by

$$\frac{I_C}{I} = 1 + \left(\frac{H}{t}\right)^2 \left(1 + \frac{4L}{C}\right) \tag{13.4}$$

For the shape shown, the stiffening factor is approximately 450. However, the section perpendicular to the pipe length has a low I; the corrugations deform like a bellows when the pipe is bent.

A twin-walled corrugated pipe is shown in Fig. 1.14. The external wall is formed by applying an internal air pressure to the molten tubular extrudate (Fig. 13.2), which expands against pairs of mould sections—these move with the cooling pipe for some distance, then return on a caterpillar track to the die.

The second, interior wall is extruded, while the inner surface of outer wall is still molten, creating a twin-walled pipe. The weld line between the two layers is visible in Fig. 1.14. Such pipes have a high bending stiffness, both longitudinally and for soil loading.

13.3.2 Ribs on injection mouldings

The easiest way to stiffen injection mouldings is to use ribs on the hidden surface (Fig. 13.3a). Ribs correspond to slots in the mould, so straight ribs are easier to machine (using a milling wheel) than those with corners. The rib thickness r should be less than 2/3 of the thickness t of the surface that they support, so they complete solidification first, thereby avoiding sink marks appearing on the product surface (Chapter 6).

A basic analysis assumes that the neutral surface is planar, so it passes through the centroid or centre of gravity, of the section (Fig. 13.3b). The section repeats to create a multi-ribbed plate. Each unit of the plate, of area

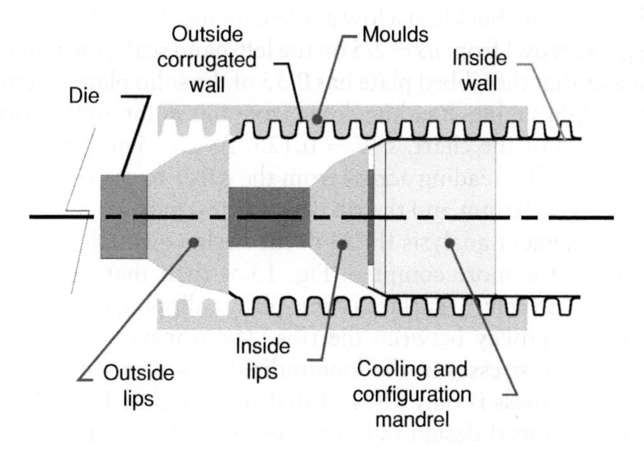

Figure 13.2 Extrusion method of making twin-walled corrugated HDPE pipe (from Diez C, Characteristics of Corrugated HDPE pipe, 7th Ed. 2005, Inc, Quebec,).

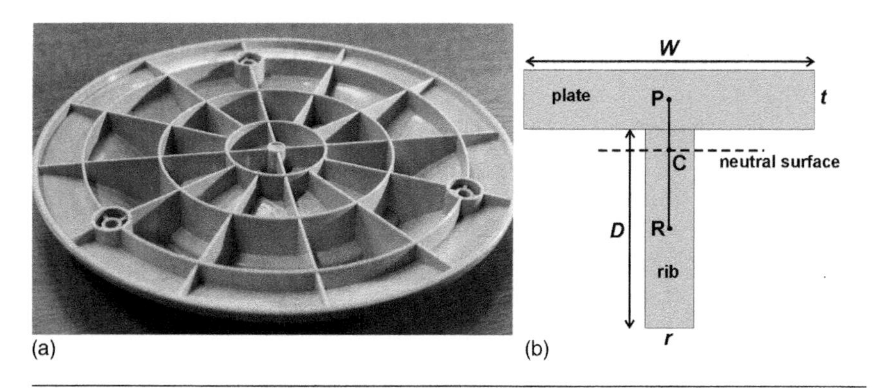

Figure 13.3 a) Ribbing on the underside of the base of a computer monitor stand, b) location of the centroid if a section of ribbed plate.

Wt, has a centroid at its centre P, while the rib of area *Dr* has a centroid at its centre R. The overall centroid is at a position C, where by the lever rule

$$WtRC = DrCR \tag{13.5}$$

This calculation neglects the taper angle of 1° or 2°, needed on the rib to allow easy ejection from the mould.

When a ribbed plate is designed, several solutions are possible. Consider the replacement of a solid plate of thickness $s = 5$ mm by a cross-ribbed plate of uniform thickness $t = 2$ mm, having the same bending stiffness (Fig. 13.4). The chart recommends one of the pairs of parameters (W, t) that meet the bending stiffness requirement. This is not the minimum mass solution. The mass can be reduced further by increasing the rib depth D and spacing W, but such ribs would buckle at a low applied moment. Reading across horizontally (the grey arrow) from $t/s = 2/5$ on the left-hand scale, the number by the curve indicates that the ribbed plate has 0.55 of the solid plate volume, i.e. there is a 45% weight saving. Reading down from this point, to the horizontal scale at the bottom of the chart, $s/W = 0.125$. As $s = 5$ mm, the rib spacing $W = 40$ mm. Finally, reading across from the curve to the right-hand scale, $T/s = 1.95$, so $T = 10$ mm and the rib depth D is 8 mm.

Finite element analysis (FEA) of the design example shows that the stress distribution is more complex (Fig. 13.5) than that assumed in the simple analysis. The neutral surface is non-planar, dipping towards the midplane of the plate, midway between the ribs. The transverse ribs also have a local effect on the stresses in the longitudinal ribs. Consequently, the computed bending stiffness is only 78% of that predicted by Fig. 13.4.

A cross-ribbed design only provides stiffening in the two rib directions. There is a low bending stiffness along a direction at 45° to the ribs, and a low torsional stiffness. An isotropic stiffened plate needs ribs in at least

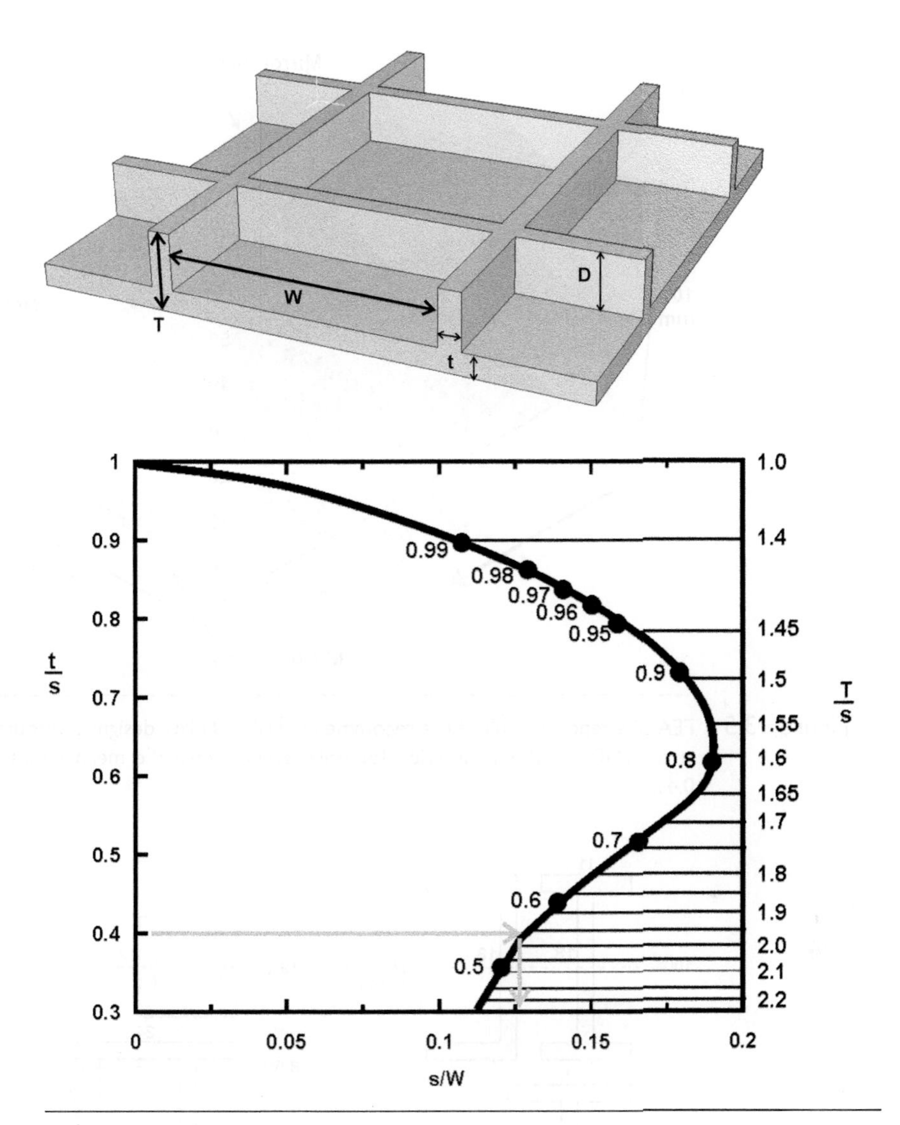

Figure 13.4 (a) Replacement of a flat plate with a cross-ribbed plate of the same bending stiffness. (b) The reduction in mass achieved by the change, and the rib dimensions required (redrawn from *Delrin Design Handbook*, Du Pont, 1980).

three directions, so the mass saving will not be quite as high as suggested by Fig. 13.4. If the direction of the main bending moments is known, ribs can be placed in the appropriate directions. The edges of a product, such as stacking polypropylene chair, are natural places to incorporate ribs.

Figure 13.6 shows the relative efficiency of different beam cross sections. Each has the same cross-sectional area and the neutral surface is horizontal. The *I* beam, of height equal to twice its width, is given a second moment of

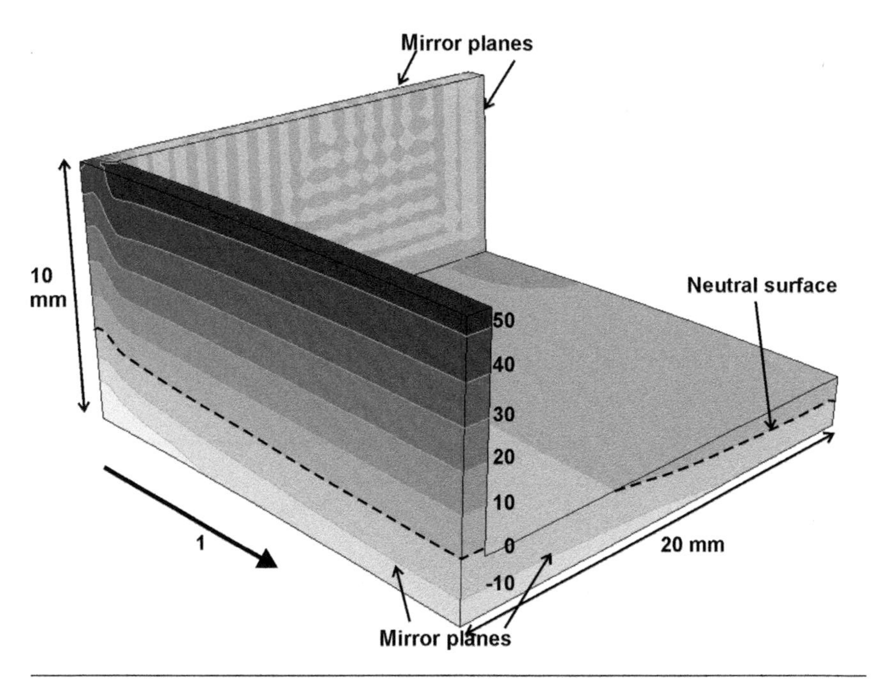

Figure 13.5 FEA of a repeating unit of the recommended ribbed plate design: Contours of longitudinal stress (MPa) in the 1 direction for polystyrene, when the mean radius of curvature is 0.4 m.

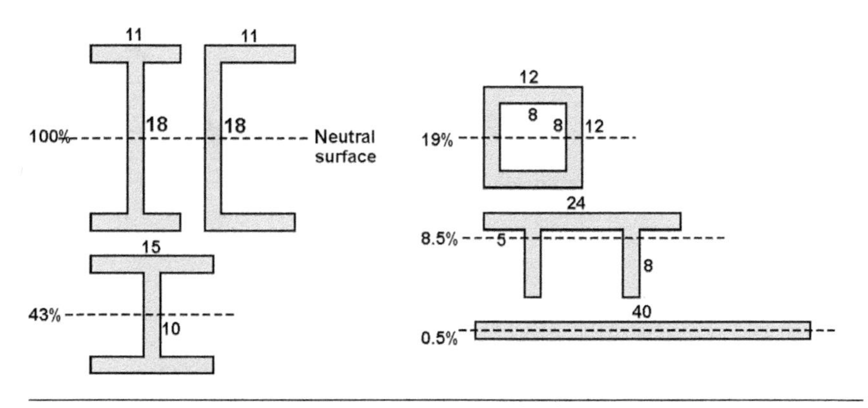

Figure 13.6 Relative bending stiffness of beams, of the same cross-sectional area, about the neutral axes shown (beam dimensions in millimeters, with 2 mm wall thickness).

area of 100%. The C beam, while apparently as efficient, will twist when placed under a bending moment. A ribbed plate, with a rib depth equal to 2/3 of the rib spacing, has a relative efficiency of only 9%. Although *I* beams and hollow tubes are efficient, they are difficult to mould in one piece. Therefore, it is common to use ribbed casings for injection-moulded products.

13.3.3 Buckling of ribs

The bending moment in service should not cause ribs put into compression, to buckle. If the rib was not connected to the plate, its critical buckling load could be calculated using Eq. (C.14) of Appendix C, with L set equal to the rib spacing W. However, as one side of the rib is supported by the plate, only the free side can buckle. FEA is necessary to predict the shape. Figure 13.7

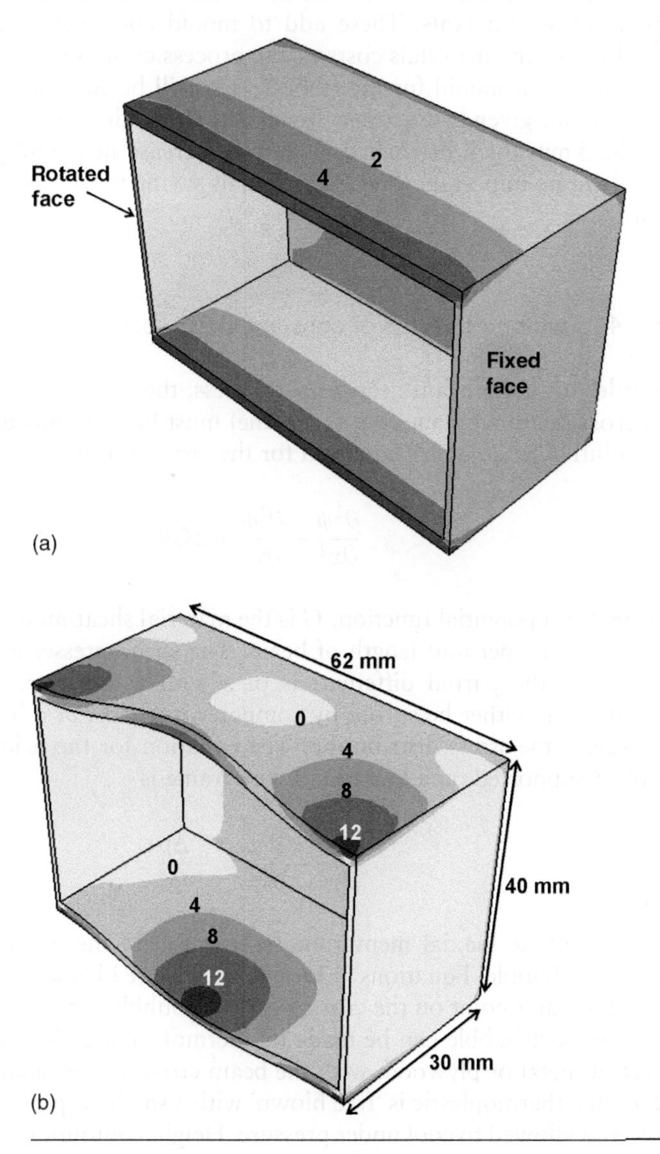

Figure 13.7 FEA of bent section of cross-ribbed PP plate: a) realatively undeformed when the radius of curvature R = 4.2 m, b) some bulked ribs when R = 1.24 m. The rib and plate thickness is 2mm. Cotours of von Mises stress (MPa).

shows a unit of a cross-ribbed plate, with deeper and longer ribs than those considered in the last section. The free sides of the longitudinal ribs are initially straight, but buckle when the plate curvature becomes large. The bending moment versus curvature relationship becomes non-linear when buckling commences, but the slope remains positive.

If FEA shows that some ribs buckle, the product can be redesigned with thicker or shorter ribs. The latter can be achieved by having cross-ribs at closer intervals. These add to mould construction costs, but add very little to the materials costs or the process cycle time. The productivity of an injection mould for the ribbed plate will be controlled by the solidification time given by Eq. (5.4). It will be ~3 s for the ribbed plate and ~19 s for the 5 mm thick original design. The increase in mould productivity is usually more important than the materials saving from product mass reduction.

13.3.4 Torsion of beams of constant cross section

In order to find a beam's torsional stiffness, the shear stress distribution in the cross-sectional plane (the x, y plane) must be determined. This requires the solution of Poisson's equation for the cross section

$$\frac{\partial^2 \phi}{\partial x^2} + \frac{\partial^2 \phi}{\partial y^2} = 2G\theta \tag{13.6}$$

where ϕ is a potential function, G is the material shear modulus and θ is the angle of twist per unit length of beam. The shear stresses in the xz and yz planes are the partial differentials of z with x and y, respectively. This equation can either be solved by computer methods, or a bubble analogue can reveal the stress distribution. The equation for the height z of a soap bubble, supported on a horizontal wire frame is

$$\frac{\partial^2 z}{\partial x^2} + \frac{\partial^2 z}{\partial y^2} = -\frac{\Delta p}{\sigma} \tag{13.7}$$

where σ is the biaxial membrane stress, and Δp the pressure differential across the bubble. Equations (13.6) and (13.7) are identical in form, and the analogues at a point on the cross section or bubble are given in Table 13.2. A permanent bubble can be made by thermoforming. A hole is cut from a sheet of metal or plywood, with the beam cross section shape. A thin sheet of molten thermoplastic is 'free blown' with a small air pressure through the hole and allowed to cool under pressure. Height contours are then drawn on the bubble.

In Appendix A, Eq. (A.14) is the differential equation for one-dimensional transient heat flow. Many finite element packages allow the

Table 13.2 Analogues between beam torsion, soap bubbles and two-dimensional steady heat flow

Torsion	*Soap bubble*	*Heat flow*
Beam boundary shape	Wire frame shape	Material boundary shape
Shear stress magnitude	Maximum slope	Heat flux magnitude
Shear stress direction	Along a height contour	Along a temperature contour
Torsional stiffness	Bubble volume	Total heat generated

solution of a two-dimensional, steady state, heat flow problem in a body with a constant heat generation rate A. The differential equation that applies is

$$\frac{\partial^2 T}{\partial x^2} + \frac{\partial^2 T}{\partial y^2} = A \tag{13.8}$$

which is another form of Poisson's equation. A boundary at a fixed temperature is the equivalent of the free surface condition in the beam torsion problem. Computed temperature contours (isotherms) can be interpreted like soap bubble height contours (Table 13.3), and heat flux magnitude as shear stress magnitude in the twisted beam. Isotherms close to free surfaces (Fig. 13.8) are parallel to the surface. Hence, the shear stress directions

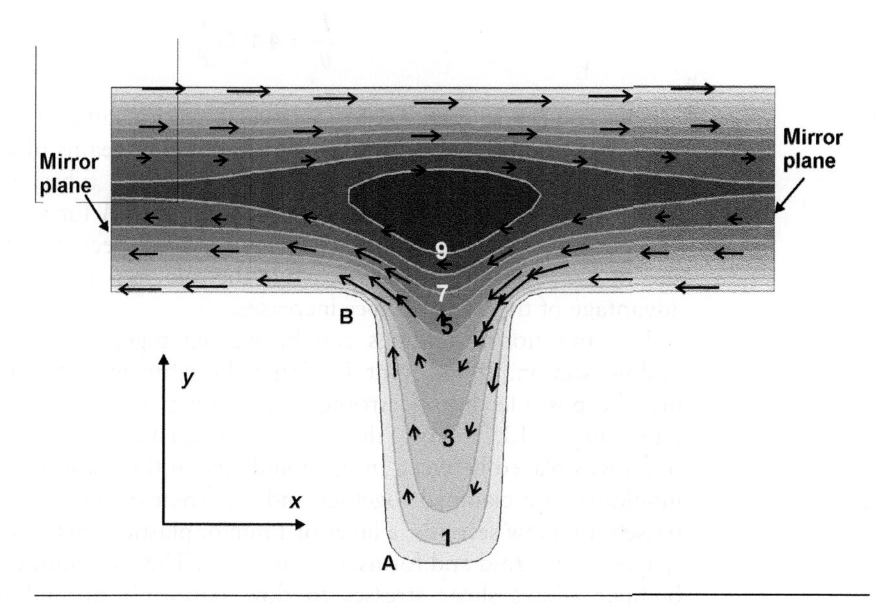

Figure 13.8 Temperature contours (arbitrary units) in a 2–D heat flow, interpreted to show the shear stress direction and magnitude (arrows) for the torsion of a multi-ribbed plate.

in these locations, are nearly parallel to the boundary. The highest temperature occurs at the mid-section, indicating that the shear stress is zero at this location. The shear stresses double back inside the cross section and the total moment of the shear forces about the axis of twist is low. Hence, the torsional stiffness of all open-section shapes is low.

Four different open sections, shown in Fig. 13.9, all have the same total width $W = 40$ mm, and thickness $t = 2$ mm. As the effects of corners in the I and ribbed sections can be ignored, the torsional stiffness is the same as for a flat plate of width W. As $W \gg t$, the ends of the rectangular section can be ignored; the linear variation of shear stress across the majority of section leads to the torsional stiffness being

$$\frac{T}{\theta} \cong GW\frac{t^3}{3} \tag{13.9}$$

To model the torsion of a hollow, thin-walled tube, a bubble is blown through a circular cutout, with a flat horizontal surface, free to rise, representing the interior of the tube. The free part of the bubble has a constant slope in the radial direction, showing that a constant shear stress acts in the tube wall, parallel to the boundary. The bubble volume is proportional to the area A inside the tube, and to the height of the central flat surface. For a constant pressure differential this height is proportional to the wall thickness t. The torsional stiffness of a hollow tube of perimeter P, wall thickness t and enclosed area A is

$$\frac{T}{\theta} \cong 4A^2G\frac{t}{P} \tag{13.10}$$

This reduces to $2\pi R^3Gt$ for a circular section tube of mean radius R. Figure 13.9 ranks the torsional stiffness of some beams of constant cross-sectional area. The preferred designs are hollow tubes; the best has a circular section, as this includes the greatest area A for a given perimeter. The per cent stiffness values only apply to these specific dimensions. If the size of the section is increased, while the thickness t is kept at 2 mm, the advantage of the hollow tubes increases.

Two injection mouldings can be welded together to produce a stiff, hollow section. If the joint is planar, hotplate welding (Section 13.2.4) may be possible, but electrothermal welding is preferred for non-planar joints. Figure 13.10 shows the section through a car bumper; woven copper braid was placed between three mouldings, of thickness 3.0 mm. When the mouldings are clamped together and a current of hundreds of amperes is passed for a few seconds, a layer of 1 mm of plastic melts, flows through the spaces in the braid and forms a strong weld. The continuous weld along the bumper allows shear stresses to flow uniformly around the section. If

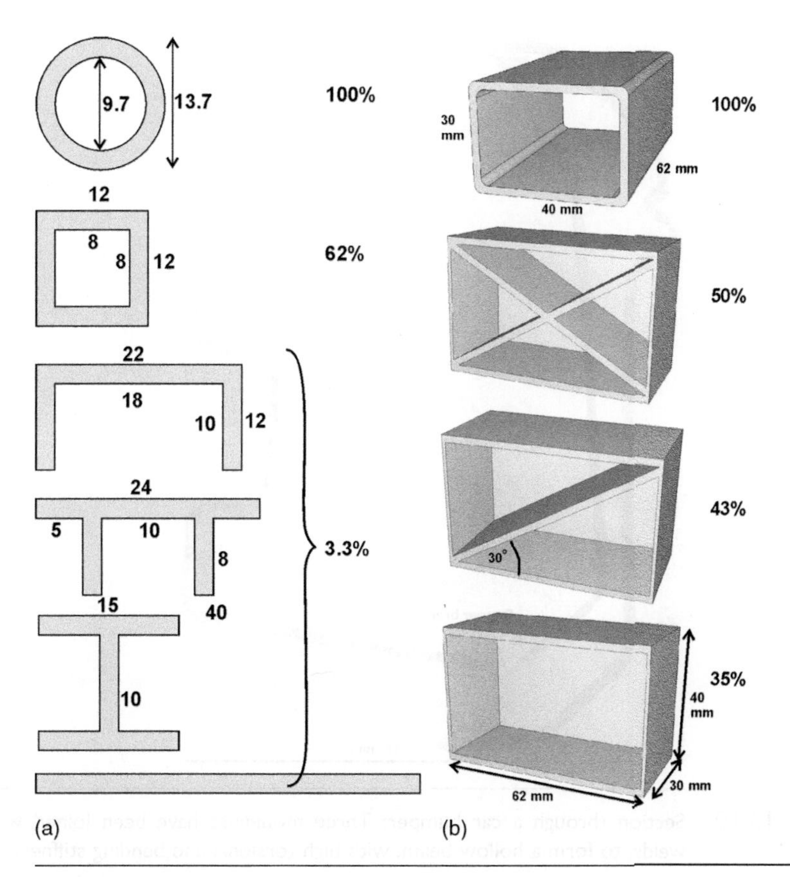

Figure 13.9 Specific torsional stiffness of beams a) the same cross-sectional area b) cross-ribbed, Dimensions in mm, all wall thicknesses 2mm wall thickness for torsional stiffness, relative to the thin-walled cylinder.

the parts were mechanically fastened together, the torsional stiffness would be lower, because shear stresses can only pass through small areas near the fasteners.

13.3.5 Torsion of beams of non-constant cross section

It is possible to design injection mouldings with higher torsional stiffness, by varying the beam cross section. Diagonal and cross-ribs, on **I** or **U** beams, take tensile or compressive loads when the beam is twisted. They, with ribs at the sides of the beam, act somewhat like a pin-jointed framework. If the

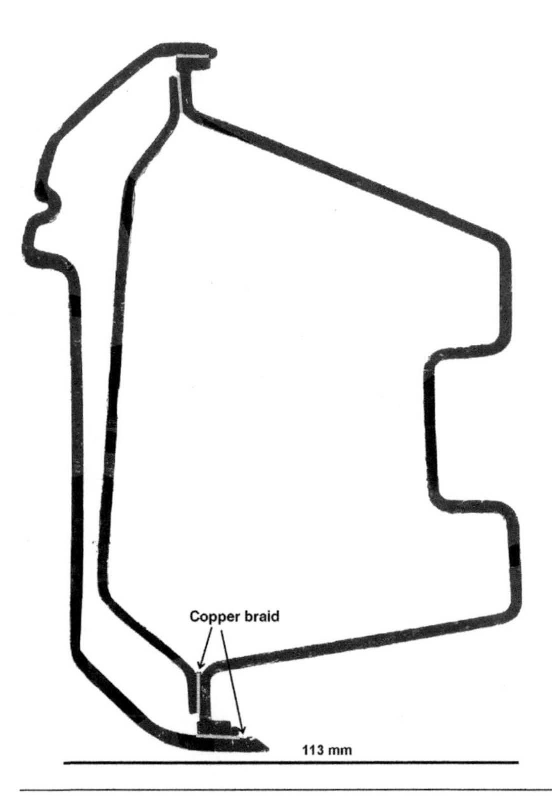

Figure 13.10 Section through a car bumper: Three mouldings have been joined with electrofusion welds, to form a hollow beam, with high torsional and bending stiffness.

geometrically equivalent framework is a *mechanism* capable of motion (Fig. 13.11), the beam torsional stiffness will be low. Try twisting the ends of the inner part of a matchbox, or a rectangular plastic tray, and note the high deflection. If the equivalent framework is rigid, the ribbed beam has a high torsional stiffness. A matchbox, with a diagonal piece of card glued in place, has a high torsional stiffness.

The specific torsional stiffness of cross-ribbed beams is lower than that of hollow tubular beams (Fig. 13.9). In the former, some parts of the structure (the cross-ribs and the back of the U—Fig. 13.12a) are lightly stressed, while in the latter there is a uniform stress. Hollow tubular beams are resistant to buckling at high torques, since they contain no unconstrained free edges. However, Fig. 13.12b shows that the free side of a diagonal rib has buckled. The twisted cross-rib has also tilted about a vertical axis.

The relationships between torque and angle of twist, for the ribbed beams of Fig. 13.9, are shown in Fig. 13.13. The hollow tube has an almost linear response up to a torque of 55 Nm before the polypropylene

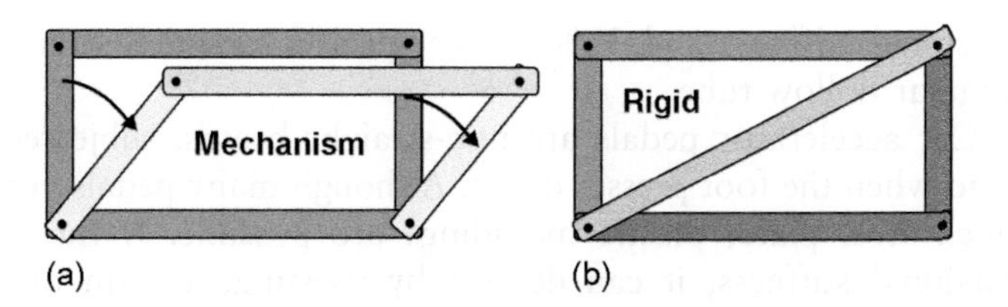

Figure 13.11 Pin-jointed frameworks: (a) mechanism; (b) simply stiff. If the rib pattern on a beam is like (a), the beam will have a low torsional stiffness; if it is like (b), it will be much stiffer.

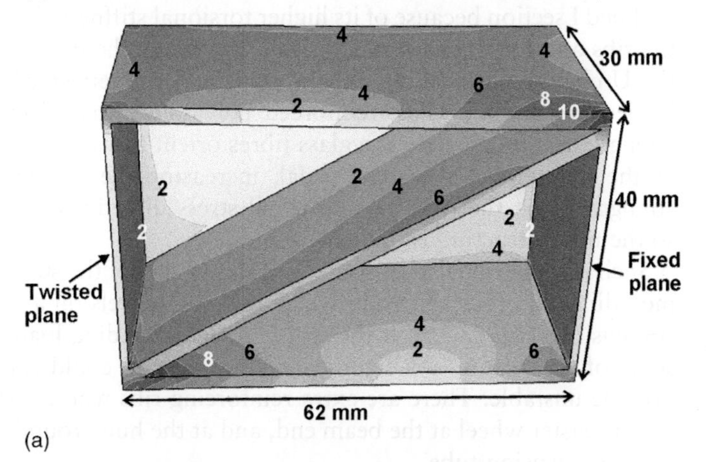

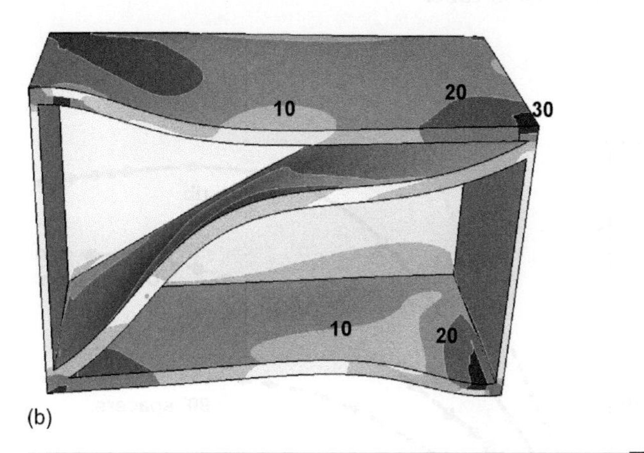

Figure 13.12 Predicted torsion of diagonally ribbed PP beam, with contours of von Mises stress (MPa): (a) At $\theta = 0.035\ \text{rad m}^{-1}$; (b) $\theta = 0.148\ \text{rad m}^{-1}$.

yields, whereas the ribbed beams have significantly lower torsional strengths, due to rib buckling. Thus, the failure modes are different. These thin-walled ribbed beams should be redesigned so that they fail in torsion by yielding. The cross-ribbed moulding design has 50% of the specific torsional stiffness and 31% of the specific torsional strength of the rectangular hollow tube.

Car accelerator pedals are non-straight beams, subjected to a bending load when the foot presses down. Although many pedals are stamped from thick steel plate, plastic mouldings are possible. If the moulding lacks torsional stiffness, it can deform by twisting, a form of buckling. It is difficult to injection mould a hollow tube with a bend by using movable cores, that retract along the axis of the tube. One design (Fig. 13.14) used a beam with a non-constant cross section. A ribbed **U** section was preferred to a ribbed **I** section because of its higher torsional stiffness. The angle between the ribs and the flange was 15°. Half way along the pedal, the direction of the **U** changed, cancelling out the slight twisting effect when the pedal is loaded in bending. Glass-reinforced polyamide was selected for its high strength and toughness. The glass fibres orient towards the flow directions as the melt flows along the pedal, increasing the Young's modulus and strength along the ribs. The principal stress directions are along the ribs, so the microstructure is optimal.

Such ribbing can also be seen on the underside of the starfish-shaped base mouldings of office chairs, in which the seat swivels on a vertical gas-strut suspension (Fig. 13.15). If the seat is rocked, bending loads are applied to some of the beams; any significant beam torsion could cause the chair to become unstable. There are extra reinforcing ribs where a shaft is inserted for the caster wheel at the beam end, and at the hub around the moulded-in metal suspension tube.

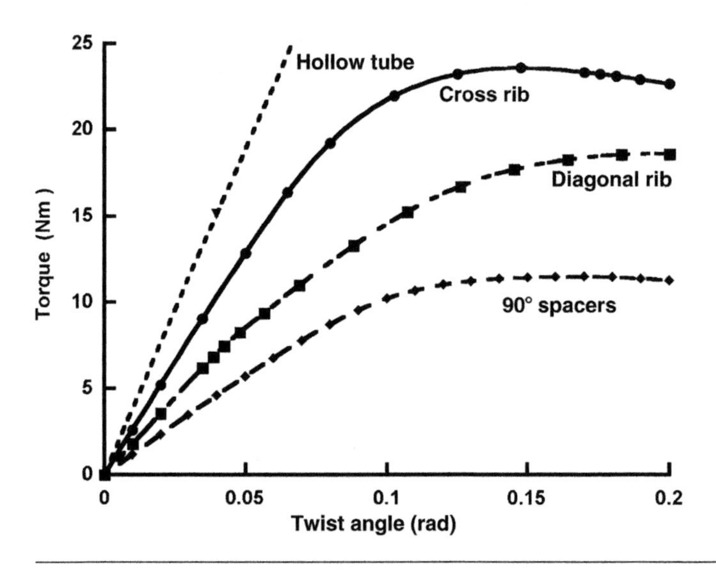

Figure 13.13 Torque vs. twist angle for the ribbed beams shown on the right of Fig. 13.9.

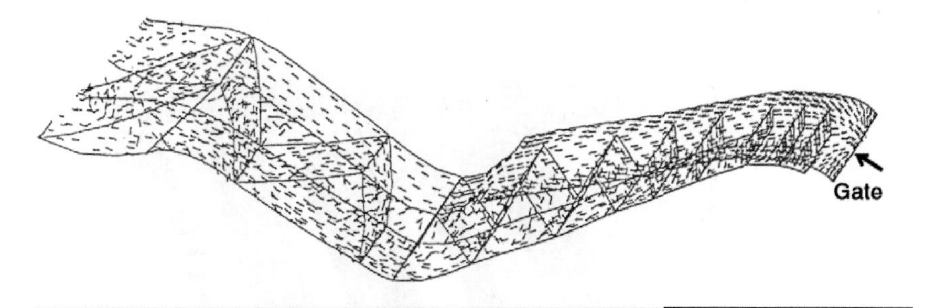

Figure 13.14 Design of a vehicle pedal. The arrows are predicted flow directions for mould filling (from Brünings W. D. *et al.*, *Kunststoffe*, **79**, 254, 1979).

Figure 13.15 Diagonal rib pattern, on the underside of office chair legs, that provides bending and torsional stiffness.

13.4 Product shapes for injection moulding

The injection moulding process allows a great variety of possible product shapes. Process optimisation, important for the economics of the product, in part determines the product design. It is important to minimise the cycle time, and to avoid shrinkage flaws, as well as using shapes that maximise the product stiffness. Two simple product design rules, considered in Sections 13.1 and 13.2, are that the product should either have uniform thickness or the thickness should decrease from the gate. The remaining sections consider how many parts can be integrated in a single moulding.

13.4.1 Uniform part thickness

Figure 1.1 shows the nearly uniform thickness of the wall of a kettle. This allows all regions of the moulding to complete solidification at nearly the

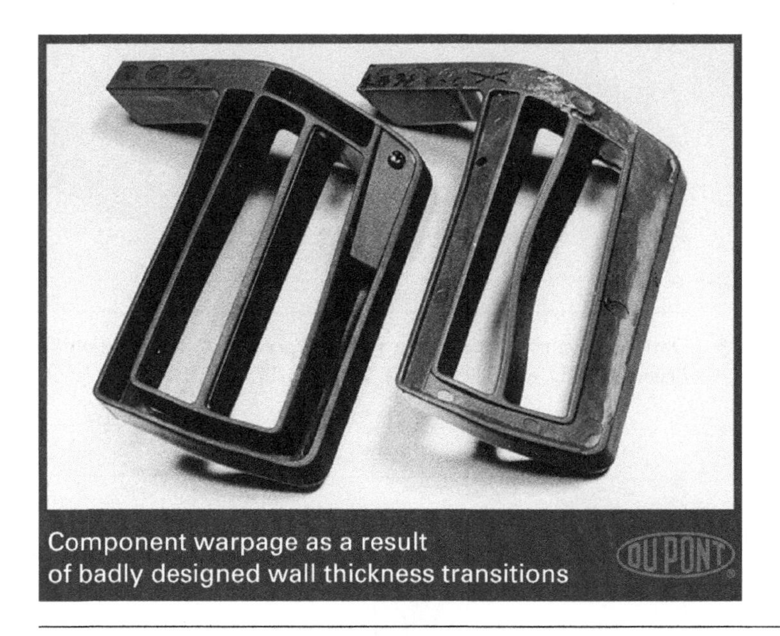

Figure 13.16 Buckled internal strut, due to shrinkage of a thin part of a moulding (Hasenauer, J. *et al.*, *Top 10 Design Tips*, Du Pont, UK, © reprinted with permission).

same time. Figure 13.16 shows one effect of ignoring this rule. The thin internal strut solidified first, then was put under compressive load by the subsequent shrinkage of the thicker outer region. As the strut had a low bending stiffness, it buckled. When the outer region was cored out to be of uniform thickness, the strut no longer buckled.

Once a mould is full, there is effectively one-dimensional heat flow in the direction perpendicular to the mould wall, while the plastic solidifies. The moulding must solidify all the way through, before it can be ejected, so the cycle time is determined mainly by this 'through-thickness solidification time' t_S. In Appendix A, Fig. A.4 is a dimensionless graph of mouldings temperature profiles versus time. If the plastic is solid at $T - T_0 = 0.4$, the melt is fully solid when the Fourier number $Fo = 0.5$. Hence, the solidification time is given by

$$t_S = \frac{FoL^2}{\alpha} \tag{13.11}$$

where L is the part half thickness in mm, and thermal diffusivity $\alpha \cong 0.1\,\text{mm}\,\text{s}^{-1}$. For a 4 mm thick part, $t_S = 20\,\text{s}$, whereas it is 5 s for a 2 mm thick moulding. The thickness chosen is a compromise between mould productivity, which is proportional to L^{-2}, and mechanical properties such as bending stiffness, which increase with L^3.

13.4.2 Part thickness that decreases away from the gate

Thin regions cool more rapidly than do thick regions, according to Eq. (13.11). Therefore, if the moulding is thicker, in a region remote from the gate, than near the gate, the remote region will have a molten core, cut off from the melt supply, at some stage during cooling. This region cannot be fed by the holding pressure, so shrinks more than other regions. Consequently, either the soft, solid skin is sucked in, leaving an unsightly sink mark, or an internal void forms (Fig. 6.13b).

Convex regions of the product surface (A in Fig. 13.8), cool more rapidly than do internal corners (B) due to the difference in heat flow geometry. Hence, the solid skin is thinnest at concave locations. This can be partly rectified by radiusing both internal and external corners, and keeping the part thickness uniform. The stress concentration factor at rounded corners is lower than at sharp corners when the product is bent (Section 9.3.1). Rounded corners make flow into the mould easier, and are unlikely to cause injury to the user of the product.

Figure 13.17 shows that, when bosses are provided for self-tapping screws, they should be separated from the main wall of the product, to avoid the creation of thick regions. Note the number of sharp internal corners in the moulding. The bosses may need to be supported by buttresses that connect to other parts of the product. An extreme example occurs in stacking chairs (Fig. 1.5).

13.4.3 Product casings that locate components

Hand-held electric tools, such as electric drills or hot-air blowers, usually have 'clam shell' mouldings. When the two casing halves close, tongues or

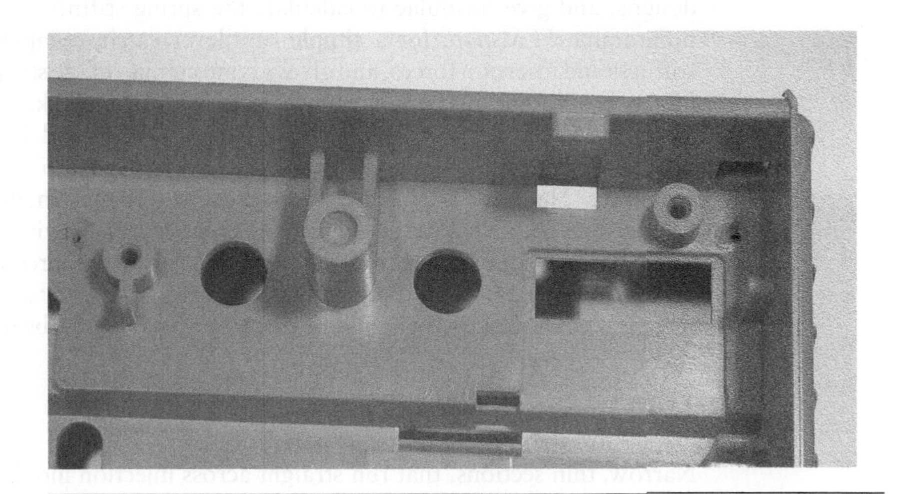

Figure 13.17 Bosses for screw assembly, seen inside a cassette player; the central one is for a screw from the product exterior.

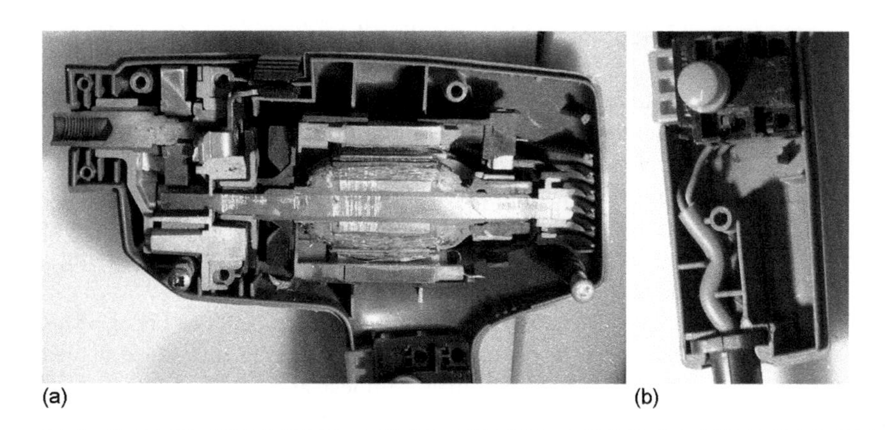

Figure 13.18 A sectioned electric drill, showing internal ribs that locate the: a) the motor and gearbox, b) the mains cable

bosses on one half locate in grooves or recesses in the other. The bending and torsional stiffness of the screwed together casing is reasonably high. Internal ribs in each half casing locate items such as an electric motor and gearbox (Fig. 13.18a). A maze of partial ribs can prevent the PVC-coated mains cable from pulling from the tool (Fig. 13.18b).

13.4.4 Integral springs and snap joints

Figure 1.4 showed separate plastic springs in a video cassette. They use the high elastic tensile strains and the low Young's modulus, of plastics to replace metal coil springs. Thin cantilever springs can also be integrated into injection mouldings. Brochures on the 'plastics.bayer.com' website describe typical designs, and give formulae to calculate the spring stiffness. An interactive programme *FEMsnap*, for a simple cantilever catch, computes the beam stiffness and insertion forces, and gives a contour map of the stresses or strains in the beam. The maximum tensile strains are usually kept below 1%. Figure 13.19 shows the principal tensile strain contours and deformed shape of such a catch, at maximum deflection during insertion into a slot in another moulding. The peak strain of 2.5% is close to the maximum allowed for polycarbonate. Such a system acts as a *snap joint*, removing the need for screw fixing. If the catch rear face is perpendicular to the direction of assembly, the product assembly is permanent. However, if angled faces are used, or there is a method of disengaging the catch, the product can be disassembled.

13.4.5 Integral hinges

Narrow, thin sections, that run straight across injection mouldings, can act as integral hinges (Fig. 1.15). These replace the metal pins, pressings and

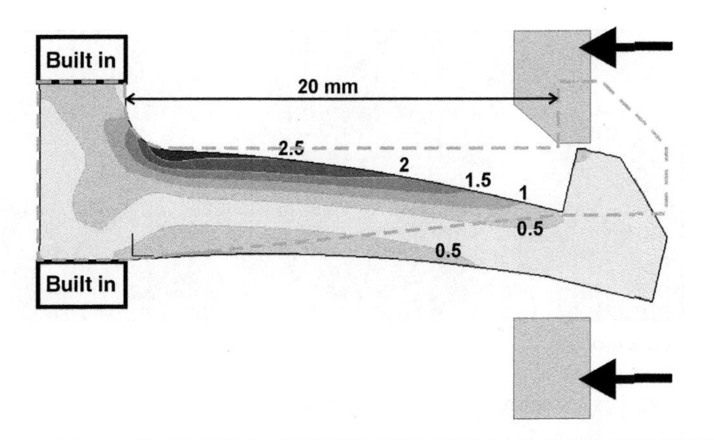

Figure 13.19 Snap joint: An integral cantilever spring at maximum deflection on insertion, compared with the undeformed shape, and principal tensile strain contours (%).

screws of conventional hinges. However, this region is a likely site for failure, unless a highly oriented semi-crystalline morphology can be achieved, with orientation direction across the hinge. Such microstructures are only easy to produce in polypropylene, because of its relatively low melt viscosity, and its propensity to crystallise in an oriented form near the surface. Figure 6.7b shows a PP hinge cross section in polarised light. An oriented region extends across the approximately 0.6 mm thick hinge. Equation (A.26) gives the thickness S of the solid layer that develops during mould filling flow in a time t, as

$$S \cong \sqrt{\alpha t} \tag{13.12}$$

As the thermal diffusivity $\alpha \cong 0.1\ \mathrm{mm^2\,s^{-1}}$, a layer $S = 0.2$ mm develops in a mould fill time of 0.4 s. Hence, the mould must be filled fast to avoid the hinge area solidifying and preventing the complete filling of the mould. When the hinge is first flexed, part of the PP stretches and whitens, but it keeps its tensile strength. Even if the PP delaminates parallel to the hinge surface, this does not detract from the hinge performance.

13.5 Instrument panel case study

13.5.1 Instrument panel shape

A car instrument panel (Fig. 5.27) is a large injection moulding with a complex shape. The panel shape was optimised, in ways discussed earlier in this chapter, to increase its bending stiffness. The visual effects of scratching are minimised by having a textured surface on the mould. Figure 13.20

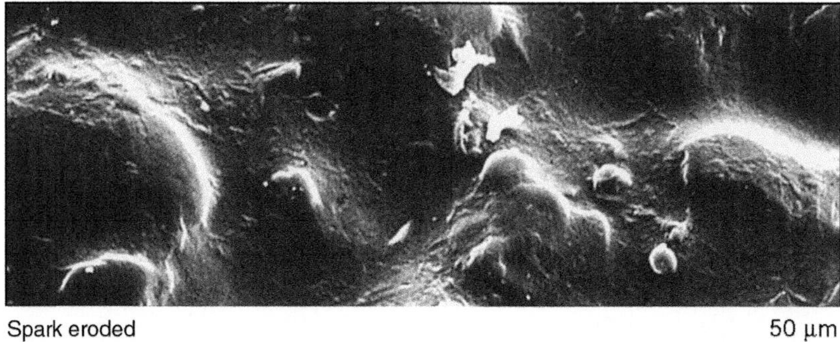

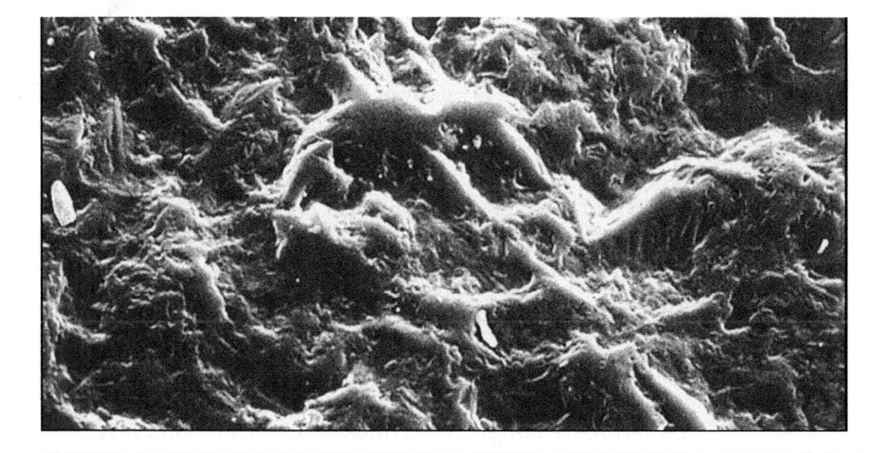

Figure 13.20 Texture on a nylon 6 moulding, due to spark erosion (top) or photo-etching (lower) mould surfaces (from Schauf D, ATI 584e, booklet, © Bayer Material Science AG, 1988).

shows a matt surface on a similar moulding that scatters the majority of incident light, so there are no direct reflections. The spark-eroded mould surface provides rounded features on the moulding, with a better appearance than those from the photo-etched mould surface. An alternative is to use a 'soft touch' finish as in the Ford Focus—the moulding has a surface layer of a low modulus ethylene copolymer, that is difficult to scratch. The effects of the differential thermal expansion coefficient between the plastic instrument panel and the steel body are disguised, by making sure that any gaps are hidden from view.

13.5.2 Free head-form impact tests

Federal Motor Vehicle Standard FMVSS 201 requires that a 4.5 kg head-form is fired at 15 mph at any interior region that could be hit by the head of an unbelted vehicle occupant. There is a triaxial accelerometer inside the head-form, which has a rubbery skin. A weighed integral of the head-form acceleration must not exceed a level that would cause moderate head injury. This is roughly equivalent to the peak acceleration being less than 140 g. Consequently, the instrument panel must absorb a kinetic energy of 100 J and it must not have any rigid projections on its surface. Figure 13.21a shows a finite element simulation of the head-form impacting an instrument panel, and figure 13.21b shows more details of the head-form impacting a ribbed plastic moulding, covering the roof rail at the top of the windscreen A pillar. For the impact on the instrument panel, the peak forces would be about 6 kN, and peak deflections about 40 mm.

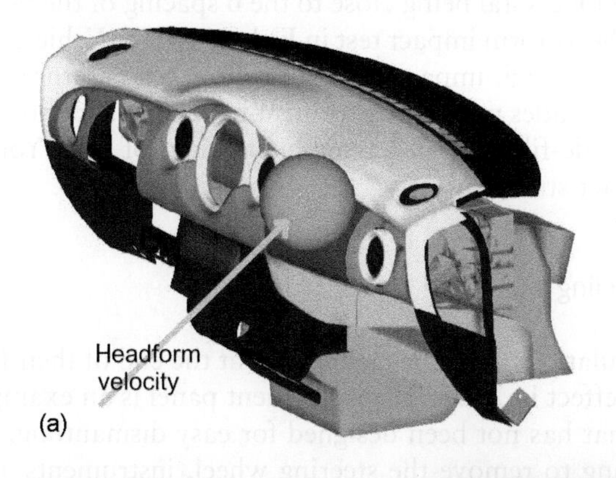

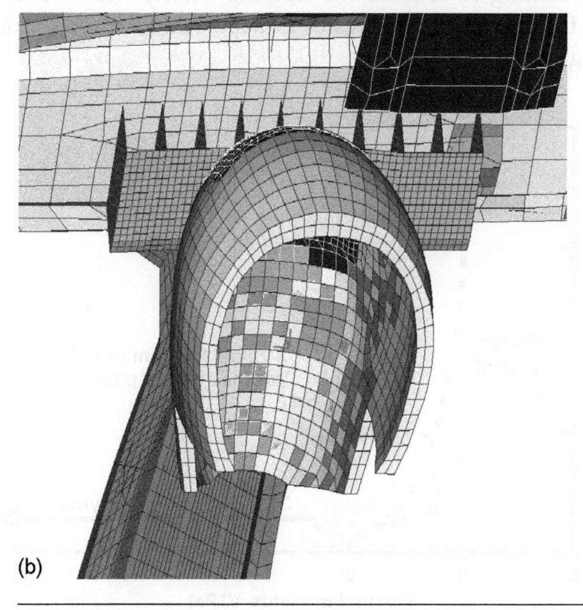

Figure 13.21 FEA of head-form impact on a) an instrument panel (Silk, G., SAE Technical Paper 2002-01-1270; Haque *et al.*, SAE Paper 2000-01-0624 for an impact b) a ribbed moulding at the top of an padded A pillar © SAE International.

13.5.3 Grade development

ABS, with an upper layer of plasticised PVC to simulate leather, was initially used. However, by the late 1980s, PP grades were developed to suit the application. Nucleated grades had increased crystallinity, hence increased yield stress (35 MPa at 23 °C) and Young's modulus. The PP was toughened by adding compatible ethylene propylene copolymer rubber, which is well bonded to the PP matrix. Talc was added to increase the Young's modulus and the temperature resistance of the instrument panel. Talc acts as a nucleating agent for PP. PP crystals grow epitaxially on the talc platelets, with the **c** spacing of the PP crystal being close to the **b** spacing of the talc crystal.

Due to the head-form impact test in Federal Motor Vehicle Standards, the PP should have a high-impact strength at low temperatures. Figure 13.22 shows how the grades used for instrument panels are superior to just rubber-toughened or talc-filled PP in having high values of both Young's modulus and Izod impact strength.

13.5.4 Recycling

European regulations about recycling cars at the end of their lives are slated to come into effect in 2007. The instrument panel is an example of a major component that has not been designed for easy dismantling. It is currently time consuming to remove the steering wheel, instruments and the wiring harness from the many places of attachment. In future, redesign will make recycling easier.

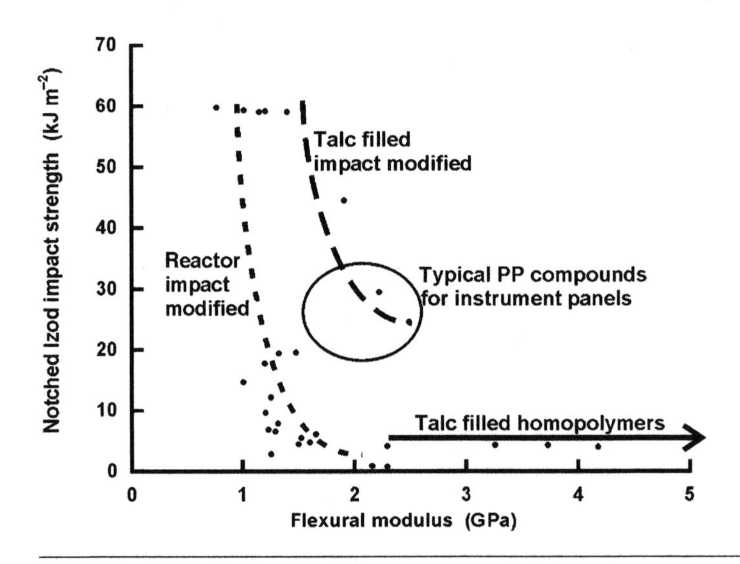

Figure 13.22 Young's modulus vs. charpy impact strength at 23 °C (redrawn from Juan, P. *et al.*, SAE paper 980067).

Chapter 14

Engineering case studies

Chapter contents

14.1 Introduction

The three case studies illustrate different areas of product design. That on gas pipes involves mechanical properties such as creep, yield and fracture, and the process technology of welding. The mechanical design loads are well known, unlike in the bicycle helmet case study, where the type of impact is unpredictable, and the impact tolerance of the brain and skull variable. This case study emphasises the loading geometry, and the difficulties in meeting conflicting requirements. The last case study on digital data storage illustrates the optical properties of the glassy polymers and the injection-moulding technology required to make compact discs.

The problems in this chapter can be tackled as the reader progresses. The answers, essential to the progress of the argument, are provided at the end of the chapter, but it is strongly recommended that the readers attempt the problem first! This is a means of revising concepts from earlier parts of the book.

14.2 Pipes for natural gas distribution

14.2.1 Introduction

The distribution of natural gas in the UK is a major enterprise. A national grid collects and distributes gas from the North Sea, and the total network, with connections to 14 million domestic and industrial consumers, has a length exceeding 200 000 km. A number of different materials are used (Table 14.1). We will consider gas distribution at the area and district level, where plastics have replaced cast iron. The plastic system was designed and installed by the British Gas Corporation, starting in the early 1970s. The successor company, responsible for the gas distribution network, is National Grid Transco (www.transco.uk.com). Gas at a pressure of 70 bar from the national grid is reduced to a pressure of 7–16 bar in the local high pressure grid, and 1–4 bar in the local distribution system.

Table 14.1 Parts of the gas distribution system

Part of system	*Gas pressure (bar)*	*Requirement*	*Material (old material)*
National grid	7–70	Maximum flow high hoop strength	High strength steel
Local distribution	0.075–4	See later	Plastic (cast iron/steel)
Inside house	<0.075	Safe durable connections	Copper (lead)

Individual housing estates take their gas supplies via pressure governors which reduce the gas pressure to 40 mbar.

The order of presentation follows the stages of determining the plastic and the pipe wall thickness, before considering installation and joining.

14.2.2 The creep rupture test

Creep rupture testing must be introduced early in the case study, because the creep rupture performance affects materials selection. Figure 14.1 shows the equipments used. Short lengths of pipe are fitted with mechanically screwed end fittings, a constant internal water pressure applied, and the time recorded when leakage occurs. There are two reasons for testing the pipe, rather than tensile specimens cut from it. Firstly, the internal pressure p produces biaxial stresses in the hoop and longitudinal directions (Section C.3 of Appendix C). The biaxial stress failure mode differs from that in tensile tests. Secondly, the pipes contain residual stresses from fabrication (Fig. 6.15b), and this, or other effects of processing, may affect the creep rupture times.

The creep rupture times for many tests are plotted versus the wall hoop stress (Fig. 14.2) on logarithmic scales. The data falls on a line of small negative slope. In the ductile creep rupture process, the pipe wall balloons out at one location then necks, with the extension being mainly in the hoop direction. The necked material has high molecular and crystalline orientation in the hoop direction, but very little orientation in the length direction, so it has a much higher tensile strength in the hoop direction. Consequently, the smaller longitudinal stress causes a split to occur in the rz plane,

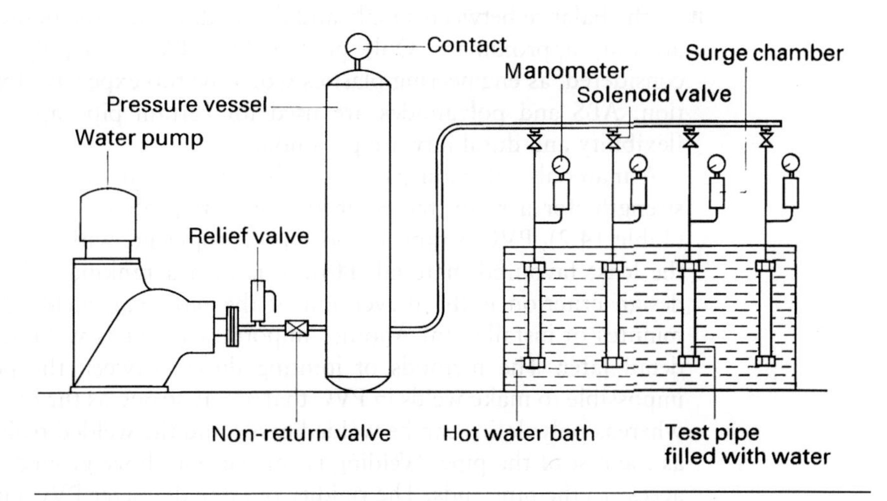

Figure 14.1 Diagram of test rig with automatic pressure control for measuring the creep strength of plastic pipes (Courtesy Hoechst AG).

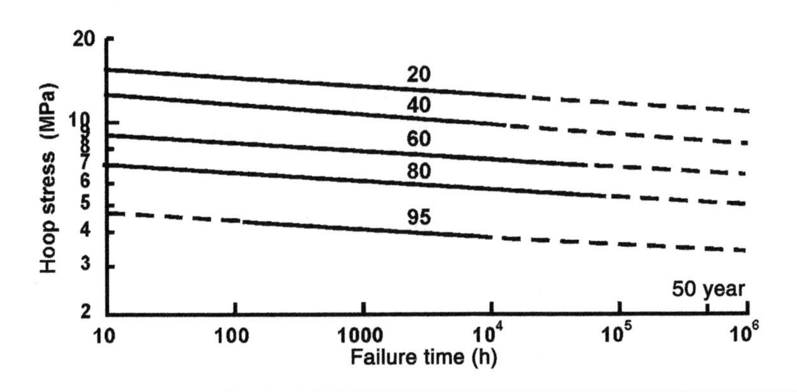

Figure 14.2 Creep rupture data for polyethylene gas pipes, made from Solvay Eltex TVB 121 at a range of temperatures (Bocker, H. *et al.*, *Kunststoffe*, **82**, 739, 1992).

producing the characteristic 'parrot's beak' fracture (Fig. 14.3a). Figure 14.3b shows a section through a welded joint in a polyethylene water pipe that has failed in an accelerated laboratory test at 80 °C, by slow crack growth. The weld bead is a stress concentrating feature.

14.2.3 Choosing a plastic

Several factors are involved in choosing a plastic, specifying the best grade for the application and deciding which additives are necessary. Three criteria—cost, strength and toughness—determined the plastics to be investigated in detail. The price of plastics fluctuates with that of oil, and depends on the balance between supply and demand. Hence, the figures in Table 1.1 are only approximate. Only polyethylene, PVC and polypropylene were considered, as engineering plastics would be too expensive for this application. ABS and polyamides are used for certain pipe applications where flexibility and durability are paramount.

A materials selection package (Chapter 13) can sort polymers by yield strength, for a room temperature test lasting about 1 min. On this basis (Table 14.2), PVC is ranked above polypropylene, with polyethylene being the least favoured material. However, such a ranking only applies if the design criterion is the prevention of the pipe wall yielding within a few minutes of installation! Another important factor is welding, discussed in detail later. The methods of jointing differ between the polymers. It is impossible to make welds in PVC that are as strong as the original material, whereas polyolefins can be welded easily and the welded region is as strong as the rest of the pipe. Welding is carried out above ground, to allow easy access to the pipe ends. The rigidity of large diameter PVC pipe means that they are difficult to manoeuvre into a trench. In contrast, medium-density polyethylene (MDPE) pipe is flexible enough to be pushed, a section at a

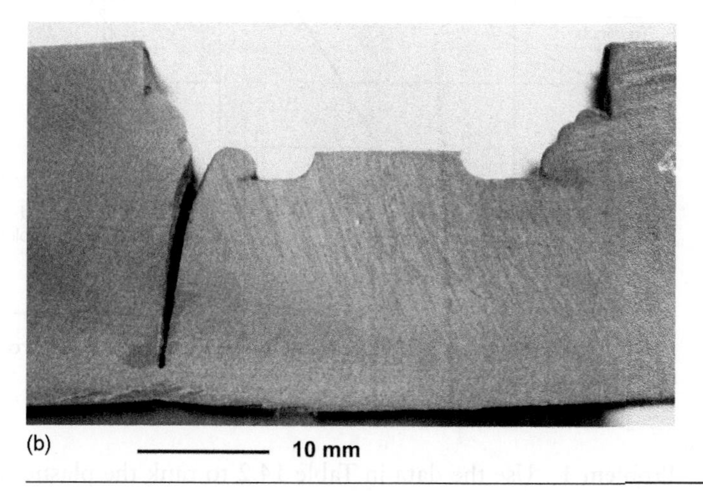

Figure 14.3 Failure modes in a pressurised MDPE pipe: (a) Ductile 'parrot's beak' fracture; (b) brittle section through a welded joint that has failed in a creep rupture test in water at 80 °C.

time, into a trench, or to be pulled into old cast iron mains in the relining process (Fig. 14.4).

Toughness is another critical material property. The explosive nature of gas/air mixtures was known from newspaper accounts of domestic fatalities in the 1970s, whereas a leak in a water pipe, although inconvenient, rarely imperils life. High-speed brittle fracture occurred in some thick-walled rigid PVC pipes used as water mains. Fortunately, the crack arrested at the mechanical seals at the ends of the 12 m pipe lengths. Plastics should not be selected on the basis of their plane strain fracture toughness K_{IC} alone (Table 9.1). The pipe wall thickness should not allow plane strain fracture. Table 14.2 gives data for the fracture toughness and yield stress, measured in tests in which the loads were applied slowly. The data for PVC illustrates the effect of incomplete particle fusion (Section 6.4.2) in thick-walled pipe.

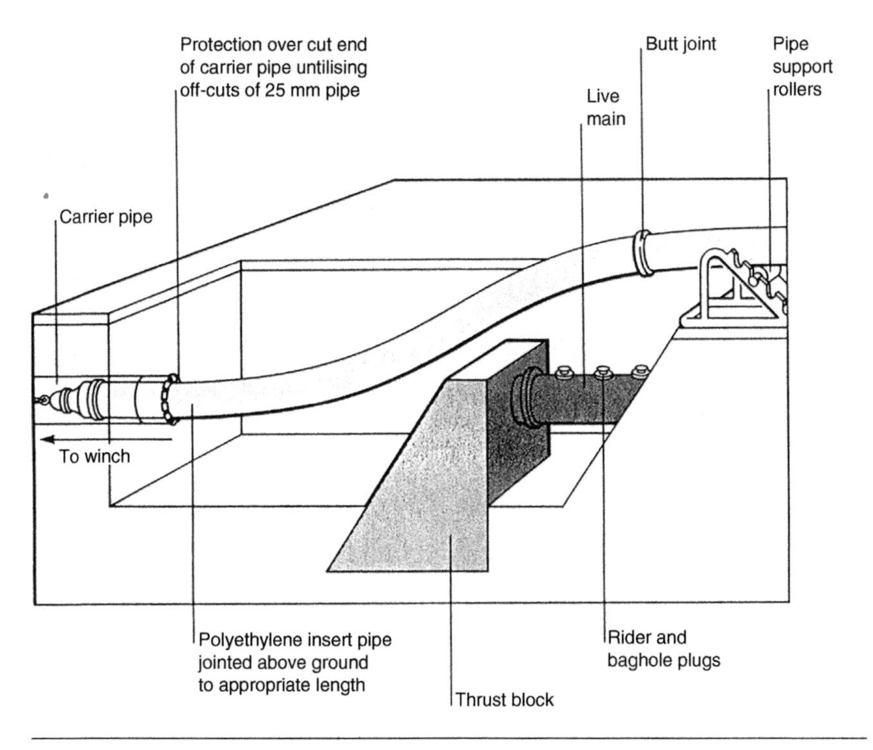

Figure 14.4 Insertion of a length of MDPE pipe into an old cast iron main (from the *Wavin Gas Handbook*, courtesy of Wavin Industrial products Ltd.).

Problem 1 Use the data in Table 14.2 to rank the plastics in terms of the transition thickness t_c for plane strain fracture (Eq. 9.21), and comment on why polyethylene is preferred to the other polymers.

Grades of PE

The grade of PE is determined mainly by its performance in pipe creep rupture tests, data not available in polymer selection packages. The tests can be carried out at 80 °C to accelerate ductile failures and possibly to induce slow crack growth (Chapter 9). For HDPE, lowering the MFI (increasing the molecular weight) increases the creep rupture strength at long times. It improves the resistance to ESC, and delays the onset of the brittle fracture by slow crack growth.

Grades of PE have been specially developed for the gas pressure pipe market. First-generation HDPEs were not used in the UK. The second generation of MDPE copolymers has superior creep rupture resistance at the pipe design lifetime of 50 years (Fig. 14.5). The data falls on one or more lines; lines of shallow slope are associated with ductile failure and lines of steeper slope with 'brittle' *plane strain* crack growth (Fig. 14.3). The

Table 14.2 Fracture toughness and yield stress of plastics

Plastic	*Density* ($kg\,m^{-3}$)	*MFI* ($g\,10\,min^{-1}$)	K_{IC} *(at °C)* $MN\,m^{-1.5}$	σ_y *(MPa)*
Polyethylene	940	0.2	3.1 (−35)	22
Polyethylene	933	2	5.0 (−60)	15.8
Polyethylene	930	8	3.0 (−60)	16.4
Polyethylene	929	16	2.1 (−60)	14.5
Polyethylene	916	18	1.1 (−60)	9.4
PP copolymer		4	3.5 (−60)	26.4
PVC well processed		$K = 68$	4.0 (−60)	57.0
PVC poorly processed		$K = 68$	2.7 (−60)	57.0

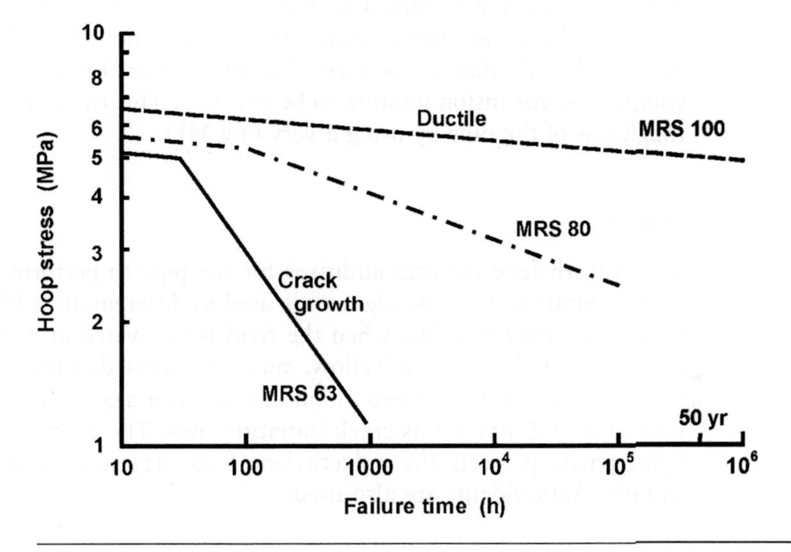

Figure 14.5 Creep rupture data at 80 °C for three generations of PE pipe (from Bocker, H. *et al.*, *Kunststoffe*, **82**, 739, 1992).

third-generation copolymers, introduced in the late 1980s, do not exhibit such brittle behaviour. In these materials, side chains produced by copolymerisation are grouped together rather than being at random along the chain. A greater percentage of comonomer can be used without reducing the crystallinity too much (Fig. 2.11 shows the effect for HDPE copolymers). This increases the number of inter-lamellar links, reinforcing the amorphous phase and suppressing the brittle fracture mode.

Links between the material requirements and the polymer microstructure are shown in Fig. 14.6. The fracture toughness depends on a number of

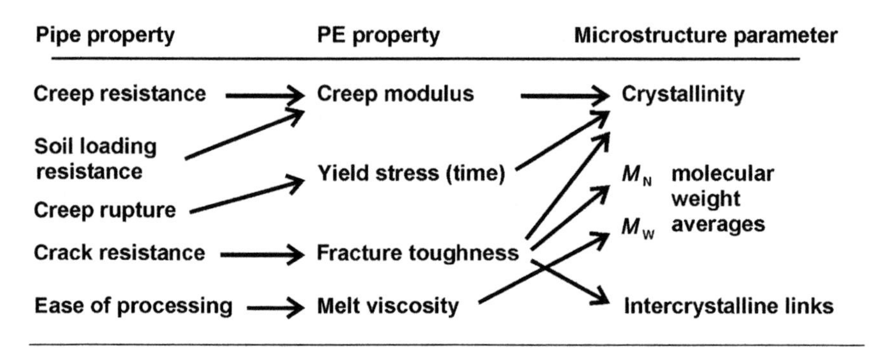

Figure 14.6 Links between pipe properties, polyethylene properties and microstructural parameters.

microstructural parameters, whereas the Young's modulus and initial yield stress are largely determined by the crystallinity. Since connecting sockets (Section 14.2.6) are manufactured by injection moulding, there is a lower limit on the MFI that can be used. The MFIs of the pipe and sockets must be comparable for fusion jointing to be effective. This rules out increasing the toughness of the pipe by using a very low MFI.

Additives

The polyethylene requires additives for the pipe to perform effectively. In many countries, a yellow pigment is used to differentiate a blue water pipe from a red electric cable, when the road is excavated at a later date. The pigment, usually cadmium yellow, must be dispersed in the polyethylene so that no agglomerates exceed 10 Ìm in size. Such agglomerates can fracture internally and may act as crack initiation sites. The pigment, by absorbing light, partly protects the polyethylene from degradation during outdoor storage. Antioxidants are also used.

Gas diffusion losses

The constituents of natural gas should not have deleterious effects on the pipe properties, nor should the gas diffuse through the pipe wall at an excessive rate. Methane and the other hydrocarbons in natural gas, diffuse through polyethylene at a very low rate, that neither causes economic loss, nor dangerous build-up of gas on the outside of the pipe. The maximum daily loss from 1 km of pipe, of 90 mm outer diameter and wall thickness 8.5 mm, pressurised to 1 bar pressure (the partial pressure of methane is 2 bar inside and 0 bar outside), is 4.4 l. This is less than 1% of the losses that occur with jointed cast iron pipe, and can be compared with a daily throughput of 2×10^5 l of gas for a pressure drop of 1 mbar km^{-1}.

14.2.4 Determining the pipe wall thickness

When the gas distribution network was planned, the number of pipe diameters was rationalised. For each pipe diameter, joints and couplings need to be made. Ideally, the pipe diameter should suit the planned flow, but a large number of pipe diameters increases the costs of manufacture and of stocks. Typical pipe dimensions are given in Table 14.3.

The standard dimension ratio (SDR) is defined as

$$\text{SDR} = \frac{\text{Minimum outside diameter}}{\text{Minimum wall thickness}} \tag{14.1}$$

The pipe is clearly identified at 1 m intervals with the manufacturer's identity, class of polyethylene (e.g. PE 80), external diameter (e.g. 90 mm), SDR and information on the date of manufacture. If faults are found with a particular pipe, the labelling system makes it possible to identify all the rest of the pipe from that particular batch, and to find out the polymer used, and the processing conditions.

The main consideration that determines the pipe wall thickness is the avoidance of creep rupture. Gas pipes are usually designed for a 50 year life. The hoop stress in the pipe is given by Eq. (C.17) of Appendix C. Since the outer diameter is the mean diameter plus the wall thickness, the hoop stress can be expressed in terms of SDR

$$\sigma_H = \frac{p}{2}(\text{SDR} - 1) \tag{14.2}$$

Since experimental creep rupture times rarely exceed 10^4 h, it is necessary to extrapolate the data, using a straight line extension of the ductile rupture line on the log–log graph. The British Gas Specification for polyethylene pipe required the 50 year creep rupture stress $\sigma_{50} > 10$ MPa. The International Standard ISO 9080 classifies polyethylene as PE80 if the lower confidence limit of the 50 year creep rupture strength lies between 8.0 and 9.9 MPa, and as PE100 if it lies between 10.0 and 11.9 MPa.

Table 14.3 PE gas pipe sizes (Wavin website, 2004)

			Diameters (mm)	
Polymer classification	*SDR*	*Pressure (bar)*	*Coil 50 or 100 m*	*Straight 6 or 12 m*
PE100	11	7	90	125, 180, 250,
PE100	17.6	4	90, 125, 180	315, 355, 400
PE80	11	5.5	20, 25, 32, 50, 63, 90, 125, 180	450, 500
PE80	17.6	3	90, 125, 180	

A design safety factor S allows for unexpected variations in the pipe properties or dimensions, or defects caused during installation. The design hoop stress σ_{DH} is given by

$$\sigma_{DH} = \frac{\sigma_{50}}{S} \tag{14.3}$$

For gas pipe of SDR = 23, made of a material with $\sigma_{50} = 10$ MPa, used at an internal pressure of 4 bar, Eqs (14.2) and (14.3) show that the safety factor $S = 2.3$.

Fig. 14.2 shows creep rupture data for a particular MDPE, for ductile failure at a range of temperatures. An Arrhenius plot, of the logarithm of the creep rupture time versus the reciprocal of the absolute test temperature, is usually a straight line graph. This can be used to estimate the creep rupture times at lower temperatures. If there is ductile failure in the higher temperature tests, it is unlikely that brittle failure will occur at long times at low temperatures.

As experience with creep rupture testing of polyolefins has been gained, elevated temperature tests have been used for quality control purposes, and standards set using such tests, i.e. the creep rupture time for pipes for natural gas distribution must exceed 170 h at 80 °C and a hoop stress of 3 MPa. Care must, however, be exercised if a polyethylene made by a different process is introduced, because the use temperature is close to 10 °C when the pipe is buried in the ground; the slope of the Arrhenius plot varies between different polyethylenes.

Problem 2 If PE100 is used for gas pipe, calculate the maximum SDR value of a pipe for 4 bar gas pressure. Allow a safety factor of 2.

Creep strains

The stresses in a pressurised pipe, free to expand in length, were analysed in Section C.3 of Appendix C. However, buried gas pipe is connected at both ends to immovable objects such as houses, and the surrounding soil prevents the pipe moving laterally. Consequently, the longitudinal strain in the pipe is zero. In this 'elastic plane strain' situation, the hoop strain is given by Eq. (C.21) as

$$\varepsilon_H = \frac{\sigma_H}{E^*} \tag{14.4}$$

where $E^* \equiv E/(1-\nu^2)$. Since Poisson's ratio ν is less than 0.4, the value of E^* is at most 16% smaller than E. The requirement, that the hoop strain should not exceed 3% after 50 years in use, relates to the need to butt weld new pipe to older material. There should not be a mismatch in dimensions. Although the polyethylene pipe begins to recover its original dimensions once the gas pressure is removed, this process is very slow.

Problem 3 An HDPE pipe, for which the tensile creep data is given in Fig. 7.6, must have a hoop creep strain less than 3% after 50 years. If the gas pressure is 4 bar, calculate the maximum SDR that can be used.

Soil loads on buried pipe

When a gas pipe is installed in a trench, the trench back filled and the road surface replaced, the soil exerts forces on the pipe. The exact magnitude of these forces depends on the degree of support of the back-fill by the trench walls, and on the relative stiffness of the pipe and the soil. An order of magnitude calculation is used to ensure that the pipe will not significantly distort in shape before the gas pressure is applied. Figure 14.7 shows two approximations to the loads experienced. In Figure 14.7a, the entire weight of the back-fill acts as a concentrated diametral load on a pipe that is free to expand laterally. Although easy to analyse, this is more severe than the real situation. A depth h of back-fill, of density ρ, exerts a force q per unit length of pipe, where

$$q = Dh\rho g$$

and g is the acceleration of gravity $= 9.8\,\mathrm{ms}^{-2}$. In Fig. 14.7b, the sides of the pipe are constrained from expanding sideways by the smooth vertical walls. This is the test situation specified for the BGC test for resistance to external loads. In reality, within about 1 year, the soil loads on the pipe surface become a uniformly distributed pressure, and creep ceases. The gas pressure also resists the deformation caused by the soil loads.

The pipe wall in Fig. 14.7a is a *statically indeterminate* structure; the bending moment depends both on the external loads and the wall bending stiffness. A distributed load $q\,\mathrm{N\,m}^{-1}$ acts at the top and bottom of the pipe of diameter D. If the pipe were cut in half horizontally and the cut ends supported on a frictionless surface (Fig. 14.7c), the bending moment m_C would be related to the angular distance θ by

$$m_C = \frac{qR}{2}(1 - \cos\theta) \tag{14.5}$$

To keep the pipe walls vertical at the cut positions, a bending moment m_0 must be applied. Its value, obtained by analysing how the stored elastic energy in the pipe wall varies with rotation of the cut points, is

$$m_0 = -qR\left(\frac{1}{2} - \frac{1}{\pi}\right)$$

In the intact pipe, the total bending moment (per unit length) $M = m_0 + m_C$; the maximum value occurs at B

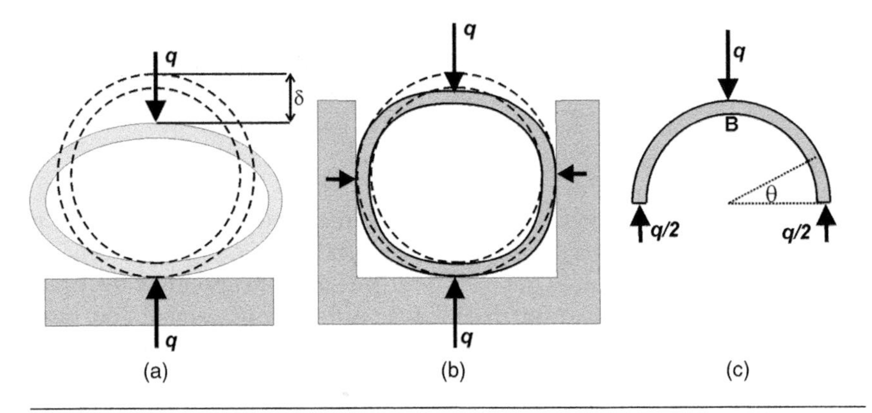

Figure 14.7 Diametral compression test on pipe: (a) Without side support; (b) smooth rigid side wall support.; (c) half pipe under three point bending.

$$M_{\max} = 0.159\ qD \tag{14.6}$$

Consequently, the elastic bending stresses reach a maximum value at the inner surface of

$$\sigma_{\max} = \frac{6M_{\max}}{t^2} = 0.95\frac{qD}{t^2} \tag{14.7}$$

where t is the pipe wall thickness. The wall curvature is calculated from the variation of M with angular position; integration gives the elastic deflection δ (the change in vertical diameter) as

$$\delta = 0.0186q\frac{D_{\mathrm{m}}^3}{EI}$$

This equation is modified for viscoelastic materials, by the methods of Chapter 7, to give the time dependent deflection $\delta(t)$. After substituting for I using Eq. (C.9), the result is

$$\delta(t) = 0.223\,q\left(\frac{D}{t}\right)^3 J(\sigma_{\mathrm{m}}, t) \tag{14.8}$$

where J is the creep compliance for the maximum stress.

Problem 4 An empty pipe in a trench is covered with a depth of 1 m of soil of density 2000 kg m^{-3}. The soil load can be considered as a concentrated force as in Fig. 14.7a. What is the maximum allowable pipe SDR if the vertical deflection after 1 h is not to exceed 10% of the pipe diameter? Use polyethylene creep data from Fig. 7.6.

Hint: Start by assuming SDR = 20. Calculate the maximum stress using Eq. (14.7); then find the creep compliance for this stress. If the deflection predicted by Eq. (14.8) is too big, revise the value of the SDR and repeat the calculation.

Such calculations show that it is unrealistic to use high strength polymers, such as biaxially oriented PET (Section 4.5.3) for gas pipe; the thin pipe walls would have insufficient resistance to soil loads in the period before the gas main is pressurised.

Fracture mechanics of the pipeline

There is a potential risk that an accidental breach of the pipe, caused by careless excavation, could propagate at high speed, down the length of the welded network. Such fast crack growth allows no time for gas flow along the pipe, hence pressure reduction. In Section 8.5.2, stress analysis of the growth of parallel cracks gave an expression for the variation of the stored elastic energy W with the crack area A. This analysis is reused, with the crack separation S replaced by the pipe circumference πD_m. Equation (8.20) becomes

$$\frac{\partial W}{\partial A} = -\frac{\pi D_m}{2E}\sigma_H^2 \tag{14.9}$$

The hoop stress is given by Eq. (C.22). In order to express this result in terms of the stress intensity factor, we need a general relationship from fracture mechanics

$$K_I^2 = -E\frac{\partial W}{\partial A} \tag{14.10}$$

derived by considering the energy release when a crack grows by a small amount. Hence, the stress intensity of the cracked pipe is

$$K_1 = \sigma_H\sqrt{\frac{\pi D_m}{2}} \tag{14.11}$$

The design hoop stress (Problems 2 and 3) is independent of the pipe diameter, so large diameter pipes are potentially more at risk from high-speed fracture. However, by using MDPEs with a low melt flow index and a K_{IC} value of $6.0\,\text{MN}\,\text{m}^{-1.5}$, and keeping the hoop stress low, the risk of high-speed brittle crack propagation is taken care of. The fracture toughness K_{IC} must be measured at the same temperature and strain rate that the pipe experiences. The data in Table 14.2 is for slow loading in polyethylene, cooled well below room temperature to ensure plane strain fracture. Experiments, in which loads were applied in the order of 1 ms to polyethylene at room temperature, show that K_{IC} appears not to change with temperature. For the latest grades of medium-density polyethylene, crack tip plasticity spreads to the far boundary of the pipe wall before the crack propagates. Consequently, to determine whether fracture occurs, it is necessary to calculate the pressure necessary for through-section yielding. The presence of a small crack no longer significantly reduces the failure pressure from that for ductile creep rupture.

14.2.5 Summary of the design requirements

When the design calculations are completed, the critical (smallest) safety factor can be found, and an upper limit on the design SDR set. In the calculations, the gas pressure was assumed to be 4 bar.

Design criterion	*Conclusions*
Avoid ductile creep rupture at 20 °C with safety factor $S = 2$	$\sigma_H < 10$ MPa at 50 years, so SDR < 26
50 year hoop strain < 3%	SDR < 21
Soil loading of 20D kN m^{-1} causes diametral deflection <10% at 1 h	SDR < 20
Avoid slow crack growth	Select a suitable MDPE of low MFI (chapter 8)
Avoid runaway crack growth	Use a super tough grade of MDPE

Therefore, the SDR should be less than 20 to ensure safety against all the failure modes listed, and the polymer grade should have sufficient fracture toughness. Table 14.3 shows that an SDR of 17.6 is used with PE100, but SDR11 pipe is used for diameters <90 mm, probably to provide extra durability against accidental damage. The initial British Gas design used high safety factors, because of uncertainties in the size of flaws in fusion joints, in the bending stresses at junctions of service pipes to the main, and the possibility of accidental damage during pipe installation. Since there have been no failures, confidence has grown in the MDPE pipe system. Consequently, increased pressures have been allowed in new pipes. Table 14.3 shows that the recommended maximum pressures are now 7 bar for SDR11 pipes. For water, similar SDR11 pipes are used at pressures up to 12.5 bar.

14.2.6 Pipe installation and jointing

Reducing the installation cost

A major part of the total system cost lies in the restoration of road surfaces after trenching. Some ingenious systems minimise the need to cut trenches. The website www.subterra.co.uk describes a range of processes. The slip lining process, shown in Fig. 14.1, inserts a smaller polyethylene pipe inside an old cast iron or steel pipe. However, as a significant clearance is needed to minimise friction, there is a loss in diameter. There is a *roll-down* process in which the welded polyethylene pipe is dragged through two pairs of U-shaped rollers, causing plastic deformation of the pipe wall and reducing its diameter by about 20%. Once the pipe has been inserted in the old gas main, it is expanded to fit by the application of internal water pressure. The initial roll-down makes this later expansion uniform, rather than the non-uniform expansion that occurs in a creep rupture test.

Types of joints

A gas distribution system can be assembled using compression fittings, similar to those used on domestic water pipes, for pipes of diameter <63 mm. However, fusion joints which cannot leak, are more common. Socket fusion, used to join pipes of sizes up to 125 mm, involves the use of an injection-moulded socket that fits on the pipe ends. A simple socket allows straight continuation of the pipe, whereas 45 and 90° elbows allow sharp corners, and equal tees allow branches to be added. Butt fusion, used to make axial joints on pipes of diameter 63 mm and above, is preferred for diameters >180 mm.

Butt welding

The procedure for the butt fusion of pipes has five stages

(1) The two pipes are clamped into a machine which ensures that they are in axial alignment. A set of rotating blades machine the pipe ends so that they are clean and parallel.
(2) A double-sided flat heating plate at a temperature of 205 ± 8 °C is inserted into the gap between the pipes. A pneumatic RAM pushes the pipes into contact with the heater, with a compressive stress of 0.15 MPa over the end surfaces. This stress is maintained until a 2 mm wide bead of molten polymer forms in contact with the heated plates.
(3) The pressure is reduced to zero and heating is continued for the appropriate time (120 s for a 125 mm diameter pipe).
(4) The carriage is opened, the heating plate removed, then the carriage is rapidly closed to form the weld. The compressive stress of 0.15 MPa is maintained for a cooling time of 10 min (125–180 mm pipe).
(5) After removing the pipe from the machine, the bead is checked for completeness, and that its width falls in the range 7–11 mm.

Butt joints have beads at the inner and outer pipe surfaces. The outer beam may be cut off, but the inner one is difficult to remove. The bead meets the pipe wall at a relatively acute angle, and the resulting stress concentration will be exacerbated if there are differences in the wall thickness of the two pipes to be joined. It was found, by testing butt-jointed pipes at 80 °C, that steps greater than 10% of the pipe wall thickness caused an unacceptable reduction in the creep rupture life. This explains the requirement for wall thickness limits.

Welding conditions

To melt the polyethylene, the temperature must exceed 135 °C. The hotplate temperature should exceed 170 °C, to avoid the risk of the polyethylene cooling and crystallising in the time interval before the joint is made. However it should not exceed 270 °C, to avoid rapid degradation of the polyethylene. Research showed that the optimum hotplate temperature was 205 °C. Figure 14.8 shows how the temperature profile in the pipe changes with time, when the contact pressure is zero. It takes about 2 min to produce a 3 mm thick molten layer.

When the pipe ends are clamped together, the pressure causes the molten layers of high viscosity, rubbery liquid to flow towards the pipe surfaces. If the pressure is low, the hardly distorted molten layer crystallises in spherulitic form, producing a weak weld interface. If it is too high, the pressure squeezes all the melt to the sides, and the unmelted regions come into contact. The excessive crystal orientation in the direction of the flow makes the joint weak. The strongest joints, for pressures of about 0.15 MPa, have microstructures between these two extremes (Fig. 14.9).

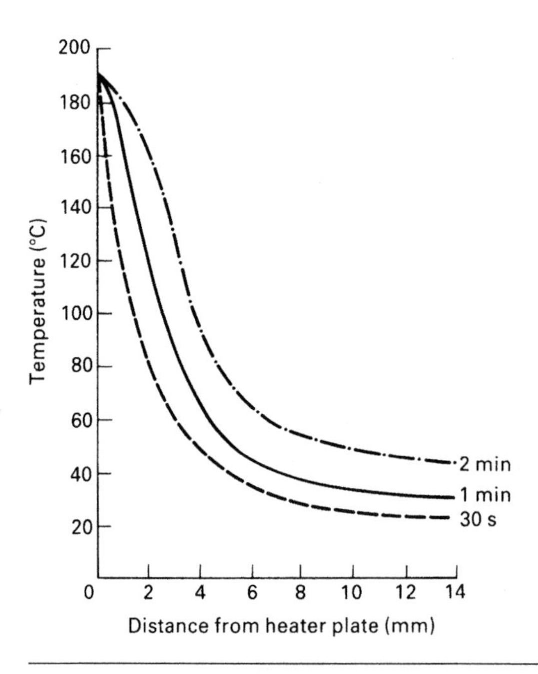

Figure 14.8 Temperature profile in MDPE after contact with a heater plate at 210 °C for various times.

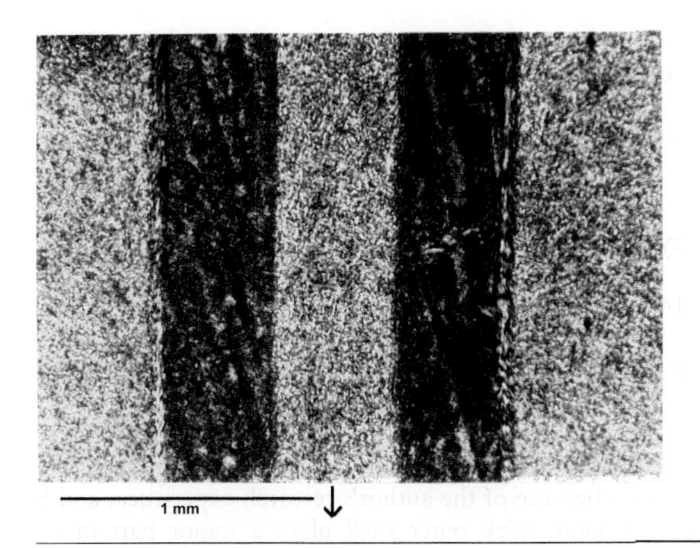

Figure 14.9 Polarised light micrograph of a butt weld in a polyethylene pipe. The arrow shows the melt flow direction during fusion.

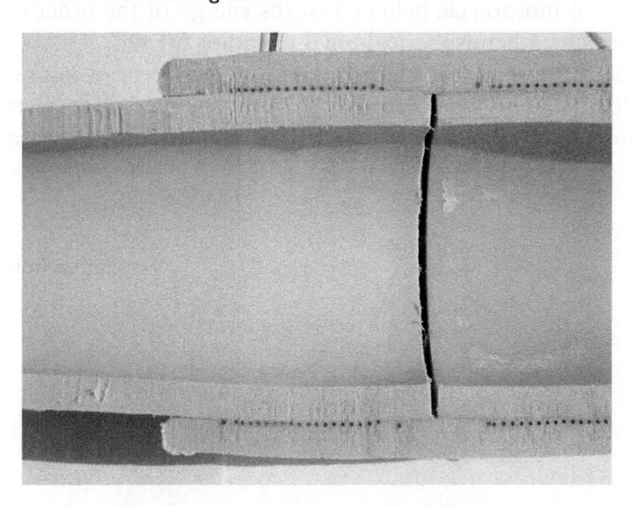

Figure 14.10 Section of an electrofusion socket joint. The embedded copper wires have been used to melt the polyethylene surfaces.

Electrofusion

Electrofusion socket joints have a copper heating coil inserted in the injection moulding (Fig. 14.10). The pipe ends and socket are slid into place, and a low voltage, high current unit is used to heat in the wire. The typical heating time of 2 min only melts a few millimetres thick region surrounding the wires. The melt expands and generates an interface pressure of about 0.6 MPa for a further 2 or 3 min. The remainder of the socket cools the weld, and the coil remains in place after welding. When tensile tests are performed on the pipe, the welded joint is loaded in shear. Examination of the failure surface shows ductile failure of the polyethylene in the regions between the copper wires.

14.3 Bicycle helmets

14.3.1 Introduction

Foamed plastics are used in helmets because of their very low density, their ability to crush and absorb energy and the possibility of economical mass production. There are many other applications of polymer foams for injury prevention, for instance the fascia padding in cars. Bicycle helmets are chosen because of the author's research experience, and because the typically 0.3 mm thick outer shell plays a minor part in energy dissipation. Consequently, the design process concentrates on the foam liner (Fig. 14.11). In contrast, the deformation of the 4–5 mm thick thermoplastic shell of a motorcycle helmet absorbs energy of the order of 40 J, when the helmet hits a hemispherical anvil of radius 50 mm.

The purpose of bicycle helmets is to reduce head injuries and deaths in 'accidents'. The word 'accident' is a misnomer as there may be culpable parties, whose behaviour could have been modified by training or by the

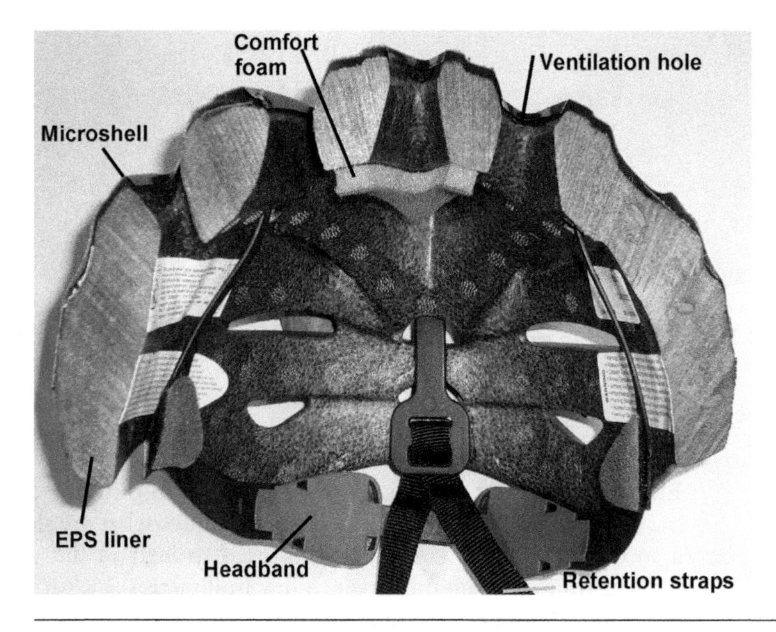

Figure 14.11 Components of a bicycle helmet, seen in cross section.

enforcement of traffic laws. The psychological factor of perceived risk may cause the user of a safety product, be it a seat belt or a protective helmet, to behave in a more reckless way than without the product. He may increase his speed until his perceived risk is the same as before. The social benefits of the product, reducing the cost of hospital treatment and supporting disabled citizens, have been evaluated by epidemiological studies. These showed that the wearing of bicycle helmets decreased the risk of serious injuries and deaths by approximately 60% (Thompson *et al.*, 2003). The compulsory wearing of bicycle helmets in Australia after July 1990, caused a significant reduction in head injuries, some of which can be attributed to a reduction in the number of cycles on the roads.

14.3.2 Biomechanics criteria for head injuries

The causes of head injuries can be classified into three types

(a) *Skull fractures:* Localised high pressures on the skull cause excessive bending stresses in the 'sandwich' structure of the skull. The penetration of convex objects into the helmet foam increases the contact area on, and spreads out the forces applied to, the skull. Skull fracture is a natural mechanism for absorbing impact energy. Some minor skull fractures do not cause brain injuries.
(b) *Linear acceleration of the brain:* In current UK bicycle and motorcycle helmet standards, the peak linear acceleration of a rigid head-form must not exceed 250 and 300 'g', respectively. The foam liner of the helmet, by crushing provides a stopping distance for the head. This reduces the peak linear acceleration and extends the time duration of the acceleration pulse. The standards assume that the injury severity correlates with peak head linear acceleration. A direct blow to the skull can cause brain swelling and bleeding (haematoma) below the impact point (a coup injury, on the opposite side of the skull; a contra-coup injury), or distributed in various parts of the brain. Experimental evidence shows that blows to the sides of the head can cause more severe brain injuries than frontal blows that cause the same acceleration levels.
(c) *Rotational acceleration of the brain:* When the heads of animals were subjected to high levels of rotational acceleration, while the skull remained undeformed, it was possible to produce concussion or permanent brain damage of a diffuse nature. However, when accident victims suffer diffuse brain damage, there is always evidence of a direct blow to the head. There are no tests for rotational acceleration in helmet standards. However, bicycle helmets reduce the peak rotational acceleration of a head-form in oblique impact tests; the peak forces (both tangential and towards the head centre) are reduced by foam crushing.

We will examine the design and materials selection for reducing linear acceleration of the brain, under the constraints of mass, size and cost.

Helmets cannot prevent all head injuries, and the aim is to minimise the social costs of injuries to the population of road users with a wearable product.

14.3.3 Geometry of the helmet/impacted object interface

The micro-shell of bicycle helmets causes negligible load spreading, so it is ignored. Initially, the foam is assumed to contain no ventilation holes; these are allowed for later. Although neither the human skull nor the outer surface of a helmet is exactly spherical, it is a reasonable approximation that the impact site is locally spherical. Both the skull and the road surface are treated as being rigid. The approximately 5 mm thick scalp is so soft and deformable that it plays little part in the impact energy absorption.

The contact geometry between a flat rigid surface (the road) and a helmet of outer radius R is shown in Fig. 14.12. Zero load spreading is assumed (Mills, 1990), which means that the boundary between the crushed and uncrushed foam is vertical. The linear crush distance x is much less than R (100–200 mm), since the linear thickness $T < 30$ mm. The foam crushes over a disc of radius a. Applying Pythagoras's theorem to the triangle gives

$$R^2 = (R - x)^2 + a^2$$

If the x^2 term is ignored in the expansion of the brackets, the contact area A is

$$A = \pi a^2 = 2\pi R x \tag{14.12}$$

Assuming that the foam has a constant yield stress σ_y while the strain is increasing, the force F transmitted by the foam is

$$F = A\sigma_y = 2\pi R \sigma_y x \tag{14.13}$$

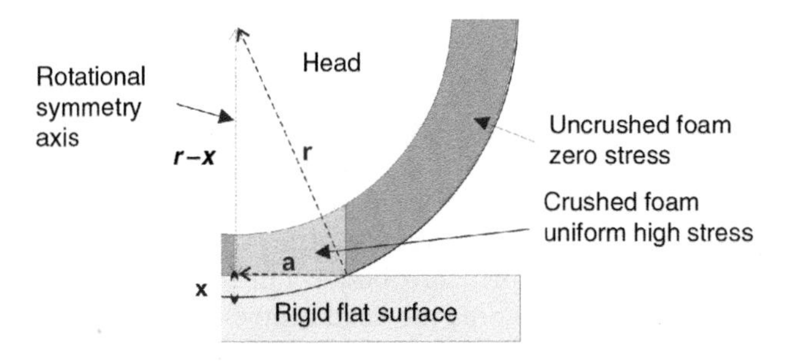

Figure 14.12 Geometry of head and helmet foam crushing, assuming zero load spreading.

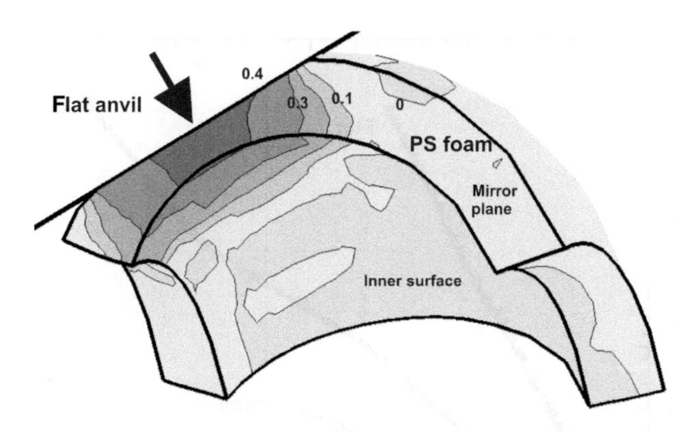

Figure 14.13 FEA of flat impact on a cycle helmet impacting a fiat rigid surface. Contours of the principal compressive stress (mPa) are shown in the rear half of the helmet.

This straight line relationship has a *loading slope k*. Substituting typical values of $R = 140\,\text{mm}$ for the front of a helmet liner, and $\sigma_y = 0.7\,\text{MPa}$, gives $k \cong 600\,\text{N}\,\text{mm}^{-1}$. Finite element analysis (FEA), using the measured stress–strain properties of the foam, predicts a linear force–distance relationship with a slope about 5% higher than that predicted by Eq. (14.13). The predicted pressure distribution (Fig. 14.13) is nearly constant across the contact area. The radius of curvature R of helmets varies from about 100 mm at the front to about 170 mm at the sides. Consequently, the loading constant predicted by Eq. (14.13), changes with the impact site, if the foam has a constant density.

For more complex impact geometries, such as on to a kerbstone, FEA is essential to predict the foam deformation geometry and the loading force versus deflection. Figure 14.14 shows that the loading slope is a function of the object hit. Consequently, helmet design (next section) will depend on the object hit. A design that is optimum for impacts on a flat road surface will be sub-optimum for an impact on a kerbstone. The loading slope for a 50 mm radius hemisphere is lower than that for other surfaces; such anvils are used in motorcycle helmet testing, but objects of this shape are rarely hit by cyclists. Any design that passes a hemispherical anvil test is likely to contain foam with a yield stress that is too high for impacts on a flat surface.

14.3.4 Design of a helmet liner for a particular impact velocity

Section 4.6.2 described the gas pressure hardening that occurs when closed-cell foams are compressed. This hardening seems to have little effect on the linearity of the loading response. However, the behaviour described by Eq. (14.8) ceases when the foam 'bottoms out'; when the compressive strain approaches $1 - R$ (R is the foam relative density), many cell faces touch and the compressive stress rises rapidly. As the foams have $R \cong 0.05$, when the

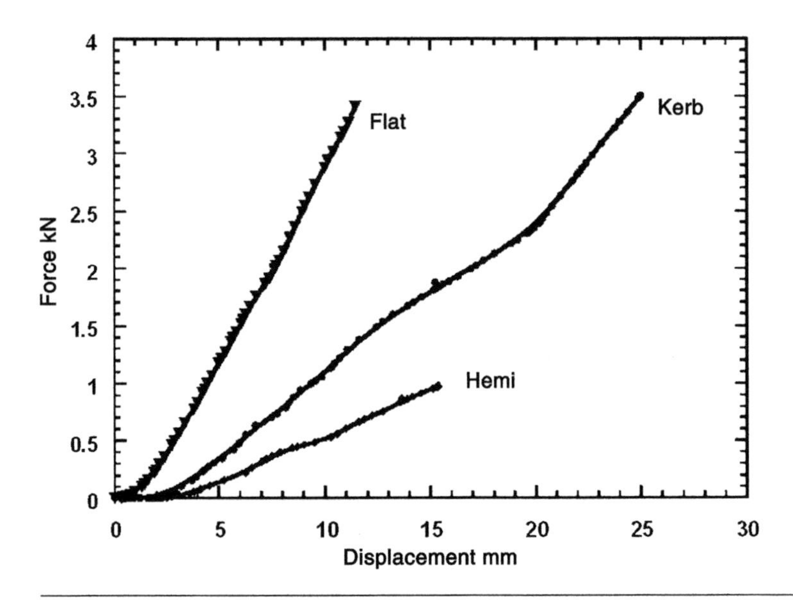

Figure 14.14 Force–deflection curves predicted for a bicycle helmet impacting flat, kerbstone and hemispherical anvils.

foam liner is compressed by more than 80% of its thickness, marked non-linearity occurs in the force–deflection relation.

Standards specify the impact velocity as the helmet and head-form strike a rigid fixed anvil. In BSEN 1078 this is $5.4\,\mathrm{m\,s^{-1}}$ for the flat anvil, corresponding to a free fall from 1.5 m, the typical height of the head above the road while riding. The mass of a typical helmet (0.25 kg) is much less than that of the head-form (4–6 kg depending on size). For an average-size head, the total kinetic energy is approximately 75 J. In BSEN 1078, the peak head-form acceleration must be $<250\,\mathrm{g}$. Helmet designers allow a margin for material variability and tests at high and low temperatures, so that the target maximum acceleration is 200 g at 20 °C. This is equivalent to the peak impact force on a 5 kg head-form $<10\,\mathrm{kN}$.

In Fig. 14.15, the energy under the force–deflection curve is equal to the energy input E. Calculations will be made for $E = 100\,\mathrm{J}$. If the force just reaches 10 kN, when the head-form decelerates to a momentary halt at deflection x_{max}

$$0.5 \times 10\,\mathrm{kN} \times x_{max}\,\mathrm{mm} = 100\,\mathrm{J} \tag{14.14}$$

To avoid bottoming out, the foam thickness T must exceed 1.25 x_{max}.

Problem 5 Determine the minimum thickness of foam in a bicycle helmet to keep the head acceleration below 200 g for an impact on a flat surface with 100 J kinetic energy, at a site where the radius of curvature is 100 mm.

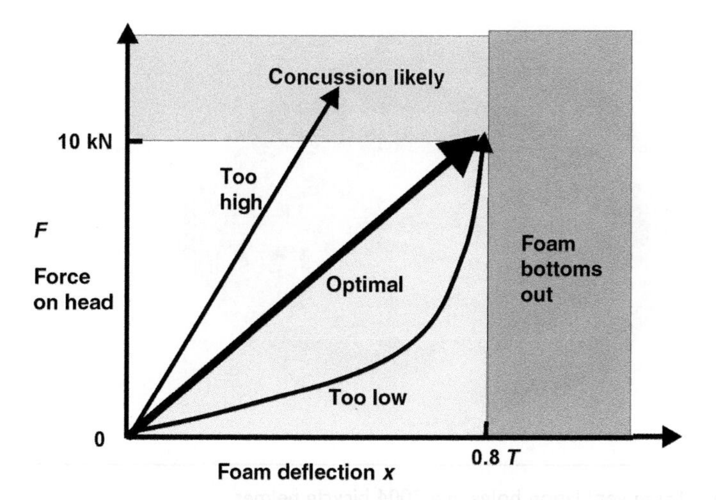

Figure 14.15 Force–deflection graph for a bicycle helmet impact—the impact energy (area under the loading curve) must be absorbed without exceeding the injury force limit, or the foam bottoming out.

What is the optimum density of expanded polystyrene beads (EPS) for the helmet?

There is no optimum design, since the shape of the surface struck is unpredictable. A foam that is ideal for an impact with a flat surface has a too low yield stress to be ideal for an impact on a kerbstone. Designers must also consider ventilation, and compensate for the lower radius of curvature R at the front and rear of the helmet. Ventilation holes reduce the contact area of the foam on the head, so the foam density must be increased to compensate. Ventilation holes are particularly important at the front and rear, to achieve the required air flow. However, since accident surveys show the front and sides of helmets are the most frequently impacted sites, it is important that there is a good thickness of foam in these regions.

As it is difficult to vary the foam density in a moulding, Eq. (14.10) predicts that the loading stiffness is lower at the front and rear of the helmet where R is the lowest. Consequently, the foam must be made thicker at these sites, to prevent bottoming out before the design impact energy is absorbed. Figure 14.16 shows the size of ventilation holes in modern helmets, and the thicker foam at the front and rear.

As the impact velocity increases, the thickness of the foam must increase with the square of the velocity, and the yield stress of the foam must decrease in proportion to keep the head acceleration below 200 g. The thickness of the foam is limited by the mass of helmet that is comfortable to wear, the necessity not to restrict the field of vision of the wearer, and increased aerodynamic drag with increasing helmet size.

Figure 14.16 Large ventilation holes in a 2004 bicycle helmet.

14.3.5 Choice of foam

The helmet liner should have minimum mass and a low production cost. There are several processes for moulding helmet liners. A self-foaming rigid polyurethane can be poured into a mould. However, this low capital cost process is slow, and the foam density is double that of polystyrene foam of the same yield stress. The main process used is the fusion of EPS or expanded polypropylene beads (EPP), using pressurised steam in a moulding process related injection moulding. EPS is used for cycle helmets, as EPP liners cost several times as much. However, EPP is less brittle and recovers better after an impact, a consideration for skate-boarding helmets which may suffer a large number of minor impacts. EPS only recovers by a small amount after an impact (Chapter 8), so the helmets should be destroyed after a crash. It has limited surface durability, but the thermoplastic micro-shell prevents abrasion of the liner exterior. EPS is brittle when bent excessively, or indented locally, but the high tensile strength micro-shell protects it against crack initiation.

Equation (7.24) indicates that, if the foam is required to have a certain compressive yield stress, yet minimum density, the yield stress of the polymer in the bulk state must be high. Polystyrene has a yield stress at high strain rates of $\sim$120 MPa whereas polypropylene has a yield stress of $\sim$60 MPa. Consequently, an EPS helmet will have a lower density than an EPP helmet designed to meet the same impact tests. Therefore, EPS is optimal for helmets that offer single-impact protection.

The optimum foam is one where the loading line meets the intersection of the 'dangerous' areas for head injury and foam bottoming out (Fig. 14.15). Lower yield stress foams bottom out before the load reaches 10 kN. If the foam has a too high yield stress, the force reaches 10 kN before the foam bottoms out.

Problem 6 Discuss whether EPS or EPP foam is preferred for the liner of a sports helmet.

Ventilation of helmets has become a selling point. Helmets are advertised on the number or size of the ventilation openings. Such openings mean that the foam density must be increased, to increase the compressive yield stress of the remaining material. Aerodynamics has been applied to increase the air flow through such ventilation holes. However, given the lack of research on heat transfer from the head, the benefit of some of the styling features has not been established

14.3.6 Summary

Bicycle helmets can be designed for impacts of up to 100 J kinetic energy, their mass can be as low as 200 g and they are comfortable to wear. Helmets designed for much higher impact energy levels would be unacceptably large, so it is impossible to protect riders from the most extreme impacts. Compromises are necessary when designing to protect people, who have variable (and unknown) tolerances to impact acceleration, from crashes with variable circumstances.

14.4 Data storage on polycarbonate discs

14.4.1 Information storage on plastic discs

Compact discs for music are one of a family of digital data storage products, which includes CD-ROMs for computer data storage and DVDs. The rewritable versions of these products use another technology for data storage. In this case study, we will concentrate on polymer selection, the required features of the moulding process, and the optical properties of the disc. Figure 14.17 shows the 'light pen' that reads the information stored in the pits in the CD surface.

The introduction of CDs onto the markets in 1982 required the development of

(a) solid-state lasers, that are small and light enough to fit into the moving pick-up head of the CD player. A focussed laser provides the required, high-intensity, polarised light beam. The Al Ga As laser, of 1 mW power, operates at a wavelength of 780 nm. Layers of this material are grown epitaxially on Ga As crystal substrates, then doped to be n- or p-type. The laser light emerges from a thin layer in the centre of a Ga As sandwich, in plane-polarised form. The laser, less than 1 mm long, forms a negligible part of the total mass of the light pen;
(b) servomechanisms that can keep the light pen focussed on the information track, as it moves past at 1.25 m s^{-1}. The vertical position of the light pen

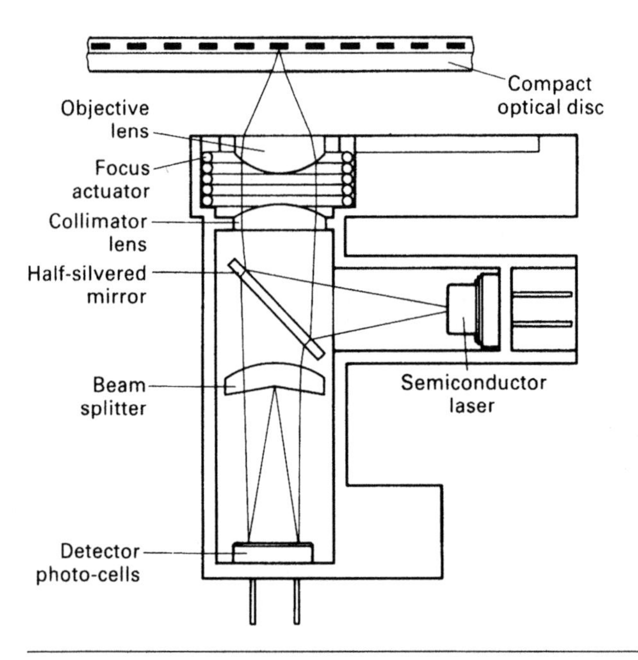

Figure 14.17 Light from a solid-state laser is focussed onto the series of pits in the compact disc. The diffracted light is detected by a series of photocells.

must be kept within ±2 μm so that the track is in focus, and the radial position within 0.2 μm of the track centre. Focussing is required over a range of 1 mm because the disc is not flat, while the radial position of the track can vary by 300 μm as the disc rotates. The focus response time is measured in ms; the resonant frequency of the suspension is 45 Hz in the focussing direction and 900 Hz in the radial direction;

(c) digital signal processing, to reduce noise in the replayed music. Error correcting codes prevent faults in the disc, or scratches or dirt on its surface, from causing the clicks that mar the response of vinyl LPs. The signal is sampled digitally as a 16-bit binary number; the 1 part in 64 500 accuracy allows a 90 dB signal-to-noise ratio, compared with 60 dB for the LP signal and only 30 dB for the channel separation. A computer buffer stores a section of the signal before replay, giving the time base stability of the computer clock. This cures wow and flutter, detectable when sustained notes are reproduced on a turntable with a variable rotation speed. There is feedback control of the CD rotation motor from the detected signal.

14.4.2 Optical design for information storage

There are two channels of analogue signals for music recorded in stereo. The human ear is not sensitive to frequencies above 20 kHz, so each channel is sampled at a frequency of 44.1 kHz. This allows some leeway above the 22 kHz at which aliasing of the signal would occur. With the sound intensity digitised to 16-bit accuracy, there are 1.41 Mbit s^{-1} of information to be recorded. Error correction coding increases the rate by a factor of 1/3. Finally each block of 8 data bits is modulated into 17 channel bits for optical recording. This modulation ensures that the pit lengths vary between 3 and 11 channel bits (0.9 and 3.3 μm) and there are at least 3 channel bits between transients (the ramp at the end of the pits of length <0.3 μm). Reading 4.332 Mbits s^{-1} of information means that the light pen reads bits of length 0.3 μm, moving past at 1.25 m s^{-1}. The SEM image of the surface of an uncoated CD (Fig. 14.18) shows pits of depth approximately 0.12 μm, that are 0.6 μm wide and 1.6 μm apart in the radial direction.

The small size of the pits, compared to the 0.78 μm wavelength of light and 1 μm diameter of the focussed spot used to read the disc, means that the disc diffracts rather than reflects light. Explanations that the light beam is not reflected when it hits a pit, are over simplifications. The regularly spaced pits in the radial direction act as a diffraction grating, therefore colours are observed when white light is reflected from the disc surface.

Signals, from four detectors that scan the two diffraction peaks at 18° on either side of the disc surface normal (Fig. 14.19), are combined to give the following signals

radial tracking error $= D_1 + D_2 - D_3 - D_4$
axial focussing error $= D_1 - D_2 - D_3 + D_4$ maximum when in focus
audio $= D_1 + D_2 + D_3 + D_4$

The pits of the CD, of equal reflectivity to the rest of the disc, form a phase object. Light reflected from a pit of depth $h = 0.12$ μm is advanced in phase by

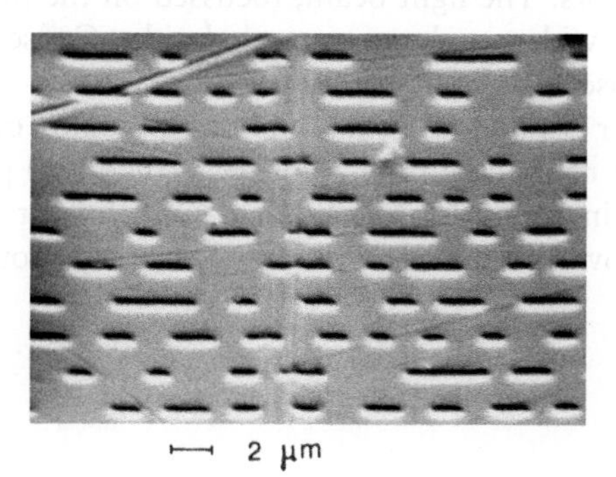

Figure 14.18 SEM micrograph of the surface of a CD showing part of the spiral track.

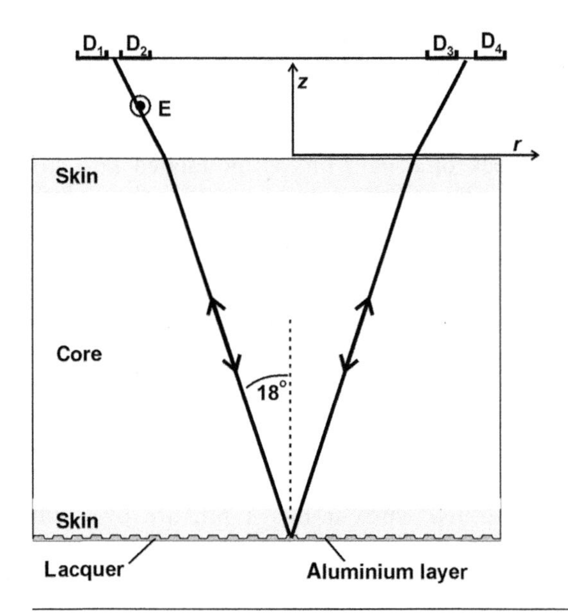

Figure 14.19 A radial section through a CD showing the rows of pits as a diffraction grating, the positioning of the detectors at the first diffraction peak, and the oblique path of the plane-polarised light.

$$\phi = 2\frac{nh}{\lambda} = 0.5 \text{ rad} \tag{14.15}$$

since the polycarbonate refractive index $n = 1.596$. The detection system is related to a phase contrast optical microscope, with the reflected light passing twice through the phase object. A transmitted light disc would require pits twice as deep. Pits of depth-to-width ratio >0.2 are very difficult to mould injection successfully, so a transmitted light disc would have wider pits. The phase shift moves the diffraction pattern laterally, relative to the detectors. The light beam, focussed on the protected side of the disc, is 0.7 mm wide on the unprotected side. Consequently, minor scratches do not cause signal deterioration.

The reflective layer is a thin layer of aluminium, vacuum evaporated onto the disc under clean room conditions (not more than 100 particles of size greater than 0.5 μm in each cubic foot of air). A subsequent lacquer coating is UV cured, then overprinted with the label, further protecting the pits underneath.

14.4.3 DVD

The digital video disc or digital versatile disc (DVD), introduced in the mid 1990s, allows the storage of a complete movie, in compressed form. A single layer, double-sided DVD has a capacity of 9.4 GB, compared to 0.64 GB of a CD, yet has the same diameter and thickness. This is achieved by reducing the pit length unit from 0.8 to 0.4 μm, reducing the track pitch from 1.6 to 0.74 μm, and having two recorded surfaces back-to-back at the mid-plane of the disc. The reading laser is in the red part of the spectrum (650 nm) rather than the infrared. As the pit spacing is reduced, the radial tilt margin is also reduced. Surface dirt and damage, now only 0.6 mm from the recorded surface, are more likely to be in focus than in a CD. However, the error correction system is more powerful, dealing with error bursts up to 6 mm long.

14.4.4 Requirements on the plastic, and its Polymer selection

Table 14.4, which lists some requirements for CDs, allows the elimination of some contending plastics. Polystyrene has too low a resistance to crazing and stress cracking, and the disc birefringence would be too high because the relatively elastic melt has a high stress-optical coefficient (defined in Eq. 9.9). Values for melts differ from values for glassy polymers given in Table 11.5. PVC has too low a heat distortion temperature, and its lack of thermal stability makes the injection moulding of high definition surfaces difficult. Silicate glass cannot be moulded with sufficient surface detail, and is brittle.

This leaves PMMA and PC as the contenders. The disc design is asymmetric, with an impermeable aluminium coating on one side, so dimensional changes caused by water diffusing into the polymer make the disc bow. The water absorption of PMMA at saturation relative humidity is 2.1 wt.%, compared with 0.4% for PC. The diffusion constants at 23 °C for water are 0.5×10^{-6} and $4.8 \times 10^{-6}\,mm^2\,s^{-1}$, respectively. The CD surface warping must be less than 0.6°, for the laser spot to focus properly. This rules out a PMMA disc with one side sealed, because it would expand over a period of tens of days as water diffuses through the 1 mm thick

Table 14.4 Qualitative comparison of glassy materials for compact disc production

Property	*Glass*	*PMMA*	*PC*	*PS*	*PVC*
Heat distortion temperature	5	3	4	3	2
Birefringence	5	4	2	1	3
Toughness	1	3	5	2	4
Solvent resistance	5	3	3	1	3
Processing	1	4	3	4	2

The scale ranges from 5 (excellent) to 1 (poor).

polymer. However DVDs, of symmetrical construction, do not bend as they expand.

PC has the required >90% transmission at a wavelength of 780 nm.

14.4.5 Optimising the processing of polycarbonate

The first process requirement is to reproduce accurately the shape of the mould surface. This requires a polymer melt of low viscosity and a high mould pressure while the surface of the disc is solidifying. The viscosity requirement could be met by using a polymer of sufficiently low molecular weight and processing the melt at a high temperature. However, the impact strength and stress cracking resistance depend on M_N exceeding 14 000 to achieve a strong entanglement network. A low molecular weight could cause these mechanical properties to fall below acceptable values. A regular polycarbonate could not be used for CDs as the melt viscosity is too high at the 340 °C processing temperature. The melt flow index of the Makrolon CD2005 grade used for CDs was 63 g 10 min^{-1} at 300 °C using a 1.2 kg mass, and 73 g 10 min^{-1} for the DP1–1265 grade used for DVDs. The Izod impact strength is 55 and 10 kJ m^{-2} for the two grades, showing the effect of reduced molecular weight on impact strength. An injection-moulding grade for automotive headlights has an MFI of 19 g 10 min^{-1} and an Izod impact strength of 70 kJ m^{-2}. However, other properties, such as yield stress and Young's modulus are unaffected by the reduction in molecular weight. The mould must be filled rapidly or the surface roughness of the type shown in Fig. 6.14b could occur. The mean surface roughness R_a value for typical injection mouldings is of the order of 1 μm, so the CD requirement that $R_a < 15$ nm is out of the ordinary. The extremely flat mould, patterned with bumps of 120 nm height, must be reproduced in detail. Dust must be kept away from the polycarbonate in all stages of transport and drying prior to moulding, and the plasticising stage of the injection-moulding machine has a special homogenising screw with starve feed.

14.4.6 Control of birefringence

The optical path difference $\Delta p_{r\theta}$ between rays polarised in the radial (r) and circumferential (θ) directions, traversing the disc once, must be less than 50 nm. As the disc is 1.2 mm thick, this means that the birefringence $\Delta n_{r\theta} < 4.2 \times 10^{-5}$. The laser light is polarised in the θ direction. Figure 14.19 shows that the light passes through the disc, oblique to the z axis. Its path length is affected both by the refractive index n_θ in the direction of polarisation, and by the value n_r in the radial direction. If the birefringence $\Delta n_{r\theta}$ is too large, the laser light cannot focus properly on the pits and the signal-to-noise ratio is reduced.

A central film gate is used so that the flow into the mould cavity has axial symmetry, in preference to a four arm gate, which starts the flow along four radii. The birefringence criterion can only be met by minimising the orientation effects discussed in Chapter 6—molecular orientation in the skin or in the core of the moulding, and residual stresses.

Problem 7 What strategies can be used to reduce the molecular orientation in the skin of injection mouldings, and to keep the skin as thin as possible?

The radial spreading flow affects the pattern of orientation. For the shear flow in the rz plane, the high shear strain rate region is close to the disc surfaces. For the extensional flow in the $r\theta$ plane (Fig. 5.6), the strain rate is tensile in the θ direction and compressive in the r direction. It is difficult to predict the orientation distribution for this complex flow from the viscous and elastic flow properties of the melt. The solution to Problem 7 suggests a method of reducing the skin orientation, but does not address the core orientation. The moulding core does not solidify until after the flow ceases, so there is some time for orientation to relax (see Eq. 5.2) before solidification. Core orientation is reduced by using a melt of low elasticity, and filling the mould relatively slowly. A high mould temperature allows relatively slow solidification, with time for the core orientation to relax. The core orientation is equally biaxial in the r and θ directions, so it is less important than the other contributions.

The residual stresses relate to the pressure history in the mould cavity (Fig. 14.20). The step at about 80 bar represents the end of the filling stage. A packing pressure of 450 bar is necessary to obtain a good impression of the bumps on the mould surface. The packing pressure reduces to zero after 1 s. The steep fall in the melt pressure at this stage shows that melt is flowing

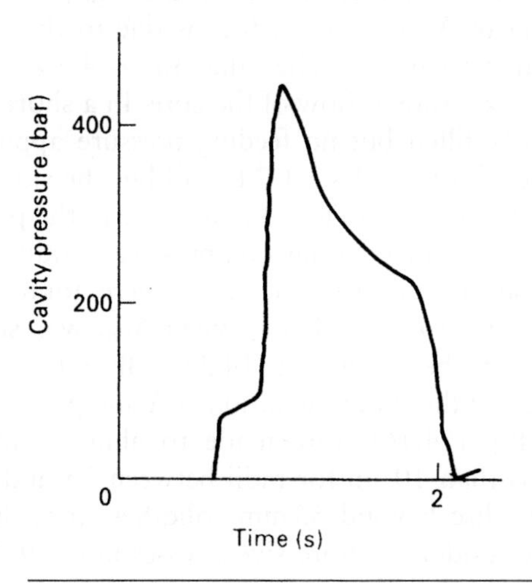

Figure 14.20 Mould pressure history for injection moulding a CD (from Anders S. and Hardt B., *Kunststoffe*, **77**, 25, 1987).

out of the cavity, and that the gate is not yet frozen. This reduction in mould pressure occurs while the solid layer at the surface of the moulding is only about 0.12 mm thick.

Problem 8 Section 5.3.4 describes the residual stresses found in injection mouldings. What is the effect of allowing the majority of the moulding to solidify while the melt pressure is zero? Assume that a 0.12 mm skin layer has solidified before pressure falls to zero.

The residual stresses in the CD are—equally biaxial compression near the surface and equally biaxial tensions in the core. According to the stress optic law (Eq. 8.9), the birefringence is proportional to the difference between the principal stresses. Therefore, the birefringence $\Delta n_{r\theta}$ due to residual stresses should be nearly zero.

The path difference for a ray travelling in the z direction is the integral

$$\Delta p_{r\theta} = \int_0^t \Delta n_{r\theta} \mathrm{d}z \tag{14.16}$$

where t is the disc thickness. To characterise birefringence, it is best to measure the path difference for an axis along which Δn does not vary. The radial flow symmetry means that Δn_{rz} does not vary along the θ axis, and $\Delta n_{z\theta}$ varies slowly along the r axis because the flow velocity vectors are along r. Wimberger-Friedl (1990) cut thin sections of CDs and measured the birefringence for light paths in the θ and r directions. Figure 14.21a shows the distributions of the birefringence Δn_{rz} as functions of radial position and the distance z from the mid-plane. The values are high, and vary markedly with position. The peak A near the surface is due to the skin solidifying during the mould filling shear flow. The value $\Delta n_{rz} \sim 4.5 \times 10^{-4}$ at the mid-plane C is due to the extensional flow of the core. In a short shot moulding (the mould was nearly filled but no feeding pressure applied), there is a nearly uniform value of $\Delta n_{rz} \sim 4 \times 10^{-4}$ for all but the skin (Fig. 14.21b). Therefore, the negative peak B in Fig. 14.21a is due to the packing pressure being applied for 0.5 s, and creating compressive residual stresses. The measured distributions of birefringence $\Delta n_{\theta z}$ were found to be almost identical to those in Fig. 14.21a. Hence, when $\Delta n_{\theta z}$ was subtracted from Δn_{rz}, $\Delta n_{r\theta}$ was found to be small (Fig. 14.21c). By adjusting the process parameters, the effects of the skin orientation at A and packing pressure at B on the integral in Eq. (14.16) were made to almost cancel. The path difference $\Delta p_{r\theta}$ is less than 10 nm for radii between 25 and 55 mm on the disc. The rim of the disc beyond 55 mm solidifies first, due to its more effective cooling; the residual compressive stresses in the θ direction in the rim cause an increased path difference. The outermost 5 mm of the disc, not used for recording, can be handled.

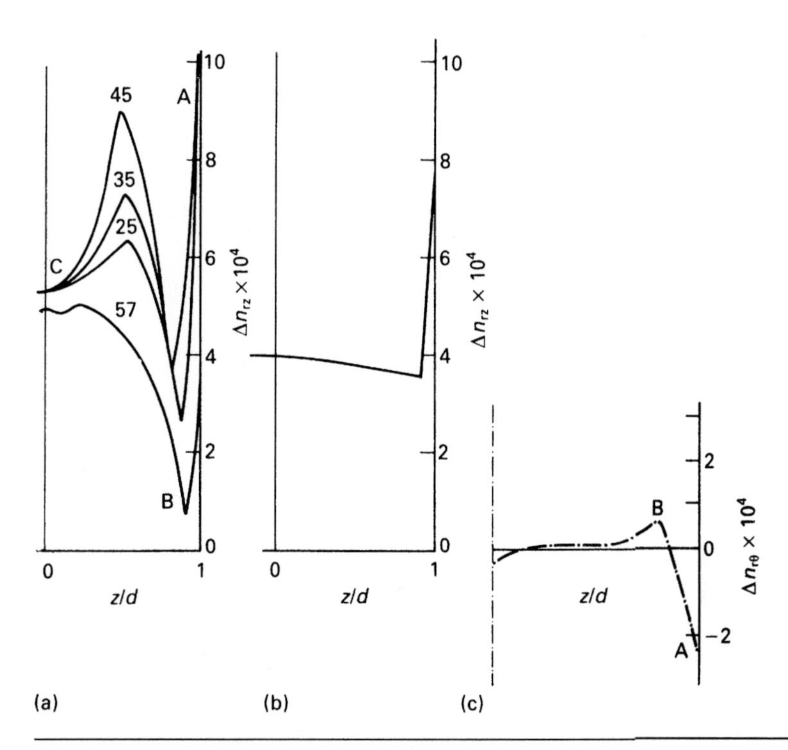

Figure 14.21 Birefringence variation with distance z from the mid-plane of a CD in the *rz* plane (*d* is the half thickness). (a) Δn_{rz} for a standard disc at various radial distances; (b) Δn_{rz} for a short shot when the mould cavity pressure <40 bar; (c) $\Delta n_{r\theta}$ for a standard disc. The integral of this distribution gives a path difference of ~10 nm (from R. Wimberger-Friedl, *Polym. Eng. Sci.*, **30**, 813, 1990).

14.5 Summary

The CD surface detail requirements and optical birefringence limits could only be satisfied by development of the polymer and the process technology. The discs are mass-produced at a low cost, using the established technology of injection moulding. Significant advances were required in process cleanliness, and excluding bubbles from the moulding. The effects of molecular orientation and residual stresses were reduced in magnitude, and counterbalanced, to meet the <50 nm path difference requirement. The discs are not subjected to wear in use, unlike LPs, or the ferrite coating of a tape passing through a tape recorder. Therefore, CDs are a durable storage medium.

Chapter 15

Sport and biomaterials case studies

Chapter contents

15.1 Introduction

This chapter contains three case studies. The first one on climbing rope illustrates the technology of *fibre* production and rope construction to meet specific mechanical property targets. There are some similarities between rope and the much smaller scale braided sutures used in surgery. Further sport case studies, on polymer foam protective gear and materials in running shoes, are mentioned in the *Further Reading* section.

The blood bag case study illustrates the use of *polymeric film* as a flexible container. It considers the permeability of polymers, plus processes for fabricating plastics film. Plasticised PVC has dominated the market for years, but there could be a changeover to flexible polyolefin films. The case study on replacement joints for implanting in the body illustrates *wear* and the effects of wear debris. Research continues on improving the wear resistance of the ultra high molecular weight polyethylene (UHMWPE) and mitigating the effects of sterilisation on the implant properties.

15.2 Dynamic climbing ropes

15.2.1 Introduction

Ropes are part of the safety equipment for rock climbing. 'Dynamic' ropes protect climbers who fall off a rock face, whereas 'static' ropes are used for ascending and descending pitches in caving, where falls are unlikely while a high tensile stiffness is required. Although bungee ropes, made from natural rubber fibres, would be strong enough for this purpose, they would cause the climber to be dragged up and down the rock face before coming to a halt.

The mass of a 50 m dynamic rope must be low (typically 3.5 kg) to enable it to be carried. The rope must be affordable and it must survive several years of use. Television series, showing expert climbers falling while attempting extreme routes, illustrate the required rope properties—tensile strength, low weight, easy bending to be clipped through belays, durability, low water absorption.

In the case study, rope mechanical factors are analysed prior to discussing rope construction. Fibre selection and optimisation are then discussed, before considering durability and performance testing.

15.2.2 Rope flexibility in bending

For a rope to provide a certain tensile load capacity, it needs to contain a minimum cross-sectional area A of polymer. The analysis of Appendix C shows that the bending stiffness EI_R of a rope, containing n fibres, is related to the EI_S of a solid rod of the same cross section by

$$EI_R = \frac{EI_S}{n} \tag{15.1}$$

Typical fibres have diameter $d = 0.031$ mm. In a typical rope, there are 41 000 fibres in the core and 18 000 in the sheath. As the Young's modulus E of polyamide fibre is about 2 GPa, the bending stiffness of the rope $EI_R = 5.3 \times 10^{-6}\,\mathrm{N\,m^2}$. The equivalent strength solid rod of diameter 7.5 mm would have a bending stiffness of $0.31\,\mathrm{N\,m^2}$. Due to the construction of dynamic ropes, there is considerable friction between the fibres, so the rope bending stiffness is somewhat higher than the value calculated above.

15.2.3 Dynamic loads in falls

The mechanics analysis of the forces on the climber's body (Smith, 1998) assumes that the rope has negligible bending stiffness, and that a length L of rope has a linear force F versus extension x relationship with slope (tensile stiffness) k_L ($\mathrm{N\,m^{-1}}$). We see later (Fig. 15.7) that this is a reasonable approximation. The rope tensile parameter k is defined as the product of a rope's tensile stiffness and its length

$$k \equiv k_L L \tag{15.2}$$

k with unit N, is independent of the rope length. It is related to the total cross section of polymer A, and effective fibre Young's modulus E^* by

$$k = E^* A \tag{15.3}$$

The peak force depends on the *fall factor*, defined by (Fig. 15.1a)

$$FF \equiv \frac{H}{L} \tag{15.4}$$

where H is the vertical distance fallen, and L the length of rope, from the belay to the climber, that is stretched. The maximum value of FF is 2, when a climber is the full extent of the rope above the belay before falling. The analysis considers the conversion of the climber's potential energy before a fall into the strain energy of the extended rope. The potential energy zero is set when the rope has its maximum elastic extension δ, when the climber's kinetic energy is again zero. The climber's potential energy before the fall is $mg(H + \delta)$, while the maximum rope strain energy is $\tfrac{1}{2}k_L\delta^2 = k\delta^2/2L$. Equating these leads to a quadratic equation in δ

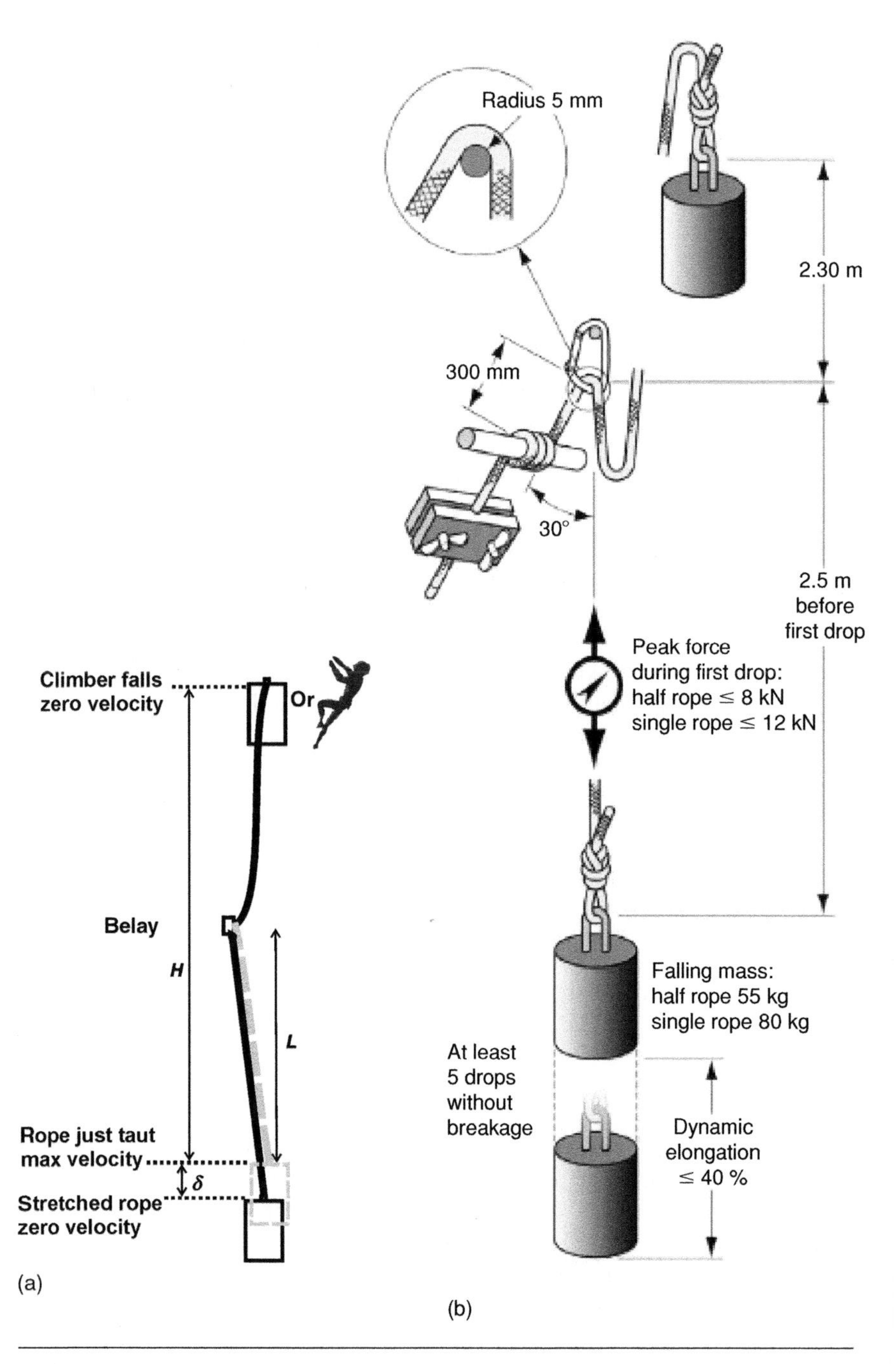

Figure 15.1 (a) Idealisation of a fall, with the definition of fall factor. (b) Dynamic rope test in the UIAA dynamic rope standard 101. © UIAA, reprinted with permission

$$\frac{k}{2L}\delta^2 - mg\delta - mgH = 0$$

The solution of this quadratic equation uses the positive square root

$$\delta = \frac{mgL}{k}\left[1 + \sqrt{1 + \frac{2k}{mg}\frac{H}{L}}\right]$$

Hence, the climber's maximum acceleration, relative to the acceleration of gravity, is

$$\frac{F_m}{mg} = 1 + \sqrt{1 + \frac{2k}{mg}FF} \qquad (15.5)$$

The kinetic energy of the falling climber increases in proportion to the rope length, so L does not appear in Eq. (15.5).

Human tolerance to whole body acceleration depends on how the loads are applied. When climbers use a harness (Fig. 15.2) to spread the load to the thighs and pelvis, a peak acceleration of 15 g is tolerable. The nineteenth century practice, of tying a hemp rope around the waist, risks severe damage to internal organs. A bungee rope would produce much lower g levels; the natural rubber filaments, of Young's modulus a few MPa, arrest bungee jumpers with about 3 g deceleration. However, bungee ropes are heavy, and climbers would not appreciate being dragged up and down a rock face.

For the peak acceleration of an 80 kg climber to be less than 15 g for a fall factor of 2, the tensile rope parameter must be $k < 39$ kN. A typical

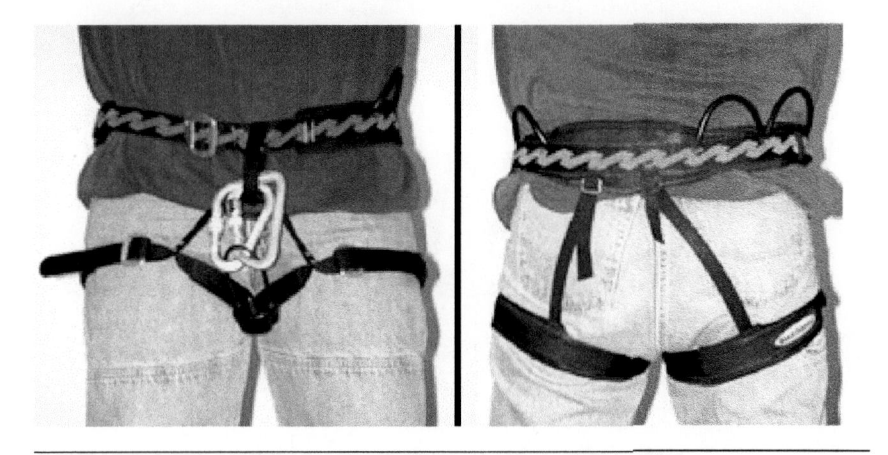

Figure 15.2 Harness: Left front view with carabiners, right rear view (from howstuffworks.com).

10.5 mm climbing rope has $k = 27$ kN. The peak acceleration limit can also be related to the rope extension; the climber falls under 1 g, and then is arrested in 1/7.5 of the fall distance; his average deceleration is 7.5 g. This means that the peak rope extension should be approximately 13%.

A 10.5 mm diameter nylon rope has a mass of 70 g m^{-1}. As the nylon fibre has density 1150 kg m^{-3}, the rope has a nylon cross-sectional area of 61 mm^2. Therefore, by Eq. (15.3), the effective rope Young's modulus E^* should be less than 0.63 GPa. However, the Young's modulus of straight, high-strength nylon fibre is 2.0 GPa. The next section shows how rope construction affects E^*.

15.2.4 Rope design and manufacture

Kurzbock (1987) patented a kern–mantel (core–sheath) rope, of the type shown in Fig. 15.3. He claimed the relatively hard sheath resisted wear from contact with rock, carabiners, etc. The braided sheath also holds the components of the core together, especially when the rope is bent around a small radius. It prevents the core strands contacting the rock. The sheath is flexible because of the helical path of the strands. The rope is a composite; the braided sheath resists abrasion and acts as a barrier to moisture, dirt and UV light, while the twisted core takes the main tensile loads.

The nylon fibres are twisted into yarn, then the yarn is twisted into strands of approximately 1.2 mm diameter, with a helix angle of approximately 50°. A machine (Fig. 15.4) feeds the strands, from bobbins on rotors, through two stages of twisting. The core of the rope contains 13 such parallel strands (four surrounding some central black fibres, then nine more surrounding these).

Figure 15.3 11 mm diameter kern mantel rope, with the core exposed.

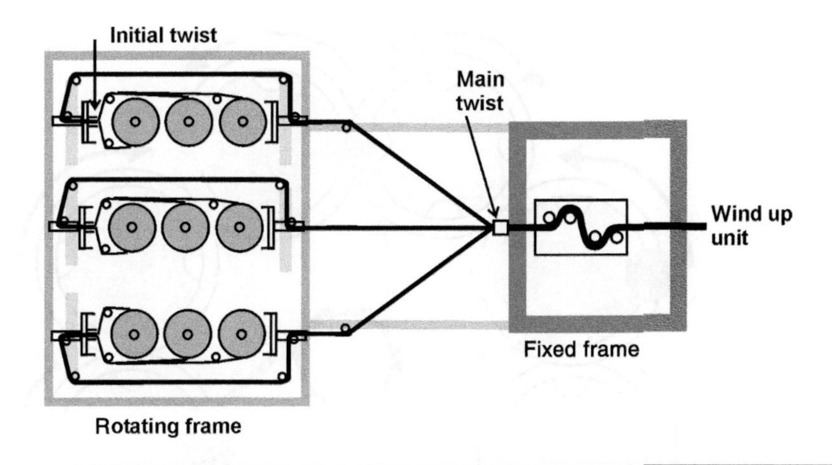

Figure 15.4 Schematic of rope twisting machine.

The sheath strands are also twisted from yarn. They are braided into a cylindrical sheath, around the core. The braiding process (Fig. 15.5) is analogous to a maypole dance. As the clockwise dancers pass in and out of the anticlockwise dancer, their ribbons braid around the pole. There are between 24 (half in clockwise and half anticlockwise) and 44 strands in the rope sheath. Manufacturers may quote the number of bobbins, from which the strands are unwound. A large number means the strands are finer. The bobbins transfer from rotor to rotor, so they make circuits of the core.

The sheath mass is typically a third of the rope mass. When the rope is under load, the sheath extends axially; the angle between the $\pm 45°$ strands decreases, so the sheath diameter decreases, compressing the core. This encourages fibre-to-fibre load transfer, important when some fibres have failed. If a section of sheath is removed from a rope, it stretches easily. The braided sheath has relatively small air gaps, making the ingress of dirt relatively difficult (see Section 11.3).

15.2.5 Polymer selection

Table 15.1 gives data on polymers used for rope. The material selection criterion for a lightweight rope of high tensile strength is σ^*/ρ, where σ^* is the tensile strength and ρ the density. Nylon 6 and polyester fibres have higher values than PP. However, factors, such as cost, high elongation at failure and low Young's modulus must also be considered. Kevlar is too expensive and has too low an elongation at failure. The elongation at failure of polyester fibres is only just sufficient for a dynamic rope. Nylon 6 is used in preference to nylon 66 because its abrasion resistance is higher, and it has good resistance against rotting and alkalis. Its equilibrium water uptake at 65% RH and 20 °C is 3.5–4.5%, similar to nylon 66, and it dries quickly.

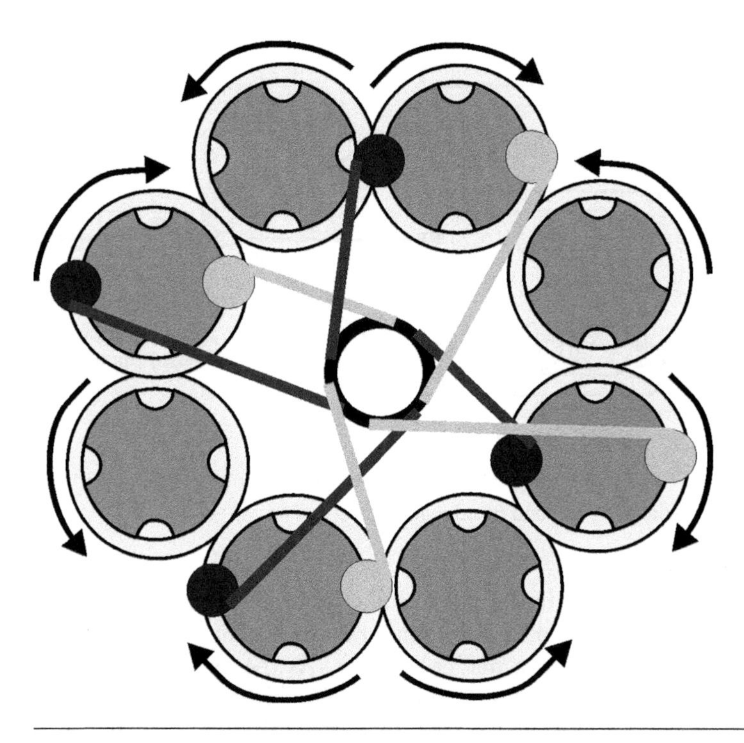

Figure 15.5 Schematic of a braiding machine, seen from above—the black bobbins circulate anticlockwise, while the grey ones circulate clockwise. The strands are drawn off towards the viewer and braid around the central core.

Table 15.1 Properties of fibres used in ropes (mainly from McKenna *et al.*, 2004)

Polymer	*Density* (ρ) ($kg\,m^{-3}$)	*Tensile strength* σ^* *(MPa)*	*Elongation at failure (%)*	*Cost ($/kg)*	σ^*/ρ
Nylon 6	1135	950	20	4	0.84
PET	1380	1130	12	4	0.82
Polypropylene	905	560	20	2	0.62
Kevlar 29	1450	2900	3.5	45	2.00

15.2.6 Optimising the rope tensile strength and Young's modulus

The tensile strength of nylon 6 was circa 80 cN/tex in 1970–1975 when the spin–draw technology was introduced. This had increased to 83–84 cN/tex by 1985, with a potential for values of 85–90 cN/tex (Club Alpino Italiano, 2002). The tex is the mass in g of 1 km of fibre, so the tensile strength is measured in units of length (this freely hanging length has

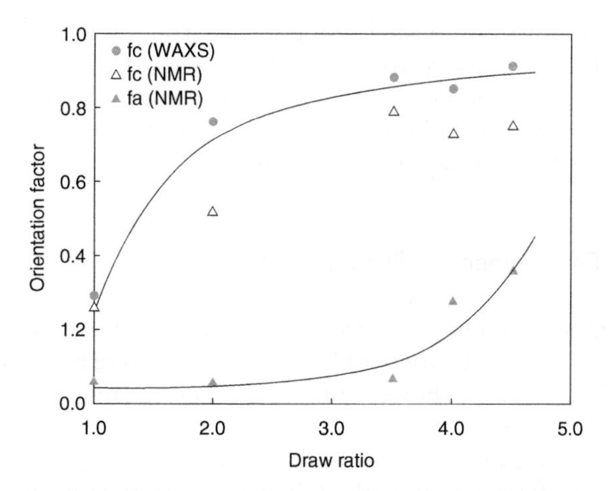

Figure 15.6 Crystal orientation and amorphous orientation increase during the cold drawing of nylon 66. (Penning, J. P. *et al.*, *Polymer,* **44**, 5869, 2003). © Elsevier

sufficient weight to cause the fibre to fail). The density of amorphous nylon 6 is $1090\,\mathrm{kg\,m^{-3}}$ while that of the α crystalline form is $1180\,\mathrm{kg\,m^{-3}}$. Therefore, for nylon 6 of 50% crystallinity, i.e. density $1135\,\mathrm{kg\,m^{-3}}$, the tensile strength in MPa is that in cN/tex multiplied by 11.35, i.e. 950 MPa for 84 cN/tex.

Figure 15.6 shows how the crystal and amorphous orientation increases with the fibre draw ratio. A fibre draw ratio of 4.5 aligns the polymer crystal c axes almost perfectly with the fibre length axis (perfect alignment would be $f_c = 1.0$). The pseudo-affine model (Section 3.4.10) for the crystalline phase predicts a rapid increase in orientation factor with draw ratio, while the affine rubber elasticity model for the amorphous phase predicts a less rapid increase.

The microstructure of nylon 6 fibres is complex, with both α form and γ crystalline forms being possible. In-line measurement of fibre birefringence shows that some crystalline orientation occurs during spinning. This increases considerably during subsequent conditioning. Hot stretching, in two stages at 140 and 185 °C, leads to a product that only contains the α crystalline form, at a volume fraction of about 50%. Subsequent heat treatment at 120 °C reduces the fibre Young's modulus, without much reduction in tensile strength.

Twisting the cords reduces the effective Young's modulus further. McKenna *et al.* (2004) analysed twisted yarn or strand. Each fibre follows a helical path of constant pitch and constant radius r. The helix angle θ increases from zero at $r = 0$ to a maximum of α at the outer radius. The Young's modulus of the strand depends on the average value of $\cos^4 \theta$. For the twisted strand, the effective Young's modulus depends on $\cos^2 \alpha$, with a typical value of 0.41 when $\alpha = 50°$. For the rope core, in which both the strand and yarn are twisted, the effective Young's modulus is reduced to about 0.6 GPa. The effective Young's modulus of the sheath is lower than that of the core. Commercial programs can compute the rope response from the fibre properties and the rope construction (Fibre Rope Modeller, from Tension Technology International, Eastbourne).

15.2.7 Environmental effects on rope durability

Absorbed water reduces the strength of nylon, but the strength returns when the rope dries out. A variety of 'dry' treatments increase the hydrophobicity of the sheath fibres—however these are not equally effective. Cutler and Lebaron (1992) investigated factors that reduce the strength of nylon ropes. Their factorial experiment showed that immersing rope in water for 8 h had the largest effect on strength, followed by stomping dirt into the rope. Signoretti (2002) found that soaking a rope in water for 48 h significantly reduced the number of rope impacts (on an EN 392 rig) before failure. This is expected, as water absorption reduces the T_g of nylon 6, hence, reducing the modulus and strength of the fibre. The equilibrium water content in air at 20 °C with 50% RH is 2.5–3%, which causes the T_g to fall from 90 °C to approximately 48 °C.

Nylon 6 is susceptible to UV damage, but the sheath, which contains UV stabilisers, protects the rope core. Some colours of the sheath begin to fade before the impact performance severely deteriorates. Dirt would reduce the core strength if it could penetrate the sheath. Fine particles of rock would get among the fibres of the core, then abrade and fracture them when the rope was flexed.

15.2.8 Rope testing standards

The load required to arrest an 80 kg mass with an acceleration of 15 *g*, is 12 kN. However, typical 10.5 mm climbing ropes have a tensile failure load of circa 30 kN. The UIAA standard 101(2004) is based on EN 892:1997. An 80 kg mass, attached to the end of a 2.8 m rope, has a guided vertical fall through 5 m (Fig. 15.1b), so the fall factor is 1.78. The rope passes over a steel edge of radius 5 mm, representing the curved surface of a karabiner, before its upper end is attached to a 30 mm diameter cylinder. In the first drop, the peak force must be less than 12 kN. The rope must survive at least five falls without failing. This does not mean however that, in mountaineering use, a rope should be discarded once it has suffered five falls. In the impact test, the rope damage occurs where it passes over the steel edge, whereas in climbing use the damage locations are more random, if near one end. The use of a rigid mass leads to dynamic loads that are higher than when a climber, with articulated limbs and deformable soft tissue, falls.

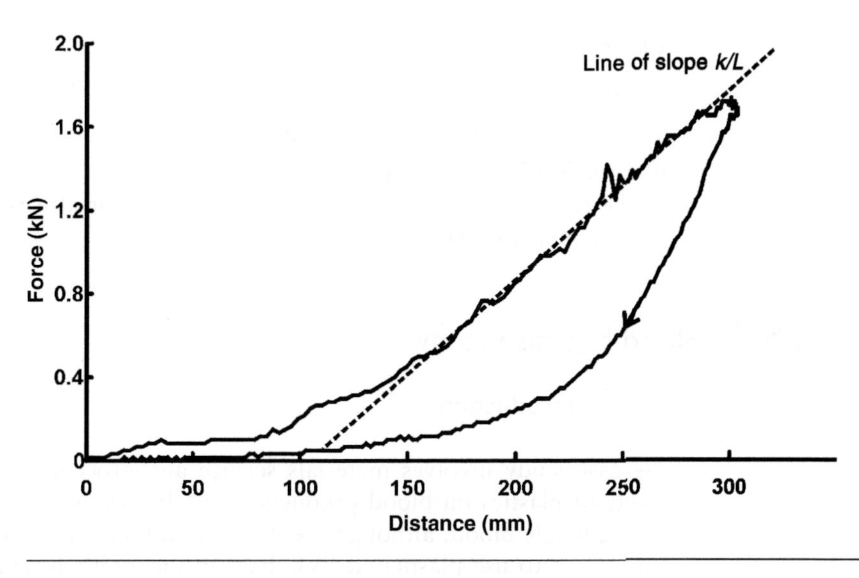

Figure 15.7 Graph of force vs. extension of a 11 mm dynamic rope.

There are very few published measurements of rope loads in falls. In some Austrian tests with a 70 kg climber and a fall factor of 0.375, the peak force was 4.0 kN when the rope was attached directly to a bolt, but only 2.3 kN when a karabiner was used for the belay. If the rope k value was a typical 27 kN, the peak force calculated using Eq. (15.5) would be 4.5 kN. However, these were low fall factor tests; the 'deformation' of the climber is expected to increase as the fall factor increases.

Figure 15.7 shows the rope force versus extension from a drop test on 1.8 m of dynamic climbing rope. A rigid mass of 11 kg was used with a fall factor of 1.0. The assumed linear loading response is a reasonable approximation once the load exceeds 0.6 kN and the strands have 'bedded-in'. There is significant energy absorption on unloading.

There is concern that ropes could fail if dynamically loaded when passing over a sharp rock. However, modifications of the rope impact test, with one side of the steel support having a 'sharp' edge with radius of curvature 0.75 mm, lead to non-reproducible results. There was evidence of melting of the nylon sheath fibres as they were dragged over the edge, and the number of falls, that the ropes could survive, decreased.

15.2.9 Further sources of information

Rope manufacturers' websites such as www.beal-planet.com detail products and give technical information on rope construction and testing.
art1.candor.com/detzel/rope.htm

Learning outcomes

Explain how

(a) high strength is achieved in polymer fibres;
(b) rope bending flexibility is achieved;
(c) the tensile stiffness of a rope and the fall factor affects the maximum force experienced.

15.3 Blood bag case study

15.3.1 Introduction

The case study involves materials science and processing, as well as the effects of plastics on blood products. Glass bottles were initially used for storing whole blood; although these are inert, they are fragile. The Americans began to use plasticised PVC bags in the 1950s Korean War for the collection of whole blood, blood processing and storage. A blood bag set (Fig. 15.8) can consist of one or three bags, with the connecting tubes and locations for blood input and egress.

The main blood products include

Red blood cells, used to replace blood loss in surgery, and for certain forms of anaemia;
Whole blood, rarely used except to counteract sudden massive blood loss;

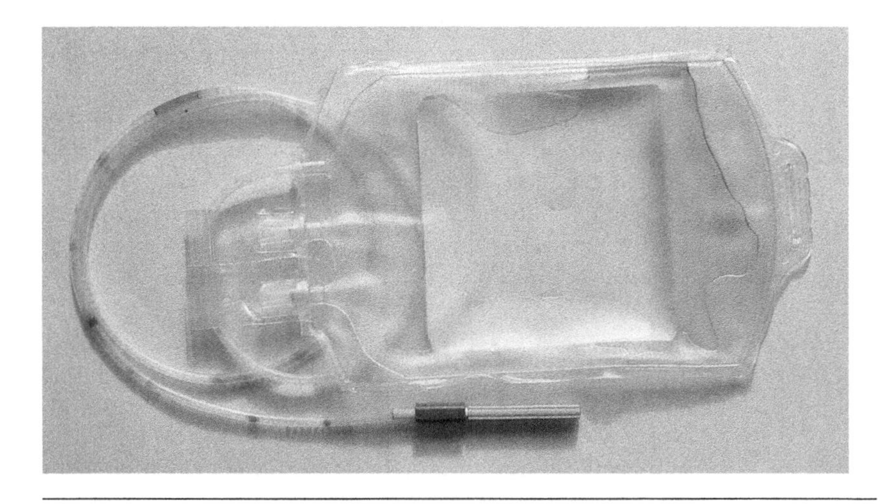

Figure 15.8 A blood bag with attached needle.

Platelets, for patients with a low platelet count (leukaemia or after bone marrow transplants);

Plasma, for processing to extract immunoglobulin, albumin (to manage severe shock and burn injuries), and clotting factors, such as factor VIII.

When a donor initially gives whole blood it is initially stored in a DEHP plasticised PVC bag (see Section 15.3.3 for abbreviation). Usually, blood is centrifuged within 8 h, and separated into red blood cells and plasma. The plasma can be centrifuged further to remove platelets. However, there is an alternative process of apheresis (removal of blood from a donor and re-infusion after components are removed), where platelets are removed continuously in disposable polycarbonate centrifuge unit.

Red blood cells are stored in refrigerators at 1–6 °C for up to 42 days in DEHP plasticised PVC bags. Plasma or platelets have a shelf life of 5 days in TETM plasticised PVC or polyolefin bags. Fresh frozen plasma, called so since it is frozen within 8 h of production, can be stored for up to 1 year, in either a DEHP-plasticised PVC or polyolefin bag.

Starr (1999) described problems that arose in the USA and Asian countries which paid donors. In the UK, where donors are volunteers, problems have arisen from pooling products such as Factor 8—notably the infection of some haemophiliacs with AIDS.

Background information on the blood transfusion service can be found at: www.blood.co.uk, www.blooddonor.org.uk and in a National Audit office report at www.blood.co.uk/pdfdocs/national_audit_2000.pdf. Bag manufacturer's websites include www.baxter.com and www.haemotronic com.

15.3.2 Polymer selection for blood bags

Carmen (1993) surveyed the requirements of blood bag materials—most of these are discussed in the following sections. Polymer selection is by elimination, as candidate materials must meet all the requirements. For cost reasons, the bag material should be made from a commodity plastic.

15.3.3 Plasticisers in PVC

Plasticisers in PVC, and antioxidants used to prevent melt degradation in polyolefins, must not be toxic, since they could enter the blood product. The ingredients of the anticoagulant/preservative solution (adenine and glucose) must be stable for autoclaving, and shelf storage for as long as 3 years.

Unplasticised PVC has far too high a modulus for use as a blood bag. The traditional plasticiser is di(ethylhexyl) phthalate (DEHP) (sometimes called di-octyl phthalate (DOP)). Addition of 100 parts per hundred of DEHP reduces the T_g of the polymer to −51 °C. This enables the material to remain flexible if it is frozen rapidly before being stored in a freezer. It turns out that DEHP plasticises the membrane of the red blood cells, giving them a longer shelf life.

DEHP plasticiser leaches from the PVC into the blood product, producing concentrations of the order of 10 mg 100 ml^{-1} of blood after 2 weeks storage. Tickner (2000), reviewing of the effects of DEHP in the body, showed that there are concerns about the metabolites of DEHP possibly causing cancer. Phthalate plasticisers, when fed in large quantities to rats, can cause cancer. This does not prove that the storage of whole blood in plasticised PVC bags is a health risk. The www.nogharm.org website contains many further documents, including some on the release of dioxin if PVC is incinerated. Koop and Juberg (1999) have also reviewed the risks of plasticisers.

Tri(2-ethylhexyl) trimellitate (TEHTM) or citrate plasticisers can be used in PVC. They have lower leaching rates and the citrate plasticisers appear not to have any potential toxicity. However, the costs (Krauskopf, 2003) are $4.1/lb for TEHTM and $1.9/lb for acetyl tris(*n*-butyl) citrate, compared with $0.60/lb for DEHP.

One typical composition for a blood bag is 59 wt.% PVC, 12% Vitamin E, 6% citrate plasticiser, 14% TEHTM and 9% epoxidised soya oil. Vitamin E suppresses haemolysis of red blood cells, while the soya oil is a thermal stabiliser for PVC processing which also has some plasticising effect. Thus, plasticised PVC can act as a reservoir for additives of a preservative nature.

There are various ways of reducing the rate at which plasticiser leaches into the body. The higher molecular weight polymeric plasticisers, such as poly 1,3 butylene adipate (PBA) are less mobile in the PVC. When DEHP is replaced by PBA in plasticiser blends, the amount extracted by kerosene at room temperature in 24 h reduces more than tenfold (Fig. 15.9).

15.3.4 Translucency, so contents can be seen

The bags should be translucent so it is easy to check if they are full, and to see layers in centrifuged bags. Section 11.4.2 explains light scattering from semi-crystalline polymers. Thin films scatter less light than thick mouldings, hence they appear more transparent. The polymer must not contain added pigments. Thermoplastics do not need pigments, which would absorb and scatter light, but most rubbers need mineral fillers for strength.

15.3.5 Flexibility, to allow blood processing

After the blood bags are centrifuged to separate out the layers of platelets, plasma and red cells, they are extracted by squeezing the bag, noting when the colour of the liquid in the exit tubing changes. It should only require a small force to bend the bag wall during such processing. Section C.3 of Appendix C gives formulae for the bending stiffness *EI* of sheet and tubing.

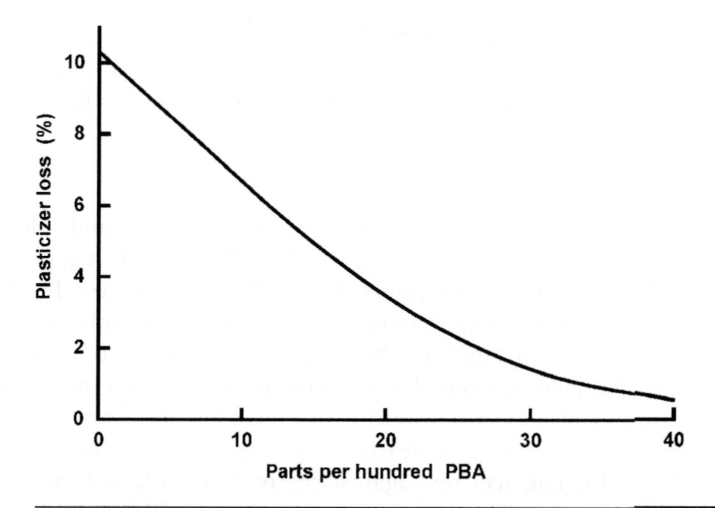

Figure 15.9 Leaching of DEHP/PBA blends from PVC (redrawn from Lakshmi S. *et al.*, *Artif. Organs*, **22**, 222, 1998).

A 0.1 mm thick bag will be flexible whatever polymer is used; consider the flexibility of a 0.13 mm thick OHP sheet, the PET has Young's modulus $E =$ 3 GPa. Thin films, made from low crystallinity PE copolymers or plasticised PVC have $E < 0.1$ GPa, and are much more flexible. The tubing, of typical inner and outer diameters 3 and 4 mm, respectively, has a higher bending stiffness than the bags.

15.3.6 Heat resistance, to allow sterilisation

Before use, the bags must be sterile. The most common sterilisation method is by steam, in an autoclave at 10 bar pressure and 121 °C for about 30 min. γ radiation may cause visible discoloration of PVC, as a result of polymer degradation. Consequently, the plastic must not melt below 121 °C, ruling out PS with T_g of 100 °C. For PP, the crystal melting temperatures of 170 °C is high enough for it to remain solid in the autoclave. The crystalline phase in PVC (roughly 10% by volume) melts at 220 °C, so at 121 °C PVC is a soft rubbery material. However, low crystallinity metallocene polyethylene (mPE), with a melting temperature of 105 °C, must be sterilised by another method.

15.3.7 Tensile strength, to survive centrifugation and handling

Centrifugation is used to separate out the white and red cells, which are slightly denser than the plasma. The high-speed centrifuge generates $5000 \times g$ of linear acceleration, where g is the acceleration of gravity. Several bags are placed in a strong 'bucket'. When this is rotated at the end of an arm, the 0.5 kg unit of blood experiences a centripetal force of 25 kN. The hydrostatic pressure p, at the base of the bag of depth $h = 0.2$ m, is

$$p = \rho a h \qquad (15.6)$$

For a linear acceleration $a = 50\,000\,\text{m s}^{-2}$ and blood of density $\rho = 1000\,\text{kg m}^{-3}$, the pressure $p = 10$ MPa. Initially, it appears that a strong polymer, such as biaxially stretched PET, is necessary. If, at the base of a rigid container, there is an unsupported corner of radius $r = 2$ mm and wall thickness $t = 0.5$ mm, the hoop stress in the wall would, by Eq. (C.18), be 40 MPa. Hence, a rigid blood container would need to be made of a strong material. This stress would cause most thermoplastics to yield and fail. However, by using a flexible bag, the strength requirement is reduced. A flexible bag will rest against the bucket wall, and the maximum stress will be of the order of pressure p, i.e. about 10 MPa.

While handling, full blood bags are sometimes supported by the tubing. Allowing a handling acceleration of 4 g, the peak load on the tubing is four times the full bag weight, e.g. 20 N. If the plastic has a tensile strength of 10 MPa, the wall thickness of a 5 mm diameter tube would need to be a minimum of 0.06 mm.

British Standard BS EN ISO 3826-1:2003 Plastic collapsible containers for human blood and blood components requires that

(a) a bag can be emptied of blood in 2 min if a relative pressure of 50 kPa is applied by two flat plates. This replaces the design-restrictive requirement in BS 2463:1990 that the minimum tube inner diameter was 2.7 mm;
(b) a tensile force of 20 N applied to the tubing for 15 s must not cause leakage.

Ko and Odegaard (1997) and Tickner (2000) discuss alternative materials to plasticised PVC. mPE with a low crystallinity (density 900–905 kg m^{-3}) allows the downsizing of the bag gauge, since the biaxially oriented polyethylene film is stronger than plasticised PVC, and it has a higher tear strength (Fig. 15.10). The typical in-plane tensile strength of the mPE film is 29 MPa, compared with 16–24 MPa for plasticised PVC film.

Some blood components are shipped on dry ice, at −70°C. If a bag is dropped at this temperature, the impact can cause plasticised PVC bags to fracture.

15.3.8 Permeability

Platelets need oxygen to survive. The gas flow rate Q through a plastic film of area A and thickness L is given by Eq. (11.5). Table 15.2 gives transmission rates for particular film thickness, for a specific gas pressure.

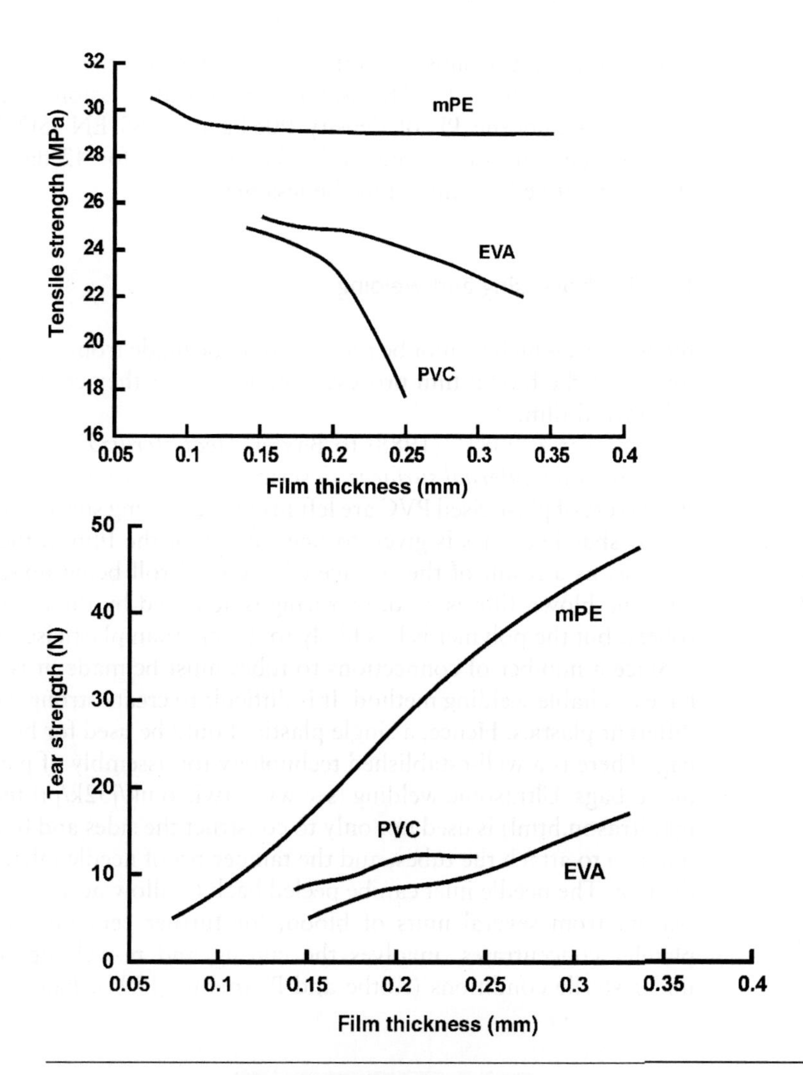

Figure 15.10 Tensile strength and tear strength of various films as a function of thickness (redrawn from Lipsitt, B., *Med. Plastics Biomater.*, **Sept.**, 1998).

Table 15.2 Transmission rates for blood bag film (Lipsitt, 1998)

Polymer	*Film thickness (mm)*	*Oxygen transmission rate ($cm^3\ m^{-2}\ day^{-1}$)*	*Water vapour transmission rate ($g\ m^{-2}\ day^{-1}$)*
Metallocene PE	0.35	1100	3
EVA	0.25	1200	14
Plasticised PVC	0.25	550	20

The oxygen transmission rates are high since these low crystallinity polymers are above T_g. The water vapour transmission rates are lowest for the metallocene PE of density 905 kg m^{-3}. BS EN ISO 3826-1:2003 requires the water loss, when a full bag is stored for 42 days at $4 \pm 2\,^{\circ}C$ at 55% relative humidity, must be less than 2%.

15.3.9 Processing and welding

Blood bags, which cannot be reused, could be made from a length of tubular film from the blown film process (Chapter 5), or they could be made from calendered film.

US patent 4790815 (1988) to Baxter, *Heat sterilizable plastic container with non-stick internal surfaces*, describes the need to prevent sticking when two layers of plasticised PVC are left in contact during sterilisation. A texture of crosshatched lines is given to one surface of the film in the calendaring process, as a result of the surface of one steel roll being textured. If polyethylene blown film is used, texturing is achieved by the use of embossing rollers, but the polymer is less likely to 'block' than plasticised PVC.

Since a number of connections to tubes must be made, it is important to have a reliable welding method. It is difficult to create strong welds between different plastics. Hence, a single plastic should be used for both tubing and bag. There is a well-established technology for assembly of plasticised PVC blood bags. Ultrasonic welding (see www.twi.co.uk/j32k/protected/band_3/pjkultrason.html) is used not only to construct the sides and base of the bag, but also to attach the tubes, and the tamper-proof needle inlet, to the top of the bag. The needle inlet can be peeled back to allow access. The pooling of plasma from several units of blood, for further centrifugation to obtain platelet concentrates, involves the cutting and re-welding of the tubing under sterile conditions (Kothe and Platmann, 1994). Figure 15.11 shows

Figure 15.11 Plasticised PVC tubing, welded with the hemotronic process, then pressure tested to failure.

a length of welded tube after pressure testing to failure. A 0.3 mm thick copper wafer at 230 °C passes through a pair of tubes, then the required ends are shifted laterally to allow the weld to form.

15.3.10 Biocompatibility

Blood bag materials should not allow large numbers of platelets to stick to their surface, since the adhesion process releases cytokines, which can lead to adverse reactions in patients who receive the platelets in the form of injection. Figure 15.12 shows a smaller number of macrophage cells on the surface of a PVC, plasticised with TEHTM, compared with the surface of a polyolefin blood bag.

15.3.11 Summary

Due to cost pressures, the market is dominated by plasticised PVC. Since 1990, there has been a search for suitable alternatives. Changes will probably be triggered by health or environmental concerns, rather than cost saving.

After completing the case study you should be able to

(a) explain the mechanical property and liquid/gas transmission, requirements of plastics used for blood bags;
(b) select from grades of PE (specify the density) or plasticised PVC (specify the plasticiser and its content) for blood bags;
(c) select processing routes for plastics film and for joining film.

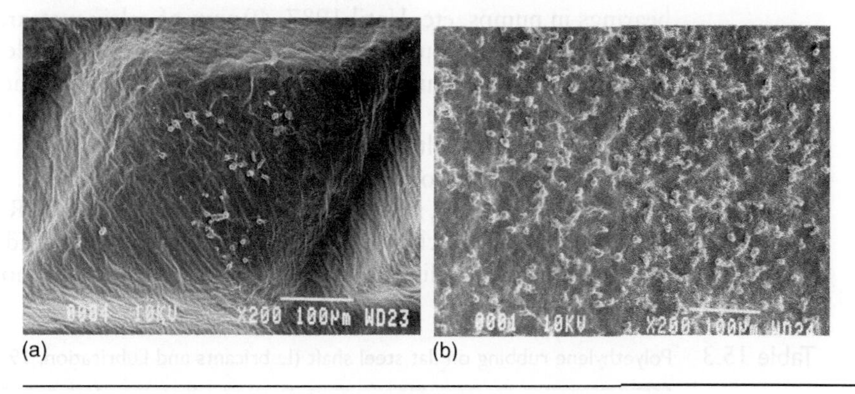

Figure 15.12 SEM micrographs of human macrophage cells on the surface of: (a) Plasticised PVC; (b) polyolefin after 5 days in an incubator at 37 °C (from Elkattan, I. *et al.*, *Clin. Diagn. Lab.*, **6**, 509, 1999) © American Society for Microbiology.

15.4 UHMWPE for hip joint implants

15.4.1 Introduction

Hip joint replacements are the most common type of joint replacements, with 300 000 operations in the USA in 2000. The aim of this case study is to highlight the wear resistance and low friction properties of UHMWPE, to discuss the effects of sterilisation on the properties of plastics, and the effects of wear debris in the body. The order of presentation is that followed by the polymer, through fabrication, sterilisation, implantation and wear, to the reactions caused by wear debris.

UHMWPE has been used as the acetabular cup of hip joint implants for more than 40 years (Kurtz, 2004). It acts as the socket for a metallic alloy or ceramic ball, connected to a tapered shaft inserted into the upper part of the femur. A high coefficient of friction in the joint could cause sufficient local heating to damage tissue (Bergmann *et al.*, 2001). PTFE has an extremely low coefficient of friction when sliding against steel in a pin-on-disc machine, because a transfer film of highly oriented PTFE builds up on the steel surface. However, PTFE implants, originally used by Charnley, had excessive wear rates. Kurtz (2004) describes the search for an alternative. First glass-fibre filled PTFE was evaluated. Then, in 1962, UHMWPE was found to have a superior wear rate. UHMWPE transfers a thin film of oriented polymer to a steel surface, if it slides in a single direction. Table 15.3 shows that the low wear rate only occurs if the PE has a very high number average molecular mass.

15.4.2 Grades of UHMWPE

UHMWPE, made by the Ziegler catalysis system, was developed for sliding bearings in pumps, etc. Until 1987, 40 ppm of calcium stearate was used as a scavenger for residual catalyst, as a lubricant and nucleating agent increasing the crystallinity. The $TiCl_3$ catalyst system became more efficient over the years, so the calcium stearate concentration was cut to 1 ppm in 1987, and removed altogether in 2002.

The polymer was originally made by Ruhrchemie AG. When Hoechst took over production, it used the trade name 'Hostalen GUR' (G granular, U ultra high, R Ruhrchemie). Ticona now produces grades GUR 1020 and 1050; the third digit determines the weight average molecular weight,

Table 15.3 Polyethylene rubbing on flat steel shaft (Lubricants and Lubrication, 1967)

M_N (1000s)	32	40	49	74	250	1000
Relative wear rate (volume)	293	21	11	8	3	1

Table 15.4 Properties of UHMWPE and irradiated version (from Pruitt, 2005)

Property	*Unit*	*GUR 1050*	*GUR 1050 + 100 kGy and 150 °C*
Crystallinity	%	50 ± 3	46 ± 0.3
Tensile yield stress at 20 °C	MPa	23.5 ± 0.3	21.4 ± 0.1
Tensile yield strain at 20 °C	%	14.4 ± 0.6	1.45 ± 0.9
Tensile ultimate stress at 20 °C	MPa	50 ± 3	37 ± 3
Tensile ultimate strain at 20 °C	%	410 ± 10	230 ± 8
Compressive Young's modulus at 37 °C	MPa	650 ± 20	570 ± 15
Compressive offset yield stress at 37 °C	MPa	9.7 ± 0.2	8.8 ± 0.1
Fracture toughness	MPa(m)	4.0 ± 0.5	3.0 ± 0.6

3.5×10^6 and 5.5 to $6 \times 10^6\,\mathrm{g\,mol^{-1}}$, respectively. The molecular weight distribution is very wide. Table 15.4 gives mechanical data; the UHMWPE orientation hardens after yield, and has a high fracture toughness. Although joint manufacture is a major business, litigation has reduced to one, the number of companies supplying UHMWPE in the USA.

The ram extrusion process for UHMWPE powder was described in Section 5.4.4, and the effects of diffusion on the strength of the particle boundaries in Section 6.4.3. Ram extruded rods are machined on a lathe into the required cup shape. It is not essential to remove all machining marks, since the smooth metal counter-face polishes the PE in contact with it.

15.4.3 Acetabular cup design

Charnley (1979), who introduced UHMWPE implants in the UK in 1962, described the 149 stages of the operation, completed in 84 min. His design (Fig. 15.13) with an alloy steel shaft and ball replacing the top of the femur, and a thick UHMWPE cup implanted in the acetabulum (part of the pelvis), is the main type used. The internal diameter of the UHMWPE cup and the method of fixing to the femur, affect the performance. Charnley argued that a 22.2 mm diameter ball, considerably smaller than the replaced femoral head, provides a lower frictional torque. However, other designs use 32 and 38 mm diameters. The shaft of the prosthesis is driven into the spongy bone of the upper part of the femur; we see later that wear debris can affect bone remodelling at this site and cause loosening, requiring a repeat operation.

The external surface of the acetabular cup is grooved to allow bone cement (ceramic-filled acrylic resin) to key mechanically to the cup. This cement also binds to the excavated part of the pelvis. Further design variables are the cup wall thickness and whether it has a metal backing. For a

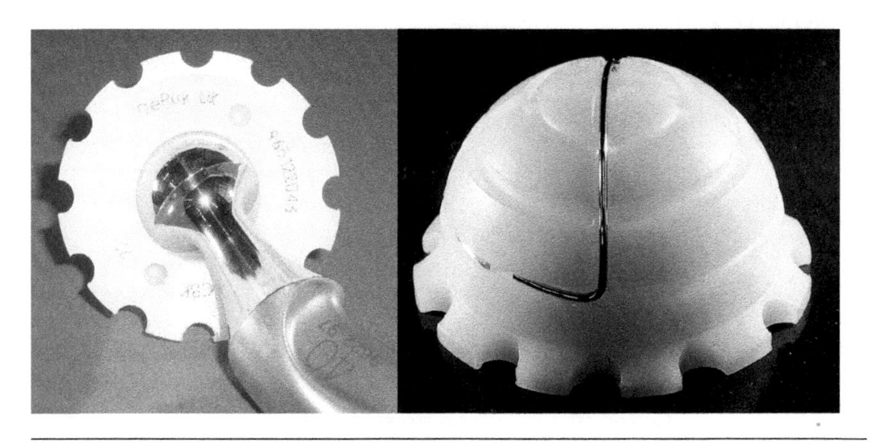

Figure 15.13 Charnley acetabular cup design showing grooves on exterior for bone cement, and the stainless-steel wire for wear assessment by X-ray.

ball diameter D, the bearing area is proportional to D^2, and the average pressure on the bearing surface changes with D^{-2}.

FEA of the stresses in the UHMWPE cup is difficult, as the stresses exceed the elastic limit. Teoh *et al.* (2002) considered an 8 mm thick cup with a metal backing, a 32 mm diameter ball and a peak load of 2.2 kN (about 2.5 × body weight) for walking. Using the unrealistic condition that the compressive stress on the ball/UHMWPE interface could not exceed the uniaxial compressive yield stress (of 8 MPa), they predicted the compressive stress to be at this level over a surface region of diameter about 8 mm. However, a von Mises type yield criterion should be used. It requires a pressure of nearly three times the uniaxial yield stress to extrude the PE to the side of the joint (Section 8.2.4).

15.4.5 Sterilising the PE before implantation

Heat sterilisation would cause cup shape to change, since the PE melting point is around 140 °C. Consequently, the machined PE joints, sealed in polyethylene bag filled with an inert gas like nitrogen, are subjected to 5 Mrad of γ rays, for sterilisation. Prior to about 1989, the irradiation was performed using air-filled bags. However, the γ rays break some PE chains, forming free radicals, which are long lasting in the bulk polymer. They lead to oxidation and chain scission, if there is dissolved oxygen in the polymer, or crosslinking in the absence of oxygen. These reactions take place in the amorphous phase, where the molecules are mobile. The end result was an increased wear rate. Crosslinking increases the fraction of the PE insoluble in a hot solvent, whereas chain scission reduces the average molecular weight (Fig. 15.14), allowing the polymer chains in the amorphous phase to

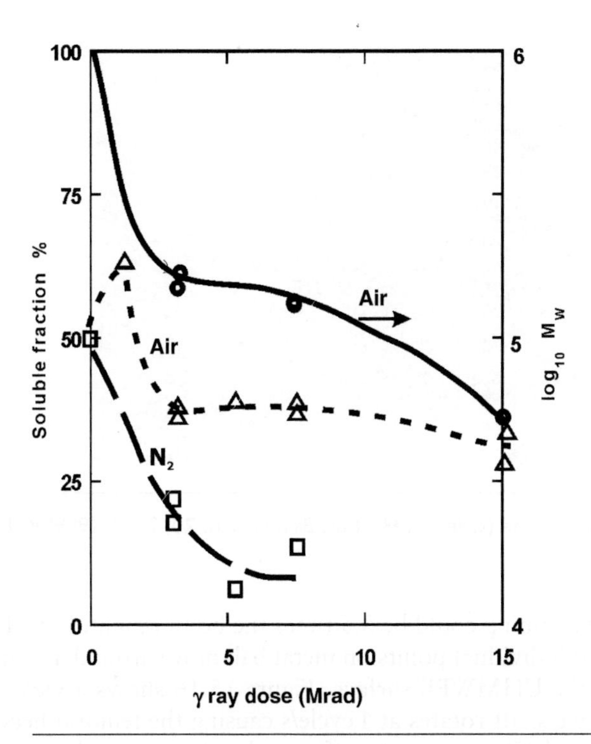

Figure 15.14 Effect of γ ray dose on the fraction of PE soluble (left scale) and molecular weight (right scale) (redrawn from Eyerer P. et al., *Kunststoffe*, **77**, 617, 1987).

rearrange and partially crystallise. Often a density profile develops at the bearing surface, indicating that a diffusion process is involved.

15.4.6 UHMWPE microstructure and mechanical properties

Olley *et al.* (1999) examined the microstructure of fabricated acetabular cups; inside the particles, there is no obvious spherulitic microstructure. Regions of about 6 μm diameter contain lamellae of width typically 0.5 μm (Fig. 15.15). These are surrounded by a looser boundary of 2 μm wide lamellae, consisting of lower molecular weight material. Such material diffuses to the boundaries to effect the bonding process.

15.4.7 Biomechanics of the patient's activities

The weight and age of patients vary, as do their physical activities. Studies using instrumented implanted hips show that the highest stresses occur if the wearer stumbles, rather than in regular walking. Monitoring a strain-gauge, attached to a prosthetic hip showed that the peak force on the joint in

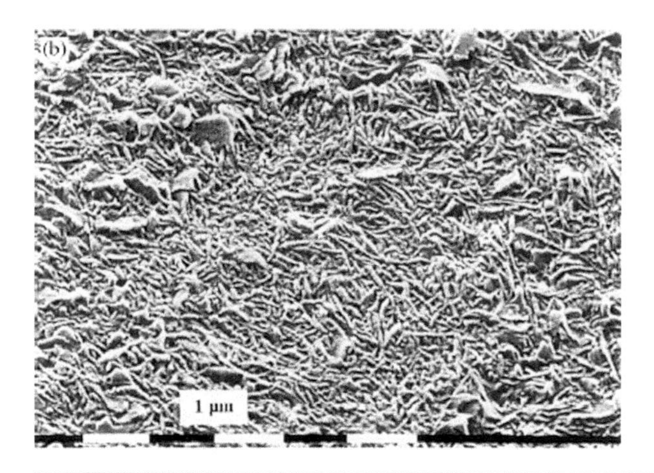

Figure 15.15 UHMWPE lamellae (Olley, R. H. *et al.*, *Biomaterials*, **20**, 2037, 1999) © Elsevier.

staircase climbing could be 3.5 times the body weight. In walking, the leg is swung, so individual points on metal ball move around a nearly rectangular path on the UHMWPE surface. Figure 15.16 shows a wear simulator; the lower drive shaft rotates at 1 cycle/s causing the femoral head to rock back and forward. There is a video of a similar rig in operation at www.machina.hut.fi/project/hip2001/. Wear tests, with a small PE pin pressing on a rotating steel wheel, do not reproduce the conditions in the body, so are not relevant. Multi-directional motion causes wear rates of the same order of magnitude as clinical wear observations. Turell *et al.* (2003) showed that a rig, producing a square sliding trajectory on the moving surface, increase the wear rate by a factor of 2.5 compared with a single-axis motion.

15.4.8 Lubrication of the joint in the body

The lubricating fluid in the body is likely to contain proteins. Wear rig experiments show that the concentration of (bovine) protein affects the wear rates. Wang (2001) found that increasing the radial clearance between the femoral head and the cup decreased both the coefficient of friction and the wear rate in a multi-axis wear test; radial clearances of 1 mm corresponded to a coefficient of friction of 0.05.

Whether thick film lubrication, with no metal to polymer contact, is possible depends on the surface roughness, the diameter of the joint and the design gap (Dowson, 2001). The average root mean square roughness of UHMWPE surfaces is circa 280 μm compared with 4 μm for the cobalt chrome alloy ball (Hutchings, 2003). For UHMWPE sockets of diameter 28 mm, the design gap of circa 0.1 mm between the ball and socket and the high roughness of the UHMWPE surface, means metal to polymer contact will occur.

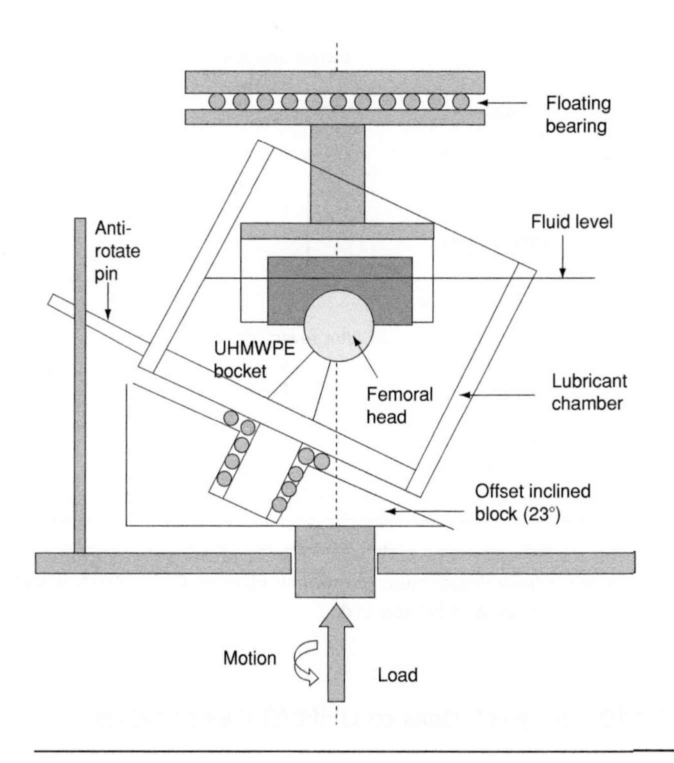

Figure 15.16 Simulator of hip motion during locomotion (Wang, A., *Wear*, **248**, 38, 2001) © Elsevier.

15.4.9 PE wear mechanisms

Studies of retrieved joints show a wide range of wear rates, with some having no noticeable wear. Data collection, to clinically evaluate the success of a particular joint design, typically considers joints up to 10 years old. In general, joint revisions are not necessitated by excessive acetabular cup wear. Rather, the side effects of wear debris (see next section) cause the implant to fail where it is attached to the femur. Hence, the wear rates must be minimised. If bone cement particles get into the joint, there will be a dramatic increase in wear. Bone cement contains hard, angular-shaped, silica particles, which can act as abrasive cutting points, if they become embedded in the metal head. Wang and Schmidig (2003) argue that the use of ceramic femoral heads would prevent the risk of runaway wear of the UHMWPE.

Observation of worn surfaces and wear debris suggest three types of wear mechanisms—adhesion, when PE that adhered to the metal surface is torn off; abrasion when bone cement particles get into the bearing and cut grooves in the soft PE; and fatigue, where surface features are deformed back and forward until they fall off. Baudriller *et al.* showed a mechanism for the detachment of a PE flake (Fig. 15.17), but they could only start the FEA modelling of microscopic wear process.

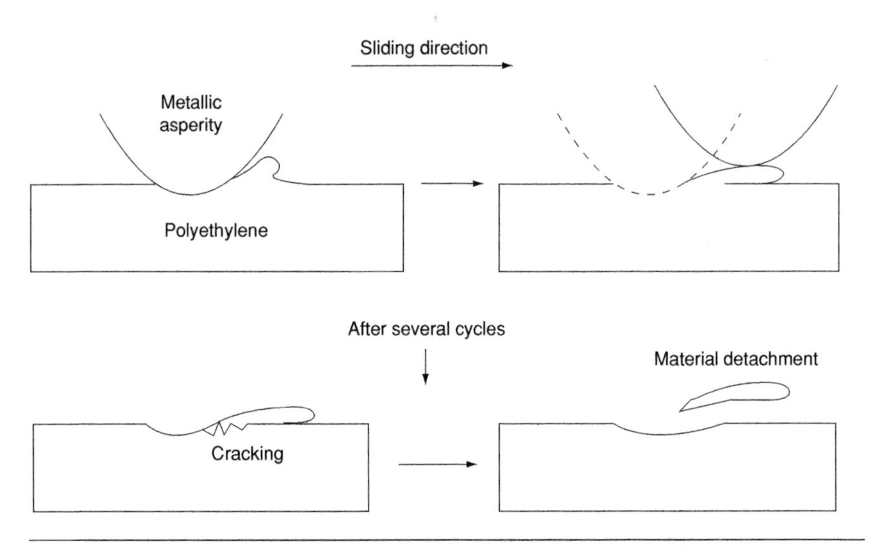

Figure 15.17 Mechanism for flake detachment (Baudriller, H. *et al.*, *Comp. Meth. Biomech Biomed. Eng.*, **7**, 227, 2004) © Taylor and Francis Ltd.

15.4.10 Body reactions to UHMWPE wear debris

The body recognises polyethylene as a foreign body, so defence mechanisms against infection come into play. *Granulocyte* cells create hydrogen peroxide, H_2O_2, a strong oxidising agent that kills foreign cells. The H_2O_2 can oxidise the PE causing chain scission and the formation of carbonyl side groups on the PE chain, which can be monitored by infrared spectroscopy. The degradation is most severe 0.5–1.5 mm below the articulating surface, where a white band forms, due to the presence of micro-cracks which scatter light. The PE in this region has a reduced tensile strength and ductility. A survey in 1995 showed that, after an average 4 years of implantation, the average crystallinity of the cups has increased by 11%, from a mean of 53% to one of 64%.

Wear debris, from retrieved UHMWPE acetabular cups, show a wide range of particle sizes from 0.1 to 1 mm, and shapes from platelets to granules to fibres (Fig. 15.18). Dowson (2001) estimated that for each of the million steps taken by a person per year, the order of 10^5 wear particles are generated. These particles migrate in the body, where m*acrophage* cells attempt to digest them, but fail. Macrophages produce cytokines, inflammatory agents that stimulate bone resorption, leading to osteolysis (bone loss). Osteolysis can cause aseptic loosening of the implanted stem in the femur, leading to failure of the implant. The website depts.washington.edu/bonebio/bonAbout/remod/remod.html has a movie of the remodelling process, which occurs continuously to keep bones strong. Green *et al.* (2000) found that UHMWPE particles of mean size 0.45 and 1.71 μm produced the

Figure 15.18 SEM of wear debris from retrieved Charnley hip prostheses (from Tipper, J. L. *et al.*, *J. Mater. Sci. Mater. Med.*, **11**, 117, 2000) with kind permission of Springer Science and Business Media.

highest rate of bone resorption, whereas 7 and 88 μm particles had no effect. Consequently, the debris size determines its effect in the body.

15.4.11 Improving the PE wear resistance

Bajaria and Bellare (1999) reviewed efforts to increase the wear resistance of UHMWPE. Du Pont-DePuy Orthopaedics used a process to increase the crystallinity of the polymer from the initial circa 50% (Du Pont, 1990). The melting temperature of polymers increases with increased pressure and the lamellae thickness increases with the temperature of crystallisation, if crystallisation is slow. The sintering of PTFE powder at high temperatures also leads to very thick lamella and a high crystallinity. In the process, machined polyethylene cups are heated in sealed cans to between 200 and 230 °C, then a pressure greater than 300 MPa applied for about 1 h. Subsequent slow cooling under pressure cause crystallisation to occur in the range 170–190 °C; further cooling to below 120 °C is necessary before the pressure is released. The typical yield stress (28 MPa) and Young's modulus (2.1 GPa) are significantly higher than pre-treatment. However, these *Hylamer* implants had an inferior wear resistance and were susceptible to oxidative degradation; use of this material has been discontinued.

In the late 1990s, highly crosslinked UHMWPE was found to have an effectively zero wear rate. Muratoglu *et al.* (2001) described how 40 mm thick discs of UHMWPE were irradiated in air using a 10 MeV electron beam, for doses up to 180 Mrad at 125 °C. After irradiation, the discs were held at 150 °C for circa 2 h until melting was complete. It is difficult to measure very low wear rates by weight loss, as the polymer absorbs some liquid.

Wang (2001) found that irradiation crosslinked UHMWPE had a significantly lower wear rate than un-crosslinked material. The radiation dose must be high to obtain the optimum effect (Fig. 15.19). Rieker *et al.* (2003) showed that the wear surfaces of highly crosslinked UHMWPE implants after 18 months in vivo, consisted of folds (Fig. 15.20). Such folds are also found in conventional UHMWPE, but fatigue leads to their detachment from the surface. The folds on the surface of the crosslinked polymer appear to stay in place. Crosslinking leads to a reduction in crystallinity, hence a

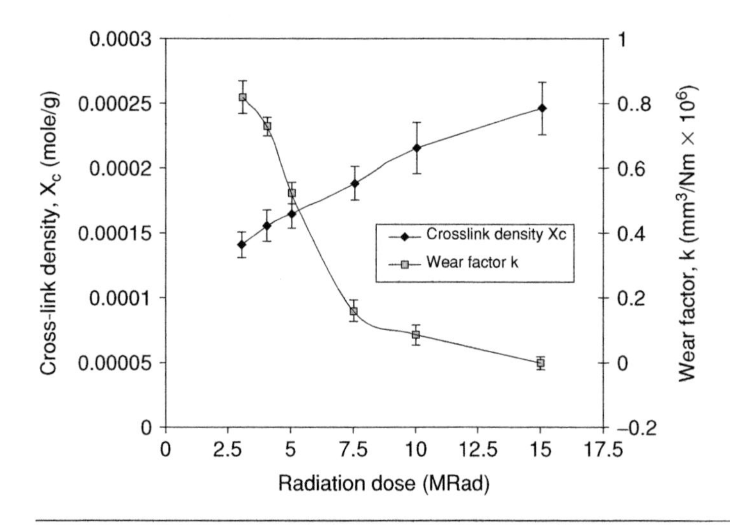

Figure 15.19 Wear rate vs. radiation dose (Wang, A., *Wear,* **248**, 38, 2001) © Elsevier.

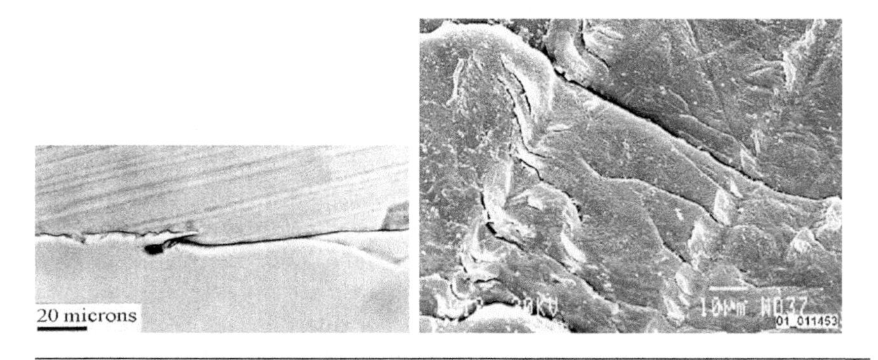

Figure 15.20 Folds on the wear surface of highly crosslinked UHMWPE (Rieker, C. B. *et al.*, *J. Arthoplasty,* **18(7)** (Suppl. 1), 48, 2003) © Elsevier.

reduction in yield stress and modulus (Table 15.4). In many applications these changes would be deleterious, but in the hip joint the effects are advantageous.

15.4.12 Summary

Research continues in many areas related to UHMWPE. The literature is much larger than that for the other two case studies in this chapter. The difficultly of establishing wear mechanisms, and the long-term nature of the 'experiments' carried out on patients, mean that progress is slow. Non-material factors, such as the skill of the surgeon, may affect success rates for particular joint designs. According to National Institute for Clinical Excellence (www.nice.org.uk), there are 64 different hip prostheses on the UK market. Their guidance suggests that implants are chosen which meet a benchmark revision rate of 10% or less after 10 years.

Appendix A

Diffusion of heat or impurities

Chapter contents

A.1 Molecular models for diffusion

The starting point for a molecular theory of diffusion is the analysis of a random walk of an atom or molecule. In Chapter 3, we dealt with the possible shapes of a one-dimensional polymer chain; Eq. (3.13) gave the number of distinguishable chains of end-to-end length r as

$$W = A\exp\left(-\frac{r^2}{2nl^2}\right) \tag{A.1}$$

where A is a constant. The probability that a chain, chosen at random, has a length r, is proportional to W. The function must be normalised, so that the sum of the probabilities of all chain lengths is unity. Since the integral

$$\int_{-\infty}^{\infty} \exp(-Ax^2)\mathrm{d}x = \sqrt{\frac{\pi}{A}} \tag{A.2}$$

the probability $P(r)$ that a chain has a length r is given by

$$P(r) = \frac{1}{\sqrt{2\pi nl^2}}\exp\left(-\frac{r^2}{2nl^2}\right) \tag{A.3}$$

This solution can be adapted for diffusion as follows: The diffusing species is assumed to take ν steps per second, so, in a time t, has traced out a walk of $n = \nu t$ steps. The step length l becomes the jump distance of the diffusing species. The one-dimensional polymer chain becomes a planar diffusion problem, in which the concentration C only varies in the x direction, and is constant in the yz plane. The equivalent of Eq. (A.3) then is

$$C(x) = \frac{M}{\sqrt{2\pi\nu tl^2}}\exp\left(-\frac{x^2}{2\nu tl^2}\right) \tag{A.4}$$

This describes the diffusion of an initially planar source of impurity atoms (Fig. A.l). M is the total amount of the impurity per unit cross-sectional area, and the diffusion distance x replaces the random walk length r. With increasing time, the Gaussian distribution spreads and the peak diminishes in magnitude.

A.2 Differential equations for diffusion

The diffusion of impurities or heat is governed by differential equations, that can be derived from the molecular models just described. We only

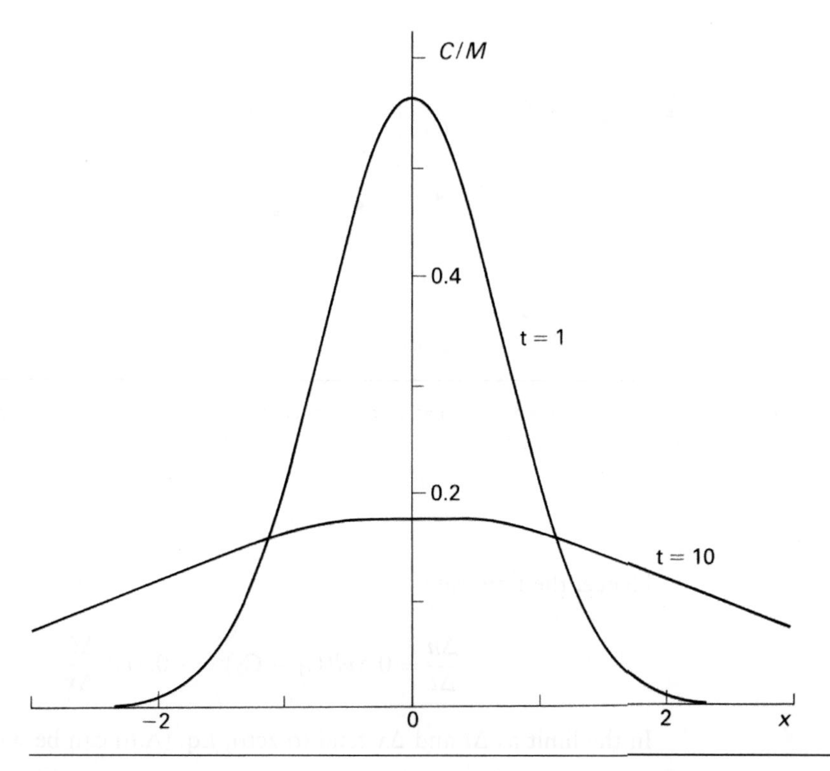

Figure A.1 Diffusion from a planar source in an infinite body at times 1 and 10 s. The distance x is in units of $2\sqrt{D}$ or $\nu\sqrt{l}$.

consider one-dimensional diffusion; two- and three-dimensional diffusion problems are analysed in texts such as Crank (1975).

Consider two layers in a solid a distance Δx apart (Fig. A.2). Let the concentration of impurity atoms be C_1 and C_2 in the two layers, so the concentration gradient is

$$\frac{\Delta C}{\Delta x} = \frac{C_2 - C_1}{\Delta x} \tag{A.5}$$

If Δx is chosen to be equal to the diffusion step length l then the numbers of impurity atoms per unit area in the layers are $C_1 l$ and $C_2 1$, respectively. In a time interval $\Delta t = l/\nu$, half of these will jump to the left and half to the right, so the net flow of atoms from layer 1 to layer 2 is

$$\Delta n = 0.5l\,(C_1 - C_2)$$

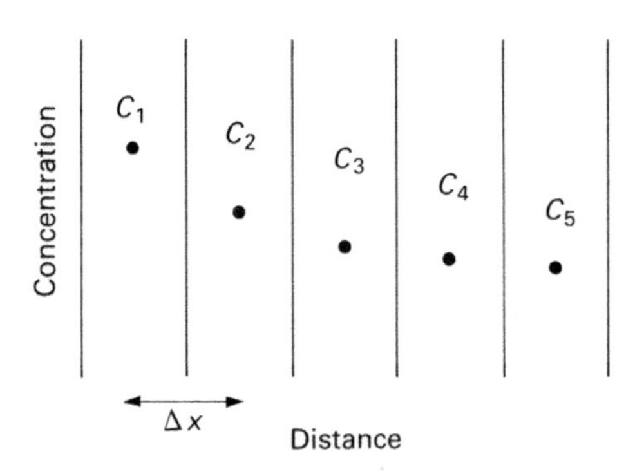

Figure A.2 Finite difference approximation to a concentration profile. The concentrations are C_1, C_2, C_3, in layers Δx thick.

Hence, the flow rate

$$\frac{\Delta n}{\Delta t} = 0.5\nu l(C_1 - C_2) = -0.5\nu l^2 \frac{\Delta C}{\Delta x} \tag{A.6}$$

In the limit as Δt and Δx tend to zero, Eq. (A.6) can be written as

$$F = -D\frac{\mathrm{d}C}{\mathrm{d}x} \tag{A.7}$$

known as Fick's first law. F is the flow rate of atoms per unit cross-sectional area, and the diffusion coefficient

$$D = 0.5\nu l^2 \tag{A.8}$$

gathers together the three constants (reciprocal of the number of step directions, frequency, step length). The heat flow equivalent of Fick's first law, used as the definition of thermal conductivity k, is

$$\frac{Q}{A} = -k\frac{\mathrm{d}T}{\mathrm{d}x} \tag{A.9}$$

where Q is the heat flow in watts down a temperature gradient $\mathrm{d}T/\mathrm{d}x$, and A is the cross-sectional area.

A second differential equation is needed for the analysis of non-steady impurity or temperature distributions. It can be derived from Eq. (A.7) or (A.9) on making the assumption that D is independent of the concentration, or k is independent of the temperature. Figure A.2 shows a solid divided into layers of thickness $\sim x$. In the finite difference heat transfer analysis, each layer is assumed to be at a uniform temperature T_j. The temperature gradient between the ith and $(i + 1)$th layer can be approximated by

$$\frac{dT}{dx} \cong \frac{T_{i+1} - T_i}{\Delta x} \tag{A.10}$$

In a time interval Δt, the increase in the thermal energy stored in layer i is the difference between the heat flows across the left- and right-hand boundaries. The calculation is made for unit cross-sectional area, and yields infinite differences from the equation

$$\Delta x \rho C_p (T_i^* - T_i) = \Delta t \left[k \frac{(T_{i-1} + T_{i+1} - 2T_i)}{\Delta x} - k \frac{(T_{i-1} + T_{i+1} - 2T_i)}{\Delta x} \right] \tag{A.11}$$

where ρ is the density and C_p is the specific heat capacity. T_j^* is the layer temperature at the end of the time interval. The equation can be rearranged to yield

$$\frac{T_i^* - T_i}{\Delta t} = \alpha \frac{(T_{i-1} + T_{i+1} - 2T_i)}{\Delta x^2} \tag{A.12}$$

in which the thermal diffusivity α is defined by

$$\alpha \equiv \frac{k}{\rho C_p} \tag{A.13}$$

Equation (A.12) can be used as a recurrence relation for finite difference calculations, or it can be expressed in differential form by going to the limit as Δt and $\Delta x \to 0$.

$$\frac{dT}{dt} = \alpha \frac{d^2 T}{dx^2} \tag{A.14}$$

A.3 Solutions to the differential equations

The steady-state solution to Fick's first law is a constant concentration or temperature gradient. Only non-steady-state solutions, having simple boundary conditions will be discussed.

A.3.1 Constant surface concentration C_0 on a semi-infinite body

Equation (A.13) is a linear differential equation because it contains only the first power of the differentials. Consequently, any two solutions can be combined to provide a further solution. Let us start with Eq. (A.4) for an initially planar source of impurity atoms. This can be re-expressed in terms of the diffusion coefficient using Eq. (A.8)

$$C(x) = \frac{M}{2\sqrt{\pi D t}} \exp\left(-\frac{x^2}{4Dt}\right) \tag{A.15}$$

The problem is modelled as an infinite body, in which there is initially a constant concentration $2C_0$ for $x < 0$. The initial impurity is split up into layers (planar sources) each of strength $2C_0\, d\zeta$ (Fig. A.3). Impurity reaching

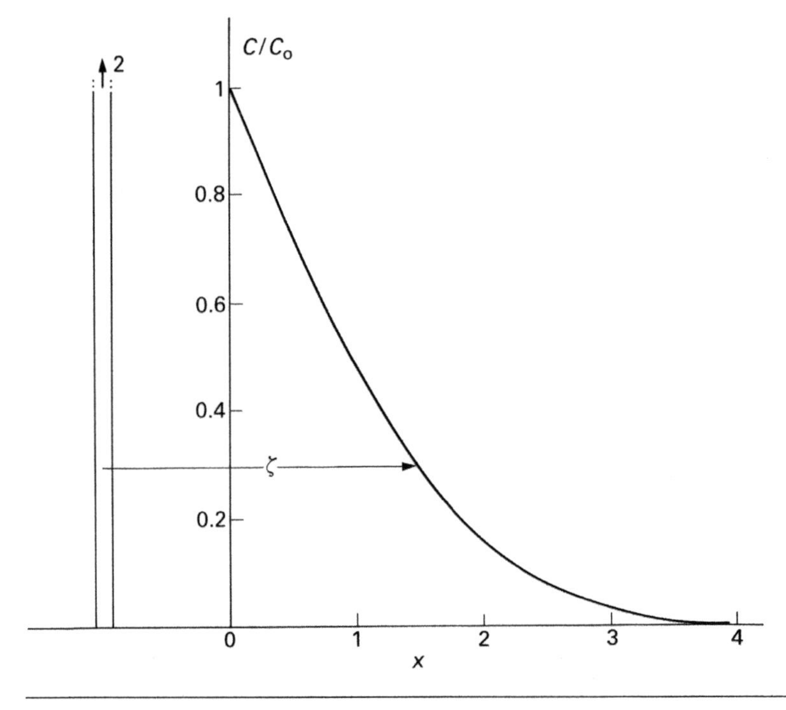

Figure A.3 Superposition of the concentration profiles from a set of planar sources from $-\infty < x < 0$, produces the erf c profile in the semi-infinite body $x > 0$. ζ is the diffusion distance from a typical source, and x is in units of $2\Delta t$.

x on the right-hand side has diffused a distance of at least x from one of the planar sources, and the total concentration is given by summing the individual contributions as

$$C(x) = \frac{2C_0}{2\sqrt{\pi Dt}} \int_x^\infty \exp\left(-\frac{\zeta^2}{4Dt}\right) d\zeta \tag{A.16}$$

The integral in Eq. (A.16) can be written in a standard form by the change of variable $\eta^2 = \zeta^2/4Dt$. The integral, known as the error function, is defined as

$$\mathrm{erf}(z) = \frac{2}{\sqrt{\pi}} \int_0^\infty \exp(-\eta^2) d\eta \tag{A.17}$$

It is the area under the normalised Gaussian function (Fig. A.l) between the ordinates $-z$ and z. The error function complement $\mathrm{erfc}(z) = 1 - \mathrm{erf}(z)$ represents the remainder of the area under the Gaussian curve. Equation (A.16) can be written as

$$C(x) = C_0 \,\mathrm{erfc} \frac{x}{2\sqrt{Dt}} \tag{A.18}$$

Tables of erfc are available (erfc 0 = 1.0, erfc 0.2 = 0.777, erfc 0.4 = 0.572, erfc 0.6 = 0.396...). This solution for an infinite body maintains $C{=}C_0$ at $x = 0$, and so is a solution to the semi-infinite body problem with a constant surface concentration C_0.

The equivalent of Eq. (A.18) for heat diffusion is

$$T - T_P = (T_0 - T_P)\mathrm{erfc}\left(x/2\sqrt{\alpha t}\right) \tag{A.19}$$

where T_0 is the temperature of the metal mould in contact with the semi-infinite polymer, and T_P is the temperature of the polymer. For typical conditions for the melting of low-density polyethylene in an extruder—barrel temperature $T_0 = 220\,°\mathrm{C}$, initial polymer temperature $T_P = 20\,°\mathrm{C}$ and melting complete at $T = 120\,°\mathrm{C}$, $(T - T_P)/(T_0 - T_P) = 0.5$ the melt front is at a position where $\mathrm{erfc}(x/2v\alpha t) = 0.5$, i.e. where

$$x = 0.94\sqrt{\alpha t} \cong \sqrt{\alpha t} \tag{A.20}$$

A.3.2 Constant surface concentration(s) on a plane sheet

A sheet of thickness $2L$ has an initially constant temperature T_m, and the surfaces $x = 0, 2L$ are held a constant zero temperature. Equation (A.13) can be solved analytically by assuming that the variables are separable, i.e.

$$T = X(x)\theta(t) \tag{A.21}$$

Substituting this in Eq. (A.13) and dividing by θX gives

$$\frac{1}{\theta}\frac{\mathrm{d}\theta}{\mathrm{d}t} = \frac{\alpha}{X}\frac{\mathrm{d}^2X}{\mathrm{d}x^2} \tag{A.22}$$

in which the variables θ and X occur on separate sides. Both sides are equated to a constant ($-\alpha\lambda^2$ for convenience) and solved separately to give

$$T = (A\sin\lambda x + B\cos\lambda x)\exp(-\lambda^2\alpha t)$$

The constants A and B are evaluated from the boundary conditions. That at $z = 0$ means that $B = 0$, and that at $x = 2L$ requires that

$$\sin 2\lambda L = 0 \quad \text{so} \quad 2\lambda L = \pi, 3\pi, 5\pi \ldots$$

The even terms are omitted because the temperature distribution has mirror symmetry about the mid-plane. Consequently, the solution is

$$T = \sum_{m=1,3,5,} A_m \sin\left(\frac{m\pi x}{2L}\right)\exp\left(-\frac{m^2\pi^2\alpha t}{4L^2}\right) \tag{A.23}$$

The constants A_m are evaluated by a Fourier transform of the initial temperature distribution. The higher the value of m, the more rapidly does the sine term die away, so that the temperature distribution eventually becomes a single half sine wave.

Alternatively, an approximate solution can be obtained using the finite difference method. The largest time interval Δt between solutions, that gives a stable solution, is

$$\Delta t = 0.5\frac{\Delta x^2}{\alpha} \tag{A.24}$$

Equation (A.12) reduces to the simple recurrence relation

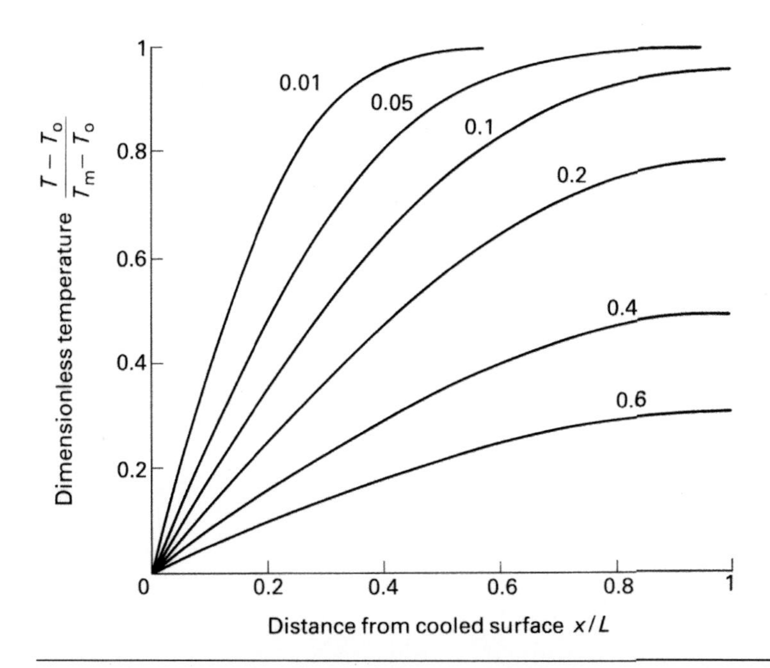

Figure A.4 Temperature profiles in a sheet for different values of the Fourier number. The initial uniform melt temperature is T_m.

$$T_i^* = 0.5(T_{i-1} + T_{i+1}) \tag{A.25}$$

for the computer program. The predictions (Fig. A.4) are displayed in dimensionless form; x/L is the dimensionless distance from the centre line, and $(T - T_0)/(T_m - T_0)$ the dimensionless temperature. T_0 is the mould temperature. The curves are labeled with the dimensionless time or *Fourier number*

$$Fo \equiv \frac{\alpha t}{L^2} \tag{A.26}$$

Reference

Crank, J., *The Mathematics of Diffusion*, 2nd Ed., Oxford University Press, 1975.

Appendix B

Polymer melt flow analysis

Chapter contents

B.1 Strain rates in channel flows

We need to be able to convert data from a melt rheometer into a flow curve, and to use such a curve to estimate pressure drops in simple melt processing equipment. Figure B.1 shows four types of flow in channels. These are assumed to be steady laminar flows. The first task is to quantify the shear strain rates. Polymer melts adhere to metal surfaces so the melt velocity is zero at the stationary channel walls.

(a) *Rectangular slot of breadth $b \gg$ height h.* With the xyz axes shown, the only non-zero strain rate is the shear strain rate $\dot{\gamma}_{yz}$ in the yz axes. Consequently, we can drop the subscripts yz on this and on the shear stress τ_{yz} without causing any confusion. There will be a strain rate $\dot{\gamma}_{xz}$ near the sides of the slot, but this will be ignored. The analysis also applies to annular channels of radius $r \gg h$. The slot is bent until the sides meet and an annulus formed.

(b) *Cylindrical channel.* The only non-zero strain rate is $\dot{\gamma}_{rz}$ in the cylindrical axes r, θ, z. Consequently, the subscripts on this and the corresponding shear stress may be dropped.

(c) *Spreading disc flow between parallel plates.* This occurs at in an injection mould cavity of constant thickness. There is a shear strain rate $\dot{\gamma}_{rz}$ in the cylindrical axes. There are also extensional strain rates in the $r\ \theta$ plane, with $\dot{e}_z = 0$ and $\dot{e}_\theta = -\dot{e}_r$ because the melt extends in the hoop direction while contracting in the radial direction.

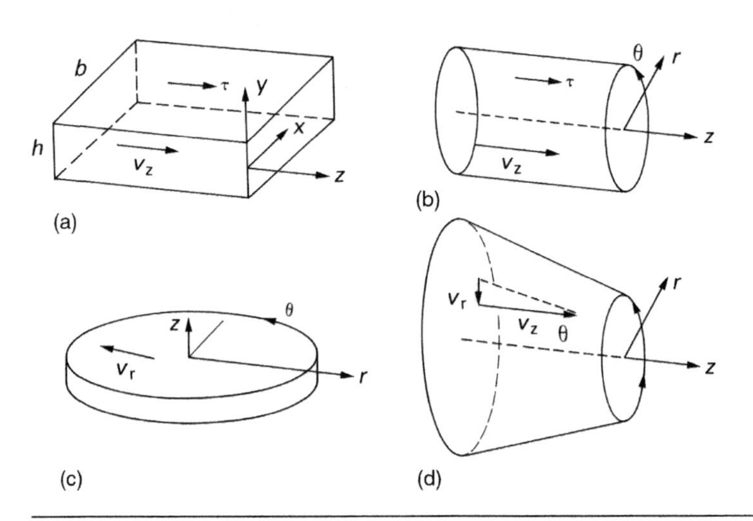

Figure B.1 Flows in channels used in polymer processing. (a) and (b) are shear flows in a rectangular slot and a cylindrical die, respectively. (c) and (d) are combinations of shear and extensional flow in a spreading disc and a tapering cylinder. The non-zero velocity components are shown.

(d) *A tapering cylindrical channel.* As well as the shear strain rate $\dot{\gamma}_{rz}$ as in (b), there is a uniaxial extension strain rate, due to the melt extending in the z direction, with

$$\dot{e}_\theta = \dot{e}_r = 0.5\dot{e}_z$$

Flows (c) and (d), although common in polymer processing, are not easy to instrument or analyse. Consequently, we concentrate on the simple shear flows (a) and (b). Even with these, tensile stresses can arise as a result of the melt elasticity.

B.2 Shear flow outputs from a slot or cylindrical die

The analysis proceeds in three stages. Details will be given for the rectangular slot, with the cylindrical slot result given in square brackets.

(a) *Shear stress variation with position.* The section of channel is assumed to be remote from any sudden changes of cross section, so that there are no elastic entrance or exit effects. The pressure p is constant across the cross section and varies linearly along the length

$$\frac{dp}{dz} = -\frac{\Delta p}{\Delta z} \tag{B.1}$$

Δp is the pressure drop between the section entrance and exit. Next, we consider the forces on the slab of liquid between $\pm y$. For the steady flow, the forces are in equilibrium, so

$$2yb\,\Delta p + 2b\,\Delta L\,\tau = 0$$

Hence

$$\tau = -y\frac{\Delta p}{\Delta z}$$

So, from Eq. (B.1)

$$\tau = -y\frac{dp}{dz}\left[= \frac{r}{2}\frac{dp}{dz}\right] \tag{B.2}$$

This linear variation of shear stress with position is a consequence of constant pressure gradient in channel.

(b) *Use the melt flow relationship to find the shear strain rate.* The relationship between the shear stress and shear strain rate is referred to as the *flow curve*. We need to assume a form for the flow curve before analysing the data from a pressure-flow rheometer. For drag-flow rheometers, in which one surface of the channel moves relative to the other, this assumption is not necessary. The melt is usually assumed to be *power law fluid*, for which

$$\tau = k\dot{\gamma}_a^n \tag{B.3}$$

where k is a constant that decreases with increasing temperature, and n is a constant that changes with the polymer and the width of its molecular mass distribution. If the pressure drops in the channel are less than 10 MPa, the flow is effectively isothermal and k is a constant. At very low strain rates, n tends to 1, and the behaviour reduces to that of a Newtonian fluid with

$$\tau = \eta\dot{\gamma} \tag{B.4}$$

The constant η is the shear viscosity. We introduce the value of the shear stress at the channel wall

$$\tau_w = -\frac{h}{2}\frac{\Delta p}{\Delta L}\left[= -\frac{R}{2}\frac{\Delta p}{\Delta L}\right] \tag{B.5}$$

so Eq. (B.2) becomes

$$\frac{\tau}{\tau_w} = -\frac{2y}{h}$$

Substituting this in Eq. (B.3) gives

$$\frac{\dot{\gamma}}{\dot{\gamma}_w} = \left(\frac{2y}{h}\right)^{\frac{1}{n}}\left[= \left(\frac{2r}{R}\right)^{\frac{1}{n}}\right] \tag{B.6}$$

(c) *Integrate to find the velocities and the flow rate.* The shear strain rate in a simple shear flow is defined by

$$\dot{\gamma} \equiv \frac{dV_z}{dy}\left[= \frac{dV_z}{dr}\right] \tag{B.7}$$

where V_z is the z component of velocity. The volume flow rate Q in the channel is given by the integral

$$Q = 2b \int_0^{h/2} V_z \mathrm{d}y$$

$$= 2b\,[V_z \mathrm{d}y]_0^{h/2} - 2b \int_0^{h/2} y \frac{\mathrm{d}V}{\mathrm{d}y} \mathrm{d}y$$

The first term is zero, since $V_z = 0$ at the channel wall. Using Eq. (B.7) in the second term we have

$$Q = -2b \int_0^{h/2} y\,\dot{\gamma}\,\mathrm{d}y \tag{B.8}$$

Equation (B.6) for the variation of the strain rate can now be substituted to give

$$Q = -\frac{2b\dot{\gamma}_{\mathrm{w}}}{(h/2)^{1/n}} \int_0^{h/2} y^{1+1/n}\ \mathrm{d}y = -\frac{b\dot{\gamma}_{\mathrm{w}} h^2}{4 + 2/n} \tag{B.9}$$

Equation (B.9) is more useful in its inverted form

$$\dot{\gamma}_{\mathrm{w}} = -\frac{(4 + 2/n)Q}{bh^2} \left[= -\frac{(3 + 1/n)Q}{\pi r^3} \right] \tag{B.10}$$

Figure B.2 shows the linear variation of shear stress and the non-linear variation of shear strain rate across a rectangular channel, for a power law fluid with $n = 0.5$. The variables have same sign. For a Newtonian fluid, the velocity variation would be parabolic, but for the power law fluid the velocity is more constant in the central region.

B.3 Presentation of melt flow data

Melt rheometers either impose a fixed flow rate and measure the pressure drop across a die, or, as in the melt flow indexer, impose a fixed pressure and measure the flow rate. Equation (B.5) gives the shear stress, but Eq. (B.10) requires knowledge of n to calculate the shear strain rate. It is conventional to plot shear stress data against the *apparent* shear rate $\dot{\gamma}_{\mathrm{wa}}$, calculated using $n = 1$ (assuming Newtonian behaviour). If the data is used subsequently to compute the pressure drop in a cylindrical die, there will be no error. However, if a flow curve determined with a cylindrical die is used to predict

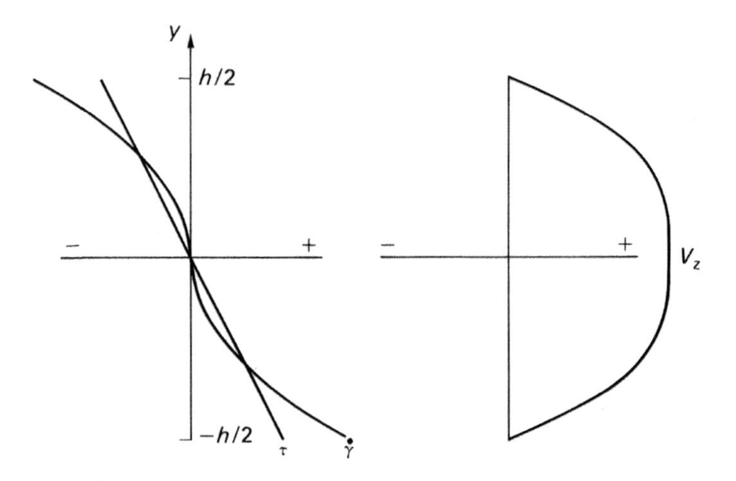

Figure B.2 Variation of shear stress, shear strain rate and velocity V_z for the flow in Fig. B.1a, with the coordinate y.

the pressure drop in a rectangular channel, there will be a slight error. The apparent shear rate at the channel wall can be calculated from the mean velocity in the channel using

$$\dot{\gamma}_{\text{wa}} = -\frac{6\overline{V}}{h}\left[= -\frac{4\overline{V}}{r}\right] \tag{B.11}$$

The *apparent viscosity* of a melt is defined by

$$\eta_{\text{wa}} \equiv \frac{\tau_{\text{w}}}{\dot{\gamma}_{\text{wa}}} \tag{B.12}$$

The apparent shear viscosity decreases as the shear strain rate increases.

Appendix C

Mechanics concepts

Chapter contents

C.1 Beam bending

C.1.1 Bending strains and stresses

In the analysis, it is assumed that the beams are much longer than they are thick, so that end effects can be neglected. If the separation of the loading points is much greater than the beam depth, the contribution of beam shear to the total deflection can also be neglected. Such long slender beams are often called 'Timoshenko beams'. Consider a beam that is bent into a circular arc of radius R (or a non-uniformly bent beam with a local radius of curvature R). The *neutral surface* is a layer in the beam with length L equal to that in the undeformed beam (L_0). Hence, the longitudinal tensile strain

$$e_x = \frac{L - L_0}{L_0}$$

is zero in the neutral surface. Figure C.1 shows a fibre in a layer a distance y above the neutral surface. A comparison of similar triangles gives $R/L_0 = (R - y)/L$. Hence, the tensile strain along the beam, in the direction of the x-axis, is

$$e_x = -\frac{y}{R} \tag{C.1}$$

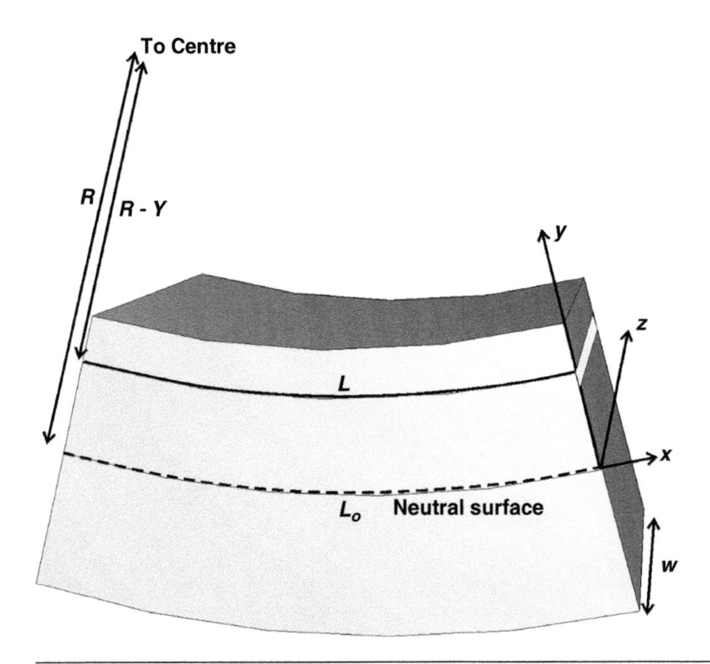

Figure C.1 Geometry of a bent beam, showing a 'fibre', a distance y above the neutral surface.

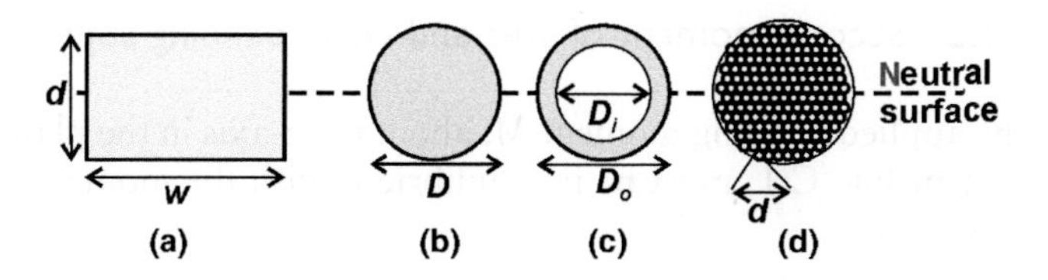

Figure C.2 Cross sections of common beam shapes: (a) Rectangular; (b) solid rod; (c) hollow tube; (d) multiple wires in a cable.

Consequently, e_x varies linearly through the thickness of the beam. The sign of the strain (whether it is tensile or compressive) depends on the direction of bending.

For beams of symmetrical cross section (Fig. C.2), the neutral surface is at the mid-depth. However, for asymmetric cross sections (Section 13.3.2), the neutral surface goes through the centroid of the cross section. For symmetrical cross section beams, the maximum and minimum strains, at the top and bottom surfaces, respectively, are given by

$$e_{\mathrm{m}} = \pm \frac{d}{2R} \tag{C.2}$$

where d is the beam depth. If, at this location, the strain reaches the yield strain e_{y}, the beam radius of curvature for initial yield R_{y} is given by

$$R_{\mathrm{y}} = \frac{d}{2e_{\mathrm{y}}} \tag{C.3}$$

We calculate the tensile stresses assuming the beam is made of an elastic material. First, we consider the elastic contractions in directions perpendicular to the stress. If there is just a uniaxial tensile stress σ_1 in a material, the lateral strains are

$$e_2 = e_3 = -\nu e_1$$

where ν is the Poisson's ratio. In a beam with a circular cross section, these contractions can occur freely, but in a rectangular beam of width$\gg$ depth, they would cause lateral curvature of opposite sign to the longitudinal curvature. Such an 'anti-clastic' curvature is observed when an eraser of rectangular cross section is bent with the fingers. However, it does not occur when a wide beam is bent over metal rollers, which means there must be lateral stresses (see Section C.2, where the plain-strain Young's modulus E^* is defined). Therefore, the longitudinal stress is given by

$$\sigma_x = E^* e_x = -\frac{E^* y}{R} \tag{C.4}$$

C.1.2 Second moment of area and beam bending stiffness

The applied bending moment M, about the z-axis in the plane shown on the right of Fig. C.1, must be in equilibrium with the moments of the internal stresses, so

$$M - \int y w \sigma_x \, dy = 0$$

where w is the beam width (which can in general vary). When Eqs (C.1) and (C.3) are substituted, this becomes

$$MR = E \int w y^2 dy \tag{C.5}$$

The *second moment of area* I of the cross section of a single-material beam is defined by

$$I \equiv \int w y^2 dy \tag{C.6}$$

Hence, the beam bending equation can be written as

$$EI = MR \tag{C.7}$$

The combination EI of a material property and beam cross section geometry is referred to as the beam *bending stiffness*. It determines the curvature for a given applied moment. The maximum stress σ_m in a symmetric beam can therefore also be expressed by substituting in Eq. (C.3)

$$\sigma_m = \frac{y_m M_m}{I} \tag{C.8}$$

where y_m is the maximum distance from the neutral surface and M_m is the maximum bending moment along the length of the beam. If the material is linearly viscoelastic, the stress variation through the beam is still linear, and the concept of the second moment of area I is still valid.

C.1.3 Second moments of area of various cross sections

For a beam of constant width w, w can be taken outside the integral in Eq. (C.6). If the beam depth is d, the integral has value $y^3/3$, evaluated at $d/2$ and $-d/2$, giving

$$I = \frac{wd^3}{12} \tag{C.9}$$

A solid cylindrical rod of diameter D has a second moment of area I given by

$$I_R = \frac{\pi}{64} D^4 \tag{C.10}$$

For a symmetrical beam with a hollow section, it is possible to subtract the I of the hole from that of the outer shape. For the hollow cylindrical beam this gives

$$I = \frac{\pi}{64}\left(D_{\text{out}}^4 - D_{\text{in}}^4\right) \tag{C.11}$$

When a multi-core cable of n wires, or a rope of n fibres, replaces a cylindrical rod with the same cross-sectional area (Fig. C.2c and d), each wire has diameter $d = D/\sqrt{n}$, so has a second moment of area

$$I_W = \frac{\pi}{64} d^4 = \frac{\pi}{64} \frac{D^4}{n^2} \tag{C.12}$$

When the cable is bent, the wires can slide relative to their neighbours, so they act as separate beams. Hence, the total second moment of area of the cable is the sum of that of individual wires

$$I_c = nI_W = \frac{\pi}{64} \frac{D^4}{n} = \frac{I_R}{n} \tag{C.13}$$

C.1.4 Beam deflection

The analysis is for the deflection of an elastic beam. In Eq. (C.7), $1/R$ can be replaced by the second differential of the deflection v

$$\frac{1}{R} = \frac{d^2 v}{dx^2} \tag{C.14}$$

so long as v is small. We take the example of a cantilever beam of length L, point loaded at one end by a force F. The bending moment is a function

$$M = F(L - x)$$

of the distance x from the clamped end. From Eq. (C.14) we have

$$\frac{d^2 v}{dx^2} = \frac{F}{EI}(L - x)$$

Integrating, with the condition that the slope is zero at the clamped end, gives

$$\frac{\mathrm{d}v}{\mathrm{d}x} = \frac{F}{EI}\left(Lx - \frac{x^2}{2}\right)$$

A second integration, with the condition of zero deflection at the clamped end, gives the deflection v as

$$v = \frac{F}{EI}\left(\frac{Lx^2}{2} - \frac{x^3}{6}\right)$$

Hence, the deflection Δ at the free end, relative to the fixed end, is

$$\Delta = \frac{FL^3}{3EI} \tag{C.15}$$

C.1.5 Beam buckling under axial compression

Buckling means a change of deformation mode, for example from uniform compressive strain to bending, when a certain deflection is exceeded. It occurs because the new deformation mode has a lower stored elastic energy than the original deformation mode.

Figure C.3 shows half of a *strut* (a beam loaded axially in compression) of length L. Its ends are *built-in* to the rest of the moulding. Consequently, they cannot rotate or move sideways when the compressive forces F are applied. If the strut were to bend, so that the lateral deflection was v at a point P, the bending moment at P would be

$$M = -Fv + M_0$$

where M_0 is an unknown moment at the strut ends that prevents rotation. Substituting this in the beam bending Eq. (C.7), and using the low deflection approximation (C.14), gives

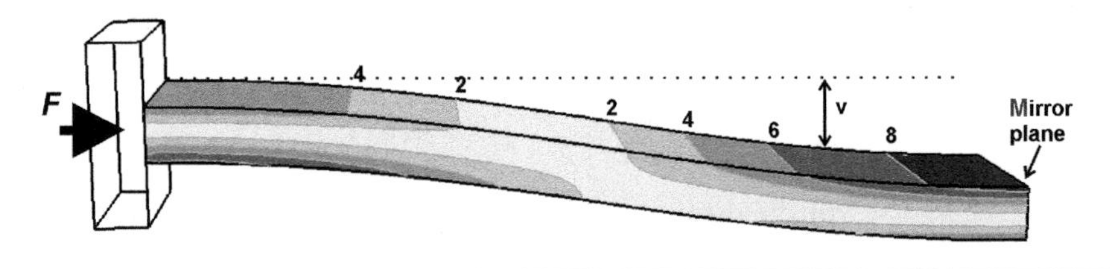

Figure C.3 Elastic buckling of a strut with built-in ends, and length to depth ratio 33:1, due to axial compressive forces F. Contours of von Mises stress (MPa).

$$EI\frac{d^2v}{dx^2} + Fv = M_0$$

The solutions of this differential equation are

$$v = \frac{M_0}{F}\left[\cos\left(\frac{2n\pi x}{L}\right) - 1\right]$$

where the mode number n is 1, 2, 3K, and the x-axis origin is at one end of the beam. In these solutions, the deflection v has an indeterminate magnitude. This implies that the strut collapses in the mode n shape at a critical axial force (the Euler buckling load) F_c given by

$$F_c = \left(\frac{2\pi n}{L}\right)^2 EI \tag{C.16}$$

The mode $n = 1$ gives the lowest buckling load as

$$F_c = \left(\frac{2\pi}{L}\right)^2 EI \tag{C.17}$$

C.1.6 Elastica for large beam deflections

The approximation for beam curvature in Eq. (C.14) can be replaced by an exact expression, relating the local orientation θ to the length coordinate s along the curved beam. Solutions of the resulting differential equation for large elastic deflections

$$\frac{\partial^2\theta}{\partial s^2} = \frac{M}{EI} \tag{C.18}$$

are known as *Elastica*.

C.2 Biaxial stresses and plane-strain elasticity

If principal stresses σ_1 and σ_2 act in a plane, in an elastic material, the strains in the 1 and 2 directions are given by

$$e_1 = \frac{\sigma_1}{E} - v\frac{\sigma_2}{E} \quad \text{and} \quad e_2 = \frac{\sigma_2}{E} - v\frac{\sigma_1}{E} \tag{C.19}$$

where v is the Poisson's ratio. If a constraint makes e_2 zero (a gas pipe may be constrained to have a constant length, or a wide beam may be constrained to contact loading rollers), this means that

$$\sigma_2 = \nu\sigma_1 \tag{C.20}$$

so

$$e_1 = \frac{\sigma_1}{E/(1-\nu^2)} = \frac{\sigma_1}{E^*} \tag{C.21}$$

The quantity $E^* \equiv E/(1-\nu^2)$, the 'plain-strain Young's modulus', can be used in place of E in formulae for the effect of a single stress. For beams of width $\leq$ depth, $E^* = E$.

C.3 Pressurised pipe

To find the hoop stress σ_H, in a pipe of wall thickness t and mean diameter D_m, consider a section of length L that contains the axis of the pipe (Fig. C.4). For equilibrium, the total force normal to the section is zero. The force acting on the pipe wall is the hoop stress σ_H multiplied by the wall cross-sectional area, whereas the force acting on the gas or liquid is the vertical area D_m L times the relative gas pressure p (the mean diameter is used because the pressure drops off approximately linearly through the pipe wall). Consequently,

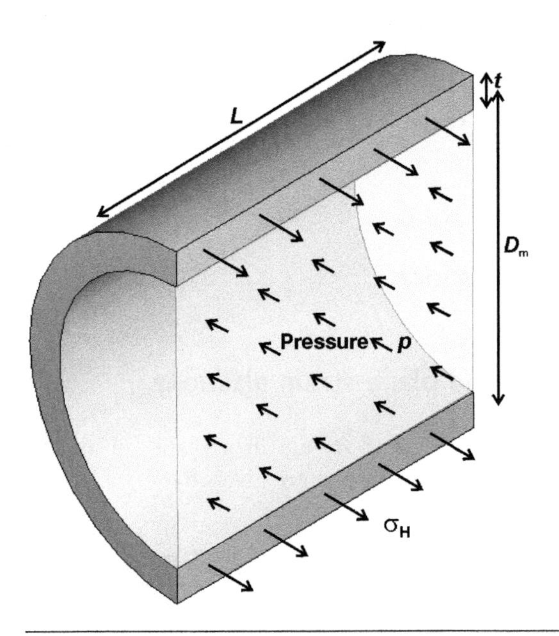

Figure C.4 Force equilibrium on a vertical section through a pressure pipe, showing the stresses and areas on which they act.

$$\sigma_{\mathrm{H}} 2Lt - pD_{\mathrm{m}}L = 0$$

so

$$\sigma_{\mathrm{H}} = \frac{pD_{\mathrm{m}}}{2t} \tag{C.22}$$

A similar method is used to consider the equilibrium of forces on a section cut perpendicular to the pipe axis, for a pipe that is free to expand in length. This leads to

$$\sigma_{\mathrm{L}} = \frac{pD_{\mathrm{m}}}{4t} \tag{C.23}$$

However, for a buried pressure pipe, which cannot change length, the elastic plane strain condition of Eq. (C.17) leads to

$$\sigma_{\mathrm{L}} = \frac{p\nu D_{\mathrm{m}}}{2t} \tag{C.24}$$

Appendix D

Questions

(For solutions, log on to the textbook website)

Chapter 2

1. Examine the data in Table 1.1, check correlation between a high T_g (data in Table 1.2) and presence of in-chain benzene rings or large side groups (structures in Tables 2.3 and 2.4). Can you explain such a correlation in terms of chain flexibility?
2. Verify Eq. (2.8) for the molecular mass averages of a step-growth polymerisation. Sum the series obtained when Eq. (2.7) is substituted in the formulae for M_N and M_W.
3. Model the stereoregularity of a PVC molecule with a 0.71 probability of a racemic conformation *r*. Generate 100 random numbers R with a calculator, and assume a racemic conformation occurs if $0.71 > R > 0$ and a meso if $1 > R > 0.71$. How many of the 100 units are contained in *rrrr*... sequences of length ~10?
4. Discuss the reasons for the relative cost of polyethylene compared to other polyamide 6,6.
5. What two main molecular parameters are controlled in the polymerisation of ethylene? What quality control tests are used on polyethylene to assess these parameters?
6. PVC is unique among the commodity plastics in that 50% is sold in a plasticised form. In what way does its microstructure differ from the other commodity plastics, to make this possible? (You may need to refer to Chapter 3.)

Chapter 3

1. Discuss the differences between the conformation of a molecule in a polymer crystal, and in a rubber. Show how the conformations can be modelled by using different sequences of C—C bond rotations.
2. Describe the nature of the disorder in a polymer glass, and discuss whether there is any evidence for the presence of micro-crystals of size 5 nm.
3. Contrast the mechanisms behind the elastic behaviour of a rubber and glassy polymer, explaining the range of shear moduli that are feasible for these materials.
4. Explain how the 2*3/1 helical conformation of polypropylene molecules spaces out the methyl side groups along the helix. Give reasons why the polypropylene crystal has a lower modulus in the c direction than the polyethylene crystal.
5. What is the condition for entanglements between two polymer chains to arise in a polymer melt, and what consequences such entanglements have on the melt flow properties?
6. Semi-crystalline polymers are isotropic and homogeneous on a scale larger than 100 μm, but anisotropic and inhomogeneous on a scale less than 0.1 μm. Discuss.

Chapter 4

1. Design a laminated steel and rubber spring for use as a bridge bearing. The compressive load on the bearing is 10^6 N and it is mounted on concrete that has a compressive design stress of 8 MPa. A shear deflection of 7 mm due to the thermal expansion of the bridge deck should not produce a shear force larger than 10^4 N. The compressive load should not produce a vertical displacement of more than 3 mm. The rubber used has a shear modulus of 0.9 MPa and is in the form of square sheets.
2. A sandwich panel (as in Fig. 4.5a) is made from a rigid polyurethane form core, of Young's modulus 40 MPa and density 92 kg m^{-3}, with equal thickness skins of glass-fibre-reinforced plastic of modulus 18 GPa and density 1900 kg m^{-3}. A one metre width of this panel must have a bending stiffness MR of 2000 N m^2. Consider core thicknesses of 5, 10 and 20 mm and see which gives the lightest panel. What will happen to the surface strains at a given load as the panel thickness is increased?
3. Discuss how the size and average separation of the rubber particles in a rubber-toughened glassy polymer influence the toughening mechanism, when a high tensile stress is applied to the material.
4. Is the mean length of the glass fibre reinforcement in injection-moulded polypropylene (fig. 4.28a) sufficient enough to give optimal stress transfer to the fibres? Explain how the stiffness anisotropy in such a moulding arises.
5. What new mechanisms of energy absorption arise when polystyrene is converted into a low density closed cell foam, and how can the compressive yield stress be controlled over a range of values?
6. Explain how an increase in the volume fraction crystallinity of polyethylene changes the moduli of a stack of lamellar crystals with amorphous inter-layers, and hence changes the macroscopic Young's modulus of the polymer.

Chapter 5

1. How has the low value of the thermal conductivity of polyethylene influenced the design of the equipment for producing polyethylene sheet?
2. A reduction in the molecular mass of a polymer reduces the melt viscosity. Why does this change not lead to a greater output rate from an extruder/die combination? What disadvantages might occur if blown film was being produced?
3. Polystyrene sheet is being extruded at $T_m = 200\,^\circ$C on to a pair of rolls of the type shown in Fig. 5.12, which are at $T_0 = 20\,^\circ$C. Calculate the Fourier number required for the whole of the sheet to be cooled below 80 °C at the end of the contact with the rolls. If the sheet is 1.25 mm thick and the roll surfaces move at 0.12 m s^{-1}, what diameter should the rolls be? The thermal diffusivity of polystyrene is 0.9×10^{-7} m^2 s^{-1}.

4. Consider the thermoforming of a domestic bath from 5 mm thick PMMA sheet. Why is it preferable to preheat the blanks in an oven rather than apply radiant heating to the sheet when it is over the mould? What are the stresses in the sheet when it is cylindrical in shape, radius 0.5 m and thickness 4 mm, when the pressure difference across it is 0.5 bar (50 kPa)? What is a typical draw ratio?
5. A 508 mm diameter cycle wheel, having five spokes, is injection moulded using a toughened nylon. It is gated at the hub and is 4 mm thick. Comment on the maximum flow lengths and whether the positions of the weld lines may cause problems.
6. Contrast the materials needed and the complexity of the moulds for conventional injection moulding, with those for RIM. What difference in cycle time would you expect for a 3 mm thick part?

Chapter 6

1. A circular, nearly-flat lid for a box is injection moulded, using polystyrene. Melt at 600 bar pressure and 200 °C. What is the volume shrinkage when the lid is at atmospheric pressure and 20 °C, assuming no feeding of the mould. Explain how feeding can partially compensate for this shrinkage, and why the diameter of the lid has a lower % shrinkage than the lid thickness.
2. Explain the design guidelines, that injection mouldings should have a uniform thickness and that there should be a 1–2° draft angle on the inside walls of boxes, in terms of the way in which the moulding solidifies, and the volume shrinkage of plastics on cooling. Does the structural foam process (Section 4.3.2) overcome these design limitations?
3. Explain how the external water cooling of extruded pipes leads to tensile residual hoop stresses at the bore of the pipe. To see whether a similar effect exists in a blow-moulded bottle, cut off the base and top, then cut down one side with a razor blade. If it curls up to a smaller diameter there are residual compressive stresses on the outer surface.
4. Discuss how increases in the mould wall temperature and the melt injection temperature would decrease the orientation in an injection moulding. What adverse effects would these changes have on the production costs?
5. The only way in which to remove orientation from an injection moulding is to anneal it at a temperature above its melting point. Explain why this process is not feasible, whereas metal components can be recrystallised at temperatures below their melting points.
6. Contrast the reasons for the persistence of particles in some PVC and UHMWPE products

Chapter 7

1. A single Voigt model, with parameters $E = 500\,\text{MPa}$ and $\eta = 2 \times 10^{12}\,\text{N s m}^{-2}$ is used to model creep. What is the retardation time? What creep strain is predicted for a creep stress of 10 MPa applied for 500 s?
2. A cantilever beam of length 400 mm and thickness 20 mm has a second moment of area of $2.0 \times 10^{-8}\,\text{m}^4$. A load of 6 N is applied to the free end. Use the data for HDPE in Fig. 7.6 to calculate the deflection of the free end after 1000 h. Suggest a cross-sectional shape for the beam that would minimise its mass if it were to be: (a) Injection moulded; (b) extruded.
3. A linearly viscoelastic polymer has a creep compliance $J = 5 \times 10^{-9}\, t^{0.1}\,\text{m}^2\,\text{N}^{-1}$, where the time t is in hours. It is subjected to an intermittent tensile stress of 5 MPa for 6 h on, 6 h off, starting at $t = 0$. Use Boltzmann's superposition principle to calculate the strain after 17 h. How does this compare with the strain for a constant 5 MPa stress applied for 11h?
4. A tensile fatigue test is carried out on a polyethylene, with the strain varying from 0 to 1% at 300 Hz. Calculate the rate of energy dissipation per unit volume if $E' = 1.0\,\text{GPa}$, $E'' = 20\,\text{MPa}$, and hence the initial heating rate, if the product of density and specific heat is $1.2 \times 10^6\,\text{J m}^{-3}\,\text{K}^{-1}$.
5. Calculate the in-phase part of complex compliance at a frequency of 0.01 Hz, of the generalised Voigt model of Fig. 7.1b.
6. Consider a washing machine, with the mass of the drum plus added weights of 20 kg, and a resonance frequency of 2.0 cycles/s (4π radians/s). Calculate the required spring constant of the suspension. If the viscous damper is such that $c/k = 0.1$, what is the vibration transmissibility at the design spin speed of 20 cycles/s? What are the disadvantages of using passive damping methods for such a product?

Chapter 8

1. A polyethylene is chosen for blow moulding of a 5 litre liquid container. When a tensile test is performed at an extension rate of $50\,\text{mm min}^{-1}$, a neck propagates down the specimen in a stable manner, yet when a prototype container is dropped through 3 m on to a hard surface, the tensile neck that forms immediately fractures. Explain.
2. Injection mouldings are usually thin walled to minimise the cooling part of the cycle time. Explain why they are more likely to fail under a compressive load by viscoelastic buckling, than by uniaxial compressive yielding.

3. Increasing the thickness of a plastics product does not necessarily make it more resistant to penetration. Discuss this statement with respect to surface indentation by a sharp object.
4. Comment on the work required to orient polyethylene in the solid state (using the data in figure 8.13) compared with that to orient the melt (as in the blown film process) then crystallize the material. Hence explain why oriented polymer film is usually made by the latter process. Comment on the ease of achieving very high levels of uniaxial orientation in the two processes.
5. How does crazing in polymers differ from crack initiation in ceramics or silicate glass, in respect of: (i) The orientation of the craze or crack plane with respect to the principal stresses; (ii) the energetics of craze or crack growth; (iii) the likelihood of multiple craze initiation?

Chapter 9

1. Make a collection of broken plastics products, then assess from the fracture appearance: (i) Whether fracture was due to yielding, fatigue or environmental stress cracking; (ii) whether the major load system was tension, bending or torsion; (iii) whether there were processing features involved such as weld lines and orientation; (iv) whether design faults could have been remedied.
2. A circular hole in a region of a product that is under tensile stress can lead to crack initiation. Contrast the mechanisms in a brittle plastic like polystyrene and a ductile plastic like polycarbonate (when the hole diameter is much less than the part thickness). How can the design of the polystyrene part be modified to avoid such crack initiation?
3. Differentiate clearly between a stress concentration factor and a stress intensity factor.
4. A surface craze in a PMMA product breaks down and becomes a crack. Treat it as an edge crack of length $a = 0.5$ mm in a body of width $w \gg a$. What tensile stress σ acting perpendicular to the crack will eventually cause the crack to grow? Use the data in Table 9.1 and the stress intensity factor $K = 1.12\sigma\sqrt{\pi a}$ for the crack geometry.
5. A cracked specimen of transparent PVC is being tensile tested. Is it realistic to expect to see the craze at the crack tip with the naked eye? The K_{IC} value is 2.0 MN m$^{-1.5}$ and the craze stress $\sigma_0 = 60$ MPa.
6. Discuss whether the notched Izod impact strengths of 3.2 mm thick plastic sheet are of any value for design purposes.

Chapter 10

1. Why does a PVC melt discolour and become more viscous when it is overheated, whereas PP melt become less viscous?
2. Explain why step-growth polymers such as PETP and PC have to be rigorously dried to prevent molecular mass reductions during melt processing, whereas this is not a problem with addition polymers.
3. Why is photo-oxidation more of a problem for the outdoor use of polyethylene film than it is for moulded or extruded products that are several millimetres thick?
4. Cracking in an ABS product was suspected to be caused by the stapling of a plasticised PVC coated cloth to the inner surface. What tests would you perform to check this hypothesis?
5. The data for the slow crack growth of polyethylene in detergent at 19 °C in Fig. 10.20 has the equation

$$\frac{da}{dt} = 9 \times 10^{-8} K^{4.0}$$

when the velocity is in $m\,s^{-1}$ and K in $MN\,m^{-1.5}$. Calculate how long it would take for an initial edge crack length $a_j = 0.1$ mm to grow to 5 mm length if the applied tensile stress $\sigma = 5$ MPa. The stress intensity factor can be calculated from $K = 1.12\sigma\sqrt{\pi a}$.

Chapter 11

1. How many grams of O_2 will pass through a 20 μm thick PVDC film in 1 day at 25 °C? The area of the film is 0.1 m^2 and the pressure differential across it is 0.1 MPa. Hence, estimate the effectiveness of such a film for wrapping meat which goes brown if it oxidises.
2. Calculate the solubility of methane at 400 kPa pressure in an MDPE at 25 °C with a volume fraction crystallinity of 0.59 using the data in Table 11.1. Calculate whether there will be a measurable loss of methane through a 6 mm thick wall MDPE pipe, filled with methane at 4 bar pressure, within 1 month of starting the experiment. Hence is the apparent loss of gas due to solution in the pipe wall or diffusion through it?.
3. Discuss why it is acceptable to use HDPE for fluid containers when HDPE has a finite permeability, i.e. for petrol through the wall of a petrol tank, or water vapour through the wall of a brake fluid container.
4. Investigate the orientation in a transparent injection moulding by examining it between polarising filters. Find out whether there is a residual stress effect by making a thin saw cut through the moulding to see if the fringe pattern alters.

5. Estimate the thickness of polyurethane foam (of thermal conductivity 0.024 $Wm^{-1}K^{-1}$) needed to insulate the inside of a brick/cavity aerated-concrete-block wall of U=1.0$Wm^{-2}K^{-1}$ to reduce the *U* value to 0.3 $W m^{-2} K^{-1}$. How much energy (in kWh) does this save in 100 days if the average temperature differential is 15° C and the wall area is 100 m^2?.

Chapter 12

1. Explain why the design electric field in a high voltage cable insulated with polyethylene is only 10 $MV m^{-1}$ when in laboratory tests the electric strength is measured as 800 $MV m^{-1}$.
2. Why is plasticised PVC suitable for insulating domestic mains cable, but not as a dielectric in a TV aerial?
3. A 2 mm wide ribbon of 1.5 μm thick PETP tape is to be used to construct a 1 μF capacitor. Calculate the length of tape necessary, given that the dielectric constant is 3.23 at 1 kHz. Calculate the resistance of the film at 50 °C using the data in Fig. 12.2, treating the geometry as a parallel plate capacitor.
4. Which thermoplastics could be used as the body of printed circuit boards, given that a soldering operation must be carried out after the components are assembled on the board? Lead tin solders melt at about 220 °C.

Chapter 13

1. Download from www.campusplastics.com the data from DuPont Europe. Select grades that meet the four conditions
   Rheological melt volume flow rate > 5
   Mechanical charpy impact strength at 23 C > 50 kJ/m^3
   Thermal vicat softening temperature > 80 C
   Processing injection mouldable *(yes)
   Rank the grades that meet these conditions in order of tensile modulus. For the grade with the highest modulus, select *multipoint* data, tick against *Creep modulus vs time*, to get a set of graphs: Magnify the E_t vs t graph by clicking on it: point the cursor at a stress of 2.4 MPa and time 10 hours, and read off the creep modulus
2. Use the data in table 13.1 to calculate the potential weight saving of replacing the steel in a nearly flat car door panel with polycarbonate, if bending stiffness is the only design criterion. Comment on the improved resistance to denting, and discuss whether the replacement of steel with polycarbonate could have any effect on safety protection such as side impact protection. (www.euroncap.com gives details of the test methods,

and www.plastics-car.org/s_plasticscar/information on car bumpers and fascia systems)

3. Use figure 13.4 to design a weight reduction of 30% when a 5 mm thick flat plate is replaced by a cross-ribbed panel. Comment on the isotropy of the bending resistance of this panel, and how the isotropy could be improved by changing the design of the ribs.
4. Acquire a defunct office chair with a plastic moulded base as in figure 13.15. With a torque wrench, measure the torsional stiffness of one of the legs, and the torsional strength (the torque at which collapse occurs). Relate these to the possible loads if a 100 kg person attempts to tip the chair back onto one leg.
5. Visit plastics.bayer.com/plastics/emea/en/femsnap/index.jsp and examine the example snap-fit joint, noting the shape and stress distribution at maximum deflection. Find some examples of snap fit joints on consumer products and compare their design with that on the Bayer site.
6. Visit plastics.bayer.com/plastics/emea/en/literature/and search for 'FEM moulded parts computation'. In the downloaded .pdf file, find the example of the parking brake pedal. Comment on the design of the ribs on this product, related to the principles explained in the chapter.

Appendix E

Solutions of problems in chapter 14

Problem 1 When Eq. (9.21) is used to calculate the transition thickness t_c, the values obtained are 40/400/66/42/27 mm for the polyethylenes, 35 mm for the PP and 10/4 mm for the PVCs. Hence, the MFI 2.0 PE of density 933 kg m^{-3} is the best.

Wall thickness ranges from 3 mm, up to 45 mm for 500 mm diameter SDR11 pipe. If PVC is not processed optimally, the wall thickness could be greater than t_c, so fast brittle fracture could be possible. The PP in the table could be used. However, the low melt stability of PP (Chapter 10) means that the inner surface of a thick-walled pipe would oxidise and suffer molecular weight loss. Consequently, PE is preferred.

Problem 2 The specification for PE 100 is for a 50 year creep rupture stress greater than 10 MPa. If a safety factor of $S = 2$ is used, the hoop stress (Eq. 14.3) must be less than 5 MPa. Substituting in Eq. 14.3, we obtain $(\text{SDR} - 1) * 0.2 < 5$, so $\text{SDR} < 26$.

Problem 3 When the pipe is placed under pressure, the hoop and longitudinal stresses in the wall are given by Eqs (C.19) and (C.21). The hoop strain for an elastic material is given using Eq. (C.18) as

$$e_{\mathrm{H}} = \frac{\sigma_{\mathrm{H}}}{E^*} = \frac{pD_{\mathrm{m}}}{2tE^*}$$

For a viscoelastic material, by the method of Section 7.3.2

$$e_{\mathrm{H}}\,(50\,\text{year}) = \frac{pD_{\mathrm{m}}\,J(50\,\text{year})}{2t}(1 - \nu^2)$$

Therefore, substituting $\nu = 0.4$, the creep hoop strain will be smaller than that in a tensile creep test by a factor $(1 - \nu^2)$, if the hoop and tensile stresses are equal. Examining Fig. 7.6, the tensile stress to cause a creep strain of 3/0.84 = 3.6% after 50 years is approximately 4 MPa. For a 4 bar pressure to cause a hoop stress of 4 MPa, the pipe SDR = 21 by Eq. (14.2). This is the maximum SDR allowed.

Problem 4 The soil load per unit length of pipe is $q = Dgh\rho = 20\,000\,D\,\text{Nm}^{-1}$. This value can be substituted in Eq. (14.7) to find the maximum stress in the pipe

$$\sigma_{\max} = 0.95\frac{qD}{t^2} = 19\,500\left(\frac{D}{t}\right)^2 = 19\,500\ \text{SDR}^2$$

Starting with a value of SDR = 20 this gives a stress of 7.6 MPa. The creep strain after 1 h from Fig. 7.6 is approximately 2.2%. Therefore, putting the condition that the diametral deflection Δ is to be $< 0.1\,D$ into Eq. (14.8) gives

$$\frac{\delta(1\ \text{h})}{D} = 0.223 \times 20\,000 \times 20^3 \times \frac{0.022}{7.6 \times 10^6} = 0.103$$

By chance this result is very close to the required one, so the maximum allowed SDR is 20. The conservative nature of the pseudo-elastic calculation provides an inbuilt safety factor.

Problem 5 The sequence of design is that

(a) given the impact energy E, Eq. (14.11) is used to find x_{max}. For $E = 100$ J, $x_{max} = 20$ mm. Allowing for bottoming out, the foam thickness $T = 25$ mm.
(b) using the helmet radius $R = 100$ mm in Eq. (14.10), a foam yield stress of 0.7 MPa gives a loading curve that passes through the point (80% of foam thickness, 10 kN)—Fig. 14.16 (EPS foam of density 65 kg m^{-3} provides this level of yield stress).

Problem 6 According to Eq. (8.20), the density of polypropylene foam must be 59% higher than the density of polystyrene foam, to achieve the same yield stress. Other factors such as materials and processing costs will be important. It was noted that PP foam recovers more than EPS after impact. Therefore, EPP is preferred for multiple-impact helmets, but EPS is preferred for bicycle helmets (single impact, minimum weight).

Problem 7 Molecular orientation during flow increases with the polymer melt elasticity and the flow rate. If the polymer molecular weight is kept to a minimum, the melt elasticity is minimised. Polycarbonate has a low melt elasticity compared with polystyrene. For CD manufacture the mould is filled in between 0.2 and 0.4 s, which is a low flow rate. The skin thickness can be reduced by having a very hot melt at 340 °C and a mould temperature of 95 °C, to reduce the solidification during mould filling.

Problem 8 Equation (6.7) involves the mould pressure p as the melt layer solidifies. While the cavity pressure is zero, all the solidifying layers have the same reference length, hence the same value of residual stress. Only the 0.12 mm thick surface layer, that solidified before the cavity pressure fell to zero, will be under compression, while the 0.76 mm core will be under a small residual tension. Consequently, the residual stress in the surface layers is kept reasonably low.

Further reading

Chapter 2

Galli, P. and Vecellio, G., Technology: Driving Force Behind Innovation and Growth of Polyolefins, *Prog. Polym. Sci.*, **26**, 1287, 2001.
Matthews, G., *PVC: Production Properties and Uses*, The Institute of Materials, London, 1996.
Saeki, Y. and Emura, T., Technical Processes for PVC Production, *Prog. Polym. Sci.*, **27**, 2055, 2002.
Scheirs, J. and Priddy, D., Eds., *Modern Styrenic Polymers*, Wiley, Chichester, 2003.

Chapter 3

Bassett, D. C., *Principles of Polymer Morphology*, Cambridge University Press, Cambridge, 1981.
Campbell, D. and White, J. R., *Polymer Characterisation: Physical Techniques*, Chapman & Hall, London, 1989.
Haward, R. N. and Young, R. J., Eds., *The Physics of Glassy Polymers*, 2nd Ed., Chapman & Hall, London.
Hemsley, D. A., Ed., *Applied Polymer Light Microscopy*, Elsevier Applied Science, Barking, 1989.
Sawyer, L. C. and Grubb, D. T., *Polymer Microscopy*, 2nd Ed. Chapman and Hall, london1995.
Schultz, J. M., *Polymer Crystallization*, Oxford University Press, Oxford 2001.
Treloar, L. R. G., *Physics of Rubber Elasticity*, 3rd Ed., Oxford University Press, Oxford, 1975.
Ward, I. M., Ed., *Structure and Properties of Oriented Polymers*, 2nd Ed., Chapman & Hall, London, 1997.

Chapter 4

Bucknell, C. B., *Toughened Plastics*, Applied Science, London, 1977.
Gent, A. N., *Engineering with Rubber*, 2nd Ed., Hanser, Gardner, Cincinnati, 2000.
Gibson, L. J. and Ashby, M. F., *Cellular Solids: Structure and Properties*, 2nd Ed., Cambridge University Press, 1988.
Hamley, I. W., *The Physics of Block Copolymers*, Oxford University Press, Oxford 1998.
Hull, D., Clyne, *Introduction to Composite Materials*, Cambridge University Press, Cambridge, 1996.

Chapter 5

Baird, D. G. and Collins, D. I., *Polymer Processing; Principles and Design*, Wiley, New York, 1998.
Bown, J., *Injection Moulding of Plastics Components*, McGraw-Hill, Maidenhead, 1979.
Chua, C. K. and Fai, L. K., *Rapid Prototyping*, Wiley, Singapore, 1997.
Macosko, C. W., RIM, *Fundamentals of Reaction Injection Moulding*, Hanser, Munich, 1989.
Michaeli, W., *Plastics Processing: An Introduction*, Carl Hanser, 1995.
Stevens, M. J. and Covas, C. A., *Extruder Principles and Operation*, 2nd Ed., Elsevier Applied Science, New York, 1995.

Chapter 6

Butters, G., Ed., *Particulate Nature of PVC*, Applied Science, Barking, 1982.
Injection Moulds, VDI, Dusseldorf, 1980.
Ogorkiewicz, R. M., Ed., *Thermoplastics: Effects of Processing*, Iliffe, London, 1969.

Chapter 7

Aklonis, J. J. and MacKnight, W. J., *Introduction to Polymer Viscoelasticity*, 2nd Ed., Wiley, New York, 1983.
Moore, D. R. and Turner, S., *Mechanical Evaluation Strategies for Plastics*, Woodhead, Cambridge 2000.
Ogorkiewicz, R. M., Ed., *Thermoplastics: Properties and Design*, Wiley, London, 1974.
Williams, J. G., *Stress Analysis of Polymers*, 2nd Ed., Ellis Horwood, Chichester, 1980.

Chapter 8

Johnson, K. L., *Contact Mechanics*, Cambridge University Press, Cambridge, 1985.
Kausch, H. H., Ed., Crazing in Polymers: 2, in Vol. 91 of *Advances in Polymer Science*, Springer, Berlin, 1990.
Pearson, R. A., Sue, H. J. and Yee, A. F., Eds., *Toughening of Plastics; Advances in Modeling and Experiments*, ACS Symposium Series 759, American Chemical Society, Washington DC 2000.
Ward, I. M., *Mechanical Properties of Solid Polymers*, 2nd Ed., Wiley, Chichester, 1983.

Chapter 9

Dugdale, D. S., Yielding of Steel Sheets Containing Slits, *J. Mech. Phys. Solids*, **8**, 100, 1960.
Engel, L., Klingele, G. W. and Schaper, H., *An Atlas of Polymer Damage*, Wolfe, London, 1981.
Kausch, H. H., *Polymer Fracture*, 2nd Ed. Springer, Berlin, 1987.
Kinloch, A. J., Ed., *Fracture Behaviour in Polymers*, Applied Science, 1983.
Williams, J. G., *Fracture Mechanics of Polymers*, Ellis Horwood, Chichester, 1984.

Chapter 10

Allen, N. S. and Edge, M., *Fundamentals of Polymer Degradation and Stablisation*, Kluwer, 1992.
Böcker, H. *et al.*, High Performance PE Provides Better Safety for Pipelines, *Kunststoffe-German Plastics*, 82, 8, 1992.
Wypych, G., *Handbook of Materials Weathering*, 3rd Ed., Chemtec, Toronto, 2003.

Chapter 11

Gilmore, M., *Fibre Optic Cabling: Theory, Design and Installation Practice*, Newnes, Oxford, 1991.
Mills, N. J., Optical Properties, in Vol. 10 of *Encyclopaedia of Polymer Science and Technology*, 2nd Ed., Wiley, New York, 1987.

Chapter 12

Blythe, A. R. and Bloor, D., *Electrical Properties of Polymers*, 2nd Ed., Cambridge University Press, Cambridge, 2005.
Riande, E. *et al.*, *Electrical Properties of Polymers*, Marcel Dekker, New York, 2004.

Chapter 13

Chow, W. W. C., *Cost Reduction in Product Design*, Van Nostrand, New York, 1978. Malloy, R. A., *Plastics Part Design for Injection Moulding: An Introduction*, Hanser, 1998.
Morton-Jones, D. H. and Ellis, J. W., *Polymer Products: Design, Materials and Manufacturing*, Chapman & Hall, London, 1986.
Rosato, D. V., Di Mattia, D. P., and Rosato, D. V., *Designing with Plastics and Composites: A Handbook*, Van Nostrand Reinhold, New York, 1991.

Chapter 14

Bicycle helmets

BSEN 1078 Helmets for Pedal Cyclists and for Users of Skateboards and Roller Skates, BSI, London.

Mills, N. J. and Gilchrist, A., The Effectiveness of Foams in Bicycle and Motorcycle Helmets, *Accid. Anal. Prev.*, **23**, 153, 1991.

Thompson, D. C., Rivara, F. P. and Thompson, R., *Helmets for Preventing Head and Facial Injuries in Bicyclists*, The Cochrane Library, 2003 Issue 1, Oxford.

Website for bicycle helmets: *www.bhsi.org*.

Optical discs

Bouwhuis, G., Ed., *Principles of Optical Disc Systems*, Adam Hilger, Bristol, 1985.

Makrolon CD2005 and DP1-1265 on *plastics.bayer.com*, 1997.

Sharpless, G., CD and DVD disc manufacturing, on *www.disctronics.com*, 2003.

Wimberger-Friedl, R., Analysis of the Birefringence Distributions in Compact Discs, *Polym. Eng. Sci.*, **30**, 813, 1990.

Chapter 15

A. Ropes

BSEN 892:1997 Mountaineering Equipment—Dynamic Climbing Ropes—Safety Requirements and Test Methods, BSI, London.

Cutler, S. and Lebaron, J., Performance of nylon climbing ropes, at *class.et.byu.edu/mfg340/qualityreports/nylon.htm*, 1992.

Kurzbock, E., US Patent 4640178, Rope, 1987.

McKenna, H. A., Searle, J. W. S. and O'Hear, N., *Handbook of Fibre Rope Technology*, Woodhead, Cambridge.

Nylon and Ropes for Mountaineering and Caving, conference, Club Alpino Italiano, Turin, 2002. See www.uiaa.ch

Smith, R. A., The Development of Equipment to Reduce Risk in Rock Climbing, *Sports Eng.*, **1**, 27–39, 1998.

UIAA Standard 101, Dynamic ropes, available at *www.uiaa.ch*, 2004.

B. Blood bags

Carmen, R., The Selection of Plastics Materials for Blood Bags. *Transfusion Med. Rev.*, 7, 1, 1993.
Ko, J. H. and Odegaard, L., Chlorine Free Blends for Flexible Medical Tubing, *Med. Plast. Biomater.*, at *www.devicelink.com/mpb/archive/97/03/004.html*, 1997.
Koop, C. E. and Juberg, D. R., A Scientific Evaluation of Health Effects of Two Plasticizers Used in Medical Devices and Toys, *www.medscape.com/viewarticle/407990_print*, 1999.
Kothe, F. C. and Platmann, G. J., The Use of the Sterile Connecting Device in Transfusion Medicine, *Transfusion Med. Rev.*, 8, 117–122, 1994.
Krauskopf, L. G., How About Alternatives to Phthalate Plasticizers? *J. Vinyl Addit. Tech.*, **9**, 159, 2003.
Lipsitt, B., Metallocene PE Films for Medical Devices, *Plast. Eng.*, 53, 25–28, 1997.
Lipsitt, B., Performance Properties of Metallocene PE, EVA and Flexible PVC Films, *Med. Plast. Biomater.*, **Sept.**, 1998.
Shah, K. *et al.*, Gas Permeability and Medical Film Products, *Med. Plast. Biomater.* at *www.devicelink.com/mpb/archive/98/09/005.html*, 1998.
Starr, D., *Blood*, Little Brown and Co., London, 1999.
Tickner, J. A., The Use of DEHP in PVC Medical Devices, Exposure, Toxicity and Alternatives, on *www.noharm.org* website, 2000.
Tickner, J. A. *et al.*, Health Risks Posed by the Use of DEHP in PVC Medical Devices, *Am. J. Ind. Med.*, **39**, 100–111, 2001.

C. Hip implants

Affatato, S., Fernandes, B. *et al.*, Isolation and Morphological Characterization of UHMWPE Wear Debris Generated in vitro, *Biomaterials*, **22**, 2325–2331, 2001.
Bajaria S. H. and Bellare A., Deformation, Morphology and Wear Behaviour of PE used in Orthopaedic Implants, *Med. Plast. Biomater.*, **March**, 1998.
Baudriller, H., Chabrand, P. *et al.*, Failure of Total Hip Arthoplasties. *Comput. Meth. Biomech Biomed. Eng.*, 7, 227, 2004.
Bennett, D. B., Orr, J. F. and Baker, R., Movement Loci of Selected Points on the Femoral Head for Individual Total Hip Arthoplasty Patients, using 3-D Computer Simulation, *J. Arthoplasty*, **15**, 909–915, 2000.
Bergmann G. *et al.*, Frictional Heating of Total Hip Implants, Part 1 Measurements in Patients, Part 2 Finite Element Study, *J. Biomechanics*, **34**, 421 and 429, 2001.
Charnley, J., *Low Friction Arthoplasty of the Hip*, Springer, New York, 1979.
Dowson, D., New Joints for the Millennium: Wear Control in Total Replacement Hip Joints, *Proc. I. Mech. Eng. Part H*, **215**, 225–357, 2001.
Du Pont, US Patent 5037928, *Process of Manufacturing UHMWPE*, 1990.
Green, T. R., Fisher, J. *et al.*, Effect of Size and Dose on Bone Resorption Activity of Macrophages by in vitro Clinically Relevant UHMWPE Particles, *J. Biomed. Mater. Res.*, **52**, 490–497, 2000.
Hutchings, I. M., Ed., *Friction, Lubrication and Wear of Artificial Joints*, Professional Engineering Publ., Bury St. Edmunds, 2003.

Kurtz, S., *The UHMWPE Handbook*, Elsevier (partly in *www.uhmwpe.org/lexicon*), 2004.
Muratoglu, O. K., Bragdon, C. R. *et al.*, A Novel Method of Crosslinking UHMWPE to Improve Wear, Reduce Oxidation and Retain Mechanical Properties, *J. Arthoplasty*, **16**, 149–160, 2001.
Olley, R. H., Hosier, I. L. *et al.*, On Morphology of Consolidated UHMWPE Resin in Hip Cups, *Biomaterials*, **20**, 2037–2046, 1999.
Pruitt, L. A., Deformation, Yield and Fracture and Fatigue of Conventional and Highly Crosslinked UHMWPE, *Biomaterials*, **26**, 905, 2005.
Rieker, C. B., Konrad, R. *et al.*, In Vivo and In Vitro Surface Changes in a Highly Crosslinked Polyethylene, *J. Arthoplasty*, **18**, 48–54, 2003.
Sedel L. and Cabanela M. E., Ed., *Hip Surgery-Materials and Development*, Martin Dunitz Ltd., London, Chapter 8, 1998.
Teoh, S. H. *et al.*, A Elasto-Plastic Finite Element Model for Polyethylene Wear in Total Hip Arthoplasty, *J. Biomechanics*, **35**, 323–330, 2002.
Turell, M., Wang, A. and Bellare, A., Quantification of the Effect of Cross-Path Motion on the Wear of UHMWPE, *Wear*, **255**, 1034–1039, 2003.
Wang, A., A Unified Theory of Wear of UHMWPE in Multi-Directional Sliding, *Wear*, **248**, 38–47, 2001.
Wang, A. *et al.*, Effect of Contact Stress on Friction and Wear of UHMWPE in Total Hip Replacements, *Proc. I. Mech. Eng. Part H*, **215**, 133, 2001.
Wang, A. and Schmidig, G., Ceramic Femoral Heads Prevent Runaway Wear for Highly Crosslinked Polyethylene Acetabular Cups by Third-Body Bone Cement Particles, *Wear*, **255**, 1057–1063, 2003.

D. Other case studies on polymers in sport

Mills, N. J., Foam Protection in Sport, in Jenkins, M. J., Ed., *Sport Materials*, Woodhead, Cambridge, pp. 9–46, Chapter 2, 2003.
Mills, N. J., Running Shoe Materials, in Jenkins, M. J., Ed., *Sport Materials*, Woodhead, Cambridge, pp. 65–99, Chapter 4, 2003.

Index

9780750651486